D0712934

HAZARDOUS AND TOXIC EFFECTS
OF INDUSTRIAL CHEMICALS

HAZARDOUS AND TOXIC EFFECTS OF INDUSTRIAL CHEMICALS

Marshall Sittig

NOYES DATA CORPORATION
Park Ridge, New Jersey, U.S.A.
1979

Copyright © 1979 by Marshall Sittig
     No part of this book may be reproduced in any form
     without permission in writing from the Publisher.
Library of Congress Catalog Card Number: 78-70739
ISBN: 0-8155-0731-3
Printed in the United States

Published in the United States of America by
Noyes Data Corporation
Noyes Building, Park Ridge, New Jersey 07656

Foreword

This handbook is intended to be a working guide for the industrial hygienist and other concerned persons. There is a continuing need to assess the status of potentially dangerous substances including those now available and those that may reach commercial availability in the future. This should be done with a predictive view to avoid or at least to ameliorate catastrophic episodes similar to those that have occurred with methyl mercury, polychlorinated biphenyls, vinyl chloride monomer, dioxin and a number of pesticides.

The information contained in this book is based upon publications issued by various governmental agencies as indicated below. While every effort has been made by the publisher and the author to correct any obvious errors in the original documents, the nature of the subject matter precludes any further representation or certainty. Research into the toxic effects of chemical substances is an evolving field, spurred by the deep concerns of science and industry and reflected in recent federal statutes but the nature of the field is such that certainty is an unobtainable goal in most cases.

The publisher and the author regard the present volume as important in providing a speedy link between the researcher and those who must be generally informed about hazardous chemicals. Every reader will recognize that this volume is not a medical text nor a substitute for medical acumen or the therapeutic knowledge needed to properly treat acute or chronic conditions encountered in practice, and the publisher and author expressly disclaim such status. This work is offered for sale upon the express understanding that further study and investigation will precede any clinical application contained herein; and the publisher and author represent only that their best efforts have been made to convey the data or information obtained from the sources cited in an accurate manner.

The basic outline of this volume is drawn mainly from summary reports by the National Institute for Occupational Safety and Health (1) and by the U.S. Environmental Protection Agency (2)(3):

> (1) *Occupational Diseases—A Guide to Their Recognition.* DHEW (NIOSH) Publication No. 77-181, Washington, D.C. (June 1977).
>
> (2) *Summary Characterization of Selected Chemicals of Near-Term Interest.* EPA report No. 560/4-70-004. Office of Toxic Substances, Washington, D.C. (April 1976).
>
> (3) *Potential Industrial Carcinogens and Mutagens.* EPA report No. 560/5-77-005. Office of Toxic Substances, Washington, D.C. (May 1977).

Also used was *TLVs® Threshold Limit Values for Chemical Substances and Physical Agents in the Workroom Environment with Intended Changes for 1978.* Cincinnati, Ohio, American Conference of Governmental Industrial Hygienists (ACGIH) (1978).

Many other sources were used in the preparation of this volume and are referenced throughout the book. Every effort has been made to include information available through November 1978.

Advanced composition and production methods developed by Noyes Data are employed to bring these durably bound books to you in a minimum of time. Special techniques are used to close the gap between "manuscript" and "completed book." Technological progress is so rapid that time-honored, conventional typesetting, binding and shipping methods are no longer suitable. We have bypassed the delays in the conventional book publishing cycle and provide the user with an effective and convenient means of reviewing up-to-date information in depth.

The book consists of about 250 individual entries arranged alphabetically by the common name of each substance described.

It should be emphasized that awareness of the toxic effects of many chemicals is both developing and being modified. Thus, while this handbook endeavors to give the latest data, the reader involved with a specific substance or material is urged to check with the above-named government agencies and others for new and revised standards and detailed guidance.

Glossary

Threshold limit values (TLV's) refer to airborne concentrations of substances and represent conditions under which it is believed that nearly all workers can be repeatedly exposed day after day without adverse effect.

Time-weighted averages (TWA's) are time-weighted average TLV concentrations for a normal 8-hour day or 40-hour workweek to which nearly all workers can be repeatedly exposed day after day without adverse effect.

Short-term exposure limits (STEL's) are maximum concentrations to which workers can be exposed for up to 15 minutes without suffering from 1) irritation, 2) chronic or irreversible tissue change or 3) narcosis of sufficient degree to materially reduce work efficiency, increase accident proneness or impair self-rescue. This limit is not cited in this volume; TWA's are usually used; the definition is given here to show the range of limits encountered.

Ceiling exposure limits (TLV-C's) are concentrations that should not be exceeded, even instantaneously. Again they are not cited here.

For particulates, mppcf means millions of particles per cubic foot of air (based on impinger samples counted by light-field techniques).

Contents

Contents

Contents

Contents **xiii**

Contents

A

ACETALDEHYDE

Description: CH_3CHO, acetaldehyde, is a flammable, volatile, colorless liquid with a characteristic odor. It is produced by oxidation of alcohol with a metallic catalyst, by hydration of acetylene, or, usually, by direct oxidation of ethylene.

Synonyms: Acetic aldehyde, aldehyde, ethanal, ethyl aldehyde.

Potential Occupational Exposures: Acetaldehyde can be reduced or oxidized to form acetic acid, acetic anhydride, acrolein, aldol, butanol, chloral, paraldehyde, and pentaerythritol. It is also used in the manufacture of disinfectants, drugs, dyes, explosives, flavorings, lacquers, mirrors (silvering), perfume, photographic chemicals, phenolic and urea resins, rubber accelerators and antioxidants, varnishes, vinegar, and yeast (1)(2).

A partial list of occupations in which exposure may occur includes:

Acetic acid makers	Paraldehyde makers
Antioxidant makers	Urea resin makers
Disinfectant makers	Rubber makers
Explosives workers	Vinegar makers
Flavoring makers	Yeast makers
Mirror silverers	

Acetaldehyde is the product of most hydrocarbon oxidations; it is a normal intermediate product in the respiration of higher plants; it occurs in traces in all ripe fruits and may form in wine and other alcoholic beverages after exposure to air. Acetaldehyde is an intermediate product in the metabolism of sugars in the body and hence occurs in traces in blood. It has been reported in fresh leaf tobacco (3) as well as in tobacco smoke (4)(5) and in automobile and diesel exhaust (6)(7).

Permissible Exposure Limits: The Federal standard (TWA) is 200 ppm (360 mg/m³); however, the ACGIH 1978 recommended TLV is 100 ppm (180 mg/m³).

Route of Entry: Inhalation of vapor.

Harmful Effects

Local—The liquid and the fairly low levels of the vapor are irritating to the eyes, skin, upper respiratory passages, and bronchi. Repeated exposure may result in dermatitis, rarely, and skin sensitization.

Systemic—Acute involuntary exposure to high levels of acetaldehyde vapors may result in pulmonary edema, preceded by excitement, followed by narcosis. It has been postulated that these symptoms may have been similar to those of alcohol, which is converted to acetaldehyde and acetic acid. Chronic effects have not been documented, and seem unlikely, since voluntary inhalation of toxicologically significant levels of acetaldehyde are precluded by its irritant properties at levels as low as 200 ppm (360 mg/m^3 of air).

Information as to the mutagenicity of aldehydes (with the exception of formaldehyde) is scant. Acetaldehyde has been found mutagenic in *Drosophila* (8).

Medical Surveillance: Consideration should be given to skin, eyes, and respiratory tract in any preplacement or periodic examinations.

Special Tests: None appear needed, but effects of exposure can be determined from blood results by gas chromatographic methods, if desired.

Personal Protective Methods: Protective clothing, gloves, goggles, and respiratory protective equipment where high concentrations of the gas or vapor are expected.

References

(1) Hayes, E.R. Acetaldehyde. *Kirk-Othmer Encyclopedia of Chemical Technology,* 2nd ed., Vol. 1, Interscience Publishers, New York (1963) pp. 75-95.
(2) Merck & Co. *The Merck Index,* Ninth ed., Merck & Co., Inc., Rahway, NJ (1976) p. 35.
(3) Shaw, W.G.J., Stephens, R.L., and Weybrew, J.A. Carbonyl constituents in the volatile oils from flue-cured tobacco. *Tobacco Sci. 4* (1960) 179.
(4) Irby, R.M., Harlow, E.S. Cigarette smoke. I-Determination of vapor constituents. *Tobacco, 148* (1959) 21.
(5) Johnstone, R.A.W. and Plimmer, J.R. The chemical constituents of tobacco and tobacco smoke. *Chem. Revs., 59* (1959) 885.
(6) Linnel, R.H., and Scott, W.E. Diesel exhaust analysis. *Arch Environ. Hlth. 5* (1962) 616.
(7) Ellis, C.F., Kendall, R.F., and Eccleston, B.H. Identification of oxygenates in automobile exhausts by combined gas-liquid chromatography and I.R. techniques. *Anal. Chem., 37* (1965) 511.
(8) Rapoport, I.A. Mutations under the influence of unsaturated aldehydes. *Dokl. Akad. Nauk SSSR, 61* (1948) 713.

Bibliography

Baker, R.N., Alenty, A.L. and Zack, J.F. 1968. Toxic volatiles in alcoholic coma. A report of

simultaneous determination of blood methanol, ethanol, isopropanol, acetaldehyde, and acetone by gas chromatography. *Bull. Los Angeles Neurol. Soc.* 33:140.

Egle, J.L. 1972. Effects of inhaled acetaldehyde and propionaldehyde on blood pressure
and heart rate. *Toxicol. Appl. Pharmacol.* 23:131.

Fairhall, L.T. 1969. *Industrial Toxicology,* 2nd ed. Hafner Publishing Company, New York
p. 138.

James, T.N., and Bear E.S. 1967. Effects of ethanol and acetaldehyde on the heart. *Am.
Heart J.* 74:243.

ACETATES

Description:
Methyl acetate: CH_3COOCH_3.
Ethyl acetate: $CH_3COOC_2H_5$.
n-Propyl acetate: $CH_3COOC_3H_7$.
Isopropyl acetate: $CH_3COOCH(CH_3)_2$.
n-Butyl acetate: $CH_3COOC_4H_9$.
Amyl acetate: $CH_3COOC_5H_{11}$.

The acetates are colorless, volatile, flammable liquids.

Synonyms:
Methyl acetate: none.
Ethyl acetate: acetic ether, vinegar naphtha.
n-Propyl acetate: acetic acid propyl ester.
Isopropyl acetate: none.
n-Butyl acetate: butyl ethanoate, acetic acid butyl ester.
Amyl acetate: isoamyl acetate, pear oil, banana oil, amyl acetate ester,
pentyl acetate.

Potential Occupational Exposures: The acetates are a group of solvents for cellulose nitrate, cellulose acetate, ethylcellulose, resins, rosin, Cumar, elemi, phenolics, oils, fats, and celluloid. They are also used in the manufacture of lacquers, paints, varnishes, enamel, perfumes, dyes, dopes, plastic and synthetic finishes (e.g., artificial leather), smokeless powder, photographic film, footwear, pharmaceuticals, food preservatives, artificial glass, artificial silk, furniture polish, odorants, and other organic syntheses.

A partial list of occupations in which exposure may occur includes:

Cellulose acetate makers	Nitrocellulose makers
Cumar makers	Paint makers
Dope makers	Perfume makers
Dye makers	Resin makers
Elemi makers	Rosin makers
Lacquer makers	Varnish makers

Permissible Exposure Limits: The Federal standards (TWA) as of 1978 are:

Methyl acetate	200 ppm	610 mg/m^3
Ethyl acetate	400 ppm	1,400 mg/m^3
n-Propyl acetate	200 ppm	840 mg/m^3
Isopropyl acetate	250 ppm	950 mg/m^3
n-Butyl acetate	150 ppm	710 mg/m^3
Isoamyl acetate	100 ppm	525 mg/m^3
n-Amyl acetate	100 ppm	530 mg/m^3
sec-Amyl acetate	125 ppm	670 mg/m^3
sec-Butyl acetate	200 ppm	950 mg/m^3
tert-Butyl acetate	200 ppm	950 mg/m^3

Route of Entry: Inhalation and ingestion.

Harmful Effects

Local—In higher concentrations, acetates are irritants to the mucous membranes. All irritate eyes and nasal passages in varying degrees. Prolonged exposure can cause irritation of the intact skin. These local effects are the primary risk in industry.

Systemic—All acetates may cause headache, drowsiness, and unconsciousness if concentrations are high enough. Those effects are relatively slow and gradual in onset and slow in recovery after exposure.

Medical Surveillance: Consider initial effects on skin and respiratory tract in any preplacement or periodical examinations, as well as liver and kidney function.

Special Tests: None in common use.

Personal Protective Methods: Barrier creams and protective clothing with gloves should be used, as well as full-face masks in areas of vapor concentration.

Bibliography

Von Oettingen, W.F. 1960. The aliphatic acids and their esters: toxicity and potential dangers. The saturated monobasic acids and their esters: aliphatic acids with three to eighteen carbons and their esters. *AMA Arch. Ind. Health* 21:100.

ACETIC ACID

Description: CH_3COOH, acetic acid, is a colorless liquid with a pungent vinegar-like odor. Glacial acetic acid contains 99% acid.

Synonyms: Ethanoic acid, ethylic acid, methane carboxylic acid, pyroligneous acid, vinegar acid.

Potential Occupational Exposures: Acetic acid is widely used as a chemical feedstock for the production of vinyl plastics, acetic anhydride, acetone, acetanilide, acetyl chloride, ethyl alcohol, ketene, methyl ethyl ketone, acetate esters, and cellulose acetates. It is also used alone in the dye, rubber, pharmaceutical, food preserving, textile, and laundry industries. It is utilized, too, in the manufacture of Paris green, white lead, tint rinse, photographic chemicals, stain removers, insecticides, and plastics.

A partial list of occupations in which exposure may occur includes:

Acetate ester makers	Plastic makers
Acetate fiber makers	Resin makers
Aspirin makers	Rubber makers
Dye makers	Stain removers
Food preservers	Textile printers
Insecticide makers	Tint rinse makers
Laundry workers	White lead makers

Permissible Exposure Limits: The Federal standard (TWA) as of 1978 is 10 ppm (25 mg/m^3).

Route of Entry: Inhalation of vapor.

Harmful Effects

Local—Acetic acid vapor may produce irritation of the eyes, nose, throat, and lungs. Inhalation of concentrated vapors may cause serious damage to the lining membranes of the nose, throat, and lungs. Contact with concentrated acetic acid may cause severe damage to the skin and severe eye damage, which may result in loss of sight. Repeated or prolonged exposure to acetic acid may cause darkening, irritation of the skin, erosion of the exposed front teeth, and chronic inflammation of the nose, throat, and bronchi.

Systemic—Bronchopneumonia and pulmonary edema may develop following acute overexposure. Chronic exposure may result in pharyngitis and catarrhal bronchitis. Ingestion, though not likely to occur in industry, may result in penetration of the esophagus, bloody vomiting, diarrhea, shock, hemolysis, and hemoglobinuria which is followed by anuria.

Medical Surveillance: Consideration should be given to the skin, eyes, teeth, and respiratory tract in placement or periodic examinations.

Special Tests: None in common use.

Personal Protective Methods: When working with glacial acetic acid, personal protective equipment, protective clothing, gloves, and goggles should be worn. Eye fountains and showers should be available in areas of potential exposure.

Bibliography

Capellini, A., and Sartorelli, E. 1967. Episodio di intossicizione collettiva da anidride acetica ed acido acetica. *Med. Lav.* 58:108.

Henson, E.V. 1959. Toxicology of the fatty acids. *J. Occup. Med.* 1:339.

Von Oettingen, W.F. 1960. The aliphatic acids and their esters: toxicity and potential dangers. The saturated monobasic acids and their esters: aliphatic acids with three to eighteen carbons and their esters. *AMA Arch. Ind. Health* 21:100.

ACETIC ANHYDRIDE

Description: $(CH_3CO)_2O$, acetic anhydride, is a colorless, strongly refractive liquid which has a strongly irritating odor.

Synonyms: Acetic oxide, acetyl oxide, ethanoic anhydride.

Potential Occupational Exposures: Acetic anhydride is used as an acetylating agent or as a solvent in the manufacture of cellulose acetate, acetanilide, synthetic fibers, plastics, explosives, resins, pharmaceuticals, perfumes, and flavorings; and it is used in the textile dyeing industry.

A partial list of occupations in which exposure may occur includes:

Acetate fiber makers	Flavoring makers
Acetic acid makers	Perfume makers
Aspirin makers	Photographic film makers
Cellulose acetate fiber makers	Plastic makers
Drug makers	Resin makers
Dye makers	Textile makers
Explosive makers	

Permissible Exposure Limits: The Federal standard (TWA) as of 1978 is 5 ppm (20 mg/m^3).

Route of Entry: Inhalation of vapor.

Harmful Effects

Local—In high concentrations, vapor may cause conjunctivitis, photophobia, lacrimation, and severe irritation of the nose and throat. Liquid acetic anhydride does not cause a severe burning sensation when it comes in contact with the skin. If it is not removed, the skin may become white and wrinkled, and delayed severe burns may occur. Both liquid and vapor may cause conjunctival edema and corneal burns, which may develop into temporary or permanent interstitial keratitis with corneal opacity due to progression of the infiltration. Contact and, occasionally, hypersensitivity dermatitis may develop.

Systemic—Immediate complaints following concentrated vapor exposure include conjunctival and nasopharyngeal irritation, cough, and dyspnea. Necrotic areas of mucous membranes may be present following acute exposure.

Medical Surveillance: Consideration should be given to the skin, eyes, and respiratory tract in any placement or periodic examinations.

Special Tests: None currently used.

Personal Protective Methods: Personal protective equipment (protective clothing, gloves, and goggles) should be used. Eye fountains and showers should be made available in areas where contact might occur.

Bibliography

Grant, W.M. 1962. *Toxicology of the Eye*. Charles C. Thomas Publishers, Springfield, Ill.
Takhirov, M.T. 1969. Hygienic standard for acetic acid and acetic anhydride. *Hyg. Sanit.* 34:122.

ACETONE (See "Ketones")

ACETONITRILE

Description: CH_3-CN, acetonitrile, is a colorless liquid with an ether-like odor.

Synonyms: Methyl cyanide, ethanenitrile, cyanomethane.

Potential Occupational Exposures: Acetonitrile is used as an extractant for animal and vegetable oils, as a solvent, particularly in the pharmaceutical industry, and as a chemical intermediate.

A partial list of occupations in which exposure may occur includes:

Animal oil processors   Vegetable oil processors
Organic chemical synthesizers

Permissible Exposure Limits: The Federal standard is 40 ppm (70 mg/m^3). This is the 1978 TWA value.

Routes of Entry: Inhalation and percutaneous absorption.

Harmful Effects

Local—At high concentrations, nose and throat irritation have been reported. Splashes of the liquid in the eyes may cause irritation. Acetonitrile may cause slight flushing of the face and a feeling of chest tightness.

Systemic—Acetonitrile has a relatively low acute toxicity, but there have been reports of severe and fatal poisonings in man after inhalation of high concentrations. Signs and symptoms may include nausea, vomiting, respiratory depression, weakness, chest or abdominal pain, hematemesis, convulsions, shock, unconsciousness, and death. In most cases there is a latent period of several hours between exposure and onset of symptoms. It has been thought that acetonitrile itself has relatively little toxic effect and that the delayed response is due to the slow release of cyanide. No chronic disease has been reported.

Medical Surveillance: Consider the skin, respiratory tract, heart, central nervous system, renal and liver function in placement and periodic examinations. A history of fainting spells or convulsive disorders might present an added risk to persons working with toxic nitriles.

Special Tests: None commonly used. Blood CN can be determined but may be of little help in evaluating low level exposures.

Personal Protective Methods: Protective clothing should be worn, and in areas of high concentration, air supplied respirators and complete skin protection are necessary. Workers in these areas must be educated to the nature of acetonitrile hazard. They should also be trained in artificial respiration and in the use of amyl nitrite antidote in emergency situations.

Bibliography

Amdur, M.L. 1959. Accidental group exposure to acetonitrile. A clinical study. *J. Occup. Med.* 1:627.
Pozzani, U.C., Carpenter, C.P., Palm, P.E., Weil, C.S. and Nair, J.H. III. 1959. An investigation of the mammalian toxicity of acetonitrile. *J. Occup. Med.* 1:634.
Rieders, F., Brieger, H., Lewis, C.E. and Amdur, M.L. 1961. What is the mechanism of toxic action of organic cyanide? *J. Occup. Med.* 3:482.

2-ACETYLAMINOFLUORENE

Description: 2-Acetylaminofluorene, is a tan crystalline solid. It has the following formula:

$$CH_3CONH-$$

Synonyms: 2-acetaminofluorene, N-2-fluorenylacetamide.

Potential Occupational Exposures: Very little 2-acetylaminofluorene is produced. It is used primarily for cancer research purposes. It was patented as a pesticide, but was never used for this purpose. Thus, occupations in which exposure may occur are those in areas of research.

Permissible Exposure Limits: 2-Acetylaminofluorene is included in the Federal standard for carcinogens; all contact with it should be avoided.

Routes of Entry: Probably by inhalation and percutaneous absorption.

Harmful Effects

Local—Unknown.

Systemic—2-Acetylaminofluorene's carcinogenic activity was first discovered in rats in which it produced nodular hyperplasia and cancer consistently in the bladder, kidney, pelvis, liver, and pancreas by ingestion. Later feeding experiments in dogs demonstrated bladder and liver tumors. Guinea pigs appear resistant to its carcinogenic effects. No human effects have been reported.

Medical Surveillance: Preplacement and periodic examinations should include history of other exposure to carcinogens, smoking history, family history, alcohol, and medications. The skin, respiratory tract, kidney, bladder, and liver should be evaluated for possible effects. Sputum and bladder cytology should be performed. Fetal effects may occur.

The scope and frequency of medical surveillance examinations can be related to the hazard, which probably is greater among research chemists or those involved in animal inhalation studies.

Special Tests: None in common use, although urinary metabolites are known.

Personal Protective Methods: Personal protective methods are designed to supplement engineering controls and to prevent all skin or inhalation exposure.

Full body protective clothing and gloves may be required. Those employed in handling operations should be provided with full-face, supplied air respirators of continuous flow or pressure demand type. On exit from a regulated area, employees should shower and change into street clothes; leaving their protective clothing and equipment at the point of exit to be placed in impervious containers at the end of the work shift for decontamination or disposal. Effective methods should be used for decontamination and changing of clothes and gloves.

Bibliography

Morris, H.P., and Eyestone, W.H. 1953. Tumors of the liver and urinary bladder of the dog after ingestion of 2-acetylaminofluorene. *J. Natl. Cancer Inst.* 13:1139.
Wilson, R.H., DeEds, F., and Cox, A.J. Jr. 1941. The toxicity and carcinogenic activity of 2-acetylaminofluorene. *Cancer Res.* 1:595.

ACETYLENE

Description: HC≡CH, acetylene, is a colorless gas with a faint ethereal odor.

Synonyms: Ethine, ethyne, narcylene.

Potential Occupational Exposures: Acetylene can be burned in air or oxygen and is used for brazing, welding, cutting, metallizing, hardening, flame scarfing, and local heating in metallurgy. The flame is also used in the glass industry. Chemically, acetylene is used in the manufacture of vinyl chloride, acrylonitrile, synthetic rubber, vinyl acetate, trichloroethylene, acrylate, butyrolactone, 1,4-butanediol, vinyl alkyl ethers, pyrrolidone, and other substances.

A partial list of occupations in which exposure may occur includes:

Acetaldehyde makers	Dye makers
Acetone makers	Foundry workers
Alcohol makers	Gougers
Braziers	Hardeners
Carbon black makers	Heat treaters
Ceramic makers	Lead burners
Copper purifiers	Rubber makers
Descalers	Scarfers
Drug makers	

Permissible Exposure Limits: No Federal standard has been established. NIOSH has recommended a ceiling limit of 2,500 ppm.

Route of Entry: Inhalation of gas.

Harmful Effects

Local—Acetylene is nonirritating to skin or mucous membranes.

Systemic—At high concentrations pure acetylene may act as a mild narcotic and asphyxiant. Most accounted cases of illness or death can be attributed to acetylene containing impurities of arsine, hydrogen sulfide, phosphine, carbon disulfide, or carbon monoxide.

Initial signs and symptoms of exposure to harmful concentrations of impure acetylene are rapid respiration, air hunger, followed by impaired mental alertness and muscular incoordination. Other manifestations include cyanosis, weak and irregular pulse, nausea, vomiting, prostration, impairment of judgment and sensation, loss of consciousness, convulsions, and death. Low order sensitization of myocardium to epinephrine resulting in ventricular fibrillation may be possible.

Medical Surveillance: No specific considerations are needed.

Special Tests: None in common use.

Personal Protective Methods: Acetylene poisoning can quite easily be prevented if 1) there is adequate ventilation and 2) impurities are removed when acetylene is used in poorly ventilated areas. General industrial hygiene practices for welding, brazing, and other metallurgical processes should also be observed.

Bibliography

Jones, A.T. 1960. Fatal gassing at an acetylene manufacturing plant. *Arch. Environ. Health* 5:417.
National Institute for Occupational Safety and Health. *Criteria for a Recommended Standard: Occupational Exposure to Acetylene.* NIOSH Doc. No. 76-195, Wash., DC (1976).
Ross, D.S. 1970. Loss of consciousness in a burner using oxyacetylene flame in a confined space. *Ann. Occup. Hyg.* 13:159.

ACRIDINE

Description: $C_{13}H_9N$, acridine, is a colorless or light yellow crystal, very soluble in boiling water.

Synonyms: Dibenzopyridine, 10-azaanthracene.

Potential Occupational Exposures: Acridine and its derivatives are widely used in the production of dyestuffs such as acriflavine, benzoflavine, and chrysaniline, and in the synthesis of pharmaceuticals such as aurinacrine, proflavine, and rivanol.

A partial list of occupations in which exposure may occur includes:

Chemical laboratory workers	Drug makers
Coal tar workers	Dye makers
Disinfectant makers	Organic chemical synthesizers

Permissible Exposure Limits: There is no Federal standard for acridine.

Route of Entry: Inhalation of vapor.

Harmful Effects

Local—Acridine is a severe irritant to the conjunctiva of the eyes, the mucous membranes of the respiratory tract, and the skin. It is a powerful photosensitizer of the skin. Acridine causes sneezing on inhalation.

Systemic—Yellowish discoloration of sclera and conjunctiva may occur. Mutational properties have been ascribed to acridine, but its effect on humans is not known.

Medical Surveillance: Evaluate the skin, eyes, and respiratory tract in the course of any placement or periodic examinations.

Special Tests: None commonly used. Can be detected in blood or urine.

Personal Protective Methods: Prevent skin, eye, or respiratory contact with protective clothing, gloves, goggles, and appropriate dust respirators. In case of spills or splashes, the skin area should be thoroughly washed and the contaminated clothing changed. Clean work clothing should be supplied on a daily basis, and the worker should shower prior to changing to street clothes.

Bibliography

Baldi, G. 1953. Patologia professionale de acridina. *Med. Lav.* 44:240.
Sawicki, E., and Engel, C.R. 1969. Fluorimetric estimation of acridine in airborne and other particulates. *Mikrochim. Acta.* 1:91.

ACROLEIN

Description: $CH_2=CHCHO$, acrolein, is a clear, yellowish liquid which is a petroleum by-product. It is commercially produced by the oxidation of propylene.

Synonyms: Acraldehyde, acrylic aldehyde, allyl aldehyde, propenal.

Potential Occupational Exposures: Acrolein is the feedstock for several types of plastics, plasticizers, acrylates, textile finishes, synthetic fibers, and methionines. It is also used in the manufacture of colloidal forms of metals, in perfumes, and, due to its pungent odor, as a warning agent in methyl chloride refrigerants. Other potential exposures may arise when acrolein vapor is given off when oils and fats containing glycerol are heated.

A partial list of occupations in which exposure may occur includes:

Acrylate makers	Refrigerant makers
Fat processors	Renderers
Methionine makers	Rubber makers
Perfume makers	Textile resin makers
Plastic makers	

Permissible Exposure Limits: The Federal standard for exposure to acrolein is 0.1 ppm (0.25 mg/m^3). This is the TWA value as of 1978.

Route of Entry: Inhalation of vapor and percutaneous absorption.

Harmful Effects

Local—In the liquid or pungent vapor form, acrolein produces intense irritation to the eye and mucous membranes of the respiratory tract. Skin burns and dermatitis may result from prolonged or repeated exposure. Sensitization in a few individuals may also occur.

Systemic—Because of acrolein's pungent, offensive odor and the intense irritation of the conjunctiva and upper respiratory tract, severe toxic effects from acute exposure are rare, as workmen will not tolerate the vapor even in minimal concentrations. Acute exposure to acrolein may cause bronchial inflammation, resulting in bronchitis or pulmonary edema.

Acrolein (as well as formaldehyde) have been found to be among the most cytotoxic aldehydes (or ketones) of a large number of organic solvents examined in short-term in vitro incubations with Ehrlich-Landschuetz diploid (ELD) ascites tumor cells (1).

In a comparison of the cytotoxicity of short-chain alcohols and aldehydes in cultured neuroblastoma cells it was found that the cytotoxicity of the alcohols increased as the number of carbons in the compound increased, whereas toxicity of the aldehydes increased with decreasing chain length (3). The marked cytotoxicity of acrolein was ascribed to the presence of both the carbonyl and the carbon-carbon double bond since propionaldehyde, having only the carbonyl group, and allyl alcohol, having only the C=C double bond, were less toxic (2).

Acrolein has been found mutagenic in *Drosophila* (3).

Medical Surveillance: Preplacement and periodic medical examinations should consider respiratory, skin, and eye disease. Pulmonary function tests may be helpful.

Special Tests: None in common use.

Personal Protective Methods: Protection in handling and transporting is advocated. Suitable ventilation and protective clothing should be provided for employees working in areas of possible exposure; protective respiratory equipment in areas of vapor concentration.

References

(1) Holmberg, B. and Malmfors, T. Cytotoxicity of some organic solvents. *Env. Res., 7* (1974) 183.
(2) Koerker, R.L., Berlin, A.J., and Schneider, F.H. The cytotoxicity of short-chain alcohols and aldehydes in cultured neuroblastoma cells. *Toxicol. Appl. Pharmacol.* 37 (1976) 281-288.
(3) Rapoport, I.A. Mutations under the influence of unsaturated aldehydes. *Dokl. Akad. Nauk SSSR* 61 (1948) 713.

Bibliography

Champeix, J., Courtial, L., Perche, E. and Catalina P. 1966. Bronchopneumopathe aigue par vapeurs d'acroleine. *Arch. Mal. Prof.* 17:794.
Gusev, M.I., Schechnikova, A.I., Dronov, I.S., Grebenskova, M.D. and Golovina, A.I. 1966. Determination of the daily average maximum permissible concentration of acrolein in the atmosphere. *Hyg. Sanit.* 31(1):8.

ACRYLAMIDE

Description: $CH_2=CHCONH_2$, acrylamide, in monomeric form consists of flake-like crystals which melt at 84.5°C. It may be stored in a cool, dark place. The monomer readily polymerizes at the melting point or under UV light.

Synonyms: Propenamide.

Potential Occupational Exposures: The major application for monomeric acrylamide is in the production of polymers as polyacrylamides. Polyacrylamides are used for soil stabilization, gel chromatography, electrophoresis, papermaking strengtheners, clarification and treatment of potable water, and foods. Approximately 70 million pounds of acrylamide were produced in 1974 in the United States. NIOSH estimates that approximately 20,000 workers in the United States are potentially exposed to acrylamide.

Permissible Exposure Limits: NIOSH recommends adherence to the present (1978) Federal standard of 0.3 mg/m^3 as a time-weighted average concentration for up to a 10-hour workday, 40-hour workweek.

Routes of Entry: Acrylamide can be absorbed through unbroken skin.

Harmful Effects

Local—Localized effects include peeling and redness of the skin of the hands and less often of the feet, numbness of the lower limbs, and excessive sweating of the feet and hands.

Systemic—The systemic effects due to acrylamide intoxication involve central and peripheral nervous system damage manifested primarily as ataxia, weak or absent reflexes, positive Romberg's sign and loss of vibration and position senses.

Medical Surveillance: Since skin contact with the substance may result in localized or systemic effects, NIOSH recommends that medical surveillance be made available to all employees working in an area where acrylamide is stored, produced, processed, or otherwise used, except as an unintentional contaminant in other materials at a concentration of less than 1% by weight.

Special Tests: None.

Personal Protective Methods: Engineering controls should be used wherever feasible to maintain airborne acrylamide concentrations below the prescribed limit, and respirators should be used only in nonroutine or emergency situations which may result in exposure concentrations in excess of the TWA environmental limit. Personal protective clothing is recommended for all workers occupationally exposed to acrylamide to further reduce the likelihood of skin contact with the substance.

Bibliography

National Institute for Occupational Safety and Health. *Criteria for a Recommended Standard: Occupational Exposure to Acrylamide,* NIOSH Doc. No. 77-112, Wash., DC (1977).

ACRYLONITRILE

Description: $CH_2=CH-CN$, acrylonitrile, is a colorless liquid with a faint acrid odor. It is both flammable and explosive.

Synonyms: Vinyl cyanide, cyanoethylene, propene nitrile.

Potential Occupational Exposures: Acrylonitrile is used in the manufacture of synthetic fibers, acrylostyrene plastics, acrylonitrile-butadiene-styrene plastics, nitrile rubbers, chemicals, and adhesives. It is also used as a pesticide.

A partial list of occupations in which exposure may occur includes:

Acrylic resin makers   Rubber makers
Organic chemical synthesizers   Synthetic fiber makers
Pesticide workers   Textile finish makers

NIOSH estimates (2) that approximately 125,000 persons are potentially exposed to acrylonitrile in the workplace.

Permissible Exposure Limits: The Federal standard has been 20 ppm (45 mg/m^3). By an emergency temporary standard (ETS), however, the Occupational Safety and Health Administration (OSHA) has amended in early 1978 (1) its present standard concerning employee exposure to acrylonitrile (AN) (also known as vinyl cyanide) and reduces the permissible exposure level from 20 parts acrylonitrile per million parts of air (20 ppm), as an 8-hour time-weighted average concentration, to 2 ppm, with a ceiling level of 10 ppm for any 15-minute period during the 8-hour day.

In addition, the standard includes an action level of 1 ppm as an 8-hour time-weighted average. Provision is also made for specific exemptions in the standard for certain operations involving the processing, use and handling of products fabricated from polyacrylonitrile (PAN). The basis for this ETS is OSHA's determination that laboratory and epidemiological data indicate that continued exposure to acrylonitrile presents a cancer hazard to workers. A grave danger therefore exists for workers exposed to AN, necessitating the issuance of an emergency standard to protect them.

In addition, the ETS requires the measurement and control of employee exposure, personal protective equipment and clothing, employee training, medical surveillance, work practices, and record keeping. It should be noted that as recently as Sept. 29, 1977 (2), the permissible exposure level was 4 ppm in air, so this limit is under continuing reevaluation. The economic impact of such changes is also being considered (3). The ETS was made permanent in late 1978 (4).

Routes of Entry: Inhalation and percutaneous absorption. It may be absorbed from contaminated rubber or leather.

Harmful Effects

Local—Acrylonitrile may cause irritation of the eyes. Repeated and prolonged exposure may produce skin irritation. When acrylonitrile is held in contact with the skin (e.g., after being absorbed into shoe leather or clothing), it may

produce blistering after several hours of no apparent effect. Unless the contaminated clothing is removed promptly and the area washed off, blistering
will occur.

Systemic—Acrylonitrile exposure may produce nausea, vomiting, headache,
sneezing, weakness, and light-headedness. Exposure to high concentrations
may produce profound weakness, asphyxia, and death.

Medical Surveillance: Consider the skin, respiratory tract, heart, central nervous system, renal and liver function in placement and periodic examinations. A history of fainting spells or convulsive disorders might present an
added risk to persons working with toxic nitriles.

Special Tests: None commonly used.

Personal Protective Methods: Leather should not be used in protective
clothing since it is readily penetrated by acrylonitrile. Rubber clothing should
be frequently washed and inspected because it will soften and swell.
Acrylonitrile should be handled with all of the same precautions as taken for
hydrogen cyanide, and workers' education should be identical. Liquid
splashed on skin should be immediately washed off. Eyes should be protected from splash (goggles), and, in areas of vapor concentration, special
cyanide masks or air supplied masks should be provided. Workers should be
trained in artificial respiration and in the use of amyl nitrite antidote in emergency situations.

References

(1) *Federal Register* 43, No. 11, 2586-2621 (Jan. 17, 1978).
(2) National Institute for Occupational Safety and Health. *Criteria for a Recommended
    Standard: Occupational Exposure to Acrylonitrile.* NIOSH Doc. No. 78-116, Wash.,
    DC, (1978).
(3) Department of Labor. *Economic Impact Assessment for Acrylonitrile.* Wash., DC, Oc-
    cupational Safety and Health Administration (Feb. 21, 1978).
(4) *Federal Register* 43, No. 192, 45762-45819 (Oct. 3, 1978).

Bibliography

Brieger H., Reiders, F. and Hodes, W.A. 1952. Acrylonitrile: spectrophotometric deter-
    mination, acute toxicity, and mechanism of action. *AMA Arch. Ind. Hyg. Occup.
    Med.* 6:128.
Hashimoto, K., and Kanai, R. 1956. Studies on the toxicology of acrylonitrile. Metabolism
    mode of action and therapy. *Ind. Health* 3:30.
Szabo, S., and Selye, H. 1971. Adrenal apoplexy and necrosis produced by acrylonitrile.
    Endokrinologie 57:405.
Wilson, R.H., and McCormick, W.E. 1949. Acrylonitrile — its physiology and toxicology.
    Ind. Med. 18:243.
Wolfsie, J.H. 1951. Treatment of cyanide poisoning in industry. *AMA Arch. Ind. Hyg.
    Med.* 4:417.

ALICYCLIC HYDROCARBONS

Description:

Cyclopropane: C_3H_6
Cyclohexane: C_6H_{12}
Cyclohexene: C_6H_{10}
Methylcyclohexane: C_7H_{14}

Alicyclic hydrocarbons are saturated or unsaturated molecules in which three or more carbon atoms are joined to form a ring structure. The saturated compounds are called cycloalkanes, cycloparaffins, or naphthenes. The cyclic hydrocarbons with one or more double bonds are called cycloalkenes or cyclo-olefins. These compounds are colorless liquids.

Synonyms: None.

Potential Occupational Exposures: Uses vary with compounds. Cyclopropane is used as an anesthetic. Cyclohexane is used as a chemical intermediate, as a solvent for fats, oils, waxes, resins, and certain synthetic rubbers, and as an extractant of essential oils in the perfume industry. Cyclohexene is used in the manufacture of adipic, maleic, and cyclohexane carboxylic acid. Methylcyclohexane is used as a solvent for cellulose ethers and in the production of organic synthetics.

A partial list of occupations in which exposure may occur includes:

Adipic acid makers	Paint removers
Benzene makers	Plastic molders
Fat processors	Resin makers
Fungicide makers	Rubber makers
Lacquerers	Varnish removers
Nylon makers	Wax makers
Oil processors	

Permissible Exposure Limits: The Federal standards (TWA) as of 1978 are:

Cyclohexane 300 ppm (1,050 mg/m^3);
Cyclohexene 300 ppm (1,015 mg/m^3);
Methylcyclohexane 400 ppm (1,600 mg/m^3).

Presently there is no standard for cyclopropane. [Note: The 1978 ACGIH lists a TLV of 300 ppm (850 mg/m^3) for cyclopentane.]

Route of Entry: Inhalation of gas or vapor.

Harmful Effects

Local—Repeated and prolonged contact with liquid may cause defatting of the skin and a dry, scaly, fissured dermatitis. Mild conjunctivitis may result from acute vapor exposure.

Systemic—Alicyclic hydrocarbons are central nervous system depressants, although their acute toxicity is low. Symptoms of acute exposure are excitement, loss of equilibrium, stupor, coma, and, rarely, death as a result of respiratory failure. The concentration of cyclopropane required to produce surgical anesthesia is low, and there is wide margin between anesthetic and toxic concentrations. The myocardium may become more sensitive to epinephrine during narcosis with cyclopropane. Severe diarrhea and vascular collapse resulting in heart, lung, liver, and brain degeneration have been reported in oral administration of alicyclic hydrocarbons to animals.

The danger of chronic poisoning is relatively slight because these compounds are almost completely eliminated from the body. Metabolism of cyclohexane, for example, results in cyclohexanone and cyclohexanol entering the bloodstream and does not include the metabolites of phenol, as with benzene. Damage to the hematopoietic system does not occur except when exposure is compounded with benzene, which may be a contaminant. Alicyclic hydrocarbons are excreted in the urine as sulfates or glucuronides, the particular content of each varying. Small quantities of these compounds are not metabolized and may be found in blood, urine, and expired breath.

Medical Surveillance: Consider possible irritant effects to the skin and respiratory tract in any preplacement or periodic examination, as well as any renal or liver complications.

Special Tests: None in common use. Some metabolites have been found in blood and urine.

Personal Protective Methods: Skin protection with barrier creams or gloves. Workers exposed to high concentrations of gas or vapor may need masks.

Bibliography

Fabre, R., Truhaut, R., and Laham, S. 1959. Etude de metabolisme du cyclohexane chez lapin. *C.R. Acad. Sci.,* Ser. D. 248:1081.

ALKANES (C_5–C_8)

Description: The alkanes have the formula C_nH_{2n+2} and are colorless, flammable liquids, such as pentane, C_5H_{12}, boiling at 36.1°C and octane, C_8H_{18}, boiling at 125.6°C.

Synonyms: Paraffins, paraffin hydrocarbons. See also entry under Hexane.

Potential Occupational Exposure: The C_5-C_8 alkanes are used in a variety of industrial applications and processes. A major use of pentane is in the formulation of gasoline. Hexane is used as a solvent in glues, varnishes, cements and other products such as inks, and as a means of extracting natural oils from seeds. Heptane and octane are used principally as solvents. NIOSH estimates that 10,000 workers in the United States are potentially exposed to pentane and heptane, 2.5 million workers are potentially exposed to hexane, and 300,000 workers are potentially exposed to octane.

Permissible Exposure Limits: NIOSH recommends that exposure to individual C_5-C_8 alkanes or mixtues of these be limited to a TWA (time-weighted average) concentration of 350 mg/m^3. In addition, a ceiling concentration of 1,800 mg/m^3, determined on the basis of a 15-minute sampling period, is recommended. These recommendations apply to exposure of workers to alkanes which are aliphatic hydrocarbons with the empirical formula C_nH_{2n+2} where n = 5, 6, 7, or 8. Occupational exposure to these alkanes has been defined as exposure above an action level of 200 mg/m^3.

Route of Entry: Inhalation of vapor.

Harmful Effects

Chronic exposure to hexane at airborne concentrations of 1,800 mg/m^3 and above has been associated with the development of polyneuropathy. Professional judgment would indicate that a TWA concentration of 350 mg/m^3 offers a sufficient margin of safety to protect against the development of chronic nerve disorders in workers exposed to hexane. Although polyneuropathy in humans has not been attributed to exposure to pentane, heptane, or octane, evidence in workers exposed to alkane mixtures suggests that these hydrocarbons are similar in toxicity to hexane.

It has been reported that similar effects result from acute exposures to hexane and to heptane. It seems likely that components of an alkane mixture may exert additive toxic effects. For these reasons, TWA concentrations of 350 mg/m^3 are recommended as environmental limits for pentane, hexane, heptane, octane, and mixtures of C_5-C_8 alkanes.

Exposure to heptane at concentrations as low as 4,100 mg/m^3 has caused slight vertigo in workers after several minutes. Marked vertigo has developed in workers exposed to hexane at concentrations several times greater than those of heptane. In order to protect workers from these acute effects, ceiling concentration limits of 1,800 mg/m^3 are recommended for the C_5-C_8 alkanes.

Medical Surveillance: It is recommended that preplacement and annual examinations be made available to all workers occupationally exposed to alkanes. These shall include physical examinations which give particular attention to tests of nervous system function and evidence of skin conditions.

Special Tests: Recommended sampling and analytical methods have been set forth in the NIOSH/OSHA Standards Completion Program. According to the procedure, a known volume of air is drawn through a charcoal tube to collect alkane vapors. The alkane vapors trapped on charcoal are desorbed with carbon disulfide. An aliquot of the desorbed sample is analyzed with a gas chromatograph. The areas of the resulting peaks are determined and compared with those of injected standards. The sampling method was chosen because it is believed to be the best available at the present time, and is expected to provide adequate collection efficiency for airborne C_5-C_8 alkanes. The gas-chromatographic method of analysis was selected because it is sensitive and relatively simple to perform. The method is not entirely specific for these alkanes.

Personal Protective Methods

Eye Protection—Full-facepiece respirators, safety glasses, or chemical safety goggles shall be provided and worn by workers during those operations in which pentane, hexane, heptane, or octane may splash into the eyes. Face shields may be used to augment chemical safety goggles and safety glasses where full facial protection is needed, but face shields are not adequate for eye protection when used alone.

Respiratory Protection—Engineering controls shall be used wherever feasible to maintain alkane concentrations below the recommended environmental limits. Compliance with the permissible exposure limit by the use of respirators is allowed only while required engineering controls are being installed or tested, when nonroutine maintenance or repair is being accomplished, or during emergencies. When a respirator is thus permitted, it shall be selected and used in accordance with the NIOSH Criteria Document (1).

Skin Protection—The employer shall provide protective apparel, including gloves, aprons, suits, boots, or face shields (8-inch minimum) with goggles, and ensure that they are worn where needed to prevent skin contact with liquid alkanes. Protective apparel shall be made of materials which most effectively prevent skin contact under the conditions for which it is deemed necessary. Rubber articles may be used provided care is taken to ensure that permeation does not occur during usage. Protective apparel should be discarded at the first sign of deterioration.

Reference

(1) National Institute for Occupational Safety and Health. *Criteria for a Recommended Standard: Occupational Exposure to Alkanes.* NIOSH Doc. No. 77-151, Wash., DC (1977).

ALLYL ALCOHOL

Description: $CH_2=CHCH_2OH$, allyl alcohol, is a colorless liquid with a pungent odor.

Synonyms: Vinyl carbinol, propenyl alcohol, 2-propenol-1, propenol-3.

Potential Occupational Exposures: Allyl alcohol is primarily used in the production of allyl esters. These compounds are used as monomers and prepolymers in the manufacture of resins and plastics. Allyl alcohol is also used in the preparation of pharmaceuticals, in organic syntheses, and as a fungicide and herbicide.

A partial list of occupations in which exposure may occur includes:

Acrolein makers   Herbicide makers
Allyl ester makers   Organic chemical synthesizers
Drug makers   Plasticizer makers
Fungicide makers   Resin makers
Glycerine makers

Permissible Exposure Limits: The Federal standard (TWA) as of 1978 is 2 ppm (5 mg/m^3).

Route of Entry: Inhalation of vapor; percutaneous absorption of liquid.

Harmful Effects

Local—Liquid and vapor are highly irritating to eyes and upper respiratory tract. Skin irritation and burns have occurred from contact with liquid but are usually delayed in onset and may be prolonged.

Systemic—Local muscle spasms occur at sites of percutaneous absorption. Pulmonary edema, liver and kidney damage, diarrhea, delirium, convulsions, and death have been observed in laboratory animals, but have not been reported in man.

Medical Surveillance: Preplacement and periodic examinations should include the eyes, skin, respiratory tract, and liver and kidney function.

Special Tests: No specific test is available.

Personal Protective Methods: Protective clothing to prevent skin contact should be made of neoprene; these must be discarded at the first sign of deterioration. The odor and irritant properties of allyl alcohol should be sufficient warning to prevent serious injury.

Bibliography

Dunlap, M.K., Kodama, D.K., Wellington, J.S., Anderson, H.H. and Hine, C.J. 1958. The toxicity of allyl alcohol. *AMA Arch. Ind. Health* 18:303.
Torkelson, T.R., Wolf, M.A., Oyen, F. and Rowe, V.K. 1959. Vapor toxicity of allyl alcohol as determined on laboratory animals. *Am. Ind. Hyg. Assoc. J.* 20:224.

ALLYL CHLORIDE

Description: Allyl chloride, $CH_2=CHCH_2Cl$, is a highly-reactive liquid halogenated hydrocarbon (boiling at 44°-45°C) with an unpleasant, pungent odor.

Synonyms: 3-chloro-1-propene, 3-chloropropylene, chlorallylene.

Potential Occupational Exposures: Allyl chloride is produced by the high-temperature chlorination of propylene, and utilized primarily in the manufacture of epichlorohydrin and glycerol. Although production in 1973 was believed to be about 300 million pounds, most of the usage of this compound occurs within divisions of Shell Oil Company and Dow Chemical USA, where it is produced. Potential exposure extends to no more than 5,000 workers.

Permissible Exposure Limits: NIOSH recommends adherence to the present Federal standard of 1 ppm of allyl chloride as a time-weighted average for up to a 10-hour workday, 40-hour workweek and proposes the addition of a 3 ppm ceiling concentration for any 15-minute period.

Route of Entry: Employees may be exposed by dermal or eye contact, inhalation, or ingestion.

Harmful Effects

The potential for explosions and for damage to the respiratory tract, liver, and kidneys from inhalation was recognized through experimental evidence during the early commercial development of the industry involving this compound and is reflected in the precautions taken during its manufacture and use. NIOSH has found no reports of known cases of either acute or chronic exposure in the United States leading to impairment of the abovementioned organs. Industrial experience in the United States has pointed to such problems as orbital pain and deep-seated aches after eye or skin contact. Both of these phenomena are believed to be transient when they occur and have been minimized through improved work practices.

Medical Surveillance: Preplacement and periodic physical examinations have been detailed by NIOSH (1). They give special attention to the respiratory system, liver, kidneys, skin and eyes.

Special Tests: None

Personal Protective Methods

Respiratory Protection—Engineering controls shall be used to maintain allyl chloride vapor concentrations below the permissible exposure limits. Compliance with the permissible exposure limits may be achieved by the use of respirators only in the following situations:

(A) During the time necessary to install or test the required engineering controls.

(B) For nonroutine operations, such as maintenance or repair activities, in which concentrations in excess of the permissible exposure limits may occur.

(C) During emergencies when air concentrations of allyl chloride may exceed the permissible limits.

When a respirator is permitted, it shall be selected and used pursuant to the detailed requirement set forth in the NIOSH Criteria Document (1).

Eye Protection—Full-facepiece respirators or chemical safety goggles shall be provided and worn for operations in which allyl chloride may splash into the eyes. Face shields may be used to augment chemical safety goggles where full facial protection is needed, but face shields, used alone, are not adequate for eye protection. Eye protection shall be selected and used in accordance with 29 CFR 1910.133.

Skin Protection—Appropriate protective apparel, including gloves, aprons, suits, boots, or face shields (8-inch minimum) shall be provided and worn where needed to prevent skin contact with liquid allyl chloride. Protective apparel shall be made of materials which most effectively prevent skin contact under the conditions for which it is deemed necessary. Since leather articles cannot be effectively decontaminated, they shall be prohibited for use as protective apparel. Rubber articles may be used provided care is taken to ensure that permeation does not occur during usage. Protective apparel should be discarded at the first sign of deterioration.

Reference

(1) National Institute for Occupational Safety and Health. *Criteria for a Recommended Standard: Occupational Exposure to Allyl Chloride.* NIOSH Doc. No. 76-204, Wash., DC, (1976).

ALUMINUM AND COMPOUNDS

Description: Al, aluminum, is a light, silvery-white, soft, ductile, malleable amphoteric metal, soluble in acids or alkali, insoluble in water. The primary sources are the ores cryolite and bauxite; aluminum is never found in the elemental state.

Synonyms: None.

Potential Occupational Exposures: Most hazardous exposures to aluminum occur in smelting and refining processes. Aluminum is mostly produced by electrolysis of Al_2O_3 dissolved in molten cryolite (Na_3AlF_6). Aluminum is alloyed with copper, zinc, silicone, magnesium, manganese, and nickel; special additives may include chromium, lead, bismuth, titanium, zirconium, and vanadium. Aluminum and its alloys can be extruded or processed in rolling mills, wireworks, forges, or foundries, and are used in the shipbuilding, electrical, building, aircraft, automobile, light engineering, and jewelry industries. Aluminum foil is widely used in packaging. Powdered aluminum is used in the paints and pyrotechnic industries. Alumina has been used for abrasives, refractories, and catalysts, and in the past in the first firing of china and pottery. Aluminum chloride is used in petroleum processing and in the rubber industry. Alkyl aluminum compounds find use as catalysts in the production of polyethylene.

A partial list of occupations in which exposure may occur includes:

Aluminum alloy grinders	Foundry workers
Aluminum workers	Petroleum refinery workers
Ammunition makers	Plastic makers
Fireworks makers	Rubber makers

Permissible Exposure Limits: There is no Federal standard specifically for metallic aluminum. It may be considered as a nuisance dust, the applicable standards being: respirable fraction, 15 mppcf or 5 mg/m^3; total dust, 50 mppcf or 15 mg/m^3. A TWA value of 10 mg/m^3 was proposed by ACGIH in 1978 for aluminum metal and oxide. A TWA value of 2 mg/m^3 was proposed by ACGIH in 1978 for soluble aluminum salts.

Route of Entry: Inhalation of dust or fume.

Harmful Effects

Local—Particles of aluminum deposited in the eye may cause necrosis of the cornea. Salts of aluminum may cause dermatoses, eczema, conjunctivitis, and irritation of the mucous membranes of the upper respiratory system by the acid liberated by hydrolysis.

Systemic—The effects on the human body caused by the inhalation of aluminum dust and fumes are not known with certainty at this time. Present data suggest that pneumoconiosis might be a possible outcome. In the majority of cases investigated, however, it was found that exposure was not to aluminum dust alone, but to a mixture of aluminum, silica fume, iron dusts, and other materials.

Medical Surveillance: Preemployment and periodic physical examinations should give special consideration to the skin, eyes, and lungs. Lung function should be followed.

Special Tests: None commonly used.

Personal Protective Methods: Workers in electrolysis manufacturing plants should be provided with respirators for protection from fluoride fumes. Dust masks are recommended in areas exceeding the nuisance levels. Aluminum workers generally should receive training in the proper use of personal protective equipment. Workers involved with salts of aluminum may require protective clothing, barrier creams, and where heavy concentrations exist, full-face air supplied respirators may be indicated.

Bibliography

Corrin, B. 1963. Aluminum pneumoconiosis. *Br. J. Ind. Med.* 20:264.

Evenshtein, Z.M. 1967. Toxicity of aluminum and its inorganic compounds. *Hyg. Sanit.* 32:244.

Kaltreider, N.E., Elder, M.J., Cralley, L.V. and Colwell, M.O. 1972. Health survey of aluminum workers with special reference to fluoride exposure. *J. Occup. Med.* 14:531.

Posner, E., and Kennedy, M.C.S. 1967. A further study of china biscuit placers in Stoke-on-Trent. *Br. J. Ind. Med.* 24:133.

AMINODIPHENYL

Description: $C_6H_5C_6H_4NH_2$, 4-aminodiphenyl, is a yellowish brown crystal.

Synonyms: Biphenyline, p-phenylaniline, xenylamine, 4-aminobiphenyl, 4-biphenylamine, p-aminobiphenyl, p-aminodiphenyl, p-biphenylamine.

Potential Occupational Exposures: It is no longer manufactured commercially and is only used for research purposes. 4-Aminodiphenyl was formerly used as a rubber antioxidant and as a dye intermediate.

A partial list of occupations in which exposure may occur includes:

> Diphenylamine workers
> Research workers

Permissible Exposure Limits: 4-Aminodiphenyl is included in the Federal standards for carcinogens; all contact with it should be avoided.

Routes of Entry: Inhalation and percutaneous absorption.

Harmful Effects

Local—None reported.

Systemic—4-Aminodiphenyl is a known human bladder carcinogen. An exposure of only 133 days has been reported to have ultimately resulted in a bladder tumor. The latent period is generally from 15 to 35 years. Acute exposure produces headaches, lethargy, cyanosis, urinary burning, and hematuria. Cystoscopy reveals diffuse hyperemia, edema, and frank slough.

Medical Surveillance: Placement and periodic examinations should include an evaluation of exposure to other carcinogens; use of alcohol, smoking, and medications; and family history. Special attention should be given on a regular basis to urine sediment and cytology. If red cells or positive smears are seen, cystoscopy should be done at once. The general health of exposed persons should also be evaluated in periodic examinations.

Special Tests: None commonly used. One urinary metabolite is 3-amino-4-hydroxydiphenyl.

Personal Protective Methods: These are designed to supplement engineering controls and to prevent all skin or respiratory contact. Full body protective clothing and gloves should be used by those employed in handling operations. Fullface, supplied air respirators of continuous flow or pressure demand type should also be used. On exit from a regulated area, employees should shower and change into street clothes, leaving their clothing and equipment at the point of exit to be placed in impervious containers at the end of the work shift for decontamination or disposal. Effective methods should be used to clean and decontaminate gloves and clothing.

Bibliography

Melick, W.F., Escue, H.M., Naryka, J.J., Mezera, R.A. and Wheeler, E.P. 1955. The first reported cases of human bladder tumors due to a new carcinogen—xenylamine. *J. Urol.* 74:760.

Melick, W.F., and Naryka, J.J. 1968. Carcinoma in situ of the bladder in workers exposed to xenylamine: diagnosis by ultraviolet light cystoscopy. *J. Urol.* 99:178.

Melick, W.F., Naryka, J.J. and Kelly, E.R. 1971. Bladder cancer due to exposure to para-aminobiphenyl: a 17-year follow-up. *J. Urol.* 106:220.

AMMONIA

Description: NH_3, ammonia, is a colorless, strongly alkaline, and extremely soluble gas with a characteristic pungent odor.

Synonyms: None.

Potential Occupational Exposures: Ammonia is used as a nitrogen source for many nitrogen-containing compounds. It is used in the production of ammonium sulfate and ammonium nitrate for fertilizers and in the manufacture of nitric acid, soda, synthetic urea, synthetic fibers, dyes, and plastics. It is also utilized as a refrigerant and in the petroleum refining, chemical, and pharmaceutical industries.

Other sources of occupational exposure include the silvering of mirrors, glue-making, tanning of leather, and around nitriding furnaces. Ammonia is produced as a by-product in coal distillation and by the action of steam on calcium cyanamide, and from the decomposition of nitrogenous materials.

A partial list of occupations in which exposure may occur includes:

Aluminum workers	Metal powder processors
Annealers	Mirror silverers
Chemical laboratory workers	Paper makers
Chemical workers	Paper pulp makers
Dye makers	Pesticide makers
Electroplate workers	Rayon makers
Fertilizer workers	Refrigeration workers
Galvanizers	Sulfuric acid workers
Glue makers	Tannery workers
Metal extractors	Water treaters

Permissible Exposure Limits: The Federal standard for ammonia is an 8-hour time-weighted average of 50 ppm (35 mg/m^3). NIOSH has recommended 50 ppm expressed as a ceiling and determined by a 5-minute sampling period. ACGIH has set TWA values of 25 ppm (18 mg/m^3) as of 1978.

Route of Entry: Inhalation of gas.

Harmful Effects

Local—Contact with anhydrous liquid ammonia or with aqueous solutions is intensely irritating to the mucous membranes, eyes, and skin. Eye symptoms range from lacrimation, blepharospasm, and palpebral edema to a rise of intraocular pressure, and other signs resembling acute-angle closure glaucoma, corneal ulceration, and blindness. There may be corrosive burns of skin or blister formation. Ammonia gas is also irritating to the eyes and to moist skin.

Systemic—Mild to moderate exposure to the gas can produce headache, salivation, burning of throat, anosmia, perspiration, nausea, vomiting, and substernal pain. Irritation of ammonia gas in eyes and nose may be sufficiently intense to compel workers to leave the area. If escape is not possible, there may be severe irritation of the respiratory tract with the production of cough, glottal edema, bronchospasm, pulmonary edema, or respiratory arrest. Bronchitis or pneumonia may follow a severe exposure if patient survives. Urticaria is a rare allergic manifestation from inhalation of the gas.

Medical Surveillance: Preemployment physical examinations for workers in ammonia exposure areas should be directed toward significant changes in the skin, eyes, and respiratory system. Persons with corneal disease, and glaucoma, or chronic respiratory diseases may suffer increased risk. Periodic examinations should include evaluation of skin, eyes, and respiratory system, and pulmonary function tests to compare with baselines established at pre-employment examination.

Special Tests: None.

Personal Protective Methods: Where ammonia hazards exist in concentrations above the standard, respiratory, eye, and skin protection should be provided. Fullface gas masks with ammonia canister or supplied air respirators, both with full facepieces, afford good protection. In areas where exposure to liquid ammonia occurs, goggles or face shields, as well as protective clothing impervious to ammonia and including gloves, aprons, and boots should be required. Where ammonia gas or concentrated ammonia solution is splashed in eyes, immediate flooding of the eyes with large quantities of water for 15 minutes or longer is advised, followed at once by medical examination.

In heavy concentrations of ammonia gas, workers should be outfitted with complete self-contained protective suits impervious to ammonia, with supplied air source, and full headpiece and facepiece. Work clothes wetted with concentrated ammonia solutions should be changed immediately, and the exposed area of the body washed thoroughly with water.

Bibliography

Highman, V.N. 1969. Early rise in intraocular pressure after ammonia burns. *Br. Med. J.* 1:359.

National Institute for Occupational Safety and Health. *Criteria for a Recommended Standard: Occupational Exposure to Ammonia.* NIOSH Doc. No. 74-136, Wash., DC (1974).

Walton, M. 1973. Industrial ammonia gassing. *Br. J. Ind. Med.* 30:78.

AMYL ALCOHOL

Description: $C_5H_{11}OH$, amyl alcohol, has eight isomers. All are colorless liquids, except the isomer 2,2-dimethyl-1-propanol, which is a crystalline solid.

Amyl alcohols are obtained from fusel oil which forms during the fermentation of grain, potatoes, or beets for ethyl alcohol. The fusel oil is a mixture of amyl alcohol isomers, and the composition is determined somewhat by the sugar source. Amyl alcohols may be prepared by acid hydrolysis of a petroleum fraction.

Synonyms: Pentanols, pentyl alcohols, fusel oil, grain oil, potato spirit, potato oil.

Potential Occupational Exposures: Amyl alcohols are used in the manufacture of lacquers, paints, varnishes, paint removers, shoe cements, perfumes, pharmaceuticals, chemicals, rubber, plastics, fruit essences, explosives, hydraulic fluids, ore flotation agents, in the preparation of other amyl derivatives, in the extraction of fats, and in the textile and petroleum refining industries.

A partial list of occupations in which exposure may occur includes:

Amyl acetate makers	Perfume makers
Amyl nitrite makers	Petroleum refiners
Explosive makers	Photographic chemical makers
Fat processors	Plastic makers
Flotation workers	Rubber makers
Lacquer makers	Shoe finishers
Mordanters	Textile workers
Oil processors	Wax processors
Painters	

Permissible Exposure Limits: The Federal standard for 3-methyl-1-butanol (isoamyl alcohol) is 100 ppm (360 mg/m^3). There are no standards for the other isomers.

Route of Entry: Inhalation of vapor, percutaneous absorption.

Harmful Effects

Local—The liquid and vapor are mild irritants to the membranes of the eyes and upper respiratory tract and skin.

Systemic—In low concentrations, amyl alcohol may cause irritation of nose and throat, nausea, vomiting, flushing, headache, diplopia, vertigo, and muscular weakness. In higher dosage, it is a narcotic.

Medical Surveillance: Consider possible irritant effects on skin and respiratory tract in any preplacement or periodic examinations.

Special Tests: None in common use. Amyl alcohol can be determined in blood.

Personal Protective Methods: Barrier creams and personal protective clothing should be used to prevent skin contact.

Bibliography

Gibel, W., Lohs, K., Wildner, G.P., Wittbrodt, S., Geibler, W., and Hilscher, H. 1969. Untersuchungen zur frage einer moglichen mutagenen wirkung von fuselol. *Arch. Geschwulstforsch.* 33:49.

Hilscher, H., Geissler, E., Lohs, K., and Gibel, W. 1969. Untersuchungen zur toxizitat und mutagenitat einzelner fuselol-komponenten an *E. coli. Acta. Biol. Med. Germ. 23:843.*

ANILINE

Description: $C_6H_5NH_2$, aniline, is a clear, colorless, oily liquid with a characteristic odor.

Synonyms: Aminobenzene, phenylamine, aniline oil, aminophen, arylamine.

Potential Occupational Exposures: Aniline is widely used as an intermediate in the synthesis of dyestuffs. It is also used in the manufacture of rubber accelerators and antioxidants, pharmaceuticals, marking inks, tetryl, optical whitening agents, photographic developers, resins, varnishes, perfumes, shoe polishes, and many organic chemicals.

A partial list of occupations in which exposure may occur includes:

Acetanilide workers	Perfume makers
Bromide makers	Photographic chemical makers
Coal tar workers	Plastic workers
Disinfectant makers	Printers
Dye workers	Rocket fuel makers
Ink makers	Rubber workers
Leather workers	Tetryl makers
Lithographers	Varnish workers

Permissible Exposure Limits: The Federal standard is 5 ppm (19 mg/m³). ACGIH (1978) has proposed TWA values of 2 ppm (10 mg/m³).

Routes of Entry: Inhalation of vapors; percutaneous absorption of liquid and vapor.

Harmful Effects

Local—Liquid aniline is mildly irritating to the eyes and may cause corneal damage.

Systemic—Absorption of aniline, whether from inhalation of the vapor or from skin absorption of the liquid, causes anoxia due to the formation of methemoglobin. Moderate exposure may cause only cyanosis. As oxygen deficiency increases, the cyanosis may be associated with headache, weakness, irritability, drowsiness, dyspnea, and unconsciousness. If treatment is not given promptly, death can occur. The development of intravascular hemolysis and anemia due to aniline-induced methemoglobinemia has been postulated, but neither is observed often in industrial practice, despite careful study of numerous cases.

Medical Surveillance: Preplacement and periodic physical examinations should be performed on all employees working in aniline exposure areas. These should include a work history to elicit information on all past exposures to aniline, other aromatic amines, and nitro compounds known to cause chemical cyanosis, and the clinical history of any occurrence of chemical cyanosis; a personal history to elicit alcohol drinking habits; and general physical examination with particular reference to the cardiovascular system. Persons with impaired cardiovascular status may be at greater risk from the consequences of chemical cyanosis. A preplacement complete blood count and methemoglobin estimation should be performed as baseline levels, also follow-up studies including periodic blood counts and hematocrits.

Special Tests: Methemoglobin levels, and other abnormal hemoglobins, and/or urine para-aminophenols, and other aniline metabolites, have been used for biologic monitoring for occupational aniline exposure.

Personal Protective Methods: In areas of vapor concentration, the use of respirators alone is not sufficient; skin protection by protective clothing should be provided even though there is no skin contact with liquid aniline. Butyl rubber protective clothing is reportedly superior to other materials. In severe exposure situations, complete body protection has been employed, consisting of air-conditioned suit with air supplied helmet and cape. Personal hygiene practices including prompt removal of clothing which has absorbed aniline, thorough showering after work and before changing to street clothes, and clean working clothes daily are essential.

Bibliography

Dutkiewicz, T., and Piotrowski, J. 1961. Experimental investigations on the quantitative estimation of aniline absorption in man. *Pure Appl. Chem.* 3:319.

Scarpa, C. 1955. The aniline test as detector of a sensitivity. *Acta. Allergol.* 9:203.

Vasilenko, N.M., Volodchenko, V.A., Khizhnyakova, L.N., Avezday, V.I., Manfanovsky, V.V., Antonovskaya, V.S., Krylova, E.V., Voskobionikova, N.A., Gnezdilova, I. and Sonkin, I.S. 1972. Data to substantiate a decrease of the maximum permissible concentration of aniline in the air of working zones. *Gig. Sanit.* 37:31.

Wetherhold, J.M., Linch, A.L., and Charsha, R.C. 1960. Chemical cyanosis—causes, effects, and prevention. *Arch. Environ. Health* 1:75.

ANTIMONY AND COMPOUNDS

Description: Sb, antimony, is a silvery-white, soft metal insoluble in water and organic solvents. The ores most often found are stibnite, valentinite, kermesite, and senarmontite.

Synonyms: None

Potential Occupational Exposures: Exposure to antimony may occur during mining, smelting or refining, alloy and abrasive manufacture, and typesetting in printing. Antimony is widely used in the production of alloys, imparting increased hardness, mechanical strength, corrosion resistance, and a low coefficient of friction. Some of the important alloys are babbitt, pewter, white metal, Britannia metal and bearing metal (which are used in bearing shells), printing-type metal, storage battery plates, cable sheathing, solder, ornamental castings, and ammunition. Pure antimony compounds are used as abrasives, pigments, flameproofing compounds, plasticizers, and catalysts in organic synthesis; they are also used in the manufacture of tartar emetic, paints, lacquers, glass, pottery, enamels, glazes, pharmaceuticals, pyrotechnics, matches, and explosives.

In addition they are used in dyeing, for blueing steel, and in coloring aluminum, pewter, and zinc. A highly toxic gas, stibine, may be released from the metal under certain conditions.

A partial list of occupations in which exposure may occur includes:

Bronzers	Miners
Ceramic makers	Paint makers
Drug makers	Rubber makers
Fireworks makers	Textile workers
Leather mordanters	Typesetters

Permissible Exposure Limits: The Federal standard for antimony and its compounds is 0.5 mg/m^3, expressed as Sb (see also Stibine). A TWA value of 2 mg/m^3 has been proposed by ACGIH (1978). Further, antimony trioxide production has been categorized as "Industrial Substances Suspect of Carcinogenic Potential for Man."

Route of Entry: Inhalation of dust or fume; percutaneous absorption.

Harmful Effects

Local—Antimony and its compounds are generally regarded as primary skin irritants. Lesions generally appear on exposed, moist areas of the body, but rarely on the face. The dust and fumes are also irritants to the eyes, nose, and throat, and may be associated with gingivitis, anemia, and ulceration of the nasal septum and larynx. Antimony trioxide causes a dermatitis known as "antimony spots." This form of dermatitis results in intense itching followed by skin eruptions. A diffuse erythema may occur, but usually the early lesions are small erythematous papules. They may enlarge, however, and become pustular. Lesions occur in hot weather and are due to dust accumulating on exposed areas that are moist due to sweating. No evidence of eczematous reaction is present, nor an allergic mechanism.

Systemic—Systemic intoxication is uncommon from occupational exposure. However, miners of antimony may encounter dust containing free silica; cases of pneumoconiosis in miners have been termed "silico-antimoniosis." Antimony pneumoconiosis, per se, appears to be a benign process.

Antimony metal dust and fumes are absorbed from the lungs into the blood stream. Principal organs attacked include certain enzyme systems (protein and carbohydrate metabolism), heart, lungs, and the mucous membrane of the respiratory tract. Symptoms of acute oral poisoning include violent irritation of the nose, mouth, stomach, and intestines, vomiting, bloody stools, slow shallow respiration, pulmonary congestion, coma, and sometimes death due to circulatory or respiratory failure. Chronic oral poisoning presents symptoms of dry throat, nausea, headache, sleeplessness, loss of appetite, and dizziness. Liver and kidney degenerative changes are late manifestations.

Antimony compounds are generally less toxic than antimony. Antimony trisulfide, however, has been reported to cause myocardial changes in man and experimental animals. Antimony trichloride and pentachloride are highly toxic and can irritate and corrode the skin. Antimony fluoride is extremely toxic, particularly to pulmonary tissue and skin.

Medical Surveillance: Preemployment and periodic examinations should give special attention to lung disease, skin disease, diseases of the nervous system, heart and gastrointestinal tract. Lung function, EKGs, blood, and urine should be evaluated periodically.

Special Tests: Blood and urine antimony levels have been suggested, but are not in common use.

Personal Protective Methods: A combination of protective clothing, barrier creams, gloves, and personal hygiene will protect the skin. Washing and showering facilities should be available and eating should not be permitted in exposed areas. Dust masks and supplied air respirators should be available in all areas where the Federal standard is exceeded.

Bibliography

Brieger, H., Semisch, C.W., III, Stasney, J., and Piatnek, D.A. 1954. Industrial antimony poisoning. *Ind. Med. Surg.* 23:521.

Chekunova, M.P., and Minkina, N.A. 1970. An investigation of the toxic effect of antimony pentafluoride. *Hyg. Sanit.* 35:30.

Cooper, D.A., Pendergrass, E.P., Vorwald, A.J., Maycock, R.L. and Brieger, H. 1968. Pneumoconiosis among workers in an antimony industry. *Am. J. Roentgenol. Radium Ther. Nucl. Med.* 103:496.

Environmental Protection Agency. *Literature Study of Selected Potential Environmental Contaminants: Antimony and Its Compounds.* Report No. EPA-560/2-76-002, Wash., DC, Office of Toxic Substances (Feb. 1976).

Levina, E.N., and Chekunova, M.P. 1964. Toxicity of antimony halides. *Gig. Tr. Prof. Zabol.* 8:608.

Sapire, D.W., and Silverman, N.H. 1970. Myocardial involvement in antimonial therapy: a case report of acute antimony poisoning with serial ECG changes. *S. Afr. Med. J.* 44:848.

Stevenson, C.J. 1965. Antimony spots. *Trans. St. Johns Hosp. Dermatol. Soc.* 51:40.

ARSENIC

Description: As, elemental arsenic, occurs to a limited extent in nature as a steel-gray metal that is insoluble in water. Arsenic in this discussion includes the element and any of its inorganic compounds excluding arsine. Arsenic trioxide (As_2O_3), the principal form in which the element is used, is frequently designated as arsenic, white arsenic, or arsenous oxide. Arsenic is present as an impurity in many other metal ores and is generally produced as arsenic trioxide as a by-product in the smelting of these ores, particularly copper. Most other arsenic compounds are produced from the trioxide.

Synonyms: None.

Potential Occupational Exposures: Arsenic compounds have a variety of uses. Arsenates and arsenites are used in agriculture as insecticides, herbicides, larvicides, and pesticides. Arsenic trichloride is used primarily in the manufacture of pharmaceuticals. Other arsenic compounds are used in pigment production, the manufacture of glass as a bronzing or decolorizing agent, the manufacture of opal glass and enamels, textile printing, tanning, taxidermy, and antifouling paints. They are also used to control sludge formation in lubricating oils. Metallic arsenic is used as an alloying agent to harden lead shot and in lead-base bearing materials. It is also alloyed with copper to improve its toughness and corrosion resistance.

A partial list of occupations in which exposure may occur includes:

Alloy makers	Dye makers
Aniline color makers	Enamelers
Arsenic workers	Fireworks makers
Babbitt metal workers	Gold refiners
Brass makers	Herbicide makers
Bronze makers	Hide preservers
Ceramic enamel makers	Insecticide makers
Ceramic makers	Lead shot makers
Copper smelters	Lead smelters
Drug makers	Leather workers

Painters	Silver refiners
Paint makers	Taxidermists
Petroleum refinery workers	Textile printers
Pigment makers	Tree sprayers
Printing ink workers	Type metal workers
Rodenticide makers	Water weed controllers
Semiconductor compound makers	Weed sprayers

Permissible Exposure Limits: The Federal standard for arsenic and its compounds was previously 0.5 mg/m^3 of air as As. In 1973, NIOSH proposed (1) the lower recommended standard of 0.05 mg As/m^3 of air determined as a time-weighted average (TWA) exposure for up to a 10-hour workday, 40-hour workweek. Then, in November 1975, OSHA proposed a workplace exposure limit for inorganic arsenic at 4 μg/m^3 (8-hour, TWA). The economic impact of such a standard has been assessed (2). The previous standard of 500 μg/m^3 for all forms of arsenic would remain in effect only for organic forms.

A 1975 NIOSH document (3) proposed that inorganic arsenic be controlled so that no worker is exposed to a concentration of arsenic in excess of 0.002 mg (2.0 μg) per cubic meter as determined by a 15-minute sampling period. Finally in 1978 a standard was promulgated (4) which limits occupational exposure to inorganic arsenic to 10 μg/m^3 (micrograms per cubic meter of air) based on an 8-hour time-weighted average.

The basis for this action is evidence that exposure to inorganic arsenic poses a cancer risk to workers. The purpose of this rule is to minimize the incidence of lung cancer among workers exposed to inorganic arsenic. Employees protected by this standard work principally in the nonferrous metal smelting, glass and arsenical chemical industries. Provisions for monitoring of exposures, record keeping, medical surveillance, hygiene facilities and other requirements are also included. The 10 μg/m^3 limit has been set because it will provide significant employee protection and is the lowest feasible level in many circumstances. ACGIH (1978) gives a TWA of 0.5 mg/m^3 for As and compounds. They go on to propose a TWA of 0.2 mg/m^3 for soluble arsenic. In addition, arsenic trioxide production is categorized under "Human Carcinogens."

Routes of Entry: Inhalation and ingestion of dust and fumes.

Harmful Effects

Local—Trivalent arsenic compounds are corrosive to the skin. Brief contact has no effect, but prolonged contact results in a local hyperemia and later vesicular or pustular eruption. The moist mucous membranes are most sensitive to the irritant action. Conjunctiva, moist and macerated areas of the skin, eyelids, the angles of the ears, nose, mouth, and respiratory mucosa are also vulnerable to the irritant effects. The wrists are common sites of dermatitis, as are the genitalia if personal hygiene is poor. Perforations of the nasal septum may occur. Arsenic trioxide and pentoxide are capable of producing skin sensitization and contact dermatitis. Arsenic is also capable of producing keratoses, especially of the palms and soles. Arsenic has been cited as a cause of skin cancer, but the incidence is low.

Systemic—The acute toxic effects of arsenic are generally seen following ingestion of inorganic arsenical compounds. This rarely occurs in an industrial setting. Symptoms develop within ½ to 4 hours following ingestion and are usually characterized by constriction of the throat followed by dysphagia, epigastric pain, vomiting, and watery diarrhea. Blood may appear in vomitus and stools. If the amount ingested is sufficiently high, shock may develop due to severe fluid loss, and death may ensue in 24 hours. If the acute effects are survived, exfoliative dermatitis and peripheral neuritis may develop.

Cases of acute arsenical poisoning due to inhalation are exceedingly rare in industry. When it does occur, respiratory tract symptoms—cough, chest pain, dyspnea—giddiness, headache, and extreme general weakness precede gastrointestinal symptoms. The acute toxic symptoms of trivalent arsenical poisoning are due to severe inflammation of the mucous membranes and greatly increased permeability of the blood capillaries.

Chronic arsenical poisoning due to ingestion is rare and generally confined to patients taking prescribed medications. However, it can be a concomitant of inhaled inorganic arsenic from swallowed sputum and improper eating habits. Symptoms are weight loss, nausea and diarrhea alternating with constipation, pigmentation and eruption of the skin, loss of hair, and peripheral neuritis. Chronic hepatitis and cirrhosis have been described. Polyneuritis may be the salient feature, but more frequently there are numbness and paresthesias of "glove and stocking" distribution. The skin lesions are usually melanotic and keratotic and may occasionally take the form of an intradermal cancer of the squamous cell type, but without infiltrative properties. Horizontal white lines (striations) on the fingernails and toenails are commonly seen in chronic arsenical poisoning and are considered to be a diagnostic accompaniment of arsenical polyneuritis.

Inhalation of inorganic arsenic compounds is the most common cause of chronic poisoning in the industrial situation. This condition is divided into three phases based on signs and symptoms.

First Phase: The worker complains of weakness, loss of appetite, some nausea, occasional vomiting, a sense of heaviness in the stomach, and some diarrhea.

Second Phase: The worker complains of conjunctivitis, and a catarrhal state of the mucous membranes of the nose, larynx, and respiratory passages. Coryza, hoarseness, and mild tracheobronchitis may occur. Perforation of the nasal septum is common, and is probably the most typical lesion of the upper respiratory tract in occupational exposure to arsenical dust. Skin lesions, eczematoid and allergic in type, are common.

Third Phase: The worker complains of symptoms of peripheral neuritis, initially of hands and feet, which is essentially sensory. In more severe cases, motor paralyses occur; the first muscles affected are usually the toe extensors and the peronei. In only the most severe cases will paralysis of flexor muscles of the feet or of the extensor muscles of hands occur.

Liver damage from chronic arsenical poisoning is still debated, and as yet the question is unanswered. In cases of chronic and acute arsenical poisoning, toxic effects to the myocardium have been reported based on EKG changes. These findings, however, are now largely discounted and the EKG changes are ascribed to electrolyte disturbances concomitant with arsenicalism. Inhalation of arsenic trioxide and other inorganic arsenical dusts does not give rise to radiological evidence of pneumoconiosis. Arsenic does have a depressant effect upon the bone marrow, with disturbances of both erythropoiesis and myelopoiesis. Evidence is now available incriminating arsenic compounds as a cause of lung cancer as well as skin cancer.

Medical Surveillance: In preemployment physical examinations, particular attention should be given to allergic and chronic skin lesions, eye disease, psoriasis, chronic eczematous dermatitis, hyperpigmentation of skin, keratosis and warts, baseline weight, baseline blood and hemoglobin count, and baseline urinary arsenic determinations. In annual examinations, the worker's general health, weight, and skin condition should be checked, and the worker observed for any evidence of excessive exposure or absorption of arsenic.

Special Tests: Chest x-rays and lung function should be evaluated; analysis of urine, hair, or nails for arsenic should be made every 60 days as long as exposure continues.

Personal Protective Methods: Workers should be trained in personal hygiene and sanitation, the use of personal protective equipment, and early recognition of symptoms of absorption, skin contact irritation, and sensitivity. With the exception of arsine and arsenic trichloride, the compounds of arsenic do not have odor or warning qualities. In case of emergency or areas of high dust or spray mist, workers should wear respirators that are supplied-air or self-contained positive-pressure type with fullface mask. Where concentrations are less than 100 x standard, workers may be able to use halfmask respirators with replaceable dust or fume filters. Protective clothing, gloves, goggles and a hood for head and neck should be provided. When liquids are processed, impervious clothing should be supplied. Clean work clothes should be supplied daily and the workers should shower prior to changing to street clothes.

References

(1) National Institute for Occupational Safety and Health. *Criteria for a Recommended Standard: Occupational Exposure to Inorganic Arsenic.* NIOSH Doc. No. 74-110, Wash., DC (1973).

(2) U.S. Department of Labor. *Inflationary Impact Statement: Inorganic Arsenic.* Wash., DC, Occupational Safety and Health Administration (Undated—presumed 1975).

(3) National Institute for Occupational Safety and Health. *Criteria for a Recommended Standard: Occupational Exposure to Inorganic Arsenic (Revised).* NIOSH Doc. No. 75-149, Wash., DC (1975).

(4) *Federal Register* 43, No. 88, 19584-19631 (May 5, 1978).

Bibliography

Air Pollution Assessment Report on Arsenic; EPA, Office of Air Quality Planning and Standards (December 1975).

Burruss, R.P. and Sargent, D.H. *Technical and Microeconomic Analysis of Arsenic and Its Compounds.* EPA, Office of Toxic Substances (performed under contract No. 68-01-2926) (September 1975).

Dinman, B.D. 1960. Arsenic; chronic human intoxication. *J. Occup. Med.* 2:137.

Elkins, H.B. 1959. *The Chemistry of Industrial Toxicology,* 2nd ed. John Wiley and Sons, New York.

Faust, S.D. and Clement, W.H. *Investigation of the Arsenic Condition at the Blue Marsh Lake Project Site, Pennsylvania.* U.S. Army Corps of Engineers, performed under contract No. DACW 67-71-C-0288 (1973).

Hazardous Waste Disposal Damage Reports. EPA, Office of Solid Waste Management Programs (publication No. EPA 530/SW-157, June 1975).

Helver, J.E. Progress on Studies on Contaminated Trout Rations and Trout Hepatoma. *NIH Report* (April 12, 1962).

Holmquist, L. 1951. Occupational arsenical dermatitis; a study among employees at a copper-ore smelting works including investigations of skin reactions to contact with arsenic compounds. *Acta. Derm. Venereol.* (Supp. 26) 31:1.

A Pilot Study on the Community Effects of Arsenic Exposure in Baltimore. EPA, Office of Toxic Sustances (performed under contract No. 68-01-2490).

Pinto, S.S., and McGill, C.M. 1953. Arsenic trioxide exposure in industry. *Ind. Med. Surg.* 22:281.

Pinto, S.S. and Nelson, K.W. 1976. Arsenic toxicology and industrial exposure. *Ann. Rev. Pharmacol. Toxicol.* 16:95.

Tseng, W.P., et al. Prevalence of Skin Cancer in an Endemic Area of Chronic Arsenicism in Taiwan. *J. National Cancer Inst.,* 40:453 (1968).

Vallee, B.L., Ulmer, D.D., and Wacker, W.E.C. 1960. Arsenic toxicology and biochemistry. *AMA Arch. Indust. Health* 21:132.

Wands, Ralph C. *Letter to APHA Panel on Arsenic Studies.* National Research Council (February 17, 1976).

ARSINE

Description: AsH_3, arsine, is a colorless gas with a slight garlic-like odor which cannot be considered a suitable warning property in concentrations below 1 ppm. Arsine's solubility is 20 ml in 100 ml of water at 20°C.

Synonyms: Hydrogen arsenide, arseniuretted hydrogen.

Potential Occupational Exposures: Arsine is not used in any industrial process but this gas is generated by side reactions or unexpectedly; e.g., it may be generated in metal pickling operations, metal drossing operations, or when inorganic arsenic compounds contact sources of nascent hydrogen. It has been known to occur as an impurity in acetylene. Most occupational exposure occurs in chemical, smelting, and refining industries. Cases of exposure have come from workers dealing with zinc, tin, cadmium, galvanized coated aluminum, and silicon and steel metals.

A partial list of occupations in which exposure may occur includes:

Acid dippers	Jewelers
Aniline workers	Lead burners
Bronzers	Paper makers
Dye makers	Plumbers
Etchers	Solderers
Fertilizer makers	Submarine workers
Galvanizers	Tinners

Permissible Exposure Limits: The Federal standard for arsine is 0.05 ppm. NIOSH has recommended that arsine be controlled to the same concentration as other forms of inorganic arsenic (0.002 mg/m^3).

Route of Entry: Inhalation of gas.

Harmful Effects

Local—High concentrations of arsine gas will cause damage to the eyes. Most experts agree, however, that before this occurs systemic effects can be expected.

Systemic—Arsine is an extremely toxic gas that can be fatal if inhaled in sufficient quantities. Acute poisoning is marked by a triad of main effects caused by massive intravascular hemolysis of the circulating red cells. Early effects may occur within an hour or two and are commonly characterized by general malaise, apprehension, giddiness, headache, shivering, thirst, and abdominal pain with vomiting. In severe acute cases the vomitus may be blood stained and diarrhea ensues as with inorganic arsenical poisoning. Pulmonary edema has occurred in severe acute poisoning.

Invariably, the first sign observed in arsine poisoning is hemoglobinuria, appearing with discoloration of the urine up to port wine hue (first of the triad). Jaundice (second of triad) sets in on the second or third day and may be intense, coloring the entire body surface a deep bronze hue. Coincident with these effects is a severe hemolytic-type anemia. Severe renal damage may occur with oliguria or complete suppression of urinary function (third of triad), leading to uremia and death. Severe hepatic damage may also occur, along with cardiac damage and EKG changes. Where death does not occur, recovery is prolonged.

In cases where the amount of inhaled arsine is insufficient to produce acute effects, or where small quantities are inhaled over prolonged periods, the hemoglobin liberated by the destruction of red cells may be degraded by the reticuloendothelial system and the iron moiety taken up by the liver, without producing permanent damage. Some hemoglobin may be excreted unchanged by the kidneys. The only symptoms noted may be general tiredness, pallor, breathlessness on exertion, and palpitations as would be expected with severe secondary anemia.

Medical Surveillance: In preemployment physical examinations, special attention should be given to past or present kidney disease, liver disease, and anemia. Periodic physical examinations should include tests to determine arsenic levels in the blood and urine. The general condition of the blood and the renal and liver functions should also be evaluated. Since arsine gas is a by-product of certain production processes, workers should be trained to recognize the symptoms of exposure and to use appropriate personal protective equipment.

Special Tests: None in common use.

Personal Protective Methods: In most cases, arsine poisoning cannot be anticipated except through knowledge of the production processes. Where arsine is suspected in concentrations above the acceptable standard, the worker should be supplied with a supplied air fullface respirator or a self-contained positive pressure respirator with full facepiece.

Bibliography

Conrad, M.E., Mazey, R.M. and Reed, J.E. 1976. Industrial arsine poisoning: report of three cases. *Ala. J. Med. Sci.* 13(1):65.

Elkins, H.G. and Fahy, J.P. 1967. Arsine poisoning from aluminum tank cleaning. *Ind. Med. Surg.* 36:747.

Grant, W.M. 1962. *Toxicology of the Eye.* Charles C. Thomas Publishers, Springfield, Ill.

Jenkins, G.C., Ind, J.E., Kazantizis, G. and Owen, R. 1965. Arsine poisoning: massive haemolysis with minimal impairment of renal function. *Br. Med. J.* 2:78.

Josephson, C.J., Pinto, S.S. and Petronella, S.J. 1951. Arsine: electrocardiographic changes produced in acute human poisoning. *AMA Arch. Ind. Hyg. Occup. Med.* 4:43.

Kipling, M.D. and Fothergill, R. 1964. Arsine poisoning in a slag-washing plant. *Br. J. Ind. Med.* 21:74.

Kobayashi, Y. 1956. Rapid method for determination of low concentrations of arsine by detector tubes. *J. Chem. Soc. Jap.* 59:899.

Sandell, E.G. 1942. Colorimetric microdetermination of arsenic after evolution as arsine. *Eng. Chem.* 14(1):82.

ASBESTOS

Description: Asbestos is a generic term that applies to a number of naturally occurring, hydrated mineral silicates incombustible in air and separable into filaments. The most widely used in industry in the United States is chrysotile, a fibrous form of serpentine. Other types include amosite, crocidolite, tremolite, anthophyllite, and actinolite.

Synonyms: None.

Potential Occupational Exposures: Almost one million tons per year of asbestos are used in the United States. In 1965, approximately 74% of the asbestos produced was used in the construction industry (532,300 tons) while 26% was used in nonconstruction industries (187,400 tons). Approximately 92% of the half million tons used in the construction industry is firmly bonded, i.e., the asbestos is "locked in" in such products as floor tiles, asbestos cements, and roofing felts and shingles; while the remaining 8% is friable or in powder form present in insulation materials, asbestos cement powders, and acoustical products.

As expected, these latter materials generate more airborne fibers than the firmly bonded products. The 187,400 tons of asbestos used in nonconstruction industries in 1965 were utilized in such products as textiles, friction material including brake linings, and clutch facings, paper, paints, plastics, roof coatings, floor tiles, and miscellaneous other products (1).

Asbestos was used in spray insulation in buildings between 1950 and 1972. This may become a major source of environmental discharge as buildings constructed during this period are demolished.

Asbestos minerals are found throughout the United States. Significant quantities of asbestos fibers appear in rivers and streams draining from areas where asbestos-rock outcroppings are found. Some of these outcroppings are being mined. Asbestos fibers have been found in a number of drinking water supplies, but the health implications of ingesting asbestos are not fully documented. Emissions of asbestos fibers into water and air are known to result from mining and processing of some minerals. Asbestiform fibers in the drinking water of Duluth and nearby communities at levels of 12 million fibers per liter have been attributed to the discharge of 67,000 tons of taconite tailings per day into Lake Michigan by Reserve Mining.

Exposure to asbestos fibers may occur throughout urban environments. A recent study of street dust in Washington, DC, showed approximately 50,000 fibers per gram, much of which appeared to come from brake linings. Autopsies of New York City residents with no known occupational exposure showed 24 of 28 lung samples to contain asbestos fibers, perhaps resulting from asbestos from brake linings and the flaking of sprayed asbestos insulation material (2).

Permissible Exposure Limits: When an asbestos criteria document was first published by NIOSH in 1972 (1), that agency recommended a standard of 2.0

asbestos fibers/cc of air based on a count of fibers greater than 5 micrometers (μm) in length. This standard was recommended with the stated belief that it would "prevent" asbestosis and with the open recognition that it would not "prevent" asbestos-induced neoplasms.

Furthermore, data were presented which supported the fact that technology was available to achieve that standard and that the criteria would be subject to review and revision as necessary. Since the time that the asbestos criteria were published in 1972, sufficient additional data regarding asbestos-related disease have been developed to warrant reevaluation.

On June 7, 1972, the Occupational Safety and Health Administration (OSHA) promulgated a standard for occupational exposure to asbestos containing an 8-hour time-weighted average (TWA) concentration exposure limit of 5 fibers longer than 5 μm/cc of air, with a ceiling limitation against any exposure in excess of 10 such fibers/cc. The standard further provided that the 8-hour TWA was to be reduced to 2 fibers/cc on July 1, 1976.

As the result of a court case, OSHA decided that to achieve the most feasible occupational health protection, a reexamination of the standard's general premises and general structure was necessary. To this end, on October 9, 1975, OSHA announced a proposed rule-making to lower the exposure limit to an 8-hour TWA concentration of 0.5 asbestos fibers longer than 5 μm/cc of air with a ceiling concentration of 5 fibers/cc of air determined by a sampling of up to 15 minutes. On December 2, 1975, OSHA requested NIOSH to re-evaluate the information available on the health effects of occupational exposure to asbestos fibers and to advise OSHA on the results of this study, so further rule-making may be expected as time goes on.

Indeed NIOSH stated in 1976 (3), that the standard should be set at the lowest level detectable by available analytical techniques, an approach consistent with NIOSH's most recent recommendations for other carcinogens (i.e., arsenic and vinyl chloride). Such a standard should also prevent the development of asbestosis.

Since phase contrast microscopy is the only generally available and practical analytical technique at the present time, this level is defined as 100,000 fibers >5 μm in length/m^3 (0.1 fiber/cc), on an 8-hour-TWA basis with peak concentrations not exceeding 500,000 fibers >5 μm in length/m^3 (0.5 fiber/cc) based on a 15-minute sample period. Sampling and analytical techniques should be performed as specified by NIOSH publication USPHS/NIOSH Membrane Filter Method for Evaluating Airborne Asbestos Fibers—T.R. 84 (1976).

This recommended standard of 100,000 fibers >5 μm in length/m^3 is intended to 1) protect against the noncarcinogenic effects of asbestos, 2) materially reduce the risk of asbestos-induced cancer (only a ban can assure protection against carcinogenic effects of asbestos) and 3) be measured by techniques that are valid, reproducible, and available to industry and official agencies.

The ACGIH (1978) has categorized asbestos (all forms) as a human carcinogen. The economic impact of the proposed OSHA standard for asbestos has been assayed (4).

Route of Entry: Inhalation.

Harmful Effects

Available studies provided conclusive evidence (3) that exposure to asbestos fibers causes cancer and asbestosis in man. Lung cancers and asbestosis have occurred following exposure to chrysotile, crocidolite, amosite, and anthophyllite.

Mesotheliomas, lung and gastrointestinal cancers have been shown to be excessive in occupationally exposed persons, while mesotheliomas have developed also in individuals living in the neighborhood of asbestos factories and near crocidolite deposits, and in persons living with asbestos workers. Asbestosis has been identified among persons living near anthophyllite deposits.

Likewise, all commercial forms of asbestos are carcinogenic in rats, producing lung carcinomas and mesotheliomas following their inhalation, and mesotheliomas after intrapleural or i.p. injection. Mesotheliomas and lung cancers were induced following even 1 day's exposure by inhalation.

The size and shape of the fibers are important factors; fibers less than 0.5 μm in diameter are most active in producing tumors. Other fibers of a similar size, including glass fibers, can also produce mesotheliomas following intrapleural or i.p. injection.

There are data that show that the lower the exposure, the lower the risk of developing cancer. Excessive cancer risks have been demonstrated at all fiber concentrations studied to date. Evaluation of all available human data provides no evidence for a threshold or for a "safe" level of asbestos exposure.

Medical Surveillance: Medical surveillance is required (1), except where a variance from the medical requirements of this proposed standard have been granted, for all workers who are exposed to asbestos as part of their work environment. For the purposes of this requirement the term "exposed to asbestos" will be interpreted as referring to time-weighted average exposures above 1 fiber/cc or peak exposures above 5 fibers/cc.

The major objective of such surveillance will be to ensure proper medical management of individuals who show evidence of reaction to past dust exposures, either due to excessive exposures or unusual susceptibility. Medical management may range from recommendations as to job placement, improved work practices, cessation of smoking, to specific therapy for asbestos-related disease or its complications.

Medical surveillance cannot be a guide to adequacy of current controls when environmental data and medical examinations only cover recent work experience because of the prolonged latent period required for the development of asbestosis and neoplasms.

Required components of a medical surveillance program include periodic measurements of pulmonary function [forced vital capacity (FVC) and forced expiratory volume for one second (FEV$_1$)], and periodic chest roentgenograms (postero-anterior 14 × 17 inches). Additional medical requirement components include a history to describe smoking habits and details on past exposures to asbestos and other dusts and to determine presence or absence of pulmonary, cardiovascular, and gastrointestinal symptoms, and a physical examination, with special attention to pulmonary rales, clubbing of fingers and other signs related to cardiopulmonary systems.

Chest roentgenograms and pulmonary function tests should be performed at least every 2 years on all employees exposed to asbestos (1). Such tests should be made annually on individuals:

> a) who have a history of 10 or more years of employment involving exposure to asbestos or

> b) who show roentgenographic findings (such as small opacities, pleural plaques, pleural thickening or pleural calcification) which suggest or indicate pneumoconiosis or other reactions to asbestos or

> c) who have changes in pulmonary function which indicate restrictive or obstructive lung disease.

Preplacement medical examinations and medical examinations on the termination of employment of asbestos-exposed workers are also required.

Special Tests: None commonly used.

Personal Protective Methods: Use of respirators can be decided on the basis of time-weighted average or peak concentration. When the limits of exposure to asbestos dust cannot be met by limiting the concentration in the workplace, the employer must utilize a program of respiratory protection and furnishing of protective clothing to protect every worker exposed (1).

Respiratory Protection—For the purpose of determining the class of respirator to be used, the employer shall measure the atmospheric concentration of airborne asbestos in the workplace when the initial application for variance is made and thereafter whenever process, worksite, climate or control changes occur which are likely to affect the asbestos concentration. The employer shall test for respirator fit and/or make asbestos measurements within the respiratory inlet covering to insure that no worker is being exposed to asbestos in excess of the standard either because of improper respirator selection or fit. Details of respirator selection are contained in a NIOSH publication (1).

Protective Clothing—

1) The employer shall provide each employee subject to exposure in a variance area with coveralls or similar full body protective clothing and hat, which shall be worn during the working hours in areas where there is exposure to asbestos dust.

2) The employer shall provide for maintenance and laundering of the soiled protective clothing, which shall be stored, transported and disposed of in sealed non-reusable containers marked "Asbestos-Contaminated Clothing" in easy-to-read letters.

3) Protective clothing shall be vacuumed before removal. Clothes shall be not be cleaned by blowing dust from the clothing or shaking.

4) If laundering is to be done by a private contractor, the employer shall inform the contractor of the potentially harmful effects of exposure to asbestos dust and of safe practices required in the laundering of the asbestos-soiled work clothes.

5) Resin-impregnated paper or similar protective clothing can be substituted for fabric type of clothing.

6) It is recommended that in highly contaminated operations (such as insulation and textiles) provisions be made for separate change rooms.

References

(1) National Institute for Occupational Safety and Health. *Criteria for a Recommended Standard: Occupational Exposure to Asbestos.* NIOSH Doc. No. HSM 72-10267, Wash., DC (1972).
(2) Environmental Protection Agency. *Summary Characterizations of Selected Chemicals of Near-Term Interest.* PB-255 817. Wash., DC. (April 1976).
(3) National Institute for Occupational Safety and Health. *Revised Recommended Asbestos Standard.* NIOSH Doc. No. 77-169, Wash., DC (Dec. 1976).
(4) Arthur D. Little, Inc. *Impact of Proposed OSHA Standard for Asbestos. First Report to the U.S. Dept. of Labor.* Report PB-283 478, Springfield, VA., Nat. Tech. Info. Service (1972).

Bibliography

Asbestos, Its Sources, Uses, Associated Environmental Exposure and Health Effects. EPA, Office of Toxic Substances (Sept. 1975).
Asbestos, the Need for and Feasibility of Air Pollution Controls. National Academy of Sciences (1971).

Biological Effects of Asbestos. HEW, National Institute of Health (Feb. 1973).

Castleman, B.I., and Fritsch, A.J. *Asbestos and You.* Washington, Center for Science in the Public Interest (Feb. 1973).

Effluent guidelines for the asbestos manufacturing industry: building materials and paper. 40 *Federal Register* 1874 (Jan. 9, 1975).

Effluent guidelines for the asbestos manufacturing industry: friction materials and textiles. 40 *Federal Register* 18172 (April 25, 1975).

Haley, T.J. Asbestosis - a reassessment of the overall problem. *J. Pharm. Science* 64(9):1435 (1975).

1975 - Review. *Asbestos* 57(7):12 (Jan. 1976).

National emission standards for hazardous air pollutants: asbestos and mercury. 39 *Federal Register* 38064 (Oct. 25, 1974).

National emission standards for hazardous air pollutants: asbestos and mercury (amendment). 40 *Federal Register* 48292 (Oct. 14, 1975).

Occupational exposure to asbestos: proposed rules. 40 *Federal Register* 47652 (Oct. 9, 1975).

Occupational safety and health standards: subpart Z - toxic and hazardous materials. 40 *Federal Register* 73072 (May 28, 1975).

ASPHALT FUMES

Description: Asphalt fumes have been defined by NIOSH (1) as the nimbose effusion of small, solid particles created by condensation from the vapor state after volatilization of asphalt. In addition to particles, a cloud of fume may contain materials still in the vapor state.

The major constituent groups of asphalt are asphaltenes, resins, and oils made up of saturated and unsaturated hydrocarbons. The asphaltenes have molecular weights in the range 1,000-2,600, those of the resins fall in the range 370-500, and those of the oils in the range 290-630.

Asphalt has often been confused with tar because the two are similar in appearance and have sometimes been used interchangeably as construction materials. Tars are, however, produced by destructive distillation of coal, oil or wood whereas asphalt is a residue from fractional distillation of crude oil.

The amounts of benzo(a)pyrene found in fumes collected from two different plants that prepared hot mix asphalt ranged from 3 to 22 nanograms/m³; this is approximately 0.03% of the amount in coke oven emissions and 0.01% of that emitted from coal-burning home furnaces.

Synonyms: None.

Potential Occupational Exposures: Occupational exposure to asphalt fumes can occur during the transport, storage, production, handling, or use of asphalt. The composition of the asphalt that is produced is dependent on the refining process applied to the crude oil, the source of the crude oil, and the penetration grade (viscosity) and other physical characteristics of the asphalt required by the consumer.

The process for production of asphalt is essentially a closed-system distillation. Refinery workers are therefore potentially exposed to the fumes during loading of the asphalt for transport from the refinery during routine maintenance, such as cleaning of the asphalt storage tanks, or during accidental spills. Most asphalt is used out of doors, in paving and roofing, and the workers' exposure to the fumes is dependent on environmental conditions, work practices, and other factors. These exposures are stated to be generally intermittent and at low concentrations. Workers are potentially exposed also to skin and eye contacts with hot, cut-back, or emulsified asphalts. Spray application of cut-back or emulsified asphalts may involve respiratory exposure also.

Asphalt sales in the United States have increased yearly from approximately 3 million tons in 1926 to over 27 million tons in 1975. Of the asphalt produced in 1975, 77.9% was used for paving, 17.4% for roofing, and 4.7% for miscellaneous purposes, such as insulating and waterproofing. Because of the nature of the major uses of asphalt and asphalt products, it is not possible to determine accurately the number of workers potentially exposed to asphalt fumes in the United States, but an estimate of 500,000 can be derived from estimates of the number of workers in various occupations involved.

Permissible Exposure Limits: Occupational exposure to asphalt fumes shall be controlled so that employees are not exposed to the airborne particulates at a concentration greater than 5 mg/m^3 of air, determined during any 15-minute period.

Occupational exposure to asphalt fumes is defined as exposure in the workplace at a concentration of one-half or more of the recommended occupational exposure limit. If exposure to other chemicals also occurs, as is the case when asphalt is mixed with a solvent, emulsified, or used concurrently with other materials such as tar or pitch, provisions of any applicable standard for the other chemicals shall also be followed.

Routes of Entry: Inhalation of dusts and fumes. Skin exposure can cause thermal burns from hot asphalt.

Harmful Effects

The principal adverse effects on health from exposure to asphalt fumes are irritation of the serous membranes of the conjunctivae and the mucous membranes of the respiratory tract. Hot asphalt can cause burns of the skin. In animals, there is evidence that asphalt left on the skin for long periods of time may result in local carcinomas, but there have been no reports of such effects on human skin that can be attributed to asphalt alone. No reliable reports of malignant tumors of parenchymatous organs due to exposure to asphalt fumes have been found, but there has been no extensive study of this possible consequence of occupational exposure in the asphalt industry.

Medical Surveillance: Medical surveillance shall be made available as outlined below to all workers subject to occupational exposure to asphalt fumes.

(A) Preplacement examinations shall include at least:

> 1) Comprehensive medical and work histories with special emphasis directed towards the eye, skin, and respiratory system.

> 2) Physical examination giving particular attention to evidence of abnormalities in the eyes, skin, or respiratory system.

> 3) A judgment of the worker's ability to use positive and negative pressure respirators.

(B) Periodic examinations shall be made available at a frequency to be determined by the responsible physician. These examinations shall include at least:

> 1) Interim medical and work histories.

> 2) Physical examination as outlined in (A-2) above.

(C) During examinations, applicants or employees found to have medical conditions which would be directly or indirectly aggravated by exposure to asphalt fumes shall be counseled on the increased risk of impairment of their health by working with this substance.

(D) Initial medical examinations shall be made available to all workers as soon as practicable after the promulgation of a standard based on these recommendations.

(E) In the event of illness known or suspected to be due to asphalt fumes, a physical examination shall be made available.

(F) Pertinent medical records shall be maintained for all employees exposed to asphalt fumes in the workplace. Such records shall be kept for at least 30 years after termination of employment. These records shall be made available to the designated medical representatives of the Secretary of Health, Education, and Welfare, of the Secretary of Labor, of the employer, and of the employee or former employee.

Special Tests: A gravimetric method is recommended for estimation of the air concentration of asphalt fumes. When large amounts of dust are present in the same atmosphere in which the asphalt fume is present, which may occur in road-building operations, the gravimetric method may lead to erroneously high estimates for asphalt fumes, and to possibly undeserved sanctions and citations for ostensibly exceeding the environmental limit for asphalt fumes or nuisance particulates.

NIOSH recommends (1) that where the resolution of such problems becomes necessary, a more specific procedure which involves solvent extraction and gravimetric analysis, be employed for the determination of asphalt fumes. The best procedure now available seems to be ultrasonic agitation of the filter in benzene and weighing of the dried residue from an aliquot of the clear benzene extract. NIOSH is attempting to devise an even more specific method for asphalt fumes for use under such conditions.

Personal Protective Methods:

(A) Respiratory Protection

> 1) Engineering controls shall be used when needed to keep concentrations of asphalt fumes below the recommended exposure limit. The only conditions under which compliance with the recommended exposure limit may be achieved by the use of respirators are:

>> (a) During the time required to install or test the necessary engineering controls.

>> (b) For operations such as nonroutine maintenance or repair activities causing brief exposure at concentrations above the environmental limit.

>> (c) During emergencies when concentrations of asphalt fumes may exceed the environmental limit.

> 2) When a respirator is permitted by paragraph (A-1) of this section, it shall be selected from a list of respirators approved by NIOSH.

(B) Protective Clothing

Employees shall wear appropriate protective clothing, including gloves, suits, boots, face shields (8-inch minimum), or other clothing as needed, to prevent eye and skin contact with asphalt.

References

(1) National Institute for Occupational Safety and Health. *Criteria for a Recommended Standard: Occupational Exposure to Asphalt Fumes,* NIOSH Doc. No. 78-106; Wash. DC., (Sept. 1977).

B

BARIUM AND COMPOUNDS

Description: Ba, barium, a silver white metal, is produced by reduction of barium oxide. The primary sources are the minerals barite ($BaSO_4$) and witherite ($BaCO_3$). Barium may ignite spontaneously in air in the presence of moisture, evolving hydrogen. Barium is insoluble in water but soluble in alcohol. Most of the barium compounds are soluble in water. The peroxide, nitrate, and chlorate are reactive and may present fire hazards in storage and use.

Synonyms: None.

Potential Occupational Exposures: Metallic barium is used for removal of residual gas in vacuum tubes and in alloys with nickel, lead, calcium, magnesium, sodium, and lithium.

Barium compounds are used in the manufacture of lithopone (a white pigment in paints), chlorine, sodium hydroxide, valves, and green flares; in synthetic rubber vulcanization, x-ray diagnostic work, glassmaking, papermaking, beet-sugar purification, animal and vegetable oil refining. They are used in the brick and tile, pyrotechnics, and electronics industries. They are found in lubricants, pesticides, glazes, textile dyes and finishes, pharmaceuticals, and in cements which will be exposed to salt water; and barium is used as a rodenticide, a flux for magnesium alloys, a stabilizer and mold lubricant in the rubber and plastics industries, an extender in paints, a loader for paper, soap, rubber, and linoleum, and as a fire extinguisher for uranium or plutonium fires.

A partial list of occupations in which exposure may occur includes:

Animal oil refiners	Paint makers
Brick makers	Plastic makers
Ceramic makers	Soap makers
Glass makers	Textile workers
Ink makers	Tile makers
Linoleum makers	Wax processors

Permissible Exposure Limits: The Federal standard for soluble barium compounds is 0.5 mg/m^3.

Routes of Entry: Ingestion or inhalation of dust or fume.

Harmful Effects

Local—Alkaline barium compounds, such as the hydroxide and carbonate, may cause local irritation to the eyes, nose, throat, and skin.

Systemic—Barium poisoning is virtually unknown in industry, although the potential exists when the soluble forms are used. When ingested or given orally, the soluble, ionized barium compounds exert a profound effect on all muscles and especially smooth muscles, markedly increasing their contractility. The heart rate is slowed and may stop in systole. Other effects are increased intestinal peristalsis, vascular constriction, bladder contraction, and increased voluntary muscle tension.

The inhalation of the dust of barium sulfate may lead to deposition in the lungs in sufficient quantities to produce "baritosis"—a benign pneumoconiosis. This produces a radiologic picture in the absence of symptoms and abnormal physical signs. X-rays, however, will show disseminated nodular opacities throughout the lung fields, which are discrete, but sometimes overlap.

Medical Surveillance: Consideration should be given to the skin, eye, heart, and lung in any placement or periodic examination.

Special Tests: None have been used.

Personal Protective Methods: Employees should receive instruction in personal hygiene and the importance of not eating in work areas. Good housekeeping and adequate ventilation are essential. Dust masks, respirators, or goggles may be needed where amounts of significant soluble or alkaline forms are encountered, as well as protective clothing.

Bibliography

Pendergrass, E.P., and R.R. Greening. 1953. Baritosis. Report of a case. *AMA Arch. Ind. Hyg. Occup. Med.* 7:44.

BENZENE

Description: C_6H_6, benzene, is a clear, volatile, colorless, highly flammable liquid with a characteristic odor. The most common commercial grade contains 50-100% benzene, the remainder consisting of toluene, xylene, and other constituents which distill below 120°C.

Synonyms: Benzol, phenyl hydride, coal naphtha, phene, benxole, cyclohexatriene.

Potential Occupational Exposures: Benzene is used as a constituent in motor fuels, as a solvent for fats, inks, oils, paints, plastics, and rubber, in the extraction of oils from seeds and nuts, and in photogravure printing (1)(2). It is also used as a chemical intermediate. By alkylation, chlorination, nitration, and sulfonation, chemicals such as styrene, phenols, and maleic anhydride are produced. Benzene is also used in the manufacture of detergents, explosives, pharmaceuticals, and dyestuffs.

A partial list of occupations in which exposure may occur includes:

Adhesive makers	Furniture finishers
Asbestos product impregnators	Glue makers
Dry-battery makers	Linoleum makers
Chemists	Maleic acid makers
Benzene hexachloride makers	Nitrobenzene makers
Burnishers	Petrochemical workers
Carbolic acid makers	Putty makers
Chlorinated benzene workers	Rubber makers
Detergent makers	Styrene makers
Dye makers	Welders

Currently benzene is consumed by the chemical industry in the U.S. at the rate of 1.4 billion gallons annually and is expected to increase when additional production facilities become available (1). NIOSH estimates that approximately 2 million workers in the U.S. are potentially exposed to benzene (1)(3). Increased concern for benzene as a significant environmental pollutant arises from public exposure to the presence of benzene in gasoline and the possibility of its increased content in gasoline (1) due to requirements for unleaded fuels for automobile equipped with catalytic exhaust converters.

Permissible Exposure Limits: In 1974 NIOSH published a criteria document (4) for occupational exposure to benzene which recommended adherence to the existing Federal standard of 10 ppm as a time-weighted average with a ceiling of 25 ppm.

The Federal emergency standard for benzene effective May 21, 1977 (5), is 1 ppm for an 8-hour TWA, with 5 ppm as a maximum peak above the acceptable ceiling for a maximum duration of 15 minutes.

OSHA again has amended its permanent standard as of mid-1978 limiting worker exposure to benzene to 1 ppm to exempt workplaces where benzene levels in liquids do not exceed 0.5%. After three years the minimum allowable level would be 0.1%. When OSHA first issued the standard no exemption for benzene mixtures was allowed. Later it amended the standard to exempt workplaces where the level of benzene in mixtures was less than 0.1%. In October 1978, the U.S. Circuit Court of Appeals (New Orleans) struck down the OSHA standard because it failed to show a cost/benefit analysis. In addition, ACGIH in 1978 designated benzene as an "Industrial Substance Suspect of Carcinogenic Potential for Man."

Routes of Entry: Inhalation of vapor which may be supplemented by percutaneous absorption although benzene is poorly absorbed through intact skin.

Harmful Effects

Local— Exposure to liquid and vapor may produce primary irritation to skin, eyes, and upper respiratory tract. If the liquid is aspirated into the lung, it may cause pulmonary edema and hemorrhage. Erythema, vesiculation, and dry, scaly dermatitis may also develop from defatting of the skin.

Systemic—Acute exposure to benzene results in central nervous system depression. Headache, dizziness, nausea, convulsions, coma, and death may result. Death has occurred from large acute exposure as a result of ventricular fibrillation, probably caused by myocardial sensitization to endogenous epinephrine. Early reported autopsies revealed hemorrhages (nonpathognomonic) in the brain, pericardium, urinary tract, mucous membranes, and skin. Chronic exposure to benzene is well documented to cause blood changes. Benzene is basically a myelotoxic agent. Erythrocyte, leukocyte, and thrombocyte counts may first increase, and then aplastic anemia may develop with anemia, leukopenia, and thrombocytopenia. The bone marrow may become hypo- or hyper-active and may not always correlate with peripheral blood.

Recent epidemiologic studies along with case reports of benzene related blood dyscrasias and chromosomal aberrations have led NIOSH to conclude that benzene is leukemogenic. The evidence is most convincing for acute myelogenous leukemia and for acute erythroleukemia, but a connection with chronic leukemia has been noted by a few investigators.

Recent work has shown increases in the rate of chromosomal aberrations associated with benzene myelotoxicity. These changes in the bone marrow are stable or unstable and may occur several years after exposure has ceased. "Stable" changes may give rise to leukemic clones and seem to involve chromosomes of the G group.

A number of recent reviews on benzene toxicity have appeared (1)(4)(6)-(8). Benzene has long been suggested as a leukemogenic agent based on many individual cases of leukemia which have been linked to benzene (4)(6)-(13). It should be noted that although in the majority of cases the individuals were subjected to mixed exposures, benzene was the agent common to all cases (1). It has been also suggested that it is possible that all cases reported as "leukemia associated with benzene exposure" have resulted from exposure to rather high concentrations of benzene and other chemicals (1). Dose-response relationships in chronic exposure of humans to benzene and details of the extent of exposures are generally considered to be either lacking or inadequate (1).

Conflicting epidemiological surveys relating to a correlation between leuke-mia and benzene exposure should be cited. For example, in the first major epidemiological survey (10), a study of 28,500 shoe workers showed an annual incidence of leukemia of 13/100,000 compared to 6/100,000 in the general population. However, an epidemiologic study (14) on 38,000 petroleum work-ers who had potential exposures to benzene failed to indicate an increase of leukemia.

The role of benzene-induced chromosome aberrations is not currently defini-tive (1). Chromosomal aberrations of both the stable and unstable type have been noted (15)-(17). In general, the chromosomal aberrations were higher in peripheral blood lymphocytes of workers exposed to benzene than in the con-trols even in the absence of overt signs of bone-marrow damage. The stable type of chromosomal aberrations persisted several years after recovery from benzene hemopathy. It was suggested that benzene might induce various types of chromosomal aberrations and that leukemia may develop in cases when potentially leukemia alone with selective advantage is produced as a toxic response to benzene exposure (15).

Numerous studies involving benzene-induced lymphocyte chromosome damage and hemopathies have been cited (18)-(25). It should be stressed that no quantitative data on total benzene exposure were available on all of the above studies on chromosome aberration in humans, with all indications sug-gesting very high levels (e.g., several hundred ppm) of benzene (1). In gen-eral, no correlation was found between the persistence of chromosomal changes and the degree of benzene poisoning (1)(25).

Exposure of cultured human leukocytes and HeLa cells to 2.2×10^{-3}M ben-zene has resulted in a decrease in DNA synthesis. Cultured human leukocytes exposed to dose levels of 1.1×10^{-3}M and 2.2×10^{-3}M exhibited chromo-some aberrations consisting of breaks and gaps (26).

Chromosomal aberrations have been noted in rats exposed to 0.2g/kg/day of benzene, 0.8g/kg/day of toluene and a mixture of benzene and toluene at levels of 0.2 and 0.8g/kg/day respectively (27).

Rabbits injected subcutaneously with a dose of 0.2mg/kg/day of pure ben-zene showed a normal karyotype in 15 of 16 test animals. However, the fre-quency of mitoses with chromosomal aberrations which was initially in the range of 5.9% increased to 57.8% after an average of 18 weeks (28).

It should be noted that animal experiments have not been supportive of the view that benzene is a leukemogenic agent (1)(29)(30). However, a co-leuke-mogenic role for benzene could explain the failure to induce leukemia in ben-zene-exposed animals (1).

The role of benzene metabolism in its toxicity as well as the significance of benzene-induced chromosome aberrations appears to be undefined (1). Urinary excretion products following benzene exposure include phenol, hydro-quinone, catechol, hydroxyhydroquinone, trans,trans-muconic acid, and L-phenylmercapturic acid (31). The major route of metabolism in all species tested was conjugation which included both ethereal sulfate and glucuronide conjugates.

The rate of benzene metabolism depends on the dose administered as well as the presence of compounds which either stimulate or inhibit benzene metabolism (1)(31).

Although the mechanism of benzene hydroxylation has not been definitively determined, it has been suggested that the reactions occur via an arene oxide intermediate (31). While benzene oxide has not been found in liver microsomes (probably due to its extreme lability) it should be noted that incubation of benzene oxide with microsomes yields the metabolic products of benzene and that naphthalene oxide has been isolated from the incubation of naphthalene with microsomes (31).

In summary, it has been established that exposure to commercial benzene or benzene-containing mixtures may result in damage to the hematopoietic system (1)(31)(32) although the mechanism by which benzene acts is not known.

In advanced stages, the result can be pancytopenia due to bone marrow aplasia. DNA synthesis is reduced in bone marrow of benzene-treated animals either because of inhibition of enzymes involved in DNA synthesis or because of lesions revealed as reduced incorporation of tritiated thymidine in DNA occurs at some point in the cell cycle.

A relationship between exposure to benzene or benzene-containing mixtures and the development of leukemia is suggested by many case reports (1)(31)(32). However, it would appear that more definitive data are required to enable a more accurate assessment of the myelotoxic, leukemogenic and chromosome-damaging effects of benzene (1)(31)(32).

Medical Surveillance: Preplacement and periodic examinations should be concerned especially with effects on the blood and bone marrow and with a history of exposure to other myelotoxic agents or drugs or of other diseases of the blood. Preplacement laboratory exams should include: (a) complete blood count (hematocrit, hemoglobin, mean corpuscular volume, white blood count, differential count, and platelet estimation), (b) reticulocyte count, (c) serum bilirubin, and (d) urinary phenol.

The type and frequency of periodic hematologic studies should be related to the data obtained from biologic monitoring and industrial hygiene studies, as well as any symptoms or signs of hematologic effects. Recommendations for proposed examinations have been made in the criteria for a recommended standard. Examinations should also be concerned with other possible effects such as those on the skin, central nervous system, and liver and kidney functions.

Special Tests: Biologic monitoring should be provided to all workers subject to benzene exposure. It consists of sampling and analysis of urine for total phenol content. The objective of such monitoring is to be certain that no worker absorbs an unacceptable amount of benzene. Unacceptable absorption of benzene, posing a risk of benzene poisoning, is considered to occur at levels of 75 mg phenol per liter of urine (with urine specific gravity corrected to 1.024), when determined by methods specified in the NIOSH "Criteria for Recommended Standard—Benzene." Alternative methods shown to be equivalent in accuracy and precision may also be useful. Biological monitoring should be done at quarterly intervals. If environmental sampling and analysis are equal to or exceed accepted safe limits, the urinary phenol analysis should be conducted every two weeks. This increased monitoring frequency should continue for at least 2 months after the high environmental level has been demonstrated.

Two follow-up urines should be obtained within one week after receipt of the original results, one at the beginning and the other at the end of the work week. If original elevated findings are confirmed, immediate steps should be taken to reduce the worker's absorption of benzene by improvement in environment control, personal protection, personal hygiene, and administrative control.

Personal Protective Methods: Protective clothing should be worn at all times; benzene-wetted clothing should be changed at once. Impervious clothing and gloves to cover exposed areas of body should be worn where exposure is continuous. In areas where there is likelihood of spill or splash, face shields or goggles should be provided. In areas of elevated vapor concentration, organic vapor cartridge masks or supplied air or self-contained breathing apparatus may be required.

References

(1) National Academy of Sciences. *A Review of Health Effects of Benzene.* National Academy of Sciences, Washington, D.C., June (1976).
(2) Kay, K. Toxicologic and cancerogenic evaluation of chemicals used in the graphic arts industries. *Clin. Toxicol.*, 9 (1976) 359-390.
(3) Anon, NIOSH links benzene to leukemia, *Chemecology* (Oct. 1976) p. 5.
(4) National Institute for Occupational Safety & Health, *Criteria for a Recommended Standard: Occupational Exposure to Benzene.* NIOSH Doc. No. 74-137, Washington, D.C. (1974).
(5) *Federal Register* 42. No. 103, 27452-27478 (May 27, 1977).
(6) Berlin, M., Gage, J., and Johnson, E. Increased aromatics in motor fuels: A review of the environmental and health effects. *Work Env. Hlth.*, 11 (1974) 1-20.
(7) Saita, G. Benzene induced hypoplastic anaemias and leukaemias. Girdwood, R.H., ed. *Blood Disorders Due to Drugs and Other Agents.* Amsterdam, Excerpta Medica. (1973) p. 127-146.

(8) Benzene in the Work Environment. Considerations bearing on the question of safe concentrations of benzene in the work environment (MAK-Wert). Communication of the Working Group "Establishment of MAK-Werte" of the Senate Commission for the Examination of Hazardous Industrial Materials, prepared in cooperation with Dr. Gertrud Buttner, Boppard, Germany, Harald Boldt Verlag (1974).

(9) Eckardt, R.E. Recent development in industrial carcinogens. *J. Occup. Med.,* 15 (1973) 904-907.

(10) Aksoy, M., Erdem, S., and Dincol, G. Leukemia in shoe-workers exposed chronically to benzene. *Blood* 44 (1974) 837-841.

(11) Aksoy, M., Erdem, S., Dincol, K., Hepyuksel, T., and Dincol, G. Chronic exposure to benzene as a possible contributary etiologic factor in Hodgkin's disease. *Blut*, 28 (1974) 293-298.

(12) Aksoy, M., Dincol, K., Erdem, S., and Dincol, G. Acute leukemia due to chronic exposure to benzene. *Am. J. Med.,* 52 (1972) 160-166.

(13) Viadana, E., and Bross, I.D.J. Leukemia and occupations. *Prev. Med.,* 1 (1972) 513.

(14) Thorpe, J.J. Epidemiologic survey of leukemia in persons potentially exposed to benzene. *J. Occup. Med.,* 16 (1974) 375-382.

(15) Vigliani, E.C., and Forni, A. Benzene, chromosome changes and leukemia. *J. Occup. Med.,* 11 (1969) 148-149.

(16) Forni, A.M., Capellini, A., Pacifico, E., and Vigliani, E.C. Chromosome changes and their evolution in subjects with past exposure to benzene. *Arch. Environ. Health,* 23 (1974) 385-391.

(17) Forni, A., Pacifico, E., and Limonta, A. Chromosome studies in workers exposed to benzene or toluene or both. *Arch. Environ. Hlth.,* 22 (1971) 373-378.

(18) Forni, A., and Moreo, L. Cytogenetic studies in a case of benzene leukaemia. *Eur. J. Cancer,* 3 (1967) 251-255.

(19) Forni, A., and Moreo, L. Chromosome studies in a case of benzene-induced erythroleukaemia. *Eur. J. Cancer,* 5 (1969) 459-463.

(20) Hartwich, G., Schwantiz, G., and Becker, J. Chromosome anomalies in a case of benzene leukaemia. *Ger. Med. Monthly,* 14 (1969) 449-450.

(21) Khan, H., and Khan, M.H. Cytogenetic studies following chronic exposure to benzene. *Arch. Toxikol.,* 31 (1973) 39-49.

(22) Sellyei, M., and Kelemen, E. Chromosome study in a case of granulocytic leukaemia with "Pelgerisation" 7 years after benzene pancytopenia. *Eur. J. Cancer,* 7 (1971) 83-85.

(23) Tough, I.M., and Court Brown, W.M. Chromosome aberrations and exposure to ambient benzene. *Lancet,* 1 (1965) 684.

(24) Pollini, G., and Colombi, R. Lymphocyte chromosome damage in benzene blood dyscrasia. *Med. Lav.,* 55 (1964) 641-654.

(25) Tough, I.M., Smith, P.G., Court Brown, W.M., and Harnden, D.G. Chromosome studies on workers exposed to atmospheric benzene. The possible influence of age. *Eur. J. Cancer,* 6 (1970) 49-55.

(26) Koizumi, A., Dobashi, Y., Tachibana, Y., Tsuda, K., and Katsunuma, H. Cytokinetic and cytogenetic changes in cultured human leucocytes and HeLa cells induced by benzene. *Ind. Health* (Japan) 12 (1974) 23-29.

(27) Dobrokhotov, V.B. The mutagenic influence of benzene and toluene under experimental conditions. *Gig Sanit.,* 37 (10) (1972) 36-39 (Translated by Air Pollution Technical Information Center, Environmental Protection Agency, Research Triangle Park, NC. Translation No. HS-138).

(28) Kissling, M., and Speck, B. Chromosome aberrations in experimental benzene intoxication. *Helv. Med. Acta.,* 36 (1971) 59-66.

(29) Laerum, O.D. Reticulum cell neoplasms in normal and benzene treated hairless mice. *Acta. Pathol. Microbiol. Scand, Sect. A,* 81 (1973) 57-63.

(30) Ward, J.M., Weisburger, J.H., Yamamoto, R.S., Benjamin, T. Brown, C.A., and
    Weisburger, E.K. Long-term effect of benzene in C57BL/6N mice. *Arch. Environ.
    Health,* 30 (1975) 22-25.
(31) Snyder, R., and Kocsis, J.J. Current concepts of chronic benzene toxicity. *CRC Crit.
    Rev. Toxicol.,* 3 (1975) 265-288.
(32) *IARC,* Vol. 7, International Agency for Research on Cancer, Lyon (1974) pp. 203-221.

Bibliography

Browning, E. *Toxicity and Metabolism of Industrial Solvents.* Amsterdam, Elsevier
    (1965).
Harris, R. *The Implications of Cancer Causing Substances in Mississippi River Water.*
    Washington, Environmental Defense Fund (November 6, 1974).
Occupational Safety and Health Standards: Subpart Z—Toxic and Hazardous Substances.
    29 CFR 1910.1000, Table Z-2.
IARC Monographs on the Evaluation of the Carcinogenic Risk of Chemicals to Man
    7:203. Lyon, International Agency for Research on Cancer (1974).
Moran, J.B. *Lead in Gasoline: Impact on Current and Future Automotive Emissions.*
    presented to Air Pollution Control Association meeting (June 1974).
Preliminary Assessment of Carcinogens in Drinking Water: Report to Congress. EPA,
    Office of Toxic Substances (December 1975).
*Development of Analytical Techniques for Measuring Ambient Atmospheric Carcinogenic
    Vapors.* EPA, Office of Air Quality Planning and Standards (Publication No. EPA-
    600/2-75-076, November 1975).
Sources of Contamination, Ambient Levels, and Fate of Benzene in the Environment.
    EPA, Office of Toxic Substances (Publication No. PB 244139, December 1974).
Sherwood, R.J., and Carter, F.W.G. 1970. The measurement of occupational exposure to
    benzene vapor. *Ann. Occup. Hyg.* 13:125.
Tauber, J.B. 1965. Instant benzol death. *J. Occup. Med.* 12:520.

BENZIDINE AND SALTS

Description: $NH_2C_6H_4C_6H_4NH_2$, benzidine, is a crystalline solid with a significant vapor pressure. The salts are less volatile, but tend to be dusty.

Synonyms: 4,4'-Biphenyldiamine, para-diaminodiphenyl, 4,4'-diaminobi-
phenyl, 4,4'-diphenylenediamine, benzidine base.

Potential Occupational Exposures: Benzidine is used (1)(2) primarily in the
manufacture of azo dyestuffs; there are over 250 of these produced. Other
uses, including some which may have been discontinued, are in the rubber
industry as a hardener, in the manufacture of plastic films, for detection of oc-
cult blood in feces, urine, and body fluids, in the detection of H_2O_2 in milk, in
the production of security paper, and as a laboratory reagent in determining
HCN, sulfate, nicotine, and certain sugars. No substitute has been found for
its use in dyes.

The three identified manufacturers (Allied, GAF, and Fabricolor) estimate that they produce 45 million pounds of azo dyes annually from benzidine. The dyes are used by about 300 major manufacturers of textile, paper, and leather products. The largest manufacturer (Allied) recently announced its intention to phase out benzidine production.

Free benzidine is present in the benzidine-derived azo dyes. According to industry, quality control specifications require that the level not exceed 20 ppm, and in practice the level is usually below 10 ppm. Industry has estimated a total environmental discharge at the 300 user facility sites of 450 pounds per year or about 1.5 pounds per year per facility, assuming all of the free benzidine is discharged in the liquid effluent.

No measurements for benzidine in ambient air, surface water, or drinking water have been reported. Further, no measurements for free benzidine in finished products containing azo dyes have been reported.

Estimates of the number of people exposed to benzidine and 3,3'-dichlorobenzidine are difficult to obtain. It has been suggested that 62 people in the U.S. are exposed to the former and between 250 and 2,500 to the latter (3). It is possible that exposures could be exceeded since 1.5 million pounds of benzidine were produced in the U.S. in 1972 while 3.5 million pounds of 3,3'-dichlorobenzidine were produced domestically in addition to another 1.4 million pounds imported in 1971 (3)(4).

A partial list of occupations in which exposure may occur includes:

Biochemists	Plastic workers
Dye workers	Rubber workers
Medical laboratory workers	Wood chemists
Organic chemical synthesizers	

Permissible Exposure Limits: Benzidine and its salts are included in a Federal standard for carcinogens; all contact with them should be avoided.

Routes of Entry: Inhalation, percutaneous absorption, and ingestion of dust.

Harmful Effects

Local—Contact dermatitis to primary irritation or sensitization has been reported.

Systemic—Benzidine is a known human urinary tract carcinogen with an average latent period of 16 years. The first symptoms of bladder cancer usually are hematuria, frequency of urination, or pain.

For a number of years, the manufacture and use of benzidine have been associated with a high risk of bladder cancer among exposed workers. Many scientists believe that tumors can result from ingestion, inhalation, or skin absorption. A number of animal studies have demonstrated the carcinogenic effects of benzidine. Mice, rats, and hamsters develop liver tumors, and dogs develop bladder cancer. Such studies have many deficiencies for estimating the risk associated with the levels of exposure to carcinogens likely to be encountered in the environment.

Free benzidine has been detected in the urine of monkeys fed benzidine-derived azo dyes, establishing a potential for reconversion of azo dyes to benzidine. Metabolism of benzidine-derived azo dyes may be similar in humans. Japanese silk painters reportedly have a high incidence of bladder cancer, possibly resulting from licking brushes and spatulas coated with benzidine-derived azo dyes. However, the carcinogenicity of such dyes has not been specifically determined.

Although benzidine is a recognized bladder carcinogen in exposed workers (2)(3)(5)-(10) analogous to other aromatic amines, the nature of the precise mechanisms responsible for the induction of neoplasia following exposure to diverse aromatic amines is not known.

Evidence exists that the metabolism of these compounds is analogous to that observed with other aromatic amines, via ring hydroxylation, N-hydroxylation of the monoacetyl derivative, and conjugation with sulfate and glucuronic acid (11)-(15). It has also been suggested that the sulfate and glucuronide conjugates of the aromatic amines might be the carcinogenically active forms *in vivo* (3).

Benzidine and its analogs (e.g., 3,3'-dichlorobenzidine; 3,3',5,5'-tetrafluorobenzidine) have been shown to be frameshift mutagens in a liver mixed function oxidase system with *S. typhimurium* TA 1538 (16)(17). Other compounds tested in this study (17), 3,3'-dianisidine (3,3'-dimethoxybenzidine) and 3,3',5,5'-tetramethylbenzidine which were either weak (18) carcinogens or noncarcinogenic (19) respectively were found to have slight mutagenic activity (in the activated system only) and no mutagenic activity respectively indicating a good correlation between animal carcinogenicity experiments and the bacterial mutagenicity assay. Benzidine has also been shown to be mutagenic in the micronucleus test in rats (20) inducing high incidences of micronucleated erythrocytes following both dermal application and subcutaneous injection. 2-Amino-, and 4-aminobiphenyl have been found mutagenic in the Salmonella/microsome test (21). 4-Aminobiphenyl is carcinogenic in the mouse, rat, rabbit, and dog (2). A high incidence of bladder carcinomas has been reported in one series of workers occupationally exposed to commercial 4-aminobiphenyl (2).

Medical Surveillance: Placement and periodic examinations should include an evaluation of exposure to other carcinogens; use of alcohol, smoking, and medications; and family history. Special attention should be given on a regular basis to urine sediment and cytology. If red cells or positive smears are seen, cystoscopy should be done at once. The general health of exposed persons should also be evaluated in periodic examinations.

Special Tests: None in common use although several metabolites are known.

Personal Protective Methods: These are designed to supplement engineering controls and to prevent all skin or respiratory contact. Full body protective clothing and gloves should also be used. On exit from a regulated area employees should shower and change into street clothes, leaving their protective clothing and equipment at the point of exit to be placed in impervious containers at the end of the work shift for decontamination or disposal. Effective methods should be used to clean and decontaminate gloves and clothing.

References

(1) Radding, S.B., Holt, B.R., Jones, J.L., Liu, D.H., Mill, T., and Hendry, D.G. *Review of Environmental Fate of Selected Chemicals.* Stanford Research Inst. Rept. Jan 10 (1975).

(2) *IARC.* Monograph No. 1. International Agency for Research on Cancer. Lyon (1972) pp. 80-86; 89-91; 74-79.

(3) Haley, T.J. Benzidine revisited: A review of the literature and problems associated with the use of benzidine and its congeners. *Clin. Toxicol.,* 8 (1975) 13-42.

(4) Anon. Final Rules Set for Exposure to Carcinogens. *Chem. Eng. News.* Feb. 11 (1974) p. 12.

(5) Hueper, W.C. *Occupational and Environmental Cancers of the Urinary System.* Yale University Press, New Haven and London (1969).

(6) Case, R.A.M., Hosker, M.E., McDonald, D.B., and Pearson, J.T. 1954. Tumors of the urinary bladder in workmen engaged in the manufacture and use of certain dyestuff intermediates in the British Chemical Industry. I. The role of aniline, benzidine, α-naphthylamine and β-naphthylamine, *Brit. J. Ind. Med.,* 11 (1954) 75.

(7) Goldwater, L.J., Rosso, A.J., and Kleinfeld, M. Bladder tumors in a coal tar dye plant. *Arch. Env. Hlth.,* 11 (1965) 814.

(8) Ubelin, F. Von., and Pletscher, A. Atiologie und prophylaxe gewerblicher tumoren in der farbstoff-industrie. *Schweiz, Med. Wschr.,* 84 (1954) 917.

(9) Anon. *Federal Register,* 39 (1974) 3756 (Jan. 29, 1974).

(10) Zavon, M.R., Hoegg, U., and Bingham, E. Benzidine exposure as a cause of bladder tumors, *Arch. Env. Hlth.,* 27 (1973) 1.

(11) Miller, J.A., and Miller, E.C. The metabolic activation of carcinogenic aromatic amines and amides. *Progr. Exptl. Tumor Res.,* 11 (1969) 273.

(12) Miller, J.A., and Miller, E.C. Chemical carcinogenesis: Mechanisms and approaches to its control. *J. Natl. Cancer Inst.,* 47 (1971) V-XIV.

(13) Miller, J.A. Carcinogenesis by Chemicals: An Overview—G.H.A. Clowes Memorial Lecture. *Cancer Res.,* 30 (1970) 559-576.

(14) Gutmann, H.R., Malejka-Giganti, D., Barry, E.J., and Rydell, R.E. On the correlation between the hepato carcinogenicity of the carcinogen N-2-fluorenyl-acetamide and its metabolite activation by the rat. *Cancer Res.,* 32 (1972) 1554-1560.

(15) Arcos, J.C., and Argus, M.F. *Chemical Induction of Cancer.* Vol. IIB, Academic Press, New York (1974) pp. 23-37.

(16) Ames, B.N., Durston, W.E., Yamasaki, E., and Lee, F.D. Carcinogens are mutagens. Simple test system combining liver homogenates for activation and bacteria for detection. *Proc. Natl. Acad. Sci.* (USA) 70 (1973) 2281-2285.

(17) Garner, R.C., Walpole, A.L., and Rose, F.L. Testing of some benzidine analogs for microsomal activation to bacterial mutagens. *Cancer Letters,* 1 (1975) 39-42.
(18) Hadidian, Z., Fredrickson, T.N., et al. Tests for chemical carcinogens. Report on the activity of derivatives of aromatic amines, nitrosamines, quinolines, nitroalkanes, amides, epoxides, aziridines and purine antimetabolites. *J. Natl. Can. Inst.,* 41 (1968) 985.
(19) Holland, V.R., Saunders, B.C., Rose, F.L., and Walpole, A.L. Safer substitute for benzidine in the detection of blood. *Tetrahedron.,* 30 (1974) 3299.
(20) Urwin, C., Richardson, J.C., and Palmer, A.K. An evaluation of the mutagenicity of the cutting oil preservative Grotan BK. *Mutation Res.,* 40 (1976) 43.
(21) McCann, J., Choi, E., Yamaski, E., and Ames, B.N. Detection of carcinogens as mutagens in the Salmonella/microsome test: Assay of 300 chemicals. *Proc. Nat. Acad. Sci.,* 72 (1975) 5135.

Bibliography

Benzidine: Wastewater Treatment Technology. EPA, Office of Water and Hazardous Materials (June 1974).
Cranmer, M. Dr. General considerations in setting a benzidine standard: Testimony to EPA, Office of Water and Hazardous Materials.
EPA/SOCMA. Stipulation of fact. EPA, Office of Water and Hazardous Materials (Hearings, May 30, 1974).
Hazard Review of Benzidine. HEW, National Institute for Occupational Safety and Health (July 1973).
Laham, S., Farant, J.P., and Potvin, M. 1971. Biochemical determination of urinary bladder carcinogens in human urine. *Occup. Health Rev.* 21:14.
Occupational safety and health standards: Subpart Z—toxic and hazardous materials (benzidine). 39 *Federal Register* 20 (January 29, 1974).
Rinde, E., and Troll, W. Metabolic reduction of benzidine azo dyes to benzidine in the rhesus monkey. *J. National Cancer Inst.,* 55:181 (1975).
SOCMA. Comments on production and use of benzidine. Submitted to EPA, Office of Water and Hazardous Materials (March 9, 1973).
SOCMA. Affidavit of Mr. Kelvin H. Ferber. Submitted to EPA, Office of Water and Hazardous Materials (May 13, 1974).
SOCMA. Benzidine effluent data. Submitted to EPA, Office of Water and Hazardous Materials (June 6, 1974).
SOCMA. Industrial hygiene and environmental control. Submitted to EPA, Office of Water and Hazardous Materials (May 30, 1975).
SOCMA. Submission. Submitted to EPA, Office of Water and Hazardous Materials (June 5, 1975).
SOCMA. Second Submission. Submitted to EPA, Office of Water and Hazardous Materials (August 5, 1975).

BENZOYL PEROXIDE

Description: Benzoyl peroxide, $C_6H_5CO-O-O-COC_6H_5$, is a crystalline solid, melting at 103° to 106°C which may explode when heated.

Synonyms: Dibenzoyl peroxide, benzoyl superoxide.

Potential Occupational Exposures: A partial list of occupations in which exposure may occur include:

Automobile body repair workers	Pharmaceutical products makers
Bakers	Pharmacists
Benzoyl peroxide makers	Physicians
Cheesemakers	Plastic products makers
Dentists	Polyester makers
Dental assistants	Printers
Flour-mill workers	Silicone rubber makers
Miners	Styrene makers
Nurses	Telephone repair workers

NIOSH estimates that 25,000 workers in the United States are potentially exposed to benzoyl peroxide or its formulations.

Permissible Exposure Limits: Exposure to benzoyl peroxide shall be controlled so that employees are not exposed at a concentration greater than 5 milligrams per cubic meter (mg/cu m) of air, determined as a time-weighted average (TWA) concentration for up to a 10-hour shift in a 40-hour workweek (1).

Routes of Entry: The major concerns from occupational exposure to benzoyl peroxide are the hazards arising from its instability, flammability, and explosive properties. In addition, benzoyl peroxide may cause local irritation of the eyes and skin. Inhalation and skin contact are the main routes of entry.

Harmful Effects: The following table gives some effects of benzoyl peroxide exposure on humans:

Route of Exposure	Exposure Concentration and Duration	Effects	Reference
Dermal	20%	Irritation in 1 of 180	2
Dermal	1%, 5%, and 10% nine 24-hr applications	Severe eczematous skin reactions in 25 of 69 at end of experiment	3
Dermal	5% 12 hr	Marked erythema and burning	4
Dermal	5% 48 hr	Severe irritation	4
Dermal	Unknown	Positive patch test and dermatitis in 3 of 30	5
Dermal	20-100%	Slight skin irritation	6
Respiratory	1.34-17.0 mg/m^3	Nose and throat irritation	7
Respiratory	2.58-82.5 mg/m^3	Eye, nose, and throat irritation	7
Dermal and respiratory	Unknown	Severe dermatitis, asthmatic wheezing	8

Medical Surveillance: Preplacement and periodic medical examinations should be conducted with particular attention to skin conditions (1).

Special Tests: None in common use.

Personal Protective Methods: Protective clothing and safety glasses with side shields or safety goggles should be worn by employees to reduce the possibility of skin contact and eye irritation. Such protection is especially important where benzoyl peroxide and other powder or granular benzoyl peroxide formulations may become airborne or where liquid or paste formulations of benzoyl peroxide might be spattered or spilled.

Protective clothing should be fire resistant. Any fabric that generates static electricity is not recommended. To prevent the buildup of static electricity, appropriate conductive footwear should be worn (9). Gloves made of rubber, leather, or other appropriate material should be worn by employees for protection when they are opening shipping boxes of pure benzoyl peroxide (9) or otherwise handling pure benzoyl peroxide. Aprons made of rubber or another appropriate material are recommended for added protection when handling benzoyl peroxide and its formulations. Plastic aprons that generate static electricity should not be used (9).

Respiratory protection as specified by NIOSH (1) must be used whenever airborne concentrations of benzoyl peroxide cannot be controlled to the recommended workplace environmental limit by either engineering or administrative controls.

References

(1) National Institute for Occupational Safety and Health. *Criteria for a recommended standard: Occupational Exposure to Benzoyl Peroxide.* NIOSH Doc. No. 77-166, Wash., D.C. (1977).
(2) Morley, M.H. Decubitus ulcer management—A team approach. *Can Nurse* 69:41-43, 1973.
(3) Poole, R.L., Griffith, J.F., MacMillan, F.S.K. Experimental contact sensitization with benzoyl peroxide. *Arch Dermatol* 102:635-39, 1970.
(4) Eaglstein, W.H. Allergic contact dermatitis to benzoyl peroxide. *Arch Dermatol* 97:527, 1968.
(5) Malten, K.E. Synthetics and eczema as a problem of occupational hygiene. *Ned Tijdschr Geneeskd* 101:1319-25, 1957 (Dut).
(6) Jirasek, L. and Kalensky, J. Occupational eczema caused by epoxy resins. *Symposium Dermatologorum, Corpus Lectionum,* 1st, Prague, Oct. 12-15, 1960. Prague, Universita Karlova, 1962, vol 2, pp 203-11 (Ger).
(7) Moskowitz, S. and Burke, W.J. Lucidol Division—Novadel-Agene Corporation, report No. L-430-50. Tonawanda, NY, New York City Division of Industrial Hygiene and Safety Standards, Chemical Unit, 1950, 9 pp.
(8) Baird, K.A. Allergy to chemicals in flour—A case of dermatitis due to benzoic acid. *J. Allergy* 16:195-98, 1945.

(9) *Properties and Essential Information for Safe Handling and Use of Benzoyl Peroxide.* Chemical Safety Data Sheet SD-81. Washington, D.C., Manufacturing Chemists' Association Inc., 1960, 10 pp.

BENZYL CHLORIDE

Description: $C_6H_5CH_2Cl$, benzyl chloride is a colorless liquid with an unpleasant, irritating odor.

Synonyms: Alpha-chlorotoluene.

Potential Occupational Exposures: In contrast to phenyl halides, benzyl halides are very reactive (1). Benzyl chloride is used in production of benzal chloride, benzyl alcohol, and benzaldehyde. Industrial usage includes the manufacture of plastics, dyes, synthetic tannins, perfumes, resins, and pharmaceuticals (2).

Suggested uses of benzyl chloride include: in the vulcanization of fluororubbers (3) and in the benzylation of phenol and its derivatives for the production of possible disinfectants (4).

A partial list of occupations in which exposure may occur includes:

Drug makers	Plastic makers
Dye makers	Resin makers
Gasoline additive makers	Rubber makers
Germicide makers	Tannin makers
Perfume makers	Wetting agent makers
Photographic developer makers	

Permissible Exposure Limits: The Federal standard is 1 ppm (5 mg/m^3).

Route of Entry: Inhalation of vapor.

Harmful Effects

Local—Benzyl chloride is a severe irritant to the eyes and respiratory tract. At 160 mg/m^3 it is unbearably irritating to the eyes and nose. Liquid contact with the eyes produces severe irritation and may cause corneal injury. Skin contact may cause dermatitis.

Systemic—Benzyl chloride is regarded as a potential cause of pulmonary edema. One author has reported disturbances of liver functions and mild leukopenia in some workers, but this has not been confirmed. Benzyl chloride has been shown to induce local sarcomas in rats treated by subcutaneous injection (5)(6). Benzyl chloride was reported to be weakly mutagenic in *S. ty-phimurium* TA 100 strain (7).

Medical Surveillance: Preplacement and periodic examinations should include the skin, eyes, and an evaluation of the liver, kidney, respiratory tract, and blood.

Special Tests: None in common use.

Personal Protective Methods: Personal protective equipment should include industrial filter respirators with goggles, and protective clothing for face, hands, and arms.

References

(1) Roberts, J.D. and Caserio, M.C. *Modern Organic Chemistry.* W.A. Benjamin, Inc., New York (1967) p. 571.
(2) *IRAC,* Vol. II. International Agency for Research on Cancer, Lyon (1976) 217-223.
(3) Okada, S., and Iwa, R. Fluorolefin elastomer stocks. Japan Kokai 7,414,560, Feb. 8 (1974) *Chem. Abstr.,* 81 (1974) P50806K.
(4) Janata, V., Simek, A., and Nemeck, O. Benzyl phenols. Czech Patent 152,190, Feb. 15 (1974) *Chem. Abstr.,* 81 (1974) P25347D.
(5) Druckrey, H., Kruse, H., Preussmann, R., Wankovic, S., and Lanschutz, C. Cancerogene alkylierende substanzen. III. Alkyl-halogenide-sulfate sulfonate und ring-gespannte heterocyclen. *Z. Krebsforsch.,* 74 (1970) 241.
(6) Preussmann, R. Direct alkylating agents as carcinogens. *Food Cosmet. Toxicol.* 6 (1968) 576.
(7) McCann, J., Springarn, N.E., Kobori, J., and Ames, B.N. Carcinogens as mutagens: Bacterial tester strains with R factor plasmids. *Proc. Natl. Acad. Sci.,* 72 (1975) 979-983.

Bibliography

Mikhailova, T.V. 1965. Comparative toxicity of chloride derivatives of toluene: benzyl chloride, benzal chloride, and benzotrichloride. *Gig. Tr. Prof. Zabol.* 8:14. [Translation published in 1965. *Fed. Proc.* (Trans. Suppl.) 24:877.]

BERYLLIUM AND COMPOUNDS

Description: Be, beryllium, is a grey-metal which combines the properties of light weight and high tensile strength. Beryllium is slightly soluble in hot water and in dilute acids and alkalis. All beryllium compounds are soluble to some degree in water. Beryl ore is the primary source of beryllium, although there are numerous other sources.

Synonyms: None.

Potential Occupational Exposures: Beryllium metal is widely used in the atomic energy field as a moderator for fission reactions, as a reflector to reduce leakage of neutrons from the reactor core, and, in a mixture with uranium, as a neutron source. Beryllium foil is the window material for x-ray tubes. Beryllium may be alloyed with a number of metals to increase hardness. Beryllium-copper alloy is the most common and is used in parts subjected to abnormal wear, extreme vibration, or shock loading such as in bushings, current-carrying springs, electric contacts and switches, and radio and radar components; it is also used in nonsparking tools. Beryllium-nickel alloy has high tensile strength, increased hardness, and age-hardening characteristics which make it useful in diamond drill-bit matrices, watch-balance wheels, and certain airplane parts. Beryllium bronzes are used in nonspark tools, electrical switch parts, watch springs, diaphragms, shims, cams, and bushings. Other alloys may be formed with zinc, magnesium, iron, aluminum, gold, silver, platinum, nickel, and steel. Beryllium also has potential for use in the aircraft and aerospace industry.

Beryllium compounds are utilized in the manufacture of ceramics and refractories, as chemical reagents and gas mantle hardeners, and in atomic energy reactions. The use of phosphors produced from beryllium oxide in fluorescent lamps has been discontinued.

Hazardous exposure to beryllium is generally associated with the milling and use of beryllium and not the mining and handling of beryl ore.

A partial list of occupations in which exposure may occur includes:

Beryllium alloy workers	Gas mantle makers
Cathode ray tube makers	Missile technicians
Ceramic makers	Nuclear reactor workers
Electric equipment makers	Refractory material makers

Permissible Exposure Limits: The present Federal standard for beryllium and beryllium compounds is 2 μg/m^3 as an 8-hour TWA with an acceptable ceiling concentration of 5 μg/m^3. The acceptable maximum peak is 25 μg/m^3 for a maximum duration of 30 minutes. The standard recommended in the NIOSH Criteria Document (1) is 2 μg Be/m^3 as an 8-hour TWA with a peak value of 25 μg Be/m^3 as determined by minimum sampling time of 30 minutes. Recently (2) it was recommended that OSHA treat beryllium as a carcinogen.

Route of Entry: Inhalation of fume or dust.

Harmful Effects

Local—The soluble beryllium salts are cutaneous sensitizers as well as primary irritants. Contact dermatitis of exposed parts of the body are caused by acid salts of beryllium. Onset is generally delayed about two weeks from the time of first exposure. Complete recovery occurs following cessation of exposure. Eye irritation and conjunctivitis can occur. Accidental implantation of beryllium metal or crystals of soluble beryllium compound in areas of broken or abraded skin may cause granulomatous lesions. These are hard lesions with a central nonhealing area. Surgical excision of the lesion is necessary. Exposure to soluble beryllium compounds may cause nasopharyngitis, a condition characterized by swollen and edematous mucous membranes, bleeding points, and ulceration. These symptoms are reversible when exposure is terminated.

Systemic—Beryllium and its compounds are highly toxic substances. Entrance to the body is almost entirely by inhalation. The acute systemic effects of exposure to beryllium primarily involve the respiratory tract and are manifest by a nonproductive cough, substernal pain, moderate shortness of breath, and some weight loss. The character and speed of onset of these symptoms, as well as their severity, are dependent on the type and extent of exposure. An intense exposure, although brief, may result in severe chemical pneumonitis with pulmonary edema.

Chronic beryllium disease is an intoxication arising from inhalation of beryllium compounds, but it is not associated with inhalation of the mineral beryl. The chronic form of this disease is manifest primarily by respiratory symptoms, weakness, fatigue, and weight loss (without cough or dyspnea at the onset), followed by nonproductive cough and shortness of breath. Frequently, these symptoms and detection of the disease are delayed from five to ten years following the last beryllium exposure, but they can develop during the time of exposure. The symptoms are persistent and frequently are precipitated by an illness, surgery, or pregnancy. Chronic beryllium disease usually is of long duration with exacerbations and remissions.

Chronic beryllium disease can be classified by its clinical variants according to the disability the disease process produces.

1. Asymptomatic nondisabling disease is usually diagnosed only by routine chest x-ray changes and supported by urinary or tissue assay.
2. In its mildly disabling form, the disease results in some nonproductive cough and dyspnea following unusual levels of exertion. Joint pain and weakness are common complaints. Diagnosis is by x-ray changes. Renal calculi containing beryllium may be a complication. Usually, the patient remains stable for years, but eventually shows evidence of pulmonary or myocardial failure.
3. In its moderately severe disabling form, the disease produces symptoms of distressing cough and shortness of breath, with marked x-ray changes. The liver and spleen are frequently affected, and spontaneous pneumothorax may occur. There is generally weight loss, bone and joint pain, oxygen desaturation, increase in hematocrit, disturbed liver function, hypercalciuria, and spontaneous skin lesions similar to those of Boeck's sarcoid. Lung function studies show measurable decreases in diffusing capacity. Many people in this group survive for years with proper therapy. Bouts of chills and fever carry a bad prognosis.

4. The severely disabling disease will show all of the above mentioned signs and symptoms in addition to severe physical wasting and negative nitrogen balance. Right heart failure may appear causing a severe nonproductive cough which leads to vomiting after meals. Severe lack of oxygen is the predominant problem, and spontaneous pneumothorax can be a serious complication. Death is usually due to pulmonary insufficiency or right heart failure.

Medical Surveillance: Preemployment history and physical examinations for worker applicants should include chest x-rays, baseline pulmonary function tests (FVC and FEV_1), and measurement of body weight. Beryllium workers should receive a periodic health evaluation that includes: spirometry (FVC and FEV_1), medical history questionnaire directed toward respiratory symptoms, and a chest x-ray. General health, liver and kidney functions, and possible effects on the skin should be evaluated.

Special Tests: Beryllium can be determined in the urine, but shows poor correlation with quantitative exposures. Tissue biopsies for beryllium content have also been utilized in diagnostic procedures, but often show no relation to the severity of the disease and indicate only that exposure has occurred.

Personal Protective Methods: Work areas should be monitored to limit and control levels of exposure. Personnel samplers are recommended. Good housekeeping, proper maintenance, and engineering control of processing equipment and technology are essential. The importance of safe work practices and personal hygiene should be stressed. When beryllium levels exceed the accepted standards, the workers should be provided with respiratory protective devices of the appropriate class, as determined on the basis of the actual or projected atmospheric concentration of airborne beryllium at the worksite. Protective clothing should be provided all workers who are subject to exposure in excess of the standard. This should include shoes or protective shoe covers as well as other clothing. The clothing should be reissued clean on a daily basis. Workers should shower following each shift prior to change to street clothes.

Reference

(1) National Institute for Occupational Safety and Health. *Criteria for a Recommended Standard: Occupational Exposure to Beryllium.* NIOSH Doc. No. 72-10268, Wash., D.C. (1972).
(2) *Chemical Week*, p. 80 (Nov. 15, 1978).

Bibliography

Tepper, L.B., Hardy, H.L., and Chamberlin, R.I. 1961. *Toxicity of Beryllium Compounds.* Elsevier, New York.

BIS(CHLOROMETHYL) ETHER

Description: ClCH$_2$OCH$_2$Cl, bis(chloromethyl) ether, is a colorless, volatile liquid with a suffocating odor. This substance may form spontaneously in warm moist air by the combination of formaldehyde and hydrogen chloride.

Synonyms: BCME, sym-dichloromethyl ether.

Potential Occupational Exposures: Exposure to bis(chloromethyl) ether may occur in industry and in the laboratory. This compound is used as an alkylating agent in the manufacture of polymers, as a solvent for polymerization reactions, in the preparation of ion exchange resins, and as an intermediate for organic synthesis.

A partial list of occupations in which exposure may occur includes:

Ion exchange resin makers	Organic chemical synthesizers
Laboratory workers	Polymer makers

Haloethers, primarily alpha-chloromethyl ethers, represent a category of alkylating agents of increasing concern (1)-(13) due to the establishment of a causal relationship between occupational exposure to two agents of this class and lung cancer in the United States and abroad (3)-(13). These haloethers are bis(chloromethyl) ether (BCME, ClCH$_2$OCH$_2$Cl) and chloromethyl methyl ether (methyl chloromethyl ether, CMME, ClCH$_2$OCH$_3$).

BCME can be produced from paraformaldehyde, sulfuric acid and hydrogen chloride while CMME can be produced via the reaction of methanol, formaldehyde, and anhydrous hydrogen chloride. It should be noted that commercial grades of CMME can be contaminated with 1% to 8% BCME (6)(7).

The potential for BCME formation increases with available formaldehyde and chloride (14)-(16) (in both gaseous and liquid phases):

$$2Cl^- + 2HCHO + 2H^+ \rightarrow ClCH_2OCH_2Cl + H_2O$$

The reaction is believed to be an equilibrium much in favor of the reactants. The extent of hazard from the combination of formaldehyde and HCl to form BCME is unknown at present, and to date, the results appear scanty and disparate (10).

Potential sources of human exposure to BCME appear to exist primarily in areas including (a) its use in chloromethylating (cross-linking) reaction mixtures in anion-exchange resin production (14); (b) segments of the textile industry using formaldehyde-containing reactants and resins in the finishing of fabric and as adhesives in the laminating and flocking of fabrics (16) and (c) the nonwoven industry which uses as binders, thermosetting acrylic emulsion polymers comprising methylol acrylamide, since a finite amount of formaldehyde is liberated on the drying and curing of these bonding agents (16).

NIOSH has confirmed the spontaneous formation of BCME from the reaction of formaldehyde and hydrochloric acid in some textile plants and is now investigating the extent of possible worker exposure to the carcinogen (17). However, this finding has recently been disputed by industrial tests in which BCME was not formed in air by the reaction of textile systems employing hydrochloric acid and formaldehyde (18).

Permissible Exposure Limits: Bis(chloromethyl) ether is included in the Federal standard for carcinogens; all contact with it should be avoided.

Route of Entry: Inhalation of vapor and perhaps, but to a lesser extent, percutaneous absorption.

Harmful Effects

Local—Vapor is severely irritating to the skin and mucous membranes and may cause corneal damage which may heal slowly.

Systemic—Bis(chloromethyl) ether has an extremely suffocating odor even in minimal concentration so that experience with acute poisoning is not available. It is not considered a respiratory irritant at concentrations of 10 ppm. Bis(chloromethyl) ether is a known human carcinogen. Animal experiments have shown increases in lung adenoma incidence; olfactory esthesioneuroepitheliomas which invaded the sinuses, cranial vault, and brain; skin papillomas and carcinomas; and subcutaneous fibrosarcomas. There have been several reports of increased incidence of human lung carcinomas (primarily small cell undifferentiated) among ether workers exposed to bis(chloromethyl) ether as an impurity. The latency period is relatively short—10 to 15 years. Smokers as well as nonsmokers may be affected.

Evidence of the human carcinogenicity of BCME and CMME have been cited (3)-(13). Regulations published recently by OSHA in the United States, specifically list both BCME and CMME as human carcinogens (5). Epidemiological studies on an industry-wide basis in the United States, have disclosed some 30 cases of lung cancer in association with BCME and CMME (11).

The carcinogenicity of BCME and CMME by skin application to mice and by subcutaneous administration to mice and rats (1)(10), the induction of lung adenomas by intraperitoneal injection of BCME in newborn mice (19) and by inhalation of CMME and BCME (20), and the induction of squamous carcinomas of the lung and esthesioneuroepitheliomas in rats by inhalation exposure (21)(22) of 0.1 ppm BCME 5 hr/day, 5 days/week through their lifetime as well as in groups of rats given 10, 20, 40, 60, 80 and 100 exposures to 0.1 ppm BCME and then held until death, have all been reported.

Van Duuren et al., (1)(2)(23) suggested that the α-haloethers be classified with the biologically active alkylating agents (e.g., nitrogen mustards, epoxides,

β-lactones, etc.). The high chemical reactivity of the α-haloethers is attributed to the reactivity of the halogen atom in displacement reactions. In comparing the carcinogenicity of 11 chloroethers (2)(10), in general, bifunctional α-chloroethers are more active than their monofunctional analogs. As the chain length increases, activity decreases, and as chlorine moves further away from the ether oxygen, carcinogenic activity also decreases. It was also noted that in a general way, the more carcinogenically active compounds are the most labile; as stability increases, carcinogenicity also decreases (10).

Medical Surveillance: Preplacement and periodic medical examinations should include an examination of the skin and respiratory tract, including chest x-ray. Sputum cytology has been suggested as helpful in detecting early malignant changes, and in this connection a smoking history is of importance. Possible effects on the fetus should be considered.

Special Tests: None have been suggested.

Personal Protective Methods: These are designed to supplement engineering controls and should be appropriate for protection of all skin or respiratory contact. Full body protective clothing and gloves should be used on entering areas of potential exposure. Those employed in handling operations should be provided with full face, supplied air respirators of continuous flow or pressure demand type. On exit from a regulated area, employees should remove and leave protective clothing and equipment at the point of exit, to be placed in impervious containers at the end of the work shift for decontamination or disposal. Showers should be taken before dressing in street clothes.

References

(1) Van Duuren, B.L., Goldschmidt, B.M., Katz, C., Langseth, L., Mercado, C., and Sivak, A. Alpha-haloethers: A new type of alkylating carcinogen. *Arch. Env. Hlth.,* 16 (1968) 472-476.

(2) Van Duuren, B.L., Katz, C., Goldschmidt, B.M., Frenkel, K., and Sivak, A. Carcinogenicity of haloethers. II. Structure-activity relationships of analogs of bis(chloromethyl) ether. *J. Natl. Cancer Inst.,* 48 (1972) 1431-1439.

(3) IARC. Bis(chloromethyl) ether. Monograph No. 4, Lyon (1974), pp. 231-238.

(4) IARC. Chloromethylmethylether. Monograph No. 4, Lyon (1974) pp. 239-245.

(5) OSHA. Occupational Safety and Health Standards: Carcinogens, *Fed. Reg.,* 39 (20) (1974) 3768-3773; 3773-3776.

(6) Albert, R.E., et al. Mortality patterns among workers exposed to chloromethyl ethers. *Env. Hlth. Persp.,* 11 (1975) 209-214.

(7) Figueroa, W.G., Raszkowski, R., and Weiss, W. Lung cancer in chloromethyl methyl ether workers. *New Engl. J. Med.,* 288 (1973) 1094-1096.

(8) Weiss, W., and Figueroa, W.G. The characteristics of lung cancer due to chloromethyl ethers. *J. Occup. Med.,* 18 (1976) 623-627.

(9) Weiss, W. Chloromethyl ethers, cigarettes, cough and cancer. *J. Occup. Med.,* 18 (1976) 194-199.

(10) Nelson, N. The chloroethers—occupational carcinogens: A summary of laboratory and epidemiology studies. *Ann. NY Acad. Sci.,* 271 (1976) 81-90.

(11) Nelson, N. The carcinogenicity of chloroethers and related compounds. Meeting on Origins of Human Cancer, Cold Springs Harbor Laboratory, New York, Sept. 7-14 (1976) p. 8.

(12) Sakabe, H. Lung cancer due to exposure to bis(chloromethyl) ether. *Ind. Hlth.,* 11 (1973) 145.

(13) Thiess, A.M., Hey, W., and Zeller, H. Zue Toxikologie von dichlorodimethyl atherverdacht auf kanzerogene wirkung auch beim menschen. *Zbl. Arbeits Med.,* 23 (1973) 97.

(14) Rohm & Haas Co. News release: Reaction of formaldehyde and HCl forms bis-CME. Rohm & Haas, Phila., PA, Dec. 27 (1972).

(15) Kallos, G.J., and Solomon, R.A. Investigation of the formation of bis(chloromethyl) ether in simulated hydrogen chloride-formaldehyde atmospheric environments. *Amer. Ind. Hyg. Assoc. T.,* 34 (1973) 469-473.

(16) Hurwitz, M.D. Assessing the hazard from BCME in formaldehyde-containing acrylic emulsions. *Amer. Dyestuff Reptr.,* 63 (1974) 62-64, 77.

(17) Anon. Industry's problems with cancer aired. *Chem. Eng. News,* 53 (1975) 4.

(18) Anon. Dow says bis(chloromethyl) ether does not form during textile operations. *Toxic Materials News,* 3 (20) (1976) 157.

(19) Gargus, J.L., Reese, W.H., Jr., and Rutter, H.A. Induction of lung adenomas in new born mice by bis(chloromethyl) ether. *Toxicol. Appl. Pharmacol.,* 15 (1969) 92-96.

(20) Leong, B.K., MacFarland, H.N., and Reese, W.H., Jr. Induction of lung adenomas by chronic inhalation of bis(chloromethyl) ether. *Arch. Env. Hlth.* 22 (1971) 663-666.

(21) Laskin, S., Kuschner, M., Drew, R.T., Capiello, V.P., and Nelson, N. Tumors of the respiratory tract inducted by inhalation of bis(chloromethyl) ether. *Arch. Env. Hlth.,* 23 (1971) 125-176.

(22) Kuschner, M. Laskin, S., Drew, R.T., Capiello, V., and Nelson, N. Inhalation carcinogenicity of alpha-haloethers, III. Lifetime and limited period inhalation studies with BCME at 0.1 ppm. *Arch. Env. Hlth.,* 30 (1975) 73-77.

(23) Van Duuren, B.L. Carcinogenic epoxides, lactones and halo-ethers and their mode of action. *Ann. NY Acad. Sci.,* 163 (1969) 633-651.

Bibliography

Lemen, R.A., Johnson, W.M., Wagoner, J.K., Archer, V.E., and Saccomanno., G. 1976. Cytologic observations and incidence following exposure to BCME. *Ann. NY Acad. Sci.* 271:71.

Note: See also the entry under Chloromethyl Methyl Ether.

BISMUTH AND COMPOUNDS

Description: Bi, bismuth, is a pinkish-silver, hard, brittle metal. It is found as the free metal in ores such as bismutite and bismuthinite and in lead ores. Bismuth is soluble in some mineral acids and insoluble in water. Most bismuth compounds are soluble in water.

Synonyms: None.

Potential Occupational Exposures: Bismuth is used as a constituent of tempering baths for steel alloys, in low melting point alloys which expand on cooling, in aluminum and steel alloys to increase machinability, and in printing type metal. Bismuth compounds are found primarily in pharmaceuticals as antiseptics, antacids, antiluetics, and as a medicament in the treatment of acute angina. They are also used as a contrast medium in roentgenoscopy and in cosmetics.

A partial list of occupations in which exposure may occur includes:

Chemists	Permanent magnet makers
Cosmetic workers	Pigment makers
Disinfectant makers	Solder makers
Fuse makers	Steel alloy makers
Laboratory workers	Tin lusterers

Permissible Exposure Limits: There is no Federal standard for bismuth or its compounds. ACGIH (1978) has set TWA values of 10 mg/m^3 for bismuth telluride and 5 mg/m^3 for Se-doped bismuth telluride.

Route of Entry: Ingestion of powder or inhalation of dust.

Harmful Effects

Local—Bismuth and bismuth compounds have little or no effect on intact skin and mucous membrane. Absorption occurs only minimally through broken skin.

Systemic—There is no evidence connecting bismuth and bismuth compounds with cases of industrial poisoning. All accounts of bismuth poisoning are from the soluble compounds used previously in therapeutics. Fatalities and near fatalities have been recorded chiefly as a result of intravenous or intramuscular injection of soluble salts. Principal organs affected by poisoning are the kidneys and liver. Chronic intoxication from repeated oral or parenteral doses causes "bismuth line." This is a gum condition with black spots of buccal and colonic mucosa, superficial stomatitis, foul breath, and salivation.

Medical Surveillance: No special considerations are necessary other than following good general health practices. Liver and kidney function should be followed if large amounts of soluble salts are ingested.

Special Tests: None have been reported.

Personal Protective Methods: Personal hygiene should be stressed, and eating should not be permitted in work areas. Dust masks should be worn in dusty areas to prevent inadvertent ingestion of the soluble bismuth compounds.

Bibliography

James, J.A. 1968. Acute renal failure due to a bismuth preparation. Calif. Med. 109:317.

BORON AND COMPOUNDS

Description: Boron, B, is a brownish-black powder and may be either crystalline or amorphous. It does not occur free in nature and is found in the minerals borax, colemanite, boronatrocalcite, and boracite. Boron is slightly soluble in water under certain conditions.

Boric acid, H_3BO_3, is a white, amorphous powder. Saturated solutions at 0°C contain 2.6% acid; at 100°C, 28% acid. Boric acid is soluble 1 g/18 ml in cold water.

Borax, $Na_2B_4O_7 \cdot 5H_2O$, is a colorless, odorless crystalline solid. Borax is slightly soluble in water.

Boron trifluoride, BF_3, is a colorless gas with a pungent, suffocating odor. It decomposes in water, forming boric acid and fluoboric acid and hydrolyzes in air giving rise to dense, white fumes.

Boron oxide, B_2O_3, is a vitreous, colorless, crystalline, hygroscopic solid and slightly soluble in water.

Synonyms: B, none; boric acid, boracic acid; borax, tincal; boron trifluoride, boron fluoride; boron oxide, boric oxide.

Potential Occupational Exposures: Boron is used in metallurgy as a degasifying agent and is alloyed with aluminum, iron, and steel to increase hardness. It is also a neutron absorber in nuclear reactors.

Boric acid is a fireproofing agent for textiles, a weatherproofing agent for wood, a preservative, and an antiseptic. It is used in the manufacture of glass, pottery, enamels, glazes, cosmetics, cements, porcelain, borates, leather, carpets, hats, soaps, and artificial gems, and in tanning, printing, dyeing, painting, and photography. It is a constituent in powders, ointments, nickeling baths, electric condensers and is used for impregnating wicks and hardening steel.

Borax is used as a soldering flux, preservative against wood fungus, and as an antiseptic. It is used in the manufacture of enamels and glazes and in tanning, cleaning compounds, for fireproofing fabrics and wood, and in artificial aging of wood.

Boron trifluoride is used as a catalyst, a flux for soldering magnesium, a fumigant, for protecting molten magnesium and its alloys from oxidation and in ionization chambers to detect weak neutrons. Boric acid is used in the manufacture of glass, enamels and glazes, in metallurgy, and in the analysis of silicates to determine SiO_2 and alkalies.

A partial list of occupations in which exposure may occur includes:

Alloy makers	Nuclear instrument makers
Antiseptic makers	Organic chemical synthesizers
Enamel makers	Tannery workers
Fumigant workers	Textile fireproofers
Glass makers	Wood workers

Permissible Exposure Limits: The applicable Federal standards are: Boron trifluoride 1 ppm (3 mg/m^3) as a ceiling value; and boron oxide 15 mg/m^3. ACGIH (1978) has set 10 mg/m^3 as the TWA for boron oxide and boron tribromide. Sodium tetraborates are in the 1-5 mg/m^3 range.

Route of Entry: Inhalation of dust, fumes, and aerosols; ingestion.

Harmful Effects

Local—These boron compounds may produce irritation of the nasal mucous membranes, the respiratory tract, and eyes.

Systemic—These effects vary greatly with the type of compound. Acute poisoning in man from boric acid or borax is usually the result of application of dressings, powders, or ointment to large areas of burned or abraded skin, or accidental ingestion. The signs are: nausea, abdominal pain, diarrhea and violent vomiting, sometimes bloody, which may be accompanied by headache and weakness. There is a characteristic erythematous rash followed by peeling. In severe cases, shock with fall in arterial pressure, tachycardia, and cyanosis occur. Marked CNS irritation, oliguria, and anuria may be present. The oral lethal dose in adults is over 30 grams. Little information is available on chronic oral poisoning, although it is reported to be characterized by mild GI irritation, loss of appetite, disturbed digestion, nausea, possibly vomiting, and erythymatous rash. The rash may be "hard" with a tendency to become purpuric. Dryness of skin and mucous membranes, reddening of tongue, cracking of lips, loss of hair, conjunctivitis, palpebral edema, gastro-intestinal disturbances, and kidney injury have also been observed.

Although no occupational poisonings have been reported, it was noted that workers manufacturing boric acid had some atrophic changes in respiratory mucous membranes, weakness, joint pains, and other vague symptoms. The biochemical mechanism of boron toxicity is not clear but seems to involve action on the nervous system, enzyme activity, carbohydrate metabolism, hormone function, and oxidation processes, coupled with allergic effects. Borates are excreted principally by the kidneys.

The toxic action of the halogenated borons (boron trifluoride and trichloride) is considerably influenced by their halogenated decomposition products. They are primary irritants of the nasal passages, respiratory tract, and eyes in man. Animal experiments showed a fall in inorganic phosphorus level in blood and on autopsy, pneumonia, and degenerative changes in renal tubules. Long term exposure leads to irritation of the respiratory tract, dysproteinemia, reduction in cholinesterase activity, increased nervous system lability. High concentrations showed a reduction of acetyl carbonic acid and inorganic phosphorus in blood, and dental fluorosis.

Skin and respiratory tract irritation and central nervous system effects have been reported from animal experiments with amine and alkylboranes. The alkylboranes seem to be more toxic than the amino compounds and decaborane, but less toxic than pentaborane. No toxic effects have been attributed to elemental boron.

Medical Surveillance: No specific considerations are needed for boric acid or borates except for general health and liver and kidney function. In the case of boron trifluoride, the skin, eyes, and respiratory tract should receive special attention. In the case of the boranes, central nervous system and lung function will also be of special concern.

Special Tests: None in common use.

Personal Protective Methods: Exposed workers should be educated in the proper use of protective equipment and there should be strict adherence to ventilating provisions in work areas. Workers involved with the manufacture of boric acid should be provided with masks to prevent inhalation of dust and fumes. Where exposure is to halogenated borons, or boranes, masks and supplied air respirators are necessary in areas of dust, gas, or fume concentration. In some areas protective clothing, gloves, and goggles may be necessary.

Bibliography

British Medical Association. 1964. Boric acid poisoning. *Br. Med. J.* 5397:1558.
Environmental Protection Agency. *Preliminary Investigation of Effects on the Environment of Boron, Indium, Nickel, Selenium, Tin, Vanadium and Their Compounds, Volume 1: Boron.* Report EPA-560/2-75-005A, Wash., D.C., Office of Toxic Substances (August 1975).
Torkelson, T.R., Sadek, S.E., and Rowe, V.K. 1961. The toxicity of boron trifluoride when inhaled by laboratory animals. *Am. Ind. Hyg. Assoc. J.* 22:263.

BORON HYDRIDES

Description: Diborane: B_2H_6, boroethane, diboron hexahydride. Diborane is a colorless gas with a nauseating odor. It ignites spontaneously in moist air, and on contact with water, hydrolyzes exothermically forming hydrogen and boric acid.

Pentaborane: B_5H_9, pentaboron monohydride. Pentaborane is a colorless, volatile liquid with an unpleasant, sweetish odor. It ignites spontaneously in air, decomposes at 150°C and hydrolyzes in water.

Decaborane: $B_{10}H_{14}$, decaboron tetradecahydride. This is a white crystal with a bitter odor. It hydrolyzes very slowly in water.

Synonyms: Boranes, hydrogen borides.

Potential Occupational Exposures: Diborane is used as a catalyst for olefin polymerization, a rubber vulcanizer, a reducing agent, a flame-speed accelerator, a chemical intermediate for other boron hydrides, and as a doping agent; and in rocket propellants and in the conversion of olefins to trialkyl boranes and primary alcohols.

Pentaborane is used in rocket propellants and in gasoline additives.

Decaborane is used as a catalyst in olefin polymerization, in rocket propellants, in gasoline additives and as a vulcanizing agent for rubber.

A partial list of occupations in which exposure may occur includes:

Dope makers	Plastic makers
Gasoline additive makers	Rocket fuel makers
Gasoline makers	Rubber makers
Organic chemical synthesizers	

Permissible Exposure Limits: The applicable Federal standards are: Diborane 0.1 ppm (0.1 mg/m^3); Pentaborane 0.005 ppm (0.01 mg/m^3); Decaborane 0.05 ppm (0.03 mg/m^3) skin. These are TWA values set by ACGIH (1978) also.

Routes of Entry: Inhalation and percutaneous absorption.

Harmful Effects

Local—Vapors of boron hydrides are irritating to skin and mucous membranes. Pentaborane and decaborane show marked irritation of skin and mucous membranes, necrotic changes, serious kerato-conjunctivitis with ulceration, and corneal opacification.

Systemic—Pentaborane is the most toxic of boron hydrides. Intoxication is characterized predominantly by CNS signs and symptoms. Hyperexcitability, headaches, muscle twitching, convulsions, dizziness, disorientation, and unconsciousness may occur early or delayed for 24 hours or more following excessive exposure. Slight intoxication results in nausea and drowsiness. Moderate intoxication leads to headache, dizziness, nervous excitation, and hiccups. There may be muscular pains and cramps, spasms in face and extremities, behavioral changes, loss of mental concentration, incoordination, disorientation, cramps, convulsions, semicoma, and persistent leukocytosis after 40 to 48 hours. Liver function tests and elevated nonprotein nitrogen and blood urea levels suggest liver and kidney damage.

Decaborane's toxic effects are similar to pentaborane. Symptoms of CNS damage predominate; however, they are not as marked as the pentaborane.

Diborane is the least toxic of the boron hydrides. In acute poisoning, the symptoms are similar to metal fume fever: tightness, heaviness and burning in chest, coughing, shortness of breath, chills, fever, pericardial pain, nausea, shivering, and drowsiness. Signs appear soon after exposure or after a latent period of up to 24 hours and persist for 1 to 3 days or more. Pneumonia may develop later. Reversible liver and kidney changes were seen in rats exposed to very high gas levels. This has not been noted in man. Subacute poisoning is characterized by pulmonary irritation symptoms, and if this is prolonged, CNS symptoms such as headaches, dizziness, vertigo, chills, fatigue, muscular weakness, and only infrequent transient tremors, appear. Convulsions do not occur. Chronic exposure leads to wheezing, dyspnea, tightness, dry cough, rales, and hyperventilation which persist for several years.

Medical Surveillance: Preemployment and periodic physical examinations to determine the status of the workers' general health should be performed. These examinations should be concerned especially with any history of central nervous system disease, personality or behavioral changes, as well as liver, kidney, or pulmonary disease of any significant nature. Chest x-rays and blood, liver, and renal function studies may be helpful.

Special Tests: None in common use.

Personal Protective Methods: Constant vigilance in the storage and handling of boron hydrides is required. Continuing worker education in the use of personal protective equipment is necessary even when maximum engineering safety measures are applied.

Adequate sanitation facilities including showers and facilities for eating away from exposure area should be provided. Workers should wash thoroughly when leaving exposure areas. Protective clothing impervious to the liquid and gas compounds are necessary. When skin is contaminated by splash or spill, immediate clothes change with thorough washing of the skin area is necessary. Showering after the shift and before changing to street clothes should be required. Masks, either dust, vapor or supplied air type depending on the compound being used in the work place, should be used by all exposed personnel and should be fullface type.

Bibliography

Lowe, H.J. and Freeman, G. 1957. Boron hydride (borane) intoxication in man. *AMA Arch. Ind. Health* 16:523.

Merritt, J.H. 1966. Pharmacology and toxicology of propellant fuels. Boron hydrides, review 3-66. USAF School of Aerospace Medicine, Aerospace Medical Division (AFSC), Brooks Air Force Base, Texas.

Roush, G., Jr. 1959. The toxicology of the boranes. *J. Occup. Med.* 46:46.

Weir, F.W. and Meyers, F.H. 1966. The similar pharmacologic effects of pentaborane, decaborane, and reserpine. *Ind. Med. Surg.* 35:696.

BORON TRIFLUORIDE

Description: Boron trifluoride, BF_3, is a colorless gas with a pungent, suffocating odor. It is corrosive to the skin and fumes in moist air.

Synonyms: None.

Potential Occupational Exposures: Boron trifluoride is a highly reactive chemical used primarily as a catalyst in chemical syntheses. It is stored and transported as a gas but can be reacted with a variety of materials to form both liquid and solid compounds. The magnesium industry utilizes the fire-retardant and antioxidant properties of boron trifluoride in casting and heat treating. Nuclear applications of boron trifluoride include neutron detector instruments, boron-10 enrichment and the production of neutron-absorbing salts for molten-salt breeder reactors. Boron trifluoride is known to be produced at the present time as a specialty chemical by only one United States' company whose production figures are not available. NIOSH estimates that 50,000 employees are potentially exposed to boron trifluoride in the United States.

Permissible Exposure Limits: Sufficient technology exists to prevent adverse effects in workers, but techniques to measure airborne levels of boron trifluoride for compliance with an environmental limit are not adequate. Therefore, an environmental limit is not recommended by NIOSH (1), in part because of the unavailability of adequate monitoring methods. ACGIH (1978) has set TWA values of 1 ppm (10 mg/m^3).

Route of Entry: Inhalation of vapors on skin and eye contact.

Harmful Effects

Boron trifluoride gas, upon contact with air, immediately reacts with water vapor to form a mist which, if at a high enough concentration provides a visible warning of its presence. The gas or mist is irritating to the skin, eyes, and respiratory system.

Medical Surveillance: In the absence of a suitable monitoring method, NIOSH recommends that medical surveillance, including comprehensive preplacement and annual periodic examinations be made available to all workers employed in areas where boron trifluoride is manufactured, used, handled, or is evolved as a result of chemical processes.

Special Tests: None in common use.

Personal Protective Methods: Engineering controls should be used to maintain boron trifluoride concentrations at the lowest feasible level. Air supplied respirators should only be used in certain nonroutine situations or in emergency situations when air concentrations of boron trifluoride are sufficient to form a visible mist.

Proper impervious protective clothing, including gloves, aprons, suits, boots, goggles, and face shields, shall be worn as needed to prevent skin and eye contact with boron trifluoride.

Reference

(1) National Institute for Occupational Safety and Health. *Criteria for a Recommended Standard: Occupational Exposure to Boron Trifluoride*. NIOSH Doc. No. 77-122, Wash., D.C. (1977).

BRASS

Description: Brass is a term used for alloys of copper and zinc. The ratio of the two compounds is generally 2 to 1, although different types of brass may have different proportions. Brass may contain significant quantities of lead. Bronze is also a copper alloy, usually with tin; however, the term bronze is applied to many other copper alloys, some of which contain large amounts of zinc.

Synonyms: None.

Potential Occupational Exposures: Brass may be cast into bearings and other wearing surfaces, steam and water valves and fittings, electrical fittings, hardware, ornamental castings, and other equipment where special corrosion-resistance properties, pressure tightness, and good machinability are required. Wrought forms of brass such as sheets, plates, bars, shapes, wire, and tubing are also widely used.

A partial list of occupations in which exposure may occur includes:

Bench molders	Junk metal refiners	Welders
Braziers	Core makers	Zinc founders
Bronzers	Galvanizers	Zinc smelters

Permissible Exposure Limits: There is no Federal standard for brass; however, there are standards for its constituents: lead (inorganic) (0.2 mg/m^3) (see the entry under Lead for further details); zinc oxide fume (5 mg/m^3); copper fume (0.1 mg/m^3).

Route of Entry: Inhalation of fume.

Harmful Effects

Local—Brass dust and slivers may cause dermatitis by mechanical irritation.

Systemic—Since zinc boils at a lower temperature than copper, the fusing of brass is attended by liberation of considerable quantities of zinc oxide. Inhalation of zinc oxide fumes may result in production of signs and symptoms of metal fume fever (see Zinc Oxide). Brass founder's ague is the name often given to metal fume fever occurring in the brass-founding industry.

Brass foundings may also release sufficient amounts of lead fume to produce lead intoxication (see Lead—Inorganic).

Medical Surveillance: See Zinc Oxide and/or Lead—Inorganic.

Special Tests: Blood lead values may be useful if lead fume or dust exposure is suspected (see Lead).

Personal Protective Methods: See Zinc Oxide and/or Lead—Inorganic.

BROMINE/HYDROGEN BROMIDE

Description: Br, bromine, is a dark reddish-brown, fuming, volatile liquid with a suffocating odor. Bromine is soluble in water and alcohol. HBr, hydrogen bromide, is a corrosive colorless gas.

Synonyms: Bromine, none; hydrogen bromide, anhydrous hydrobromic acid.

Potential Occupational Exposures: Bromine is primarily used in the manufacture of gasoline antiknock compounds (1,2-dibromoethane). Other uses are for gold extraction, in brominating hydrocarbons, in bleaching fibers and silk, in the manufacture of pharmaceuticals, military gas, dyestuffs, and as an oxidizing agent.

Hydrogen bromide and its aqueous solutions are used in the manufacture of organic and inorganic bromides, as a reducing agent and catalyst in controlled oxidations, in the alkylation of aromatic compounds, and in the isomerization of conjugated diolefins.

A partial list of occupations in which exposure may occur includes:

Drug makers   Organic chemical synthesizers
Dye makers   Petroleum refinery workers
Gasoline additive makers   Photographic chemical makers
Gold extractors   Silk and fiber bleachers

Permissible Exposure Limits: The Federal standards are: bromine 0.1 ppm (0.7 mg/m^3); and hydrogen bromide 3 ppm (10 mg/m^3).

Routes of Entry: Inhalation of vapor or gas. Bromine may be absorbed through the skin.

Harmful Effects

Local—Bromine and hydrogen bromide and its aqueous solutions are extremely irritating to eyes, skin, and mucous membranes of the upper respiratory tract. Severe burns of the eye may result from liquid or concentrated vapor exposure. Liquid bromine splashed on skin may cause vesicles, blisters, and slow healing ulcers. Continued exposure to low concentrations may result in acne-like skin lesions. These are more common in the oral use of sodium bromide as a sedation.

Systemic—Inhalation of bromine is corrosive to the mucous membranes of the nasopharynx and upper respiratory tract, producing brownish discoloration of tongue and buccal mucosa, a characteristic odor of the breath, edema and spasm of the glottis, asthmatic bronchitis, and possibly pulmonary edema which may be delayed until several hours following exposure. A measles-like rash may occur. Exposure to high concentrations of bromine can lead to rapid death due to choking caused by edema of the glottis and pulmonary edema.

Bromine has cumulative properties and is deposited in tissues as bromides, displacing other halogens. Exposures to low concentrations result in cough, copious mucous secretions, nose bleeds, respiratory difficulty, vertigo, and headache. Usually these symptoms are followed by nausea, diarrhea, abdominal distress, hoarseness, and asthmatic type respiratory difficulty.

Other effects from chronic exposure have been reported in Soviet literature, e.g., loss of corneal reflexes, joint pains, vegetative disorders, thyroid dysfunction, and depression of the bone marrow. These have not been noted in the United States literature.

Hydrogen bromide (hydrobromic acid) is less toxic than bromine, but is an irritant to the mucous membranes of the upper respiratory tract. Long term exposures can cause chronic nasal and bronchial discharge and dyspepsia. Skin contact may cause burns.

Medical Surveillance: The skin, eyes, and respiratory tract should be given special emphasis during preplacement and periodic examinations. Chest x-rays as well as general health, blood, liver, and kidney function should be considered. Exposure to other irritants or bromine compounds in medications may be important.

Special Tests: None commonly used. Blood bromides can be determined but are probably not helpful in following exposures.

Personal Protective Methods: Respiratory protection with gas masks with acceptable canister or supplied air respirators is essential in areas of excessive vapor concentration. Where aqueous solutions or liquids are used, or high vapor concentrations are present, skin and eyes should be protected against spills or splashes by impervious clothing, gloves, aprons, and face shields or goggles.

Bibliography

Degenhart, J.J. 1972. Estimation of Br in plasma with a Br selective electrode. *Clin. Chim. Acta.* 38:217.

Dunlop, M. 1967. Simple colorimetric method for the determination of bromide in urine. *J. Clin. Pathol.* 20:300.

Edmonds, A. 1966. Toxicity of vaporizing liquids. *Ann. Occup. Hyg.* 9:235.

Goodwin, J.F. 1971. Colorimetric measurement of serum bromide with a bromate-rosaniline method. *Clin. Chem.* 17:544.

Gutsche, B., and Herrmann, R. 1970. Flame-photometric determination of bromine in urine. *Analyst.* 95:805.

Leong, B.K.J., and Torkelson, T.R. 1970. Effects of repeated inhalation of vinyl bromide in laboratory animals with recommendations for industrial handling. *Am. Ind. Hyg. Assoc. J.* 31:1.

Ohno, S. 1971. Determination of iodine and bromine in biological materials by neutron-activation analysis. *Analyst.* 96:423.

1,3-BUTADIENE

Description: $H_2C=CH-CH=CH_2$, 1,3-butadiene, is a colorless, flammable gas with a pungent, aromatic odor. Because of its low flash point, 1,3-butadiene's fire and explosion hazard may be more serious than its health hazard.

Synonyms: Biethylene, bivinyl, butadiene monomer, divinyl, erythrene, methylallene, pyrrolylene, vinylethylene.

Potential Occupational Exposures: 1,3-Butadiene is used chiefly as the principal monomer in the manufacture of many types of synthetic rubber. Presently, butadiene is finding increasing usage in the formation of rocket fuels, plastics, and resins.

A partial list of occupations in which exposure may occur includes:

Organic chemical synthesizers	Rocket fuel makers
Resin makers	Rubber makers
Rocket fuel handlers	

Permissible Exposure Limits: The Federal standard for 1,3-butadiene is 1,000 ppm (2,200 mg/m^3).

Route of Entry: Inhalation of gas or vapor.

Harmful Effects

Local—Butadiene gas is slightly irritating to the eyes, nose, and throat. Dermatitis and frostbite may result from exposure to liquid and evaporating gas.

Systemic—In high concentrations, 1,3-butadiene gas can act as an irritant, producing cough, and as a narcotic, producing fatigue, drowsiness, headache, vertigo, loss of consciousness, respiratory paralysis, and death. One report states that chronic exposure may result in central nervous system disorders, diseases of the liver and biliary system, and tendencies toward hypotension, leukopenia, increase in ESR, and decreased hemoglobin content in the blood. These changes have not been seen by most observers in humans.

Medical Surveillance: No specific considerations are needed.

Special Tests: None in common use.

Personal Protective Methods: Masks are recommended in contaminated areas with high concentrations of the gas.

Bibliography

Batkina, I.P. 1966. Maximum permissible concentration of divinyl vapor in factory air (O presdel 'no dopustimoi kontsentratsii parov divinika v vozdukhe radiochikh pomeshchenii). *Hyg. Sanit.* 31:334.
Carpenter, C.P., Shaffer, C.B., and Smyth, H.F., Jr. 1944. Studies on the inhalation of 1,3-butadiene; with a comparison of its narcotic effect with benzol, toluol, and styrene, and a note on the elimination of styrene by the human. *J. Ind. Hyg. Toxicol.* 26:69.

n-BUTYL ALCOHOL

Description: $CH_3CH_2CH_2CH_2OH$, n-butyl alcohol, is a colorless volatile liquid with a pungent odor.

Synonyms: 1-Butanol, butyl hydroxide, propylcarbinol, butyric alcohol, hydroxybutane, n-butanol, n-propylcarbinol.

Potential Occupational Exposures: n-Butyl alcohol is used as a solvent for paints, lacquers, varnishes, natural and synthetic resins, gums, vegetable oils, dyes, camphor, and alkaloids. It is also used as an intermediate in the manufacture of pharmaceuticals and chemicals and in the manufacture of artificial leather, safety glass, rubber and plastic cements, shellac, raincoats, photographic films, perfumes, and in plastic fabrication.

A partial list of occupations in which exposure may occur includes:

Alkaloid makers	Photographic film makers
Detergent makers	Plasticizer makers
Drug makers	Stainers
Dye makers	Urea-formaldehyde resin makers
Lacquerers	Varnish makers

Permissible Exposure Limits: The Federal standard is 100 ppm (300 mg/m^3). [Note: the 1978 ACGIH lists a TLV of 50 ppm (150 mg/m^3).]

Route of Entry: Inhalation of vapor and percutaneous absorption.

Harmful Effects

Local—The liquid is a primary skin irritant. The vapor is an irritant to the conjunctiva and mucous membranes of the nose and throat. A mild keratitis characterized by corneal vacuoles has been noted at vapor concentrations over 200 ppm.

Systemic—Inhalation of high concentrations, in addition to the local effects, have produced transitory and persistent dizziness with Meniere's syndrome. Slight headache and drowsiness may also occur.

Medical Surveillance: Consider irritant effects on eyes, respiratory tract, and skin in any preplacement or periodic examinations.

Special Tests: None have been used. Blood levels can be determined.

Personal Protective Methods: Barrier creams and protective clothing should be used where skin contact may occur.

Bibliography

Jain, N.C. 1971. Direct blood-injection method for gas chromatographic determination of alcohols and other volatile compounds. *Clin. Chem.* 2:82.

Seitz, P.B. 1972. Vertiges graves apparus apres manipulation de butanol et d'isobetanol. *Arch. Mal. Prof.* 33:393.

Sterner, J.H., Crouch, H.C., Brockmyre, H.G., and Cusack, M. 1949. A ten year study of butyl alcohol exposure. *Am. Ind. Hyg. Assoc. Q.* 10:53.
Tabershaw, I.R., Fahy, J.P., and Skinner, J.B. 1944. Industrial exposure to butanol. *J. Ind. Hyg. Toxicol.* 26:328.

n-BUTYLAMINE

Description: $CH_3CH_2CH_2CH_2NH_2$, n-butylamine, is a flammable colorless liquid with an ammoniacal odor.

Synonyms: 1-Aminobutane.

Potential Occupational Exposures: n-Butylamine is used in pharmaceuticals, dyestuffs, rubber, chemicals, emulsifying agents, photography, desizing agents for textiles, pesticides, and synthetic agents.

A partial list of occupations in which exposure may occur includes:

Butylaminophenol makers	Insecticide makers
Chemists	Petroleum dewaxers
Drug makers	Rubber makers
Dye makers	Tanning chemical makers
Emulsifier makers	

Permissible Exposure Limits: The Federal standard is 5 ppm (15 mg/m^3) as a ceiling value.

Routes of Entry: Inhalation and percutaneous absorption.

Harmful Effects

Local—Butylamine vapor is irritating to the nose, throat, and eyes. Contact with the liquid may produce severe eye damage and skin burns.

Systemic—Inhalation of concentrations at or above the threshold limit may produce mild headaches and flushing of the skin and face.

Butylamine vapor has produced pulmonary edema in animal experiments.

Medical Surveillance: Evaluate risks of eye or skin injury and respiratory irritation in periodic or placement examinations.

Special Tests: None have been developed.

Personal Protective Methods: Protective clothing and goggles should be worn where possibility of skin or eye contact with the liquid exists. In areas of elevated vapor concentration, fullface masks with organic vapor canister or supplied air respirators and protective clothing should be worn. The odor and irritation of the mucous membranes cannot be relied upon for exposure control.

C

CADMIUM AND COMPOUNDS

Description: Cd, cadmium, is a bluish-white metal. The only cadmium mineral, greenockite, is rare; however, small amounts of cadmium are found in zinc, copper, and lead ores. It is generally produced as a by-product of these metals, particularly zinc. Cadmium is insoluble in water but is soluble in acids.

Synonyms: None.

Potential Occupational Exposures: Cadmium is highly corrosion resistant and is used as a protective coating for iron, steel, and copper; it is generally applied by electroplating, but hot dipping and spraying are possible. Cadmium may be alloyed with copper, nickel, gold, silver, bismuth, and aluminum to form easily fusible compounds. These alloys may be used as coatings for other materials, welding electrodes, solders, etc. It is also utilized in electrodes of alkaline storage batteries, as a neutron absorber in nuclear reactors, a stabilizer for polyvinyl chloride plastics, a deoxidizer in nickel plating, an amalgam in dentistry, in the manufacture of fluorescent lamps, semiconductors, photocells, and jewelry, in process engraving, in the automobile and aircraft industries, and to charge Jones reductors.

Various cadmium compounds find use as fungicides, insecticides, nematocides, polymerization catalysts, pigments, paints, and glass; they are used in the photographic industry and in glazes. Cadmium is also a contaminant of superphosphate fertilizers.

Exposure may occur during the smelting and refining of cadmium-containing zinc, lead, and copper ores, and during spraying, welding, cutting, brazing, soldering, heat treating, melting, alloying and salvage operations which require burning of cadmium-containing materials.

A partial list of occupations in which exposure may occur includes:

Alloy makers	Pesticide workers
Battery makers	Solder workers
Dental amalgam makers	Textile printers
Engravers	Welders
Metalizers	Zinc refiners
Paint makers	

NIOSH estimates that 100,000 U.S. workers are exposed to Cd or its compounds.

Permissible Exposure Limits: The Federal standard for cadmium fume is 0.1 mg (100 μg)/m^3 (as Cd) as an 8-hour TWA with an acceptable ceiling of 0.3 mg/m^3. For cadmium dust, the standard is 0.2 mg/m^3 (Cd) as an 8-hour TWA with an acceptable maximum ceiling of 0.6 mg/m^3. NIOSH has recommended (1) a TWA limit of 40 μg/m^3 with a ceiling limit of 200 μg in a 5-minute sampling period. ACGIH (1978) set a TWA value of 0.05 mg/m^3 for Cd dust, fume and salts. Beyond that, cadmium oxide production has been categorized under "Industrial Substances Suspect of Carcinogenic Potential for Man."

Routes of Entry: Inhalation or ingestion of fumes or dust.

Harmful Effects

Local—Cadmium is an irritant to the respiratory tract. Prolonged exposure can cause anosmia and a yellow stain that gradually appears on the necks of the teeth. Cd compounds are poorly absorbed from the intestinal tract, but relatively well absorbed by inhalation. Skin absorption appears negligible. Once absorbed Cd has a very long half-life and is retained in the kidney and liver.

Systemic—Acute toxicity is almost always caused by inhalation of cadmium fumes or dust which are produced when cadmium is heated. There is generally a latent period of a few hours after exposure before symptoms develop. During the ensuing period, symptoms may appear progressively. The earliest symptom is slight irritation of the upper respiratory tract. This may be followed over the next few hours by cough, pain in the chest, sweating, and chills which resemble the symptoms of nonspecific upper respiratory infection. Eight to 24 hours following acute exposure severe pulmonary irritation may develop, with pain in the chest, dyspnea, cough, and generalized weakness. Dyspnea may become more pronounced as pulmonary edema develops. The mortality rate in acute cases is about 15%. Patients who survive may develop emphysema and cor pulmonale; recovery can be prolonged.

Chronic cadmium poisoning has been reported after prolonged exposure to cadmium oxide fumes, cadmium oxide dust, cadmium sulfides, and cadmium stearates. Heavy smoking has been reported to considerably increase tissue Cd levels. In some cases, only the respiratory tract is affected. In others the effects may be systemic due to absorption of the cadmium. Lung damage often results in a characteristic form of emphysema which in some instances is not preceded by a history of chronic bronchitis or coughing. This type of emphysema can be extremely disabling. Some studies have not shown these effects.

Systemic changes due to cadmium adsorption include damage to the kidneys with proteinuria, anemia, and elevated sedimentation rate. Of these, proteinuria (low molecular weight) is the most typical. In advanced stages of the disease, there may be increased urinary excretion of amino acids, glucose, calcium, and phosphates. These changes may lead to the formation of renal calculi. If the exposure is discontinued, there is usually no progression of the kidney damage. Mild hypochromic anemia is another systemic condition sometimes found in chronic exposure to cadmium.

In studies with experimental animals, cadmium has produced damage to the liver and central nervous system, testicular atrophy, teratogenic effects in rodents after intravenous injection of cadmium, decrease in total red cells, sarcomata, and testicular neoplasms. Hypertensive effects have also been produced. None of these conditions, however, has been found in man resulting from occupational exposure to cadmium. Heavy smoking would appear to increase the risk of cumulative toxic effects.

Medical Surveillance: In preemployment physical examinations, emphasis should be given to a history of or the actual presence of significant kidney disease, smoking history, and respiratory disease. A chest x-ray and baseline pulmonary function study is recommended. Periodic examinations should emphasize the respiratory system, including pulmonary function tests, kidneys, and blood.

Special Tests: A low molecular weight proteinuria may be the earliest indication of renal toxicity. The trichloroacetic acid test may pick this up, but more specific quantitative studies would be preferable. If renal disease due to cadmium is present, there may also be increased excretion of calcium, amino acids, glucose, and phosphates.

Personal Protective Methods: Most important is the requirement that each worker be adequately protected by the use of effective respiratory protection: either by dust masks, vapor canister respirators, or supplied air respirators. Clothing should be changed after each shift and clean work clothing issued each day. Food should not be eaten in contaminated work areas. Workers should shower after each shift before changing to street clothes.

References

(1) National Institute for Occupational Safety and Health. *Criteria for a Recommended Standard: Occupational Exposure to Cadmium.* NIOSH Doc. No. 76-192, Wash., D.C. (1976).

Bibliography

Beton, D.C., Andrews, G.S., Davies, H.J., Howells, L., and Smith, G.F. 1966. Acute cadmium fume poisoning: five cases with one death from renal necrosis. *Br. J. Med.* 23:292.

British Industrial Biological Research Association. 1972. "Itai-itai byo" and other views on cadmium. *Food Cosmet. Toxicol.* 10:249.

Chauve, S., et al. Zinc and cadmium in normal human embryos and fetuses. *Archiv Environmental Health* 26:237 (1973).

Cadmium and the Environment: Toxicity, Economy, Control. Paris, Environment Directorate, Organization for Economic Cooperation and Development (1975).

Environmental Protection Agency. *Technical and Microeconomic Analysis of Cadmium and Its Compounds.* Report EPA 560/3-75-005, Wash., D.C. Office of Toxic Substances (June 1975).

Environmental Protection Agency. *Scientific and Technical Assessment Report on Cadmium.* Report EPA 600/6-75-003, Research Triangle Park, N.C. National Environmental Research Center (July 1975).

Fassett, D.W. 1972. Cadmium. p. 97. In: D.H.K. Lee, ed. *Metallic Contaminants and Human Health.* Academic Press, New York.

Fassett, D.W. 1975. Cadmium. Biological effects and occurrence in the environment. *Ann. Rev. Pharmacol* 15:425.

Ferm, V.H. Developmental malformations induced by cadmium. *Biol Neonate* 19:101 (1971).

Fleischer, M., Sarofim, A.F., Fassett, D.W., et al. 1974. Environmental impact of cadmium: A review by the panel on hazardous trace substances. *Environ. Health Perspect.* 7:253.

Friberg, L. 1959. Chronic cadmium poisoning. *AMA Arch Ind. Health* 20:401.

Friberg, L., Piscator, M., and Nordberg, G.F. 1974. *Cadmium in the Environment.* 2nd ed. CRC Press, Cleveland, Ohio.

Fulkerson, William et al. *Cadmium: The Dissipated Element.* Oak Ridge National Laboratory (report no. ORNL NSF-EP-21, January 1973).

Lane, R.F., and Campbell, A.C.P. 1954. Fatal emphysema in two men making a copper cadmium alloy. *Br. J. Ind. Med.* 11:118.

National Inventory of Sources and Emissions: Cadmium-1968. EPA, Office of Air and Water Programs (report no. APTD-68, February 1970).

Regulations for Rebuttable Presumption Against Registration, 40 CFR 162.11.

Tsuchiya, K. 1967. Proteinuria of workers exposed to cadmium fume—The relationship to concentration in the working environment. *Arch. Environ. Health* 14:875.

Webb, M. Cadmium. *Br. Med. Bull,* 31(3):246-250 (1975).

CALCIUM CYANAMIDE

Description: $CaCN_2$, calcium cyanamide, is a blackish-grey, shiny powder.

Synonyms: Nitrolim, calcium carbimide, cyanamide.

Potential Occupational Exposures: Calcium cyanamide is used in agriculture as a fertilizer, herbicide, defoliant for cotton plants, and pesticide. It is also used in the manufacture of dicyandiamide and calcium cyanide, as a desulfurizer in the iron and steel industry, and in steel hardening.

A partial list of occupations in which exposure may occur includes:

Ammonia makers	Herbicide workers
Cotton defoliant workers	Nitrogen compound makers
Cyanamide makers	Organic chemical synthesizers
Fertilizer workers	Steel workers

Permissible Exposure Limits: There is no Federal standard for calcium cyanamide. (Note: The 1978 ACGIH TLV is 0.5 mg/m³.)

Route of Entry: Inhalation of dust.

Harmful Effects

Local—Calcium cyanamide is a primary irritant of the mucous membranes of the respiratory tract, eyes, and skin. Inhalation may result in rhinitis, pharyngitis, laryngitis, and bronchitis. Conjunctivitis, keratitis, and corneal ulceration may occur. An itchy erythematous dermatitis has been reported and continued skin contact leads to the formation of slowly healing ulcerations on the palms and between the fingers. Sensitization occasionally develops. Chronic rhinitis and perforation of the nasal septum have been reported after long exposures. All local effects appear to be due to the caustic nature of cyanamide.

Systemic—Calcium cyanamide causes a characteristic vasomotor reaction. There is erythema of the upper portions of the body, face, and arms, accompanied by nausea, fatigue, headache, dyspnea, vomiting, oppression in the chest, and shivering. Circulatory collapse may follow in the more serious cases. The vasomotor response may be triggered or intensified by alcohol ingestion. Pneumonia or lung edema may develop. Cyanide ion is not released in the body, and the mechanism of toxic action is unknown.

Medical Surveillance: Evaluate skin, respiratory tract, and history of alcohol intake in placement or periodic examinations.

Special Tests: None commonly used.

Personal Protective Methods: In addition to personal protective equipment, waterproof barrier creams may be used to provide additional face and skin protection. Personal hygiene measures are to be encouraged, such as showering after work and a complete change of clothing. In areas of heavy dust concentrations, fullface dust masks are recommended.

Bibliography

Buyske, D.A., and Downing, V. 1960. Spectrophotometric determination of cyanamide. *Anal. Chem.* 32:1798.

CALCIUM OXIDE

Description: CaO, calcium oxide, occurs as white or grayish-white lumps or granular powder. The presence of iron gives it a yellowish or brownish tint. It is soluble in water and acids.

Synonyms: Lime, burnt lime, quicklime, calx, fluxing lime.

Potential Occupational Exposures: Calcium oxide is used as a refractory material, a binding agent in bricks, plaster, mortar, stucco and other building materials, a dehydrating agent, a flux in steel manufacturing, and a laboratory agent to absorb CO_2; in the manufacture of aluminum, magnesium, glass, pulp and paper, sodium carbonate, calcium hydroxide, chlorinated lime, calcium salts, and other chemicals; in the flotation of nonferrous ores, water and sewage treatment, soil treatment in agriculture, dehairing hides, the clarification of cane and beet sugar juice, and in fungicides, insecticides, drilling fluids, and lubricants.

A partial list of occupations in which exposure may occur includes:

Brick masons	Paper makers
Fertilizer makers	Plaster makers
Fungicide workers	Steel workers
Glass makers	Sugar refiners
Insecticide makers	Tannery workers
Metal smelters	Water treaters
Mortar workers	

Permissible Exposure Limits: The Federal standard for calcium oxide is 5 mg/m^3. ACGIH (1978) has set a TWA value of 2 mg/m^3.

Route of Entry: Inhalation of dust.

Harmful Effects

Local—The irritant action of calcium oxide is due primarily to its alkalinity and exothermic reaction with water. It is irritating and may be caustic to the skin, conjunctiva, cornea, and mucous membranes of upper respiratory tract, may produce burns or dermatitis with desquamation and vesicular rash, lacrimation, spasmodic blinking, ulceration, and ocular perforation, ulceration and inflammation of the respiratory passages, ulceration of nasal and buccal mucosa, and perforation of nasal septum.

Systemic—Bronchitis and pneumonia have been reported from inhalation of dust. The lower respiratory tract is generally not affected because irritation of upper respiratory passages is so severe that workers are forced to leave the area.

Medical Surveillance: Preemployment physical examinations should be directed to significant problems of the eyes, skin, and the upper respiratory tract. Periodic examinations should evaluate the skin, changes in the eyes, especially the cornea and conjunctiva, mucosal ulcerations of the nose, mouth, and nasal septum, and any pulmonary symptoms. Smoking history should be known.

Special Tests: None in common use.

Personal Protective Methods: In areas where workers are exposed to calcium oxide levels above the standard, protection to the skin, eyes, and respiratory tract should be provided. Skin protection can be provided by protective clothing and gloves. All dusty area workers should be provided with goggles and dust masks with proper cartridges. Personal hygiene is to be encouraged, with frequent change of work clothes and showering after each shift before change to street clothes.

CARBARYL

Description: $C_{10}H_7OOCNHCH_3$, carbaryl, 1 naphthyl N-methylcarbamate, is a crystalline substance melting at 145°C. Carbaryl manufactured in the United States is produced synthetically and is used primarily as an insecticide for agricultural purposes.

Synonyms: Carbaryl is the generic name for the 1-naphthyl ester of N-methylcarbamate.

Potential Occupational Exposures: Approximately 53 million pounds of carbaryl were produced in 1972 in the United States, of which 25 million pounds were used domestically, and the remainder was exported. NIOSH estimates that approximately 100,000 workers in the United States are potentially exposed to carbaryl.

Permissible Exposure Limits: NIOSH recommends adherence to the present Federal standard of 5 milligrams of carbaryl per cubic meter of air as a time-weighted average for up to a 10-hour workday, 40-hour workweek.

Routes of Entry: Inhalation, skin contact or eye contact.

Harmful Effects

The major health problem associated with occupational exposure to carbaryl is related to its inhibition of the enzyme cholinesterase in the central, autonomic and peripheral nervous systems. The inhibition of cholinesterase allows acetylcholine to accumulate at these sites and thereby leads to overstimulation of innervated organs. The signs and symptoms observed as a consequence of exposure to carbaryl in the workplace environment are manifestations of excessive cholinergic stimulation, e.g., nausea, vomiting, mild abdominal cramping, dimness of vision, dizziness, headache, difficulty in breathing, and weakness.

Medical Surveillance: NIOSH recommends that workers subject to carbaryl exposure have comprehensive preplacement medical examinations, with subsequent annual medical surveillance.

Special Tests: None in common use.

Personal Protective Methods: Engineering controls should be used wherever feasible to maintain carbaryl concentrations below the prescribed limits, and respirators should only be used in certain nonroutine or emergency situations. During certain agricultural applications, however, respirators must be used.

Any employee whose work involves likely exposure of the skin to carbaryl or carbaryl formulations, e.g., mixing or formulating, shall wear full-body coveralls or the equivalent, impervious gloves, i.e, highly resistant to the penetration of carbaryl, impervious footwear, and, when there is danger of carbaryl coming in contact with the eyes, goggles or a face shield. Any employee engaged in field application of carbaryl shall be provided with, and required to wear, the following protective clothing and equipment: goggles, full-body coveralls, impervious footwear, and a protective head covering. Employees working as flaggers in the aerial application of carbaryl shall be provided with, and required to wear, full-body coveralls or waterproof rainsuits, protective head coverings, impervious gloves and impervious footwear.

Bibliography

National Institute for Occupational Safety and Health. *Criteria for a Recommended Standard: Occupational Exposure to Carbaryl.* NIOSH Doc. No. 77-107 (1977).

CARBON DIOXIDE

Description: CO_2, carbon dioxide, is a colorless, odorless, noncombustible gas, soluble in water. It is commonly sold in the compressed liquid form, and the solid form (dry ice).

Synonyms: Carbonic acid gas, carbonic anhydride.

Potential Occupational Exposures: Gaseous carbon dioxide is used to carbonate beverages, as a weak acid in the textile, leather, and chemical industries, in water treatment, and in the manufacture of aspirin and white lead, for hardening molds in foundries, in food preservation, in purging tanks and pipelines, as a fire extinguisher, in foams, and in welding. Because it is relatively inert, it is utilized as a pressure medium. It is also used as a propellant in aerosols, to promote plant growth in green houses; it is used medically as a respiratory stimulant, in the manufacture of carbonates, and to produce an inert atmosphere when an explosive or flammable hazard exists. The liquid is used in fire extinguishing equipment, in cylinders for inflating life rafts, in the manufacturing of dry ice, and as a refrigerant. Dry ice is used primarily as a refrigerant.

Occupational exposure to carbon dioxide may also occur in any place where fermentation processes may deplete oxygen with the formation of carbon dioxide, e.g., in mines, silos, wells, vats, ships' holds, etc.

A partial list of occupations in which exposure may occur includes:

Aerosol packages	Grain elevator workers
Beverage carbonators	Inert atmosphere welders
Blast furnace workers	Insecticide makers
Brewery workers	Miners
Carbonic acid makers	Refrigerating car workers
Charcoal burners	Refrigerating plant workers
Chemical synthesizers	Soda makers
Explosive makers	Tannery workers
Fire extinguisher makers	Textile workers
Firemen	Vatmen
Foundry workers	Well cleaners

Permissible Exposure Limits: The Federal standard is 5,000 ppm (9,000 mg/m^3).

Route of Entry: Inhalation of gas.

Harmful Effects

Local—Frostbite may result from contact with dry ice or gas at low temperature.

Systemic—Carbon dioxide is a simple asphyxiant. Concentrations of 10% (100,000 ppm) can produce unconsciousness and death from oxygen deficiency. A concentration of 5% may produce shortness of breath and headache. Continuous exposure to 1.5% CO_2 may cause changes in some physiological processes. The concentration of carbon dioxide in the blood affects the rate of breathing.

Medical Surveillance: No special considerations are necessary although persons with cardiovascular or pulmonary disease may be at increased risk.

Special Tests: None in common use.

Personal Protective Methods: Carbon dioxide is a heavy gas and accumulates at low levels in depressions and along the floor. Generally, adequate ventilation will provide sufficient protection for the worker. Where concentrations are of a high order, supplied air respirators are recommended.

Bibliography

Cullen, D.J., and Eger, E.I. 1974. Cardiovascular effects of carbon dioxide in man. *Anesthesiology* 41:345.

National Institute for Occupational Safety and Health. *Criteria for a Recommended Standard: Occupational Exposure to Carbon Dioxide*. NIOSH Doc. No. 76-194 (1976).

Schulte, J.H. 1964. Sealed environments in relation to health and disease. *Arch. Environ. Health* 8:438.

Williams, H.I., 1958. Carbon dioxide poisoning report of eight cases and two deaths. *Brit. Med. J.* 2:1012.

CARBON DISULFIDE

Description: CS_2, carbon disulfide, is a highly refractive, flammable liquid which in pure form has a sweet odor and in commercial and reagent grades has a foul smell. It can be detected by odor at about 1 ppm but the sense of smell fatigues rapidly and, therefore, odor does not serve as a good warning property. It is slightly soluble in water, but more soluble in organic solvents.

Synonyms: Carbon bisulfide, dithiocarbonic anhydride.

Potential Occupational Exposures: Carbon disulfide is used in the manufacture of viscose rayon, ammonium salts, carbon tetrachloride, carbanilide, xanthogenates, flotation agents, soil disinfectants, dyes, electronic vacuum tubes, optical glass, paints, enamels, paint removers, varnishes, varnish removers, tallow, textiles, explosives, rocket fuel, putty, preservatives, and rubber cement; as a solvent for phosphorus, sulfur, selenium, bromine, iodine, alkali cellulose, fats, waxes, lacquers, camphor, resins, and cold vulcanized rubber. It is also used in degreasing, chemical analysis, electroplating, grain fumigation, oil extraction, and drycleaning.

NIOSH estimates that 20,000 workers in the United States are exposed or potentially exposed to carbon disulfide.

A partial list of occupations in which exposure may occur includes:

Ammonium salt makers	Putty makers
Bromine processors	Rayon makers
Carbon tetrachloride makers	Resin makers
Degreasers	Rocket fuel makers
Drycleaners	Rubber cement makers
Electroplaters	Rubber workers
Fat processors	Sulfur processors
Flotation agent makers	Tallow makers
Iodine processors	Textile makers
Oil processors	Vacuum tube makers
Paint workers	Varnish makers
Preservative makers	Wax processors

Permissible Exposure Limits: The Federal standard is 20 ppm (60 mg/m^3) determined as an 8-hour TWA. The acceptable ceiling concentration is 30 ppm (90 mg/m^3) with a maximum peak above this for an 8-hour workshift of 100 ppm (300 mg/m^3) for a maximum duration of 30 minutes.

NIOSH recommends that exposure be limited to a TWA concentration of 3 mg/m^3 (1 ppm). In addition, a ceiling concentration of 30 mg/m^3 (10 ppm), determined on the basis of a 15-minute sampling period, is recommended. Occupational exposure to CS$_2$ is defined as exposure above an action level of 1.5 mg/m^3. ACGIH (1978) has recommended a TWA value of 10 ppm (30 mg/m^3).

Routes of Entry: Inhalation of vapor which may be compounded by percutaneous absorption of liquid or vapor.

Harmful Effects

Local—Carbon disulfide vapor in sufficient quantities is severely irritating to eyes, skin, and mucous membranes. Contact with liquid may cause blistering with second and third degree burns. Skin sensitization may occur. Skin absorption may result in localized degeneration of peripheral nerves which is most often noted in the hands. Respiratory irritation may result in bronchitis and emphysema, though these effects may be overshadowed by systemic effects.

Systemic—Intoxication from carbon disulfide is primarily manifested by psychological, neurological, and cardiovascular disorders. Recent evidence indicates that once biochemical alterations are initiated they may remain latent; clinical signs and symptoms then occur following subsequent exposure.

Following repeated carbon disulfide exposure, subjective psychological as well as behavioral disorders have been observed. Acute exposures may result in extreme irritability, uncontrollable anger, suicidal tendencies, and a toxic manic depressive psychosis. Chronic exposures have resulted in insomnia, nightmares, defective memory, and impotency. Less dramatic changes include headache, dizziness, and diminished mental and motor ability, with staggering gait and loss of coordination.

Neurological changes result in polyneuritis. Animal experimentation has revealed pyramidal and extrapyramidal tract lesions and generalized degeneration of the myelin sheaths of peripheral nerves. Chronic exposure signs and symptoms include retrobulbar and optic neuritis, loss of sense of smell, tremors, paresthesias, weakness, and, most typically, loss of lower extremity reflexes.

Atherosclerosis and coronary heart disease have been significantly linked to exposure to carbon disulfide. Atherosclerosis develops most notably in the blood vessels of the brain, glomeruli, and myocardium. Abnormal electroencephalograms and retinal hypertension typically occur before renal involvement is noted. Any of the above three areas may be affected by chronic exposure, but most often only one aspect can be observed. A significant increase in coronary heart disease mortality has been observed in carbon disulfide workers. Studies also reveal higher frequency of angina pectoris and hypertension. Abnormal electrocardiograms may also occur and are also suggestive of carbon disulfide's role in the etiology of coronary disease.

Other specific effects include chronic gastritis with the possible development of gastric and duodenal ulcers; impairment of endocrine activity, specifically adrenal and testicular; abnormal erythrocytic development with hypochromic anemia; and possible liver dysfunction with abnormal serum cholesterol. Also in women, chronic menstrual disorders may occur. These effects usually occur following chronic exposure and are subordinate to the other symptoms.

Recently human experience and animal experimentation have indicated several possible biochemical changes. Carbon disulfide and its metabolites (i.e., dithiocarbamic acids and isothiocyanates) show amino acid interference, cerebral monoamine oxidase inhibition, endocrine disorders, lipoprotein metabolism interference, blood protein, and zinc level abnormalities, and inorganic metabolism interference due to chelating of polyvalent ions. The direct relationship between these biochemical changes and clinical manifestations is only suggestive.

Medical Surveillance: Preplacement and periodic medical examinations should be concerned especially with skin, eyes, central and peripheral nervous system, cardiovascular disease, as well as liver and kidney function. Electrocardiograms should be taken.

Special Tests: CS_2 can be determined in expired air, blood, and urine. The iodine-azide test is most useful although nonspecific, and it may indicate other sulfur compounds.

Personal Protective Methods: Local exhaust, general ventilation, and personal protective equipment should be utilized. In modern manufacture, CS_2 fumes are generally controlled by closed operations. Where fumes are present in unacceptable concentrations, vapor gas mask with fullface or used air respirators should be used. In all areas where there is a likelihood of spill or splash on any skin area, protection should be afforded by protective clothing, goggles, face shields, aprons, and coats.

Bibliography

Brieger, H.H., and Teisinger, J.J., eds. 1967. International Symposium on Toxicology of Carbon Disulfide, organized by the Sub-Committee for Occupational Health in the Production of Artificial Fibers of the Permanent Commission and International Association of Occupational Health. Prague, September 15th-17th, 1966. Excerpta Medica Foundation, Amsterdam.

Davidson, M., and Feinleib, M. 1972. Disulfide poisoning: A review. *Am. Heart J.* 83:100.

Hanninen, H. 1971. Psychological picture of manifest and latent carbon disulfide poisoning. *Br. J. Ind. Med.* 28:374.

Hernberg, S., Partanen, T., Nordman, C.H., and Sumari, P. 1970. Coronary heart disease among workers exposed to carbon disulfide. *Br. J. Ind. Med.* 27:313.

Kleinfeld, M., and Tabershaw, I.R. 1955. Carbon disulfide poisoning. Report of two cases. *J. Am. Med. Assoc.* 159:677.

National Institute for Occupational Safety and Health. *Criteria for a Recommended Standard: Occupational Exposure to Carbon Disulfide.* NIOSH Doc. No. 77-156 (1977).

U.S. Department of Health, Education, and Welfare, Public Health Service, Center for Disease Control, National Institute for Occupational Safety and Health. 1974. Case Report, Occupational Health Case Report—No. 1. *J. Occup. Med.* 16:22.

CARBON MONOXIDE

Description: CO, carbon monoxide, is a colorless, odorless, tasteless gas, partially soluble in water, but one which decomposes.

Synonyms: None.

Potential Occupational Exposures: Carbon monoxide is used in metallurgy as a reducing agent, particularly in the Mond process for nickel; in organic synthesis, especially in the Fischer-Tropsch process for petroleum products and in the oxo reaction; and in the manufacture of metal carbonyls.

It is usually encountered in industry as a waste product of incomplete combustion of carbonaceous material (complete combustion produces CO_2). The major source of CO emission in the atmosphere is the gasoline-powered internal combustion engine. Specific industrial processes which contribute significantly to CO emission are iron foundries particularly the cupola; fluid catalytic crackers, fluid coking, and moving-bed catalytic crackers in petroleum refining; lime kilns and Kraft recovery furnaces in Kraft paper mills; furnace, channel, and thermal operations in carbon black plants; beehive coke ovens, basic oxygen furnaces, sintering of blast furnace feed in steel mills; and formaldehyde manufacture. There are numerous other operations in which a flame touches a surface that is cooler than the ignition temperature of the gaseous part of the flame where exposure to CO may occur, e.g., arc welding, automobile repair, traffic control, tunnel construction, fire fighting, mines, use of explosives, etc.

A partial list of occupations in which exposure may occur includes:

Acetylene workers   Brewery workers
Blast furnace workers   Carbon black makers
Boiler room workers   Coke oven workers

Diesel engine operators	Organic chemical synthesizers
Garage mechanics	Petroleum refinery workers
Metal oxide reducers	Pulp and paper workers
Miners	Steel workers
Mond process workers	Water gas workers

Permissible Exposure Limits: The present Federal standard is 50 ppm (55 mg/m³). The standard recommended by NIOSH is 35 ppm with a ceiling value of 200 ppm. This latter value is to limit carboxyhemoglobin formation to 5% in a nonsmoker engaged in sedentary activity at normal altitude.

Route of Entry: Inhalation of gas.

Harmful Effects

Local—None.

Systemic—Carbon monoxide combines with hemoglobin to form carboxyhemoglobin which interferes with the oxygen carrying capacity of blood, resulting in a state of tissue hypoxia. The typical signs and symptoms of acute CO poisoning are headache, dizziness, drowsiness, nausea, vomiting, collapse, coma, and death. Initially the victim is pale; later the skin and mucous membranes may be cherry-red in color. Loss of consciousness occurs at about the 50% carboxyhemoglobin level. The amount of carboxyhemoglobin formed is dependent on concentration and duration of CO exposure, ambient temperature, health, and metabolism of the individual. The formation of carboxyhemoglobin is a reversible process. Recovery from acute poisoning usually occurs without sequelae unless tissue hypoxia was severe enough to result in brain cell degeneration.

Carbon monoxide at low levels may initiate or enhance deleterious myocardial alterations in individuals with restricted coronary artery blood flow and decreased myocardial lactate production.

Severe carbon monoxide poisoning has been reported to permanently damage the extrapyramidal system, including the basal ganglia.

Medical Surveillance: Preplacement and periodic medical examinations should give special attention to significant cardiovascular disease and any medical conditions which could be exacerbated by exposure to CO. Heavy smokers may be at greater risk. Methylene chloride exposure may also cause an increase of carboxyhemoglobin. Smokers usually have higher levels of carboxyhemoglobin than nonsmokers (often 5 to 10% or more).

Special Tests: Carboxyhemoglobin levels are reliable indicators of exposure and hazard.

Personal Protective Methods: Under certain circumstances where carbon monoxide levels are not exceedingly high, gas masks with proper canisters can be used for short periods but are not recommended. In areas with high concentrations, self-contained air apparatus is recommended.

Bibliography

National Institute for Occupational Safety and Health. *Criteria for a Recommended Standard: Occupational Exposure to Carbon Monoxide.* NIOSH Doc. No. 73-11,000 (1973).

CARBON TETRACHLORIDE

Description: CCl_4, carbon tetrachloride, is a colorless, nonflammable liquid with a characteristic odor. Oxidative decomposition by flame causes phosgene and hydrogen chloride to form.

Synonyms: Tetrachloromethane, perchloromethane.

Potential Occupational Exposures: Carbon tetrachloride is used as a solvent for oils, fats, lacquers, varnishes, rubber, waxes, and resins. Fluorocarbons are chemically synthesized from it. It is also used as an azeotropic drying agent for spark plugs, a dry cleaning agent, a fire extinguishing agent, a fumigant, and an anthelmintic agent. The use of this solvent is widespread, and substitution of less toxic solvents when technically possible is recommended.

A partial list of occupations in which exposure may occur includes:

Chemists	Lacquer makers
Degreasers	Metal cleaners
Fat processors	Propellant makers
Firemen	Refrigerant makers
Fluorocarbon makers	Rubber makers
Grain fumigators	Solvent workers
Ink makers	Wax makers
Insecticide makers	

Permissible Exposure Limits: The Federal standard is 10 ppm (65 mg/m^3) as an 8-hour TWA with an acceptable ceiling concentration of 25 ppm; acceptable maximum peaks above the ceiling of 200 ppm are allowed for one 5-minute duration in any 4-hour period. NIOSH has recommended a ceiling limit of 2 ppm based on a 1-hour sampling period at a rate of 750 ml/min.

Routes of Entry: Inhalation of vapor. Percutaneous absorption has been demonstrated in animals.

Harmful Effects

Local—Carbon tetrachloride solvent removes the natural lipid cover of the skin. Repeated contact may lead to a dry, scaly, fissured dermatitis. Eye contact is slightly irritating, but this condition is transient.

Systemic—Excessive exposure may result in central nervous system depression, and gastrointestinal symptoms may also occur. Following acute exposure, signs and symptoms of liver and kidney damage may develop. Nausea, vomiting, abdominal pain, diarrhea, enlarged and tender liver, and jaundice result from toxic hepatitis. Diminished urinary volume, red and white blood cells in the urine, albuminuria, coma, and death may be consequences of acute renal failure. The hazard of systemic effects is increased when carbon tetrachloride is used in conjunction with ingested alcohol.

Carbon tetrachloride has produced liver tumors in the mouse, hamster, and rat following several routes of administration including inhalation and oral (1)(2). A number of cases of hepatomas appearing in men several years after carbon tetrachloride poisoning have also been described (3)(4).

The chemical pathology of CCl_4 liver injury is generally viewed as an example of lethal cleavage (5), e.g., the splitting of the CCl_3-Cl bond which takes place in the mixed function oxidase system of enzymes located in the hepatocellular endoplasmic reticulum. While two major views of the consequences of this cleavage have been suggested, both views take into account the high reactivity of presumptive free radical products of a homolytic cleavage of the CCl_3-Cl bond (3). One possibility is the direct attack (via alkylation) by toxic free radical metabolites of CCl_4 metabolism on cellular constituents, especially protein sulfhydryl groups (6). In homolytic fission, the two odd-electron fragments formed would be trichloromethyl and monatomic chlorine free radicals. Fowler (7) detected hexachloroethane (CCl_3CCl_3) in tissues of rabbits following CCl_4 intoxication.

An alternative view has emphasized peroxidative decomposition of lipids of the endoplasmic reticulum as a key link between the initial bond cleavage and pathological phenomena characteristic of CCl_4 liver injury (3).

Similarly to chloroform, information as to the mutagenicity of carbon tetrachloride is scant. Carbon tetrachloride gave negative results when tested in *E. coli* and *Salmonella typhimurium* (8)(9). The synergistic effect of CCl_4 on the mutagenic effectivity of cyclophosphamide in the host-mediated assay *S. typhimurium* has been reported (10). CCl_4 did not effect the mutagenicity of cyclophosphamide when tested in vitro with *S. typhimurium* strains G46 and TA1950. CCl_4 was nonmutagenic when assayed in a spot-test with the above strains of *S. typhimurium*.

Medical Surveillance: Preplacement and periodic examinations should include an evaluation of alcohol intake and appropriate tests for liver and kidney functions. Special attention should be given to the central and peripheral nervous system, the skin, and blood.

Special Tests: Expired air and blood levels may be useful as indicators of exposure.

Personal Protective Methods: Barrier creams, gloves, protective clothing, and masks should be used as appropriate where exposure occurs.

References

(1) *IARC.* Vol. 1, International Agency for Research on Cancer, Lyon (1972) pp. 53-60.
(2) Warwick, G.P. *Liver Cancer.* International Agency for Research on Cancer, Lyon (1971), pp. 121-157.
(3) Tracey, J.P., and Sherlock, P. Hepatoma following carbon tetrachloride poisoning. *New York St. J. Med.* 68 (1968) 2202.
(4) Rubin, E., and Popper, H. The evolution of human cirrhosis deduced from observations in experimental animals, chlorofluorocarbons in the atmosphere. *Medicine,* 46 (1967) 163.
(5) Rechnagel, R.O., and Glende, E.A., Jr. Carbon tetrachloride: An example of lethal cleavage. *CRC Crit. Revs. In Toxicology* (1973) 263-297.
(6) Butler, T.C. Reduction of carbon tetrachloride in vivo and reduction of carbon tetrachloride and chloroform in vitro by tissues and tissue constituents. *J. Pharmacol. Exp. Therap.,* 134 (1961) 311.
(7) Fowler, S.S.L. Carbon tetrachloride metabolism in the rabbit. *Brit. J. Pharmacol.* 37 (1969) 773.
(8) Uehleke, H., Greim, H., Krämer, M., and Werner, T. Covalent binding of haloalkanes to liver constituents, but absence of mutagenicity on bacteria in a metabolizing test system. *Mutation Res.,* 38 (1976) 114.
(9) McCann, J., Choi, E., Yamasaki, E., and Ames, B.N. Detection of Carcinogens as mutagens in the Salmonella/microsome test: Assay of 300 chemicals. *Proc. Natl. Acad. Sci.,* 72 (1975) 5135-5139.
(10) Braun, R., and Schöneich, J. The influence of ethanol and carbon tetrachloride in the host-mediated assay with *Salmonella typhimurium. Mutation Res.* 31 (1975) 191-194.

Bibliography

Barnes, R., and Jones, R.C. 1967. Carbon tetrachloride poisoning. *Am. Ind. Hyg. Assoc. J.* 28:557.
Lewis, C.E. 1961. The toxicology of carbon tetrachloride. *J. Occup. Med.* 3:82.
Luse, S.A., and Wood, W.G. 1967. The brain in fatal carbon tetrachloride poisoning. *Arch. Neurol.* 17:304.
National Institute for Occupational Safety and Health. *Criteria for a Recommended Standard: Occupational Exposure to Carbon Tetrachloride.* NIOSH Doc. No. 76-133 (1976).
Nielson, V.K., and Larsen, J. 1965. Acute renal failure due to carbon tetrachloride poisoning. *Acta. Med. Scand.* p. 178.

CARBONYLS

Description: Metal carbonyls have the general formula $Me_x(CO)_y$ in which Me is the metal and x and y are whole numbers. They are generally produced by direct reaction between carbon monoxide and the finely divided metal; however, chromium, molybdenum, and tungsten carbonyls can be produced by the Grignard method, and platinum metals, iron and rhenium carbonyls may be obtained from metal sulfides, halides, or oxides. The carbonyls react with oxidizing agents and may ignite spontaneously. Reaction with water or steam results in the liberation of carbon monoxide; and on heating, the carbonyls decompose forming carbon monoxide and the finely divided metal powder which may ignite. Some of the more important carbonyls are:

Chromium carbonyl: $Cr(CO)_6$. Colorless crystals.
Cobalt tricarbonyl: $[Co(CO)_3]_4$. Black crystal.
Cobalt tetracarbonyl: $[Co(CO)_4]_2$. Orange crystals or dark brown microscopic crystals.
Cobalt carbonyl hydride: $HCo(CO)_4$. Below $-26.2°C$, exists as light yellow solid. At room temperature, a gas. It begins to decompose in air above $-26°C$.
Cobalt nitrosocarbonyl: $Co(CO)_3(NO)$. Cherry liquid.
Iron tetracarbonyl: $[Fe(CO)_4]_3$. Dark green lustrous crystals.
Iron pentacarbonyl: $Fe(CO)_5$. Viscous yellow liquid.
Iron nonacarbonyl: $Fe_2(CO)_9$. Yellow to orange crystals.
Iron carbonyl hydride: $H_2Fe(CO)_4$. A gas. Begins to decompose at $-10°C$.
Iron nitrosyl carbonyl: $Fe(NO)_2(CO)_2$. Dark red crystals.
Molybdenum hexacarbonyl: $Mo(CO)_6$. White crystals.
Nickel carbonyl: $Ni(CO)_4$. Colorless liquid.
Osmium carbonyl chloride: $Os(CO)_2Cl_3$. Dark brown. Deliquescent.
Ruthenium pentacarbonyl: $Ru(CO)_5$. Colorless liquid. Very volatile.
Tungsten carbonyl: $W(CO)_6$. Colorless crystals.

Synonyms:

Chromium carbonyl: None.
Cobalt tricarbonyl: Tetracobalt dodecacarbonyl.
Cobalt tetracarbonyl: Dicobalt octacarbonyl.
Cobalt carbonyl hydride: Cobalt tetracarbonyl hydride.
Cobalt nitrosocarbonyl: None.
Iron tetracarbonyl: None.
Iron pentacarbonyl: None.
Iron nonacarbonyl: Enneacarbonyl.
Iron carbonyl hydride: None
Iron nitrosyl carbonyl: None.

Molybdenum hexacarbonyl: Molybdenum carbonyl.
Nickel carbonyl: Nickel tetracarbonyl.
Osmium carbonyl chloride: None.
Ruthenium pentacarbonyl: None.
Tungsten carbonyl: None.

Potential Occupational Exposures: Metal carbonyls are used in isolating certain metals from complex ores, in the preparation of high purity metals, for the production of carbon steel and metallizing, and as catalysts in organic synthesis. Pure metal powders from carbonyls are used in the electronics industry for radiofrequency transformers. $Fe(CO)_5$ is used as a gasoline additive in Europe and as an antidetonator.

Metal carbonyls may be formed during other processes: in the Bessemer converter in the steel industry; inadvertent introduction of carbon monoxide onto metal catalyst beds; storage of carbon monoxide in steel cylinders producing $Fe(CO)_5$; slowly flowing water or gas in an iron pipe generating $Fe(CO)_5$; and in the Fischer-Tropsch process for the liquefaction of coal.

A partial list of occupations in which exposure may occur includes:

Acetylene welders	Nickel refiners
Blast furnace workers	Organic chemical synthesizers
Metal refiners	Petroleum refinery workers
Mond process workers	

Permissible Exposure Limits: There are no specific standards for the metal carbonyls, other than nickel carbonyl. (See Nickel Carbonyl.)

Routes of Entry: Inhalation of vapor or dust. Percutaneous absorption of liquids may occur.

Harmful Effects

Local—Aside from skin irritation caused by the specific metal liberated when the metal carbonyl decomposes, no local effects have been reported.

Systemic—Metal carbonyls as a group have somewhat similar toxicological effects, although there are differences in degrees of toxicity which range from moderate to extremely mild. Nickel carbonyl is the best known and is highly toxic, capable of causing pulmonary edema. Exposures during the Mond process have been associated with an increased incidence of lung and nasal sinuses cancer. Cancer has been produced in rats in the lung, liver, and kidneys.

The toxicity of carbonyls depends in part on the toxic character of the metal component and in part on the volatility and stability of the carbonyl itself. $Ni(CO)_4$ has a very high vapor pressure, plus stability at room temperature.

There are no reports of human injury following exposure to cobalt carbonyls. Cobalt tetracarbonyl has an odor so offensive at low levels of concentration that it provides an effective warning against toxic exposure. Iron pentacarbonyl may cause similar pulmonary symptoms to those of nickel carbonyl. Animal studies indicate that the inhalation of fumes and dusts of carbonyls causes respiratory irritation and disturbances to the central nervous system.

Medical Surveillance: (See Nickel Carbonyl.) Preemployment physical examinations should give particular attention to the respiratory tract and skin. Periodic examinations should include the respiratory tract and nasal sinuses, smoking history as well as general health. A baseline chest x-ray should be available and pulmonary function followed.

Special Tests: Urinary nickel level determinations for a few days after an acute exposure may be useful. Little information is available as to the value of biochemical studies in the case of the other carbonyls.

Personal Protective Methods: In areas where either dust or vapors of the metal carbonyls are encountered, the worker should wear appropriate supplied air respirators. Where the danger of splash or spill of liquids exists, impervious protective clothing should be used.

Bibliography

Brief, R.S., Blanchard, J.W., Scala, R.A., and Blacker, J.H. 1971. Metal carbonyls in the petroleum industry. *Arch. Environ. Health* 23:273.
McDowell, R.S. 1971. Metal carbonyl vapors—Rapid quantitative analysis of infrared spectrophotometry. *Am. Ind. Hyg. Assoc. J.* 32:621.

CEMENT DUST (See "Portland Cement")

CERIUM AND COMPOUNDS

Description: Ce, cerium, a soft, steel-gray metal, is found in the minerals monazite, cerite, and orthite. It may form either tri- or tetravalent compounds. The cerous salts are usually white and the ceric salts are yellow to orange-red. Cerium decomposes in water and is soluble in dilute mineral acids.

Synonyms: None.

Potential Occupational Exposures: Cerium and its compounds are used as a catalyst in ammonia synthesis, a deoxidizer to improve the mechanical quality and refine grain size of steel, an opacifier in certain enamels, an arc-stabilizer in carbon arc lamps, an abrasive for polishing mirrors and lenses, a sedative and as a medicinal agent for vomiting during pregnancy. It is used in the manufacture of topaz yellow glass, spheroidal cast iron, incandescent gas mantles and in decolorizing glass, to prevent mildew in textiles, and to produce a vacuum in neon lamps and electronic tubes. Alloyed with aluminum, magnesium, and manganese, it increases resistance to creep and fatigue. Ferrocerium is the pyrophoric alloy in gas cigarette lighters, and an alloy of magnesium, cerium, and zirconium is utilized for jet engine parts.

A partial list of occupations in which exposure may occur includes:

Alloy makers	Lighter flint makers
Ammonia makers	Metal refiners
Enamel makers	Phosphor makers
Glass (vitreous) makers	Rocket fuel makers
Ink makers	Textile workers

Permissible Exposure Limits: There is no Federal standard for cerium or its compounds.

Route of Entry: Inhalation of dust.

Harmful Effects

Local—No local effects have been reported due to cerium and its compounds.

Systemic—There are no records of injury to human beings from either the industrial or medicinal use of cerium. The main risk to workers is from dust in mining and production areas. Recent reports in the literature describe "Cer-pneumoconiosis," a condition found in a group of graphic arts workers who use carbon arc lights in their work. Chest x-rays reveal small, miliary, homogeneously distributed infiltrates. Cer-pneumoconiosis cannot be considered a dust disease of the lung similar to silicosis. In the later stages of the reaction to the dust of carbon arc lamps, perifocal emphysema, and slight fibrosis of lungs are noted. It has been speculated that these changes may have been due to inhalation of substances containing radioactive elements of the thorium chain. To date, these views have not been confirmed by animal experimentation, autopsy, or human biopsy. Animal experimentation has demonstrated increased coagulation time from organic preparations of cerium, disturbance of lipid metabolism from cerium and its nitrates, and profound effects on metabolism and intestinal muscle causing loss of motility from cerium chloride.

Medical Surveillance: Chest x-rays should be taken as a part of preemployment and periodic physical examinations.

Special Tests: None in common use.

Personal Protective Methods: In areas of carbon arc lights, workers should wear effective dust filters or respirators. In mining and production areas, workers should wear effective dust filters or respirators suitable for the particulate size of air borne dust.

Bibliography

Heuck, F., and Hoschek, R. 1968. Cer-pneumoconiosis. *Am. J. Roentgen.* 104:777.

CHLORINATED BENZENES

Description: Chlorinated benzenes are aromatic rings with one or more chlorines substituted for a hydrogen. Compounds with only a few chlorines are usually colorless liquids at room temperature and have an aromatic odor. The more highly substituted compounds are crystals (typically monoclinic).

Synonyms:

Chlorobenzene: phenyl chloride, monochlorobenzene, chlorobenzol.
o-dichlorobenzene: 1,2-dichlorobenzene.
m-dichlorobenzene: 1,3-dichlorobenzene.
p-dichlorobenzene: 1,4-dichlorobenzene.
1,2,3-trichlorobenzene: None.
1,2,4-trichlorobenzene: None.
1,3,5-trichlorobenzene: None.
1,2,4,5-tetrachlorobenzene: None.
Hexachlorobenzene: perchlorobenzene.

Potential Occupational Exposures: Chlorobenzene is used as a solvent and as an intermediate in dyestuffs. o-Dichlorobenzene is used as a solvent, fumigant, insecticide, and chemical intermediate. p-Dichlorobenzene finds use as an insecticide, chemical intermediate, disinfectant and moth preventative. Other chlorinated benzenes are not as widely used in industry but find use as chemical intermediates, and to an even lesser extent, as insecticides and solvents.

A partial list of occupations in which exposure may occur includes:

Cellulose acetate workers	Insecticide makers and workers
Deodorant makers	Lacquer workers
Disinfectant workers	Organic chemical synthesizers
Dyers	Paint workers
Dye makers	Resin makers
Fumigant workers	Seed disinfectors

Permissible Exposure Limits: The Federal standards are:

chlorobenzene	75 ppm	350 mg/m^3
o-dichlorobenzene	50 ppm	300 mg/m^3
p-dichlorobenzene	75 ppm	450 mg/m^3

Threshold limit values for the other compounds have not as yet been established.

Routes of Entry: Inhalation of vapor, percutaneous absorption of the liquid.

Harmful Effects

Local—Chlorinated benzenes are irritating to the skin, conjunctiva, and mucous membranes of the upper respiratory tract. Prolonged or repeated contact with liquid chlorinated benzenes may cause skin burns.

Systemic—In contrast to aliphatic halogenated hydrocarbons, the toxicity of chlorinated benzenes generally decreases as the number of substituted chlorine atoms increases. Basically, acute exposure to these compounds may cause drowsiness, incoordination, and unconsciousness. Animal exposures have produced liver damage.

Chronic exposure may result in liver, kidney, and lung damage as indicated by animal experiments.

Medical Surveillance: Preplacement and periodic examinations should consider skin, liver, lung, and kidney.

Special Tests: None commonly used. Urinary excretion of 2,5-dichlorophenol may be useful as an index of exposure.

Personal Protective Methods: Barrier creams, protective clothing, and good personal hygiene are good preventive measures. Respirators in areas of vapor concentrations are advised.

Bibliography

Brown, V.K.H., Muir, C., and Thorpe, E. 1969. The acute toxicity and skin irritant properties of 1,2,4-trichlorobenzene. *Ann. Occup. Hyg.* 12:209.

Girard, R., Tolot, F., Martin, P., and Bourret, J. 1969. Hemopathies graves et exposition a des derives chlores du benzene (a propos de 7 cas). *J. Med. Lyon* 50:771.

Hollingsworth, R.L., Rowe, V.K., Oyen, F., Tokelson, T.R., and Adams, E.M. 1956. Toxicity of o-dichlorobenzene. Studies on animals and industrial experience. *AMA Arch. Ind. Health* 17:180.

Pagnotto, L.D., and Walkley, J.E. 1965. Urinary dichlorophenol as an index of para-dichlorobenzene exposure. *Am. Ind. Hyg. Assoc. J.* 26:137.

Tolot, F., Soubrier, B., Bresson, J.R., and Martin, P. 1969. Myelose proliferative devoultion rapide. Role etiologique possible des derives chlores du benzene. *J. Med. Lyon* 50:761.

Varshavskaya, S.P. 1968. Comparative toxicological characteristics of chlorobenzene and dichlorobenzene (ortho- and para-isomers) in relation to the sanitary protection of water bodies. *Hyg. Sanit.* 33:17.

CHLORINATED LIME

Description: Chlorinated lime is a white or grayish-white hygroscopic powder with a chlorine odor. It is a relatively unstable chlorine carrier in solid form and is a complex compound of indefinite composition. Chemically, it consists of varying proportions of calcium hypochlorite, calcium chlorite, calcium oxychloride, calcium chloride, free calcium hydroxides, and water. The commercial product generally contains 24-37% available chlorine. On exposure to moisture, chlorine is released.

Synonyms: Chloride of lime, bleaching powder.

Potential Occupational Exposures: Chlorinated lime is a bleaching agent, i.e., it has the ability to chemically remove dyes or pigments from materials. It is used in the bleaching of wood pulp, linen, cotton, straw, oils, and soaps, and in laundering, as an oxidizer in calico printing, a chlorinating agent, a disinfectant, particularly for drinking water and sewage, a decontaminant for mustard gas, and as a pesticide for caterpillars.

A partial list of occupations in which exposure may occur includes:

Disinfectant makers	Straw bleachers
Dyers	Textile bleachers
Laundry workers	Textile printers
Oil bleachers	Water treaters
Sewage treaters	Wood pulp bleachers
Soap bleachers	

Permissible Exposure Limits: There is no Federal standard for chlorinated lime. (See Chlorine.)

Routes of Entry: Inhalation of dust. Inhalation of vapor and ingestion.

Harmful Effects

Local—The toxic effects of chlorinated lime are due to its chlorine content. The powder and its solutions have corrosive action on skin, eyes, and mucous membranes, can produce conjunctivitis, blepharitis, corneal ulceration, gingivitis, contact dermatitis, and may damage the teeth.

Systemic—The dust is irritating to the respiratory tract and can produce laryngitis and pulmonary edema. Chlorinated lime is extremely hygroscopic and with the addition of water evolves free chlorine. Inhalation of the vapor is extremely irritating and toxic. (See Chlorine.) Ingestion of chlorinated lime causes severe oral, esophageal, and gastric irritation.

Medical Surveillance: Consider possible effects of skin, teeth, eyes, or respiratory tract. There are no specific diagnostic tests.

Special Tests: None commonly used.

Personal Protective Methods: In dusty areas, the worker should be protected by appropriate respirators. Simple dust masks should not be used since the moisture present in expired air will release the chlorine. Skin effects can be minimized with protective clothing. Most important is the fact that free chlorine is liberated when chlorinated lime comes in contact with water. All precautions should be followed to protect the worker under these circumstances. (See Chlorine.)

CHLORINATED NAPHTHALENES

Description: $C_{10}H_{8-x}Cl_x$, the chlorinated naphthalenes, are naphthalenes in which one or more hydrogen atoms have been replaced by chlorine to form wax-like substances, beginning with monochloronaphthalene and going on to the octochlor derivatives. Their physical states vary from mobile liquids to waxy-solids depending on the degree of chlorination.

Synonyms: Chloronaphthalenes.

Potential Occupational Exposures: Industrial exposure from individual chlorinated naphthalenes is rarely encountered; rather it usually occurs from mixtures of two or more chlorinated naphthalenes. Due to their stability, thermoplasticity, and nonflammability, these compounds enjoy wide industrial application. These compounds are used in the production of electric condensers, in the insulation of electric cables and wires, as additives to extreme pressure lubricants, as supports for storage batteries, and as a coating in foundry use.

A partial list of occupations in which exposure may occur includes:

Cable coaters	Rubber workers
Condenser impregnators	Solvent workers
Electric equipment makers	Transformer workers
Insecticide workers	Wire coaters
Petroleum refinery workers	Wood preservers
Plasticizer makers	

Permissible Exposure Limits: The Federal standards are:

Trichloronaphthalene	5.0 mg/m^3
Tetrachloronaphthalene	2 mg/m^3

| Pentachloronaphthalene | 0.5 mg/m^3 |
| Hexachloronaphthalene | 0.2 mg/m^3 |

Routes of Entry: Inhalation of fumes and percutaneous absorption of liquid.

Harmful Effects

Local—Chronic exposure to chlorinated naphthalenes can cause chloracne, which consists of simple erythematous eruptions with pustules, papules, and comedones. Cysts may develop due to plugging of the sebaceous gland orifices.

Systemic—Cases of systemic poisoning are few in number and they may occur without the development of chloracne.

It is believed that chloracne develops from skin contact and inhalation of fumes, while systemic effects result primarily from inhalation of fumes. Symptoms of poisoning may include headaches, fatigue, vertigo, and anorexia. Jaundice may occur from liver damage. Highly chlorinated naphthalenes seem to be more toxic than those chlorinated naphthalenes with a lower degree of substitution.

Medical Surveillance: Preplacement and periodic examinations should be concerned particularly with skin lesions such as chloracne and with liver function.

Special Tests: None are in common use.

Personal Protective Methods: Skin contact should be avoided whenever possible. Barrier creams, protective clothing, and good personal hygiene are all good preventive measures. Use of respirators in areas of vapor concentration is advised.

Bibliography

Kleinfeld, M., Messite, J. Swencicki, R. 1972. Clinical effects of chlorinated naphthalene exposure. *J. Occup. Med.* 14:377.

Mayers, M.R., and Silverberg, M.G. 1938. Effects upon the skin due to exposure to some chlorinated hydrocarbons. *Ind. Bull. N.Y. State Dep. Labor* 17: 358 and 425.

Mayers, M.R., and Smith, A.R. 1942. Systemic effects from exposure to certain chlorinated hydrocarbons. *Ind. Bull. N.Y. State Dep. Labor* 21:30.

CHLORINE

Description: Cl_2, chlorine, is a greenish-yellow gas with a pungent odor. It is slightly soluble in water and is soluble in alkalis. It is the commonest of the four halogens which are among the most chemically reactive of all the elements.

Synonyms: None.

Potential Occupational Exposures: Gaseous chlorine is a bleaching agent in the paper and pulp and textile industries for bleaching cellulose for artificial fibers. It is used in the manufacture of chlorinated lime, inorganic and organic compounds such as metallic chlorides, chlorinated solvents, refrigerants, pesticides, and polymers, e.g., synthetic rubber and plastics; it is used as a disinfectant, particularly for water and refuse, and in detinning and dezincing iron. NIOSH estimated in 1973 that 15,000 workers had potential occupational exposure to chlorine.

A partial list of occupations in which exposure may occur includes:

Aerosol propellant makers	Paper bleachers
Bleachers	Pesticide makers
Chlorinated solvent makers	Plastic makers
Disinfectant makers	Rayon makers
Dye makers	Refrigerant makers
Flour bleachers	Silver extractors
Iron workers	Swimming pool maintenance workers
Laundry workers	Tin recovery workers

Permissible Exposure Limits: The Federal standard is 1 ppm (3 mg/m^3). NIOSH has recommended a ceiling limit of 0.5 ppm for a 15-minute sampling period. The basis for the NIOSH-recommended environmental limit is the prevention of irritation of the skin, eyes, and respiratory tract.

Harmful Effects

Local—Chlorine reacts with body moisture to form acids. It is itself extremely irritating to skin, eyes, and mucous membranes, and it may cause corrosion of teeth. Prolonged exposure to low concentrations may produce chloracne.

Systemic—Chlorine in high concentrations acts as an asphyxiant by causing cramps in the muscles of the larynx (choking), swelling of the mucous membranes, nausea, vomiting, anxiety, and syncope. Acute respiratory distress including cough, hemoptysis, chest pain, dyspnea, and cyanosis develop, and later tracheobronchitis, pulmonary edema, and pneumonia may supervene.

Medical Surveillance: Special emphasis should be given to the skin, eye, teeth, cardiovascular status in placement and periodic examinations. Chest x-rays should be taken and pulmonary function followed.

Special Tests: None in common use.

Personal Protective Methods: Whenever there is likelihood of excessive gas levels, workers should use respiratory protection in the form of fullface gas mask with proper canister or supplied air respirator. The skin effects of chlorine can generally be controlled by good personal hygiene practices. Where very high gas concentrations or liquid chlorine may be present, full protective clothing, gloves, and eye protection should be used. Changing work clothes daily and showering following each shift where exposures exist are recommended.

Bibliography

Chasis, H., Zapp, J.A., Bannon, J.H., Whittenburger, J.L., Helm, J., Doheny, J.J., and MacLeod, C.M. 1947. Chlorine accident in Brooklyn. *Occup. Med.* 4:152.

Dixon, W.M., and Drew, D. 1968. Fatal chlorine poisoning. *J. Occup. Med.* 10:249.

Henefer, D. 1969. *Disease of the Occupations,* 4th Ed. Little, Brown and Co., Boston.

Joyner, R.E., and Durel, E.G. 1947. Accidental liquid chlorine spill in a rural community. *J. Occup. Med.* 4:152.

Kaufman, J., and Burkons, D. 1971. Clinical, roentgenologic, and physiologic effects of acute chlorine exposure. *Arch. Environ. Health* 23:29.

Kramer, C.G. 1967. Chlorine. *J. Occup. Med.* 9:193.

National Institute for Occupational Safety and Health. *Criteria for a Recommended Standard: Occupational Exposure to Chlorine.* NIOSH Doc. No. 76-170 (1976).

o-CHLOROBENZYLIDENE MALONITRILE

Description: $ClC_6H_4CH=C(CN)_2$ o-chlorobenzylidene malonitrile (OCBM), is a white crystalline solid.

Synonyms: OCBM, CS.

Potential Occupational Exposures: OCBM is used as a riot control agent.

A partial list of occupations in which exposure may occur includes:

Riot controllers

Permissible Exposure Limits: The Federal standard is 0.05 ppm (0.4 mg/m³).

Route of Entry: Inhalation.

Harmful Effects

Local—OCBM is extremely irritating and acts on exposed sensory nerve endings (primarily in the eyes and upper respiratory tract). The signs and symptoms from exposure to the vapor are conjunctivitis and pain in the eyes, lacrimation, erythema of the eyelids, blepharospasms, irritation and running of the nose, burning in the throat, coughing and constricted feeling in the chest, and excessive salivation. Vomiting may occur if saliva is swallowed. Most of the symptoms subside after exposure ceases. Burning on the exposed skin is increased by moisture. With heavy exposure, vesiculation and erythema occur. Photophobia has been reported.

Systemic—Animal experiments indicate that OCBM has a relatively low toxicity. The systemic changes observed in human experiments are nonspecific reactions to stress. OCBM is capable of sensitizing guinea pigs; there also appears to be a cross-reaction in guinea pigs previously sensitized to 1-chloro-acetophenone.

Medical Surveillance: Consideration should be given to the eyes, skin, and respiratory tract in any placement or periodic evaluations.

Special Tests: None have been proposed.

Personal Protective Methods: Because of its extremely irritant properties, those using OCBM in high concentrations should wear respirators and eye protection.

Bibliography

Beswick, F.W., Hollan, P., and Kemp, K.H. 1972. Acute effects of exposure to ortho-chlorobenzylidene malonitrile (CS) and the development of tolerance. *Brit. J. Ind. Med.* 29:298.
Chung, C.W., and Giles, A.L. 1972. Sensitization of guinea pigs to alphachloroaceto-phenone (CN) and ortho-chlorobenzylidene malonitrile (CS), tear gas chemicals. *J. Immunol.* 109:284.
Gass, S., Fisher, T.L., Jascot, M.J., and Herban, J. 1971. Gas-liquid chromatography of some irritants at various concentrations. *Anal. Chem.* 43:462.
Punte, C.L., Owens, E.J., and Gutentag, P.J. 1963. Exposures to orthochlorobenzylidene malonitrile. *Arch. Environ. Health.* 6:366.

CHLORODIPHENYLS AND DERIVATIVES

Description: $C_{12}H_{10-x}Cl_x$, Chlorodiphenyls, are diphenyl rings in which one or more hydrogen atoms are replaced by a chlorine atom. Most widely used are chlorodiphenyl (42% chlorine), containing 3 chlorine atoms in unassigned positions, and chlorodiphenyl (54% chlorine) containing 5 chlorine atoms in unassigned positions. These compounds are light, straw-colored liquids with typical chlorinated aromatic odors; 42% chlorodiphenyl is a mobile liquid and 54% chlorodiphenyl is a viscous liquid.

Chlorinated diphenyl oxides are ethers of chlorodiphenyls and are included in this group. They range from clear, oily liquids to white to yellowish waxy solids, depending on the degree of chlorination.

Halogenated biphenyls, as are typical of aryl halides, generally are quite stable to chemical alteration. Polychlorinated biphenyls (PCBs) (first introduced into commercial use more than 45 years ago) are one member of a class of chlorinated aromatic organic compounds which are of increasing concern because of their apparent ubiquitous dispersal, persistence in the environment, and tendency to accumulate in food chains, with possible adverse effects on animals at the top of food webs, including man (1)-(7).

Polychlorinated biphenyls are prepared by the chlorination of biphenyl and hence are complex mixtures containing isomers of chlorobiphenyls with different chlorine contents (8). It should be noted that there are 209 possible compounds obtainable by substituting chlorine for hydrogen on from one to ten different positions on the biphenyl ring system. An estimated 40 to 70 different chlorinated biphenyl compounds can be present in each of the higher chlorinated commercial mixtures (9)(10). For example, Arochlor 1254 contains 69 different molecules, which differ in the number and position of chlorine atoms (10).

It should also be noted that certain PCB commercial mixtures produced in the United States and elsewhere (e.g., France, Germany, and Japan) have been shown to contain other classes of chlorinated derivatives, e.g., chlorinated naphthalenes and chlorinated dibenzofurans (7)(11)-(13). The possibility that naphthalene and dibenzofuran contaminate the technical biphenyl feedstock used in the preparation of the commercial PCB mixtures cannot be excluded.

Synonyms: Chlorobiphenyls, polychlorinated diphenyl, PCB.

Potential Occupational Exposures: Chlorinated diphenyls are used alone and in combination with chlorinated naphthalenes. They are stable, thermoplastic, and nonflammable, and find chief use in insulation for electric cables and wires in the production of electric condensers, as additives for extreme pressure lubricants, and as a coating in foundry use.

A partial list of occupations in which exposure may occur includes:

Cable coaters	Plasticizer makers
Dye makers	Resin makers
Electric equipment makers	Rubber workers
Herbicide workers	Textile flameproofers
Lacquer makers	Transformer workers
Paper treaters	Wood preservers

Permissible Exposure Limits: The Federal standards for chlorodiphenyl (42% Cl) and chlorodiphenyl (54% Cl) are 1 mg/m^3 and 0.5 mg/m^3 respectively.

Routes of Entry: Inhalation of fume or vapor and percutaneous absorption of liquid.

Harmful Effects

Local—Prolonged skin contact with its fumes or cold wax may cause the formation of comedones, sebaceous cysts, and pustules, known as chloracne. Irritation to eyes, nose, and throat may also occur. The above standards are considered low enough to prevent systemic effects, but it is not known whether or not these levels will prevent local effects.

Systemic—Generally, toxic effects are dependent upon the degree of chlorination; the higher the degree of substitution, the stronger the effects. Acute and chronic exposure can cause liver damage. Signs and symptoms include edema, jaundice, vomiting, anorexia, nausea, abdominal pains, and fatigue.

Studies of accidental oral intake indicate that chlorinated diphenyls are embryotoxic, causing stillbirth, a characteristic grey-brown skin, and increased eye discharge in infants born to women exposed during pregnancy.

The rates and routes of transport of the PCBs in the environment (1)(2)(5)(6) and their accumulation in ecosystems (1)(2)(15)-(20), and toxicity (1)(3)(5)(7) (21)-(23) have been reviewed.

It is generally acknowledged that the toxicological assessment of commercially available PCBs has been complicated by the heterogeneity of the isomeric chlorobiphenyls and by marked differences in physical and chemical properties that influence the rates of absorption, distribution, biotransformation and excretion (1)(7)(21)-(29).

The lower chlorine homologs of PCBs (either examined individually per se or in Aroclor mixtures) are reported to be more rapidly metabolized in the rat than the higher homologs (30)-(34). Sex-linked differences were also disclosed (e.g., the biological half-life of Aroclor 1254 in adipose tissue of rats fed 500 ppm was 8 and 12 weeks in males and females respectively) (32).

The lowest PCB homolog found from Aroclor 1254 in human fat was pentachlorobiphenyl (35).

The metabolism of many PCB isomers have consistently shown the formation of various hydroxylated urinary excretion products. For example, the metabolism of 4,4'-dichlorobiphenyl in the rat yielded four monohydroxy-, four dihydroxy- and two trihydroxy metabolites (36). The structures of the major metabolites in the rat are consistent with epoxidation of the biphenyl nucleus followed by epoxide ring opening accompanied by a 1,2-chlorine shift (NIH shift). The formation of minor rat metabolites, such as 4-chloro-3'-biphenylol, appeared to occur via reductive dechlorination (36).

Urinary metabolites of 2,5,2',5'-tetrachlorobiphenyl (TCB) in the nonhuman primate included among others: TCB, monohydroxy-TCB, and dihydroxy-TCB (37).

In studies involving the metabolism of 2,2',4,4',5,5'-hexachlorobiphenyl by rabbits, rats and mice (38)(39), it was shown that the rabbit excreted hexachlorobiphenylol-, pentachlorobiphenylol-, and methoxypentachlorobiphenylol compounds (39)(40), while rats and mice excreted only a hexachlorobiphenylol (38).

These results and previous studies by Gardner and co-workers (41) as well as in vitro metabolism studies with 4-chlorobiphenyl (42) indicate that PCBs are metabolized via metabolically activated arene oxide intermediates.

It is also of potential importance to note the presence of methyl sulfone metabolites of PCB (as well as DDE) recently found in seal blubber (43). The toxicological significance of these metabolites has not been elucidated to date.

Methyl Sulfone of PCB
(x + y = 3 to 7)

Methyl Sulfone of DDE

Increasing evidence indicates that not all chlorobiphenyl congeners produce the same pharmacologic effects (7)(29)(44)-(51). Morphological alterations in both acute and chronic toxicity have been studied in rats, monkeys, mice and cows (7)(52)-(57); the organ consistently affected was the liver. For example, when male Sprague-Dawley rats were fed a diet containing mixtures of PCB isomers (Aroclor 1248, 1254 and 1262) at a concentration of 100 ppm in the diet for 52 weeks, there was a decided increase in their total serum lipids and cholesterol and a transient increase in triglycerides accompanied by distinct morphological changes in the liver (57). Generalized liver hypertrophy and focal areas of hepatocellular degeneration were followed by a wide spectrum of repair processes. The tissue levels of PCB were greater in the animals receiving the high chlorine mixtures and high levels persisted in these tissues even after the PCB treatment had been discontinued.

Indirect effects of PCB exposure are related to increased microsomal enzyme activity and include alteration in metabolism of drugs, hormones, and pesticides (7)(44)(58)-(60). A large portion of the human population has detectable levels of PCB in adipose tissue (2)(63) and recent preliminary reports have revealed that PCBs have been found in 48 of 50 samples of mothers' milk in 10 states (61)-(63) (the average levels in the 48 samples was 2.1 ppm). The 1971-74 adipose levels of PCB in the United States showed that of 6,500 samples examined, 77% contained PCB (e.g., 26% contained <1 ppm; 44% contained 1 to 2 ppm and 7% contained >2 ppm) (63). There were no sex differences and the levels of PCB increased with age (63).

The 1971-74 PCB ambient water levels were as follows: of 4,472 water samples, 130 were in the range of 0.1 to 4.0 ppb (detection limit: 0.1 to 1.0 ppb) (63); of 1,544 sediment samples, 1,157 were in the range of 0.1 to 13,000 ppb (limit of detection 0.1 to 1.0 ppb) (63). In a recent accidental plant discharge episode, Hudson River sediments near Fort Edward, New York were found to contain 540 to 2,980 ppm PCBs (64). Soil levels of PCBs in the 1971-74 period ranged from 0.001 to 3.33 ppm (average 0.02 ppm) in 1,434 samples taken from 12 of 19 metropolitan areas (63). In limited surveys in ambient air, an average of 100 ng/cm^3 of PCB was found for each of 3-24 hour samples from Miami, Florida, Jacksonville, Florida and Fort Collins, Colorado (63). Where PCBs have been found in food, it has generally been in fish samples from the Great Lakes area (64) with contamination arising mainly through the environment, whereas in the past, milk and dairy products, eggs, poultry, animal feeds, infant foods as well as paper food packaging received PCBs principally from agricultural and industrial applications (64).

Although there has been a sharp curtailment of PCB production and dispersive use applications from a record high of 70 million lb in 1969, it is believed that it will take several years for ecosystems such as Lake Michigan to cleanse itself of the compounds even if no new input is made (65). Due to its inertness and high adsorption coefficient, the PCBs have accumulated in the bottom sediments. The final sink for PCB is predicted to be degradation in the atmosphere, with some fraction being buried in underlying sediments of lakes (65).

It is important to note that even with the cessation of PCB production per se, other environmental sources of PCB may exist. For example, it has been reported that some PCBs are products of DDT photolysis (4)(66)(67). Uyeta et al (68) recently reported the photoformation of PCBs from the sunlight irradiation of mono-, di-, tri-, tetra-, and hexachlorobenzenes.

Gaffney (68a) recently reported the formation of various mono-, di-, and tri-chlorobiphenyls resulting from the final chlorination of municipal wastes containing biphenyl. Laboratory chlorination of influent and effluent from a municipal waste treatment facility also resulted in the formation of these and other chloroorganic substances such as di- and trichlorobenzenes.

Previous clinical aspects of human poisoning in Japan ("Yusho" disease) involving at least 1,000 people consuming rice bran oil contaminated with Kanechlor 400 (a PCB containing 48% chlorine with 2,4,3',4'-, 2,5,3',4'-, 2,3,5,5'-tetrachlorobiphenyl, and 2,3,5,3',4'-pentachlorobiphenyl) (69) are well documented (1)(7)(70)(71). It has also been claimed that there are an estimated

15,000 victims of "Yusho" disease although only 1,081 persons have been officially diagnosed as such (72).

It has been very recently reported by Hirayama (73) that five of the Yusho victims died of liver cancer within 5 years after consuming the contaminated cooking oil.

Recent reports of high cancer rates among Mobil Oil employees at its Paulsboro, N.J. refinery exposed to PCBs (Aroclor 1254) have suggested a possible link between PCB exposure and skin (melanoma) or pancreatic cancer (74)-(76). The Mobil study indicated that 8 cancers developed between 1957 and 1975 among 92 research and development and refinery workers exposed for 5 or 6 years in the late 1940s and early 1950s to varying levels of Aroclor 1254. Of the 8 cancers, 3 were malignant melanomas and two were cancers of the pancreas. NIOSH said "this is significantly more skin cancer (melanoma) and pancreatic cancer than would be expected in a population of this size, based on the Third National Cancer Survey" (75).

It should be noted that Monsanto Co. could find no causal relationship between cancer and PCB exposure at its plant in Sauget, Ill. The Monsanto study was based on a review of the records of more than 300 current and former employees at the Illinois plant which had been engaged in PCB production since 1936 (77).

Earlier indications of the carcinogenicity of PCBs were reported in 1972 by Nagasaki et al (78) who cited the hepatocarcinogenicity of Kaneclor-500 in male dd mice fed 500 ppm of the PCB. The hepatomas appeared similar to those induced by the gamma-isomer of benzene hexachloride (BHC) (79)(80), whereas Kaneclor-400 and Kaneclor-300 had no carcinogenicity activity in the liver of mice. The Kaneclor-500 sample contained 55.0% pentachlorobiphenyl, 25.5% tetrachlorobiphenyl, 12.8% hexachlorobiphenyl and 5.0% trichlorobiphenyl. Later studies by Ito et al (81) also demonstrated that Kaneclor-500 not only induced hepatic neoplasms in mice when fed at levels of 500 ppm in the diet for 32 weeks but also promoted the induction of tumors by alpha-BHC and beta-BHC. Kimbrough et al (82) reported the induction of liver tumors in Sherman strain female rats fed 100 ppm of Aroclor 1260 in their diet for approximately 21 months. Recent studies also suggest that PCBs exert a potent promoting action in experimental azo dye hepatocarcinogenesis (83).

Conflicting evidence to date exists concerning mutagenic effects of mixtures of PCBs (63)(42)(84)-(89). Tests on *Drosophila* with PCB of mixed degrees of chlorination did not indicate any chromosome-breaking effects (85). However, it was suggested by Ramel (84) that PCBs may have an indirect bearing on mutagenicity and carcinogenicity since they induce enzymatic detoxification enzymes in liver microsomes.

No chromosomal aberrations have been observed in human lymphocyte cultures exposed to Aroclor 1254 at 100 ppm levels (86). Keplinger et al (87) employing a dominant lethal assay, reported no evidence of mutagenic effects of Aroclors. Green et al (88)(89) reported a lack of mutagenic activity as measured by dominant lethal test for male Osborne-Mendel rats subjected to 4 different regimens of Aroclor 1242 and 1254. While the above studies of Green et al (88)(89) were negative in regard to chromosomal mutations, they do not entirely rule out the possibility that PCBs may induce point mutations. However, to date there are no known reports in the literature concerning the induction of point mutations by PCBs in laboratory model systems.

A recent comparison of the mutagenic activity of Aroclor 1254 (average chlorine content, 4.96 Cl/molecule), 2,2',5,5'-tetrachlorobiphenyl (4 Cl/molecule), Aroclor 1268 (average chlorine content, 9.7 Cl/molecule), Aroclor 1221 (average chlorine content, 1.15 Cl/molecule) and 4-chlorobiphenyl showed that as the degrees of chlorination decreased, the mutagenicity to *Salmonella typhimurium* TA 1538 strain [in the presence of liver homogenate (90)] increased (42). This strain is sensitive to frameshift mutagens (90). The influence of the degree of chlorination on the mutagenicity to the mutant strain TA 1538 also complements the observations that as the chlorine content of the PCB substrate increases, the metabolic rate decreases (31)(91).

It is also important to note the recent report that the in vitro metabolism of 4-chlorobiphenyl proceeds via an arene oxide intermediate and is accompanied by binding to the endogenous microsomal RNA and protein (42). Preliminary results also suggest binding of PCB to exogenous DNA (42) which confirms an earlier report of Allen and Norback (92). Covalent binding of the 2,5,2',5'-tetrachlorobiphenyl metabolites (e.g., trans dihydrodihydroxy) to cellular macromolecules was suggested by Allen and Norback (93) to be a possible pathway for the carcinogenic action of the PCBs.

Teratogenic studies appear to be thus far nondefinitive (7). However, while the PCBs have not exhibited known or clearly defined teratogenic effects in mammals, their easy passage across the placenta suggests the potential for some form of fetal toxicity (7)(72)(94). Placental transport of PCBs has been reported for the rabbit, rat (95), mouse and cow as well as observed among "Yusho" patients (7)(96)(97).

No account of the toxicity of the polychlorinated biphenyls can be complete without stressing the possible role of trace contaminants (7)(11)-(13), e.g., the chlorinated dibenzofurans. For example, embryotoxicity of the PCBs Clophen A069 and Phenoclor DP-6 has been attributed to chlorinated dibenzofurans present as trace contaminants in the commercial preparations (11)(98). Subsequently, tetra-, penta-, and hexachlorodibenzofurans were detected in a number of American preparations of PCBs (e.g., Aroclor 1248, 1254, 1260), concentrations of the individual chlorodibenzofurans were in the order of 0.1 μg/kg of the PCB. Chlorinated dibenzofurans have been considered as possible causes of embryonic mortality and birth-defects observed in PCB-feeding experiments in birds (99)(100). The chlorinated dibenzofurans are structurally related to the chlorinated dibenzo-p-dioxins some of which are both highly toxic and teratogenic (101).

A number of possibilities exist to account for the presence of chlorodibenzo-furans in commercial PCB mixtures. One explanation considers the presence of the parent compound (dibenzofuran) in the technical grade biphenyl sub-jected to the chlorination process. It is also conceivable that chlorinated di-benzofurans may be produced from PCBs in the environment. Two possible mechanisms for such a transformation exist, both of which involve hydroxy derivatives.

As cited earlier, hydroxylation is a route of metabolism of the PCBs. Polar oxygenated compounds have also been found as photolytic products of the PCBs. It should be stressed that the transformation of only 0.002% of a major constituent of an Aroclor mixture to the corresponding chlorinated dibenzo-furans would produce concentrations in the mixture corresponding to the values reported by Vos et al as toxicologically significant (102)(103).

To date, there have been no published findings of chlorinated dibenzofurans in aquatic samples or in foods. The extremely low levels of these trace con-taminants in the original organic chemicals and/or complex mixtures (e.g., PCBs, chlorinated phenols) would stress the requirement for analytical pro-cedures permitting the sampling, concentration and detection in the parts-per-billion to parts-per-trillion range.

Medical Surveillance: Placement and periodic examinations should include an evaluation of the skin, lung, and liver function. Possible effects on the fetus should be considered.

Special Tests: None in common use.

Personal Protective Methods: Protection of exposed skin should be encour-aged, since the above standards may not be low enough to prevent chloracne. Barrier creams, protective clothing, and good personal hygiene are good pro-tective measures. Respirators should be used in areas of vapor concentration.

References

(1) Panel on hazardous substances, polychlorinated biphenyls—environmental im-pact. *Environ. Res.* (1972) 249.
(2) Hammond, A.L. Chemical Pollution: Polychlorinated Biphenyls. *Science,* 175 (1972) 155.
(3) National Swedish Environment Protection Board, PCB Conference, Wenner-Gren Center, Stockholm, Sweden, Sept. 29 (1970).
(4) Peakall, D., and Lincer, J. Polychlorinated biphenyls: Another long-life wide-spread chemical in the environment. *Bioscience,* 20 (1970) 958.
(5) Ahmed, A.H. PCBs in the Environment. *Environment* 18 (1976) 6.
(6) Nisbet, I.C.T., and Sarofim, A.F. Rates and Routes of Transport of PCBs in the En-vironment. *Env. Hlth. Persp.,* 1 (1972) 21.
(7) Dept. HEW. Final report of the subcommittee on the health effects of polychlo-rinated biphenyls and polybrominated biphenyls. Washington, D.C. July (1976).

(8) Hubbard, H.L. Chlorinated biphenyl and related compounds. In Kirk-Othmer's *Encyclopedia of Chemical Technology*, 2nd ed. 5 (1964) 289.

(9) Hutzinger, O., Safe, S., and Zitko, V. *The Chemistry of PCBs*. CRC Press, Cleveland, Ohio (1974).

(10) Zitko, V., Hutzinger, O., and Safe, S. Retention times and electron-capture responses of some individual chlorobiphenyls. *Bull. Env. Contam. Toxicol.,* 6 (1971) 160.

(11) Vos, J.G., Koeman, J.J., Vandermaas, H.L., TenNoever de Brauw, M.C., and DeVos. Identification and toxicological evaluation of chlorinated dibenzofuran and chlorinated naphthalene in two commercial polychlorinated biphenyls. *Food Cosmet. Toxicol.,* 8 (1970) 625.

(12) Bowes, G.W. Identification of chlorinated dibenzofurans in American polychlorinated biphenyls. *Nature,* 256 (1975) 305.

(13) Roach, J.A.G., and Pomerantz, I.H. The findings of chlorinated dibenzofurans in Aroclor PCBs of recent manufacture. Paper No. 53, presented at 88th Annual Meeting of Assn. of Official Analytical Chemists, Washington, D.C. October 14-17 (1974).

(15) Risebrough, R.W., and deLappe, R.W. Polychlorinated biphenyls in ecosystems. *Env. Hlth. Persp.,* 1 (1972) 39.

(16) Gustafson, C.G. PCBs—Prevalent and persistent. *Env. Sci. Technol.,* 4 (1970) 814.

(17) Ruopp, D.J., and deCarlo, V.J. Environmental levels of PCBs. Int. Conf. on Environ. Sensing and Assessment, Las Vegas, Nev., Sept. 14-19 (1975).

(18) Maugh, T.H. Chemical pollutants: Polychlorinated biphenyls still a threat. *Science,* 190 (1975) 1189.

(19) Kolbye, A.C. Food exposures to polychlorinated biphenyls. *Environ. Hlth. Persp.,* 1 (1972) 85-88.

(20) Anon. The Rising Clamor About PCBs. *Env. Sci. Technol.,* 10 (1976) 122.

(21) Fishbein, L. Toxicity of Chlorinated Biphenyls. *Ann. Rev. Pharmacol.,* 14 (1974) 139.

(22) Kimbrough, R.D. The Toxicity of Polychlorinated Polycyclic Compounds and Related Compounds. *Crit. Revs. Toxicol.,* 2 (1974) 445.

(23) Allen, J.R. Response of the non-human primate to polychlorinated biphenyl exposure, *Fed. Proc.,* 34 (1975) 1675.

(24) Grant, D.L., Phillips, W.E.J., and Villeneuve, D. Metabolism of polychlorinated biphenyl (Aroclor 1254) mixture in the rat. *Bull. Env. Contam. Toxicol.,* 6 (1971) 102.

(25) Bush, B., Tumasonis, C.F., and Baker, F.D. Toxicity and persistence of PCB homologs and isomers in the avian system. *Arch. Env. Con. Tox.,* 2 (1974) 195.

(26) Sissons, D., and Welti, D. Structural identification of polychlorinated biphenyls in commercial mixtures by gas-liquid chromatography, nuclear magnetic resonance and mass spectrometry. *J. Chromatog.,* 60 (1971) 15.

(27) Webb, R.G., and McCall, A.C. Identities of polychlorinated biphenyl isomers in Aroclors. *J. Assoc. Off. Anal. Chem.,* 55 (1972) 746.

(28) Ecobichon, D.J., and Comeau, A.M. Comparative effects of commercial Aroclors on rat liver enzyme activities. *Chem. Biol. Interactions,* 9 (1974) 314.

(29) Ecobichon, D.J., and Comeau, A.M. Isomerically pure chlorobiphenyl congeners and hepatic function in the rat: Influence of position and degree of chlorination. *Toxicol. Appl. Pharmacol.,* 33 (1975) 94.

(30) Matthews, H.B., and Anderson, M.W. The distribution and excretion of 2,4,5,2',5'-pentachlorobiphenyl in the rat. *Drug Metab. Dispos.,* 3 (1975) 211-219.

(31) Matthews, H.B., and Anderson, M.W. Effect of chlorination on the distribution and excretion of polychlorinated biphenyls. *Drug Metab. Dispos.,* 3 (1975) 371-380.

(32) Braunberg, R.C., Dailey, R.E., Brouwer, E.A., Kasza, L., and Blaschka, A.M. Acute, subacute, and residual effects of polychlorinated biphenyl (PCB) in rats. Biological half-life in adipose tissue. *J. Toxicol. Env. Hlth.,* 1 (1976) 683-688.

(33) Mehendale, H.M. Uptake and disposition of chlorinated biphenyls by isolated perfused rat liver. *Drug Metab. Dispos.,* 4 (1976) 124-132.

(34) Peterson, R.E., Seymour, J.L., and Allen, J.R. Distribution and biliary excretion of polychlorinated biphenyls in rats. *Tox. Appl. Pharmacol.*, 38 (1976) 609-619.

(35) Biros, F.J., Walker, A.C., and Medbery, A. Polychlorinated biphenyls in human adipose tissue. *Bull. Env. Contam. Toxicol.*, 5 (1970) 317-323.

(36) Tulp, M.T.M., Sundström, G., and Hutzinger, O. The metabolism of 4,4'-dichloro-biphenyl in rats and frogs. *Chemosphere*, 6 (1976) 425-432.

(37) Hsu, I.C., Van Miller, J.P., Seymour, J.L., and Allen, J.R. Urinary metabolites of 2,5,2',5'-tetrachlorobiphenyl in the nonhuman primate. *Proc. Soc. Exp. Biol. Med.*, 150 (1975) 185-188.

(38) Jensen, S., and Sundström, G. Metabolic hydroxylation of a chlorobiphenyl containing only isolated unsubstituted positions—2,2',4,4',5,5'-hexachlorobiphenyl. *Nature*, 251 (1974) 219-220.

(39) Hutzinger, O., Jamieson, W.D., Safe, S., Paulmann, L., and Ammon, R. Identification of metabolic dechlorination of highly chlorinated biphenyl in rabbit. *Nature*, 252 (1974) 698-699.

(40) Sundström, G., Hutzinger, O., and Safe, S. The metabolism of 2,2'4,4',5,5'-hexachlorobiphenyl by rabbits, rats and mice. *Chemosphere*, 4 (1976) 249-253.

(41) Gardner, A.M., Chen, J.T., Roach, J.A.G., and Ragelis, E.P. *Biochem. Biophys. Res. Comm.*, 55 (1973) 1377.

(42) Wyndham, C., Devenish, J., and Safe, S. The in vitro metabolism, macromolecular binding and bacterial mutagenicity of 4-chlorobiphenyl, a model PCB substrate. *Res. Communs. Chem. Pathol. Pharmacol.*, 15 (1976) 563-570.

(43) Jensen, S., and Jansson, B. Methyl sulfone metabolites of PCB and DDE. *Ambio*, 5-6 (1976) 257-260.

(44) Vos, J.G., and Notenboom-Ram, E. Comparative toxicity study of 2,4,5,2',4',5'-hexachlorobiphenyl and a polychlorinated biphenyl mixture in rabbits. *Toxicol. Appl. Pharmacol.*, 23 (1972) 563.

(45) Figita, S., Tsuji, Kato, K., Saeki, S., and Tsukamoto, H. Effects of biphenyl chlorides on rat liver microsomes. *Fukuoka Acta. Med.*, 62 (1971) 30.

(46) Ecobichon, D.J., and MacKenzie, D.O. The Uterotrophic Activity of Commercial and Isomerically Pure Chlorobiphenyls in the Rat. *Res. Commun. Chem. Pathol. Pharmacol.*, 9 (1974) 85.

(47) Chen, P.R., Mehendale, H.M., and Fishbein, L. Effect of two isomeric tetrachloro-biphenyls on rat and their hepatic microsomes. *Arch. Env. Contam. Tox.*, 1 (1973) 36.

(48) Johnstone, G.J., Ecobichon, D.J., and Hutzinger, O. The influence of pure poly-chlorinated biphenyl compounds on hepatic function in the rat. *Toxicol. Appl. Pharmacol.*, 28 (1974) 66.

(49) Jao, L.T., Hass, J.R., and Mathews, H.B. In vivo metabolism of radioactive 4-chloro- and 4,4'-dichlorobiphenyl in rats. *Toxicol. Appl. Pharmacol.*, 37 (1976) 147.

(50) Mathews, H.B., and Tuey, D.B. The distribution and excretion of 3,5,3',5'-tetrachlo-robiphenyl in the male rat. *Toxicol. Appl. Pharmacol.*, 37 (1976) 148.

(51) Seymour, J.L., Peterson, R.E., and Allen, J.R. Tissue distribution and biliary excretion of 2,5,2',5'-tetrachlorobiphenyl (TCB) and 2,4,5,2',4',5'-hexachlorobiphenyl (HCB) in male and female rats. *Toxicol. Appl. Pharmacol.*, 37 (1976) 171.

(52) Kimbrough, R.D., Linder, R.E., and Gaines, T.B. Morphological changes in livers of rats fed polychlorinated biphenyls: Light microscopy and ultrastructure. *Arch. Env. Hlth.*, 25 (1972) 354.

(53) Nagasaki, H., Tomii, S., Mega, T., Marugami, M., and Ito, N. Hepatocarcinogenicity of polychlorinated biphenyls in mice. *Gann*, 63 (1972) 805.

(54) Nishizumi, M. Light and electron microscope study of chlorobiphenyl poisoning in mouse and monkey liver. *Arch. Env. Hlth.*, 21 (1970) 620.

(55) Platonow, N.S., and Chen, N.Y. Transplacental transfer of polychlorinated biphenyls (Aroclor 1254) in cow. *Vet. Rec.,* 92 (1973) 69.

(56) Kasza, L., Weinberger, M.A., Carter, C., Hinton, D.E., Trump, B.F., and Brouwer, E.A. Acute, subacute and residual effects of polychlorinated biphenyl (PBC) in rats. II. Pathology and electron microscopy of liver and serum enzyme study. *J. Toxicol. Env. Hlth.,* 1 (1976) 689.

(57) Allen, J.R., Carstens, L.A., and Abrahamson, L.J. Responses of rats exposed to polychlorinated biphenyls for 52 weeks, 1. Comparison of tissue levels of PCB and biological changes. *Arch. Env. Contam.,* 4 (1976) 404-419.

(58) Grant, D.L., Moodie, C.A., and Phillips, W.E.J. Toxicodynamics of Aroclor 1254 in the male rat. *Env. Physiol. Biochem.,* 4 (1974) 214.

(59) Fuhremann, T.W., and Lichtenstein, E.P. Increase in the toxicity of organophosphorus insecticides to house flies due to polychlorinated biphenyl compounds. *Toxicol. Appl. Pharmacol.,* 22 (1972) 628.

(60) Villeneuve, D.C., Grant, D.L., and Phillips, W.E.J. Modification of pentobarbital sleeping times in rats following chronic PCB ingestion. *Bull. Env. Contam. Toxicol.,* 7 (1972) 264.

(61) Anon. Worrisome PCB levels found in mothers' milk in 10 states. *Toxic Mat. News,* 3 (1976) 141.

(62) Anon. PCB contamination of mothers' milk spurs inquiry. *Env. Hlth. Letter,* 15 (1976) 4.

(63) DHEW/EPA. Meeting to review and evaluate EPA's sampling of PCBs in the fat of mothers' milk, DHEW committee to coordinate toxicology and related programs in cooperation with the Environmental Protection Agency, Bethesda, MD, Sept. 23, 1976.

(64) Anon. The rising clamor about PCBs. *Env. Sci. Technol.,* 10 (1976) 122-123.

(65) Neeley, W.B. A material balance study of polychlorinated biphenyls in Lake Michigan. *Sci. Total Environ.,* 7 (1977) 117-129.

(66) Plimmer, J.R., and Kligebiel, U.L. PCB formation. *Science* 181 (1973) 994-995.

(67) Moilanen, K.W., and Crosby. Amer. Chem. Soc. Meeting, 165th, Dallas, TX. April (1973).

(68) Uyeta, M., Taue, S., Chikasawa, K., and Mazaki, M. Photoformation of polychlorinated biphenyls from chlorinated benzenes. *Nature,* 264 (1976) 583-584.

(68a) Gaffney, P.E. Chlorobiphenyls and PCBs: Formation during chlorination. *J. Water Pollut. Control,* 49 (1977) 401-404.

(69) Saeki, S., Tsutsui, A., Oguri, K., Yoshimura, H., and Hamona, M. Isolation and structure elucidation of the amine component of KC-400 (chlorobiphenyls). *Fukuoka Acta. Med.,* 62 (1971) 20.

(70) Kuratsune, M., et al, Morikawa, Y., Hirohata, T., Nishizumi, M., Kochi, S. An epidemiologic study on "Yusho" or chlorobiphenyl poisoning. *Fukuka Acta Med.,* 60 (1969) 513-532.

(71) Kuratsune, M., Yoshimura, T., Matsuzaka, J., et al. Yusho, a poisoning caused by rice oil contaminated with chlorobiphenyls. *Env. Hlth. Persp.,* 1 (1972) 119.

(72) Umeda, G. Kanemi yusho: PCB poisoning in Japan. *Ambio* (1972) 132-134.

(73) Hirayma, T. Meeting on Origins of Human Cancer, Cold Spring Harbor, NY. Sept. 7-14 (1976).

(74) Bahn, A.K., Rosenwaike, I., Herrmann, N., et al. Melanoma after exposure to PCBs. *New Engl. J. Med.,* 295 (1976) 450.

(75) Anon. Mobil study suggests possible link between PCBs and human cancer. *Occup. Hlth. Safety Letter,* 6 (1976) 4.

(76) Anon. High cancer rates among workers exposed to PCBs. *Chem. Eng. News.* 54 (1976) 26.

(77) Anon. Monsanto finds no link between cancer and PCBs in workers. *Occup. Hlth. Safety Letter,* 5 (1976) pp. 5-6.

(78) Nagasaki, H., Tomii, S., and Mega, T. Hepatocarcinogenicity of polychlorinated biphenyls in mice. *Gann,* 63 (1972) 805.

(79) Nagasaki, H., Tomii, S., Mega, T., Marugami, M., and Ito, N. Development of hepatomas in mice treated with benzene hexachloride. *Gann,* 64 (1971) 431.

(80) Kimura, N.T., Baba, T. Neoplastic changes in the rat liver induced by polychlorinated biphenyl. *Gann,* 64 (1973) 105.

(81) Ito, N., Nagasaki, H., Arai, M., Makiura, S., Sugihara, S., and Hirao, K. Histopathologic studies on liver tumorigenesis induced by mice by technical polychlorinated biphenyls and its promoting effect on liver tumors induced by benzene hexachloride. *J. Natl. Cancer Inst.,* 51 (1973) 1637.

(82) Kimbrough, R.D., Squire, R.A., Linder, R.E., Strandberg, J.D., Montali, R.J., and Burse, V.W. Induction of liver tumors in Sherman female rats by polychlorinated biphenyl Aroclor 1260. *J. Natl. Cancer Inst.,* 55 (1975) 1453-1459.

(83) Kimura, N.T., Kanematsu, I., and Baba, T. Polychlorinated biphenyl(s) as a promotor in experimental hepatocarcinogenesis in rats. *Z. Krebsforsch. Klin. Onkol.,* 87 (1976) 257-266.

(84) Ramel, C. Mutagenicity research and testing in Sweden. *Mutation Res.,* 33 (1975) 79-86.

(85) Nilsson, B., and Ramel, C. Genetic Tests on *Drosophila Melanogaster* with Polychlorinated Biphenyls (PCB). *Hereditas,* 77 (1974) 319.

(86) Hoopingarner, R., Samule, A., and Krause, D. Polychlorinated biphenyl interactions with tissue culture cells. *Env. Hlth. Persp.,* 1 (1972) 155.

(87) Keplinger, M.L., et al. Toxicological studies with polychlorinated biphenyls. PCB Conference, Quail. Roost. Conf. Center, Rougemont, NC, Dec. (1971).

(88) Green, S., Sauro, F.M., and Friemann, L. Lack of dominant lethality in rats treated with polychlorinated biphenyls. *Food Cosmet. Toxicol.,* 13 (1975) 507.

(89) Green, S., Carr, J.V., Palmer, K.A., and Oswald, E.J. Lack of cytogenetic effects in bone marrow spermatogonial cells in rats treated with polychlorinated biphenyls. *Bull. Env. Contam. Toxicol.,* 13 (1975) 14.

(90) Ames, B.N., Darston, W.E., Yamasaki, E., and Lee, F.D. Carcinogens are mutagens: A simple test for combining liver homogenates for activation and bacteria for detection. *Proc. Natl. Acad. Sci.,* 70 (1973) 2281-2285.

(91) Ghiasuddin, S.M., Nenzer, R.E., and Nelson, J.O. Metabolism of 2,5,2'-trichloro-, 2,5,2',5'-tetrachloro- and 2,4,5,2',5'-pentachlorobiphenyl in rat hepatic microsomal systems, *Toxicol. Appl. Pharmacol.,* 36 (1976) 187-194.

(92) Allan, J.R. and Norback, D.H. Pathobiological responses of primates to polychlorinated biphenyl exposure. In *National Conference on Polychlorinated Biphenyls, Proceedings,* 43-49 (1975).

(93) Allan, J.R., and Norback, D.H. Carcinogenic potential of the polychlorinated biphenyls. Origins of Human Cancer Meeting, Cold Spring Harbor, NY, Sept. 7-14. (1976) Abstract p. 75.

(94) Grant, D.L., Villeneuve, D.C., McCully, K.A., and Phillips, W.E.J. Placental transfer of polychlorinated biphenyls in the rabbit. *Env. Physiol.,* 1 (1971) 61.

(95) Takagi, Y., Otake, T., Kataoka, M., Murata, Y., Aburada, S., Akasaka, S., Hashimoto, K., Uda, H., and Kitaura, T. Studies on the transfer and distribution of ^{14}C-polychlorinated biphenyls from maternal to fetal and suckling rats. *Toxicol. Appl. Pharmacol.,* 38 (1976) 549-558.

(96) Kojima, T., Fukumoto, H., Makisumi, S. Chloro-biphenyl poisonings, gas-chromatographic detection of chloro-biphenyls in the rice oil and biological materials. *Jap. J. Legal Med.,* 23 (1969) 415.

(97) Inagami, K., Koga, T., Tomita, Y. Poisoning of chlorobiphenyls. *Shokuhin. Eisei Gaku Zasshi,* 10 (1969) 415.

(98) Vos, J.G., and Koeman, J.H. Comparative toxicologic study with polychlorinated biphenyls in chickens with special reference to porphyria, edema formation, liver necrosis, and tissue residues. *Toxicol. Appl. Pharm.,* 17 (1970) 565.

(99) Tumasonis, C.F., Bush, B., and Baker, F.D. PCB levels in egg yolks associated with embryonic mortality and deformity of hatched chicks. *Arch. Env. Contam. Toxicol.,* 1 (1973) 312.

(100) Gilbertson, M., and Hale, R. Early embryonic mortality in a herring gull colony in Lake Ontario. *Canad. Field-Naturalist,* 88 (1974) 354.

(101) Schwetz, B.A., Norris, J.M., Sparschu, G.L., Rowe, V.K., Gehring, P.J., Emerson, J.L., and Gerbig, C.G. Toxicology of chlorinated dibenzo-p-dioxins. *Env. Hlth. Persp.,* 5 (1973) 87.

(102) Vos, J.G., Koeman, J.H., Vandermaas, H.S., Debraun, M.C., and DeVos, H. Identification and toxicological evaluation of chlorinated dibenzofuran and chlorinated naphthalene in two commercial polychlorinated biphenyls. *Food Cosmet. Toxicol.,* 8 (1970) 625-633.

(103) Vos, J.G., and Koeman, J.H. Dermal toxicity studies of technical polychlorinated biphenyls and fractions thereof in rabbits. *Toxicol. Appl. Pharmacol.,* 19 (1971) 617.

Bibliography

Meigs, J.W., Albom, J.J., and Kartin, B.L. 1954. Chloracne from an unusual exposure to Arochlor. *J. Am. Med. Assoc.* 154:1417.

National Institute for Occupational Safety and Health. *Criteria for a Recommended Standard: Occupational Exposure to Polychlorinated Biphenyls,* NIOSH Doc. No. 77-225 (1977).

Peakall, D.B. 1972. Polychlorinated diphenyls: occurrence and biological effects. *Residue Rev.* 44:1.

CHLOROFORM

Description: $CHCl_3$, chloroform, is a clear, colorless liquid with a characteristic odor. Though nonflammable, chloroform decomposes to form hydrochloric acid, phosgene, and chlorine upon contact with a flame.

Synonyms: Trichloromethane, methenyl chloride.

Potential Occupational Exposures: Chloroform was one of the earliest general anesthetics, but its use for this purpose has been abandoned because of toxic effects. Chloroform is widely used as a solvent (especially in the lacquer industry); in the extraction and purification of penicillin and other pharmaceuticals; in the manufacture of artificial silk, plastics, floor polishes, and fluorocarbons; and in sterilization of catgut.

A partial list of occupations in which exposure may occur includes:

Chemists	Polish makers
Drug makers	Silk synthesizers
Fluorocarbon makers	Solvent workers
Lacquer workers	

NIOSH estimates that 40,000 people in the United States may be exposed to chloroform in their working environment (1).

Chloroform is widely distributed in the atmosphere (2)(3) and water (4)(5) (including municipal drinking water primarily as a consequence of chlorination) (4)(5). A survey of 80 American cities by EPA found chloroform in every water system in levels ranging from <0.3 to 311 ppb (4).

Permissible Exposure Limits: The Federal standard is 50 ppm (240 mg/m^3). The ACGIH recommended 1976 TLV was 25 ppm. NIOSH's recommended limit is a ceiling of 2 ppm based on a 1-hour sample collected at 750 l/min. ACGIH (1978) set 10 ppm as a TWA with the notation that chloroform was an "Industrial Substance Suspect of Carcinogenic Potential for Man."

Route of Entry: Inhalation of vapors.

Harmful Effects

Local—Chloroform may produce burns if left in contact with the skin.

Systemic—Chloroform is a relatively potent anesthetic at high concentrations. Death from its use as an anesthetic has resulted from liver damage and from cardiac arrest. Exposure may cause lassitude, digestive disturbance, dizziness, mental dullness, and coma. Chronic overexposure has been shown to cause enlargement of the liver and kidney damage. Alcoholics seem to be affected sooner and more severely from chloroform exposure. Disturbance of the liver is more characteristic of exposure than central nervous system depression or renal injury. There is some animal experimental evidence that suggests chloroform may be a carcinogen.

Chloroform is carcinogenic in Osborne-Mendel rats and G6C3F1 mice following long-term oral intubation at maximum tolerated and half maximum tolerated doses (6)(7). In rats, malignant and benign primary kidney tumors were found while chloroform treated mice showed significant incidences of hepatocellular carcinomas (6)(7). Chloroform and other halogenated hydrocarbons produce pathological effects by localizing in target tissues and binding covalently to cellular macromolecules (8)-(10).

Information as to the mutagenicity of chloroform is scant. Chloroform (as well as carbon tetrachloride) gave negative results when tested with microsomal incubates with *S. typhimurium* TA 1535 and *E. coli* K-12 for base pair substitution and *S. typhimurium* TA 1538 for frame-shift mutations (11).

Medical Surveillance: Preplacement and periodic examinations should include appropriate tests for liver and kidney functions, and special attention should be given to the nervous system, the skin, and to any history of alcoholism.

Special Tests: Expired air and blood levels may be useful in estimating levels of acute exposure.

Personal Protective Methods: Protective clothing and gloves should be worn to protect the skin, and masks should be worn in areas of high vapor concentration.

References

(1) Anon. Chloroform causes cancer in laboratory animals. NCI Study Report. *Toxic Materials News,* 3 (1976) 36.
(2) Yung, Y.L., McElroy, M.B., and Wofsy, S.C. Atmospheric halocarbons: A discussion with emphasis on chloroform. *Geophys. Res. Letters,* 2 (1975) 397-399.
(3) McConnell, G., Ferguson, O.M., and Pearson, C.R. Chlorinated hydrocarbons in the environment. *Endeavor,* 34 (1975) 13-27.
(4) Environmental Protection Agency. Draft Report for Congress: Preliminary Assessment of Suspected Carcinogens in Drinking Water, Office of Toxic Substances, Washington, D.C., October 17 (1975).
(5) Bellar, T.A., Lichtenberg, J.J., and Kroner, R.C. The occurrence of organohalides in chlorinated drinking waters. *J. Am. Waterworks Assoc.* 66 (1974) 703-706.
(6) Powers, M.B., and Welker, R.W. Evaluation of the oncogenic potential of chloroform long term oral administration in rodents. 15th Annual Meeting Society of Toxicology, Atlanta, GA, March 14-18 (1976).
(7) National Cancer Institute. Report on the carcinogenesis bioassay of chloroform, National Cancer Institute, Bethesda, MD, March (1976).
(8) Ilett, K.F., Reid, W.D., Sipes, I.G., and Krishna, G. Chloroform toxicity in mice: Correlations of renal and hepatic necrosis with covalent bonding of metabolites to tissue macromolecules. *Exp. Mol. Pathol.,* 19 (1973) 215.
(9) Reid, W.D., and Krishna, G. Centrolobular hepatic necrosis related to covalent binding of metabolites of halogenated aromatic hydrocarbons. *Exp. Mol. Pathol.,* 18 (1973) 80-95.
(10) Brodie, B.B., Reid, W.D., Cho, A.K., Sipes, G., Krishna, G., and Gillette, J.R. Possible mechanism of liver necrosis caused by aromatic organic compounds. *Proc. Natl. Acad. Sci.,* 68 (1971) 160.
(11) Uehceke, H., Greim, H., Kramer, M., and Werner, T. Covalent binding of haloalkanes to liver constituents, but absence of mutagenicity on bacteria in a metabolizing test system. *Mutation Res.,* 38 (1976) 114.

Bibliography

Challen, P.J.R., Hickish, D.E., and Bedford, J.R. 1958. Chronic chloroform intoxication. *Br. J. Ind. Med.* 15:243.
National Institute for Occupational Safety and Health. *Criteria for a Recommended Standard: Occupational Exposure to Chloroform.* NIOSH Doc. No. 75-114 (1975).

CHLOROMETHYL METHYL ETHER

Description: $ClCH_2OCH_3$, chloromethyl methyl ether, is a volatile, corrosive liquid. Commercial chloromethyl methyl ether contains from 1 to 7% bis(chloromethyl) ether, a known carcinogen.

Synonyms: CMME, methyl chloromethyl ether, monochloromethyl ether, chloromethoxymethane.

Potential Occupational Exposures: Chloromethyl methyl ether is a highly reactive methylating agent and is used in the chemical industry for synthesis of organic chemicals. Most industrial operations are carried out in closed process vessels so that exposure is minimized.

A partial list of occupations in which exposure may occur includes:

Organic chemical synthesizers

Permissible Exposure Limits: Chloromethyl methyl ether is included in the Federal standard for carcinogens; all contact with it should be avoided.

Route of Entry: Inhalation of vapor and possibly percutaneous absorption.

Harmful Effects

Local—Vapor exposure results in severe irritation of the skin, eyes, and nose. Rabbit skin tests using undiluted material resulted in skin necrosis.

Systemic—Chloromethyl methyl ether is only moderately toxic given orally. Acute exposure to chloromethyl methyl ether vapor may result in pulmonary edema and pneumonia.

Several studies of workers with CMME manufacturing exposure have shown an excess of bronchiogenic cancer predominately of the small cell-undifferentiated type with relatively short latency period (typically 10 to 15 years). Therefore, commercial grade chloromethyl methyl ether must be considered a carcinogen. At present it is not known whether or not chloromethyl methyl ether's carcinogenic activity is due to bis(chloromethyl) ether (BCME) contamination, but this may be a moot question inasmuch as two of the hydrolysis products of CMME can combine to form BCME. Animal experiments to determine chloromethyl methyl ether's ability to produce skin cancer indicated marginal carcinogenic activity; highly pure CMME was used. Inhalation studies, using technical grade CMME showed only one bronchiogenic cancer and one esthesioneuroepithelioma out of 79 animals exposed.

Medical Surveillance: Preplacement and periodic medical examinations should include an examination of the skin and respiratory tract, including a chest x-ray. Sputum cytology has been suggested as helpful in detecting early malignant changes. and in this connection a detailed smoking history is of importance. Possible effects on the fetus should be considered.

Special Tests: None have been suggested.

Personal Protective Methods: These are designed to supplement engineering controls and to prevent all skin or respiratory contact. Full body protective clothing and gloves should be used on entering areas of partial exposure. Those employed in handling operations should be provided with fullface, supplied air respirators of continuous flow or pressure demand type. On exit from a regulated area, employees should be required to remove and leave protective clothing and equipment at the point of exit, to be placed in impervious containers at the end of the work-shift for decontamination or disposal. Showers should be taken prior to dressing in street clothes.

Bibliography

DeFonso, L.R., and Kelton, S.C., Jr. 1976. Lung cancer following exposure to chloromethyl methyl ether. *Arch. Environ. Health* 31:125.
Figueroa, W.G., Raszowski, R., and Weiss, W. 1973. Lung cancer in chloromethyl methyl ether workers. *N. Eng. J. Med.* 288:1096.
Gargus, J.L., Reese, W.H., and Rutter, H.A. 1969. Induction of lung adenomas in newborn mice by bis(chloromethyl) ether. *Toxicol. Appl. Pharmacol.* 15:92.
Laskin, S., Drew, R.T., Cappiello, V., Kuschner, M., and Nelson, N. 1975. Inhalation carcinogenicity of alpha halo ethers: II. Chronic inhalation studies with chloromethyl methyl ether. *Arch. Environ. Health* 30:70.
Nelson, N. 1976. The chloroethers—occupational carcinogens: a summary of laboratory and epidemiology studies. *Ann. NY Acad. Sci.* 271:81.
Van Duuren, B.L., Sivak, A., Goldschmidt, B.N., Katz, C., and Melchionne, S. 1969. Carcinogenicity of halo-ethers. *J. Nat. Cancer Inst.* 43:481.
Weiss, W., Boucot, K.R. 1975. The respiratory effects of chloromethyl methyl ether. *J. Am. Med. Assoc.* 234:1139.

Note: See also the entry under Bis(Chloromethyl) Ether.

CHLOROPRENE

Description: $H_2C=CCl-CH=CH_2$, chloroprene, is a colorless, flammable liquid possessing a pungent odor.

Synonyms: 2-chloro-1,3-butadiene.

Potential Occupational Exposures: The only major use of chloroprene is in the production of artificial rubber (neoprene, duprene) (1)(2)(4).

A partial list of occupations in which exposure may occur includes:

Neoprene makers
Rubber makers

Chloroprene is extremely reactive, e.g., it can polymerize spontaneously at room temperatures, the process being catalyzed by light, peroxides and other free radical initiators. It can also react with oxygen to form polymeric peroxides and because of its instability, flammability and toxicity, chloroprene has no end product uses as such. An estimated 2,500 workers are exposed to chloroprene in the United States.

Permissible Exposure Limits: The Federal standard is 25 ppm (90 mg/m^3).

NIOSH in 1977 specified that the employer shall control exposure to chloroprene so that no employee is ever exposed at a concentration greater than 3.6 milligrams per cubic meter (mg/m^3) of air (1 ppm), determined as a ceiling concentration for any 15-minute sampling period during a 40-hour workweek. ACGIH (1978) has proposed TWA values of 10 ppm (45 mg/m^3).

Routes of Entry: Inhalation of vapor and skin absorption.

Harmful Effects

Local—Chloroprene acts as a primary irritant on contact with skin, conjunctiva, and mucous membranes and may result in dermatitis, conjunctivitis, and circumscribed necrosis of the cornea. Temporary hair loss has been reported during the manufacture of polymers.

Systemic—Inhalation of high concentrations may result in anesthesia and respiratory paralysis. Chronic exposure may produce damage to the lungs, nervous system, liver, kidneys, spleen, and myocardium.

Chloroprene which has been used since 1930 in the manufacture of synthetic rubber has recently been suggested to be responsible for an increased incidence of skin and lung cancer in exposed workers in the USSR (5)(6). During the period of 1956-70, epidemiological studies of industrial workers in the Yerevan region revealed 137 cases of skin cancer among 24,989 persons over 25 years of age. Three percent of the workers exposed to chloroprene and 1.6% of those working in industries using chloroprene derivatives developed skin cancer, compared to only 0.4% for persons working in nonchemical industries. The chloroprene workers who developed skin cancer had an average age of 59.6 years and an average duration of employment of 9.5 years (5).

During the same period, 87 cases of lung cancer were identified among 19,979 workers in the same region. The group exposed to chloroprene or its derivatives had the highest incidence of lung cancer (1.16%). These workers' average age was 44.5 years with an average duration of employment of 8.7 years.

Of the 34 cases of lung cancer in this group, 18 were among persons having a direct and prolonged exposure to chloroprene monomer, the remaining 16 involved individuals exposed to chloroprene latexes (6). High rates of skin and lung cancer have also recently been reported among workers in a United States plant who had handled chloroprene and its derivatives (7).

Chromosome aberrations in lymphocytes from peripheral blood of workers exposed to chloroprene has been reported (8). A significant rise in the number of chromosome aberrations in blood cultures of those exposed to an average chloroprene concentration of 18 ppm for 2 to more than 10 years was noted (8).

Exposure of S. typhimurium TA 100 strain to 0.5% to 8% of chloroprene vapor in air in the absence of any metabolic activation system caused a linear increasing mutagenic response, (reaching 3 times the spontaneous mutation rate at a concentration of 8%) (9). Exposure to a higher concentration (20%) caused a strong toxicity in the bacteria. This mutagenic and/or toxic effect could be caused by a direct action of chloroprene or more likely, by one of its enzymic (bacteria) or nonenzymic breakdown products. Up to 3-fold increased mutagenic response was found when a fortified 9000 g liver supernatant from either phenobarbitone-treated or untreated mice was added to such assays, supporting an enzymic formation of mutagenic metabolite(s) from chloroprene (9) [probably an oxirane (epoxide) in analogy with vinyl chloride, vinylidene chloride, and trichloroethylene] (9)(10).

Medical Surveillance: Preplacement and periodic examinations should include an evaluation of the skin, eyes, respiratory tract, and central nervous system. Liver and kidney function should be evaluated.

Special Tests: None commonly used.

Personal Protective Methods: Protective clothing, chemical safety goggles, air-supplied or self-contained respirators, and safety harnesses should be worn where there is exposure to the liquid or high concentrations of the vapor.

References

(1) Bauchwitz, P.S. Chloroprene. R.E. Kirk and D.F. Othmer (eds). *Encyclopedia of Chemical Technology*, 2nd ed., Vol. 5. John Wiley and Sons, New York, 1964, pp. 215-231.
(2) Bellringer, F.J., and Hollis, C.E. Make chloroprene from butadiene. *Hydrocarbon Processing*, 47 (1968) 127-130.
(3) Lloyd, J.W., Decoufle, P., and Moore, R.M. Background information on chloroprene. *J. Occup. Med.*, 17 (1975) 263-265.
(4) VanOss, J.F. *Chemical Technology: An Encyclopedic Treatment*. Vol. 5. Barnes and Noble Books, New York (1972) pp. 482-483.
(5) Khachatryan, E.A. The role of chloroprene in the process of skin neoplasm formation. *Gig. Tr. Prof. Zab.*, 18 (1972) 54-55.
(6) Khachatryan, E.A. The occurrence of lung cancer among people working with chloroprene. *Vop. Oncol.*, 18 (1972) 85-86.

(7) Anon. Industry's problems with cancer aired. *Chem. Eng. News,* 53 (1975) 4.
(8) Katosova, L.D. Cytogenetic analysis of peripheral blood of workers engaged in the production of chloroprene. *Gig. Tr. Prof. Zab.,* 10 (1973) 30-33.
(9) Bartsch, H., Malaveille, C., Montesano, R., and Tomatis, L. Tissue-mediated mutageneity of vinylidene chloride and 2-chlorobutadiene in *Salmonella typhimurium. Nature,* 255, (1975) 641-43.
(10) Bartsch, H., and Montesano, R. Mutagenic and carcinogenic effects of vinyl chloride. *Mutation Res.* 32 (1975) 93-114.

Bibliography

Infante, P.F., Wagoner, J.K., and Young, R.J. 1977. Chloroprene: observations of carcinogenesis and mutagenesis. In: H.H. Hiaff, J.D. Watson, and J.A. Winston, eds. *Origins of Human Cancer.* Cold Spring Harbor Press, Cold Spring Harbor, New York.
National Institute for Occupational Safety and Health. *Criteria for a Recommended Standard: Occupational Exposure to Chloroprene.* NIOSH Doc. No. 77-210 (1977).
Ritter, W.L., and Carter, A.S. 1948. Hair loss in neoprene manufacture. *J. Ind. Hyg. Toxicol.* 30:192.

CHROMIC ACID (See "Chromium and Compounds")

CHROMIUM AND COMPOUNDS

Description: This group includes chromium trioxide (CrO_3), and its aqueous solutions (1). Chromium may exist in one of three valence states in compounds, +2, +3, and +6. Chromic acid, along with chromates, is in the hexavalent form. Chromium trioxide is produced from chromite ore by roasting with alkali or lime, (calcium oxide) leaching, crystallization of the soluble chromate or dichromate followed by reaction with sulfuric acid.

Synonyms: Chromium trioxide is also known as chromic anhydride, chromic acid and chromium(VI) oxide.

Potential Occupational Exposures: Chromium trioxide is used in chrome plating, copper stripping, aluminum anodizing, as a catalyst, refractories, in organic synthesis, and photography.

A partial list of occupations in which exposure may occur includes:

Anodizers	Metal workers	Stainless steel workers
Copper etchers	Oil purifiers	Textile workers
Electroplaters	Photoengravers	Welders
Glass workers	Photographers	
Lithographers	Process engravers	

An estimated 15,000 industrial workers are potentially exposed to chromic acid (2).

NIOSH estimates that 175,000 workers are potentially exposed to chromium(VI) in various forms (3).

Permissible Exposure Limits: The Federal standard for chromic acid and chromates is 0.1 mg/m^3 as a ceiling concentration (1). The NIOSH Criteria for a Recommended Standard (2) would set work place limits for chromic acid of 0.05 mg/m^3 as chromium trioxide as a TWA with a ceiling concentration of 0.1 mg/m^3 as chromium trioxide determined by a sampling time of 15 minutes.

More recently, however (3) certain forms of chromium(VI) have been found to cause increased respiratory cancer mortality among workers. NIOSH has found that certain other forms of chromium(VI) are noncarcinogenic; they are the monochromates and bichromates (dichromates) of hydrogen, lithium, sodium, potassium, rubidium, cesium, and ammonium, and chromium(VI) oxide (chromic acid anhydride).

NIOSH (3) recommends that chromium(VI) in the workplace be considered to be carcinogenic unless an employer can demonstrate that only the noncarcinogenic chromium(VI) compounds mentioned above are present. If this can be demonstrated, NIOSH has less restrictive recommendations appropriate for noncarcinogenic materials.

In those instances where the recommendations for carcinogenic chromium(VI) apply, NIOSH recommends that exposure to chromium(VI) be not greater than 1 μg Cr(VI)/cu m. If only the noncarcinogenic chromium(VI) materials described above are present in the workplace, NIOSH recommends workplace environmental limits of 50 μg chromium(VI)/cu m as a 15-minute ceiling and a TWA of 25 μg chromium(VI)/cu m for up to a 10-hour workday, 40-hour workweek. The present Federal workplace environmental limit for chromium(VI) materials is a ceiling of 52 μg chromium(VI)/cu m or 100 μg chromium(VI) oxide/m^3.

Routes of Entry: Percutaneous absorption, inhalation, and ingestion.

Harmful Effects

Local—In some workers, chromium compounds act as allergens which cause dermatitis to exposed skin. They may also produce pulmonary sensitization. Chromic acid has a direct corrosive effect on the skin and the mucous membranes of the upper respiratory tract: and although rare, the possibility of skin and pulmonary sensitization should be considered.

Systemic—Chromium compounds in the +3 state are of a low order of toxicity. In the +6 state, chromium compounds are irritants and corrosive, and can enter the body by ingestion, inhalation, and through the skin. Typical industrial hazards are: inhalation of the dust and fumes released during the manufacture of dichromate from chromite ore; inhalation of chromic acid mist during the electroplating and surface treatment of metals; and skin contact in various manufacturing processes.

Acute exposures to dust or mist may cause coughing and wheezing, headache, dyspnea, pain on deep inspiration, fever, and loss of weight. Tracheobronchial irritation and edema persist after other symptoms subside. In electroplating operations, workers may experience a variety of symptoms including lacrimation, inflammation of the conjunctiva, nasal itch and soreness, epistaxis, ulceration and perforation of the nasal septum, congested nasal mucosa and turbinates, chronic asthmatic bronchitis, dermatitis and ulceration of the skin, inflammation of laryngeal mucosa, cutaneous discoloration, and dental erosion. Hepatic injury has been reported from exposure to chromic acid used in plating baths, but appears to be rare.

Working in the chromate-producing industry increases the risk of lung cancer.

A table of differentiation between noncarcinogenic and carcinogenic chromium(VI) compounds has been presented by NIOSH (3) as follows:

Evident Noncarcinogens	**Inferred Noncarcinogens**
Sodium bichromate	Lithium bichromate
Sodium chromate	Lithium chromate
Chromium(VI) oxide	Potassium bichromate
	Potassium chromate
	Rubidium bichromate
	Rubidium chromate
	Cesium bichromate
	Cesium chromate
	Ammonium bichromate
	Ammonium chromate

Evident Carcinogens	**Inferred Carcinogens**
Calcium chromate	Alkaline earth chromates and
Sintered calcium chromate	bichromates
Alkaline lime roasting	Chromyl chloride
process residue	t-Butyl chromate
Zinc potassium chromate	Other chromium(VI) materials
Lead chromate	not listed in this table

Medical Surveillance: Preemployment physical examinations should include: a work history to determine past exposure to chromic acid and hexavalent chromium compounds, exposure to other carcinogens, smoking history, history of skin or pulmonary sensitization to chromium, history or presence of dermatitis, skin ulcers, or lesions of the nasal mucosa and/or perforation of the septum, and a chest x-ray. On periodic examinations an evaluation should be made of skin and respiratory complaints, especially in workers who demonstrate allergic reactions. Chest x-rays should be taken yearly for workers over age 40, and every five years for younger workers. Blood, liver, and kidney function should be evaluated periodically.

Special Tests: Urinary chromate values have been studied in relation to exposure, but their value is questionable.

Personal Protective Methods: Full body protective clothing should be worn in areas of chromic acid exposure, and impervious gloves, aprons, and footwear should be worn in areas where spills or splashes may contact the skin. Where chromic acid may contact the eyes by spills or splashes, impervious protective goggles or face shield should be worn. All clothing should be changed at the end of the shift and showering encouraged prior to change to street clothes. Clean clothes should be reissued at the start of the shift. Respirators should be used in areas where dust, fumes, or mist exposure exceeds Federal standards or where brief concentrations exceed the TWA, and for emergencies. Dust, fumes and mist filter-type respirators or supplied air respirators should be supplied all workers exposed, depending on concentration of exposure.

References

(1) National Institute for Occupational Safety and Health. *Occupational Diseases: A Guide to Their Recognition, Revised edition.* NIOSH Doc. No. 77-181, Wash., D.C. (June 1977).
(2) National Institute for Occupational Safety and Health. *Criteria for a Recommended Standard: Occupational Exposure to Chromic Acid.* NIOSH Publ. No. 73-11021 (1973).
(3) National Institute for Occupational Safety and Health. *Criteria for a Recommended Standard: Occupational Exposure to Chromium(VI).* NIOSH Doc. No. 76-129 (1976).

Bibliography

Barborik, M. 1970. The problem of harmful exposures to chromium compounds. *Ind. Med. Surg.* 39:45.
Davidson, I.W.F., and Secrest, W.L. 1972. Determination of chromium in biological materials by atomic absorption spectrometry using a graphite furnace atomizer. *Anal. Chem.* 44:1808.
Henning, H.F. 1972. Chromium plating. *Ann. Occup. Hyg.* 15:93.
Kazantzis, G. 1972. Chromium and nickel. *Ann. Occup. Hyg.* 15:25.

COAL TAR PRODUCTS

Description: The term "coal tar products," as used by NIOSH (1), includes coal tar and two of the fractionation products of coal tar, creosote and coal tar pitch, derived from the carbonization of bituminous coal. Coal tar, coal tar pitch, and creosote derived from bituminous coal often contain identifiable components which by themselves are carcinogenic, such as benzo[a]pyrene, benzanthracene, chrysene, and phenanthrene. Other chemicals from coal tar products such as anthracene, carbazole, fluoranthene, and pyrene may also cause cancer, but these causal relationships have not been adequately documented.

Synonyms: None.

Potential Occupational Exposures: The coke-oven plant is the principal source of coal tar. The hot gases and vapors produced during the conversion of coal to coke are collected by means of a scrubber, which condenses the effluent into ammonia, water, crude tar, and other by-products. Crude tar is separated from the remainder of the condensate for refining and may undergo further processing.

Employees may be exposed to pitch and creosote in metal and foundry operations, when installing electrical equipment, and in construction, railway, utility, and briquette manufacturing. A list of primary employment in which the various types of pitch and creosote are encountered is as follows:

Product	User Industry	% of Tar Processed	No. of Jobs Affected
Electrode	Aluminum	43.2	28,000
binder pitch	Steel	3.0	50,000
	Graphite	9.2	10,000
Core pitch	Foundry	2.2	2,000
Refractory pitch	Steel	2.4	50,000
Fiber pitch	Electrical	3.5	-
Misc. pitch	Various	3.4	-
Roofing pitch	Construction	8.8	-
Other tars and fuel residue	Fuel	24.3	-
Creosote	Railway, utility, construction	-	5,000

Total 145,000

Permissible Exposure Limits: In 1967, the American Conference of Governmental Industrial Hygienists (ACGIH) adopted a threshold limit value (TLV) of 0.2 mg/cu m for coal tar pitch volatiles (CTPV), described as a "benzene-soluble" fraction, and listed certain carcinogenic components of CTPV. The TLV was established to minimize exposure to the listed substances believed to be carcinogens, viz, anthracene, benzo[a]pyrene, phenanthrene, acridine, chrysene, and pyrene. This TLV was promulgated as a federal standard under the Occupational Safety and Health Act of 1970 (29 CFR 1910.1000). No foreign standards were found for exposure to coal tar pitch or creosote.

In 1973, NIOSH published the "Criteria for a Recommended Standard—Occupational Exposure to Coke Oven Emissions," recommending work practices to minimize the harmful effects of exposure to coke-oven emissions and inhalation of coal tar pitch volatiles. In 1974, OSHA established a Standards Advisory Committee on Coke Oven Emissions to study the problem of the exposure of coke-oven workers to CTPV and to prepare recommendations for an effective standard in the assigned area. In 1975, the Committee recommended a limit of 0.2 μg/cu m for benzo[a]pyrene (Federal Register, 41:46742-46787, October 22, 1976).

According to a recent NIOSH publication (1), occupational exposure to coal tar products shall be controlled so that employees are not exposed to coal tar, coal tar pitch, creosote or mixtures of these substances at a concentration greater than 0.1 milligram/cu m of the cyclohexane-extractable fraction of the sample, determined as a time-weighted average (TWA) concentration for up to a 10-hour work shift in a 40-hour workweek.

Routes of Entry: Inhalation and skin contact.

Harmful Effects

Based on a review of the toxicologic and epidemiologic evidence presented, it has been concluded (1) that some materials contained in coal tar pitch, and therefore in coal tar, can cause lung and skin cancer, and perhaps cancer at other sites. Based on a review of experimental toxicologic evidence, it is also concluded that creosote can cause skin and lung cancer. While the evidence on creosote is not so strong as that on pitch (in part because of difficulties in chemical characterization of such mixtures), the conclusion on the carcinogenic potential of creosote is supported by information on the presence of polynuclear aromatic hydrocarbons and imputations and evidence of the carcinogenicity of such hydrocarbons.

Medical Surveillance: Medical surveillance shall be made available, as specified below, to all employees occupationally exposed to coal tar products.

  (a) Preplacement medical examinations shall include:
      1. Comprehensive initial medical and work histories, with special emphasis directed toward identifying preexisting disorders of the skin, respiratory tract, liver, and kidneys.
      2. A physical examination giving particular attention to the oral cavity, skin, and respiratory system. This shall include posteroanterior and lateral chest x-rays (35 x 42 cm). Pulmonary function tests, including forced vital capacity (FVC)

and forced expiratory volume at one second (FEV 1.0), and a sputum cytology examination shall be offered as part of the medical examination of exposed employees. Other tests, such as liver function and urinalysis should be performed as considered appropriate by the responsible physician. In addition, the mucous membranes of the oral cavity should be examined.
3. A judgment of the employee's ability to use positive pressure respirators.

(b) Periodic examinations shall be made available at least annually. These examinations shall include:
1. Interim medical and work histories.
2. A physical examination as outlined in (a) 2 above.

(c) Initial medical examinations shall be made available to all workers as soon as practicable after the promulgation of a standard based on these recommendations.

(d) Pertinent medical records shall be maintained for at least 30 years after termination of employment. They shall be made available to medical representatives of the government, the employer or the employee.

Special Tests: None commonly used.

Personal Protective Methods: Employers shall use engineering controls when needed to keep the concentration of airborne coal tar products at or below the specified limit. Employers shall provide protective clothing and equipment impervious to coal tar products to employees whenever liquid coal tar products may contact the skin or eyes. Emergency equipment shall be located at well-marked and clearly identified stations and shall be adequate to permit all personnel to escape from the area or to cope safely with the emergency on reentry.

Protective equipment shall include a) eye and face protection, b) protective clothing and c) respiratory protection as spelled out in detail by NIOSH (1).

Reference

(1) National Institute for Occupational Safety and Health. *Criteria for a Recommended Standard: Occupational exposure to coal tar products.* NIOSH Doc. No. 78-107, Wash., D.C. (Sept. 1977).

COBALT AND COMPOUNDS

Description: Co, cobalt, is a silver-grey, hard, brittle, magnetic metal. It is relatively rare; the important mineral sources are the arsenides, sulfides, and oxidized forms. It is generally obtained as a by-product of other metals, particularly copper. Cobalt is insoluble in water, but soluble in acids.

Synonyms: None.

Potential Occupational Exposures: Nickel-aluminum-cobalt alloys are used for permanent magnets. Alloys with nickel, aluminum, copper, beryllium, chromium, and molybdenum are used in the electrical, automobile, and aircraft industries. Cobalt is added to tool steels to improve their cutting qualities and is used as a binder in the manufacture of tungsten carbide tools.

Various cobalt compounds are used as pigments in enamels, glazes, and paints, as catalysts in afterburners, and in the glass, pottery, photographic, electroplating industries. Radioactive cobalt (^{60}Co) is used in the treatment of cancer.

A partial list of occupations in which exposure may occur includes:

Alloy makers	Nickel workers
Catalyst workers	Paint dryer makers
Ceramic workers	Porcelain colorers
Drug makers	Rubber colorers
Electroplaters	Synthetic ink makers
Glass colorers	

Permissible Exposure Limits: The Federal standard for cobalt, metal fume and dust, is 0.1 mg/m^3. ACGIH (1978) has proposed a TWA value of 0.05 mg/m^3.

Route of Entry: Inhalation of dust or fume.

Harmful Effects

Local—Cobalt dust is mildly irritating to the eyes and to a lesser extent to the skin. It is an allergen and has caused allergic sensitivity type dermatitis in some industries where only minute quantities of cobalt are used. The eruptions appear in the flexure creases of the elbow, knee, ankles, and neck. Cross sensitization occurs between cobalt and nickel, and to chromium when cobalt and chromium are combined.

Systemic—Inhalation of cobalt dust may cause an asthma-like disease with cough and dyspnea. This situation may progress to interstitial pneumonia with marked fibrosis. Pneumoconiosis may develop which is believed to be reversible. Since cobalt dust is usually combined with other dusts, the role cobalt plays in causing the pneumoconiosis is not entirely clear.

Ingestion of cobalt or cobalt compounds is rare in industry. Vomiting, diarrhea, and a sensation of hotness may occur after ingestion or after the inhalation of excessive amounts of cobalt dust. Cardiomyopathy has also been reported, but the role of cobalt remains unclear in this situation.

Medical Surveillance: In preemployment examinations, special attention should be given to a history of skin diseases, allergic dermatitis, baseline allergic respiratory diseases, and smoking history. A baseline chest x-ray should be taken. Periodic examinations should be directed toward skin and respiratory symptoms and lung function.

Special Tests: None are in common use.

Personal Protective Methods: Where dust levels are excessive, dust respirators should be used by all workers. Protective clothing should be issued to all workers and changed on a daily basis. Showering after each shift is encouraged prior to change to street clothes. Gloves and barrier creams may be helpful in preventing dermatitis.

Bibliography

Barborik, M., and Dusek, J. 1972. Cardiomyopathy accompanying industrial cobalt exposure. *Br. Heart J.* 34:113.

Camarasa, J.M.C. 1967. Cobalt contact dermatitis. *Acta. Derm. Venereol.* 47:287.

Miller, C.W., Davis, M.W., Goldman, A., and Wyatt, J.P. 1953. Pneumoconiosis in the tungsten-carbide tool industry. *AMA Arch. Ind. Hyg. Occup. Med.* 8:453.

COPPER AND COMPOUNDS

Description: Cu, copper, is a reddish-brown metal which occurs free or in ores such as malachite, cuprite, and chalcopyrite. It may form both mono- and divalent compounds. Copper is insoluble in water, but soluble in nitric acid and hot sulfuric acid.

Synonyms: None.

Potential Occupational Exposures: Metallic copper is an excellent conductor of electricity and is widely used in the electrical industry in all gauges of wire for circuitry, coil, and armature windings, high conductivity tubes, commutator bars, etc. It is made into castings, sheets, rods, tubing, and wire, and is used in water and gas piping, roofing materials, cooking utensils, chemical and pharmaceutical equipment, and coinage. Copper forms many important alloys: Be-Cu alloy, brass, bronze, gun metal, bell metal, German silver, aluminum bronze, silicon bronze, phosphor bronze, and manganese bronze.

Copper compounds are used as insecticides, algicides, molluscicides, plant fungicides, mordants, pigments, catalysts, and as a copper supplement for pastures, and in the manufacture of powdered bronze paint and percussion caps. They are also utilized in analytical reagents, in paints for ships' bottoms, in electroplating, and in the solvent for cellulose in rayon manufacture.

A partial list of occupations in which exposure may occur includes:

Asphalt makers	Pigment makers
Battery makers	Rayon makers
Electroplaters	Solderers
Fungicide workers	Wallpaper makers
Gem colorers	Water treaters
Lithographers	Wood preservative workers

Permissible Exposure Limits: The Federal standard for copper fume is 0.1 mg/m^3, and for copper dusts and mists, 1 mg/m^3.

Route of Entry: Inhalation of dust or fume.

Harmful Effects

Local—Copper salts act as irritants to the intact skin causing itching, erythema, and dermatitis. In the eyes, copper salts may cause conjunctivitis and even ulceration and turbidity of the cornea. Metallic copper may cause keratinization of the hands and soles of the feet, but it is not commonly associated with industrial dermatitis.

Systemic—Industrial exposure to copper occurs chiefly from fumes generated in welding copper-containing metals. (See Brass.) The fumes and dust cause irritation of the upper respiratory tract, metallic taste in the mouth, nausea, metal fume fever, and in some instances discoloration of the skin and hair. Inhalation of dusts, fumes, and mists of copper salts may cause congestion of the nasal mucous membranes, sometimes of the pharynx, and on occasions, ulceration with perforation of the nasal septum. If the salts reach the gastrointestinal tract, they act as irritants, producing salivation, nausea, vomiting, gastric pain, hemorrhagic gastritis, and diarrhea. It is unlikely that poisoning by ingestion in industry would progress to a serious point as small amounts induce vomiting and empty the stomach of copper salts.

Chronic human intoxication occurs rarely and then only in individuals with Wilson's disease (hepatolenticular degeneration). This is a genetic condition caused by the pairing of abnormal autosomal recessive genes in which there is abnormally high absorption, retention, and storage of copper by the body. The disease is progressive and fatal if untreated.

Medical Surveillance: Consider the skin, eyes, and respiratory system in any placement or periodic examinations.

Personal Protective Methods: In areas where copper dust or fume is excessive, workers should be provided with proper dust or fume filters or supplied air respirator with full facepiece.

Bibliography

Davenport, S.J. 1953. Review of literature on health hazards of metals—1. Copper, Bureau of Mines Information Circular 7666. U.S. Department of Interior, Wash., D.C.
Gleason, R.P. 1968. Exposure to copper dust. *Am. Ind. Hyg. Assoc. J.* 29:461.

COTTON DUST

Description: "Cotton dust" is defined as dust generated into the atmosphere as a result of the processing of cotton fibers combined with any naturally occurring materials such as stems, leaves, bracts, and inorganic matter which may have accumulated on the cotton fibers during the growing or harvesting period. Any dust generated from processing of cotton through the weaving of fabric in textile mills and dust generated in other operations or manufacturing processes using new or waste cotton fibers or cotton fiber by-products from textile mills is considered cotton dust.

Synonyms: None.

Potential Occupational Exposures: The Occupational Safety and Health Administration has estimated that 800,000 workers are involved in work with cotton fibers and thus are potentially exposed to cotton dust in the work place.

Permissible Exposure Limits: The current Federal standard consists only of an environmental limit, 1 mg/m^3 of air. Available epidemiologic data do not demonstrate a safe concentration of cotton dust, but suggest that a safe concentration may exist below 0.1 mg/m^3, which is close to the background dust concentration in many locations. As a consequence, NIOSH has not recommended a specific environmental limit but recommends that the Department of Labor establish the lowest feasible limit, but in no case at an environmental concentration as high as 0.2 mg lint-free cotton dust/m^3. In addition, NIOSH recommends a strong program of medical monitoring and management, administrative controls, and work practices to limit the incidence and severity of cotton dust disease.

Route of Entry: Inhalation of dust.

Harmful Effects

The processing of raw cotton is associated with byssinosis, a disease characterized by pulmonary dysfunction. There is evidence, but not proof, that cotton dust itself is not the cause of byssinosis, but may carry materials into the respiratory tract, possibly foreign protein or proteolytic enzymes.

Medical Surveillance:

a) Preplacement—A comprehensive physical examination shall be made available to include as a minimum: medical history, baseline forced vital capacity (FVC), and forced expiratory volume at 1 second (FEV_1). The history shall include administration of a questionnaire designed to elicit information regarding symptoms of chronic bronchitis, byssinosis, and dyspnea. If a positive personal history of respiratory allergy, chronic obstructive lung disease, or other diseases of the cardiopulmonary system are elicited, or where there is a positive history of smoking, the applicant shall be counseled on his increased risk from occupational exposure to cotton dust. At the time of this examination, the advisability of the workers using negative or positive pressure respirators shall be evaluated.

b) Each newly employed person shall be retested for ventilatory capacity (FVC and FEV_1) within 6 weeks of employment. This retest shall be performed on the first day at work after at least 40 hours' absence from exposure to cotton dust and shall be performed before and after at least 6 hours of exposure on the same day.

c) Periodic—Each current employee exposed to cotton dust shall be offered a medical examination at least yearly that shall include administration of a questionnaire designed to elicit information regarding symptoms of chronic bronchitis, byssinosis, and dyspnea.

d) Each current employee exposed to cotton dust shall have measurement of forced vital capacity (FVC) and of forced expiratory volume at one second (FEV_1). These tests of ventilatory function should be performed on the first day of work following at least 40 hours of absence from exposure to cotton dust, and shall be performed before and after at least 6 hours of exposure on the same day.

e) Ideally, the judgment of the employee's pulmonary function should be based on preplacement values (values taken before exposure to cotton dust). When preplacement values are not available, then reference to standard pulmonary function value tables may be necessary. Note that these tables may not reflect normal values for different ethnic groups. For example, the average healthy black male may have an approximately 15% lower FVC than a caucasian male of the same body build. A physician shall consider, in cases of significantly decreased pulmonary function, the impact of further exposure to cotton dust and evaluate the relative merits of a transfer to areas of less exposure or protective measures. A suggested plan for the management of cotton workers was proposed as a result of a conference on cotton workers' health.

f) Medical records, including information on all required medical examinations, shall be maintained for persons employed in work involving exposure to cotton dust. Medical records with pertinent supporting documents shall be maintained at least 20 years after the individual's termination of employment. These records shall be available to the medical representatives of the Secretary of Health, Education, and Welfare, of the Secretary of Labor, of the employee or former employee, and of the employer.

Special Tests: Ventilatory capacity tests (see Medical Surveillance above).

Personal Protective Methods: Engineering control shall be used wherever feasible to maintain cotton dust concentrations below the prescribed limit. Administrative controls can also be used to reduce exposure.

Respirators shall also be provided and used for nonroutine operations (occasional brief exposures above the environmental limit and for emergencies) and shall be considered for use by employees who have symptoms even when exposed to concentrations below the established environmental limit. Appropriate respirators shall be used pursuant only to the following requirements:

a) For the purpose of determining the type of respirator to be used, the employer shall initially measure the atmospheric concentration of cotton dust in the workplace and thereafter whenever process, work site, climate, or control changes occur which are likely to alter the cotton dust concentration. This requirement shall not apply when only atmosphere-supplying positive pressure respirators are used. The employer shall ensure that all workers are supplied with respirators appropriate for the concentration of dust to which they are exposed.

b) A respiratory protective program meeting the requirements of the Occupational Safety and Health Administration Standards, part 1910.134, shall be established and enforced by the employer (29 CFR, Part 1910.134 published in the Federal Register, volume 39, page 23671, dated June 24, 1974).

References

(1) National Institute for Occupational Safety and Health, *Criteria for a Recommended Standard: Occupational Exposure to Cotton Dust.* NIOSH Doc. No. 75-118 (1975).

CREOSOTE

Description: Creosote is a flammable, heavy, oily liquid with a characteristic sharp, smoky smell, and caustic burning taste. In pure form it is colorless, but the industrial product is usually brownish. It is produced by the destructive distillation of wood or coal tar at temperatures above 200°C. The chemical composition is determined by the source and may contain guaiacol, creosols, phenol, cresols, pyridine, and numerous other aromatic compounds.

Synonyms: Creosotum, creosote oil, brick oil.

Potential Occupational Exposures: Creosote is used primarily as a wood preservative, and those working with the treated wood may be exposed. It is also used as a waterproofing agent, an animal dip, a constituent in fuel oil, a lubricant for die molds, as pitch for roofing, and in the manufacture of chemicals and lampblack. In the pharmaceutical industry, it is used as an antiseptic, disinfectant, antipyretic, astringent, styptic, germicide, and expectorant.

A partial list of occupations in which exposure may occur includes:

Coal tar workers	Pitch workers
Fuel oil blenders	Waterproofers
Lampblack makers	Wood preservers
Organic chemical synthesizers	

Permissible Exposure Limits: There is no Federal standard for creosote.

Route of Entry: Skin absorption.

Harmful Effects

Local—The liquid and vapors are strong irritants producing local erythema, burning, itching, pigmentation (grayish yellow to bronze), vesiculation, ulceration, and gangrene. Eye injuries include keratitis, conjunctivitis, and permanent corneal scars. Contact dermatitis is reported in industry. Photosensitization has been reported. Skin cancer may occur.

Systemic—Symptoms of systemic illness include salivation, vomiting, vertigo, headache, loss of pupillary reflexes, hypothermia, cyanosis, convulsions, thready pulse, respiratory difficulties, and death.

Medical Surveillance: Consider the skin, eyes, respiratory tract, and central nervous system in placement and periodic examination.

Special Tests: None commonly used.

Personal Protective Methods: Protective clothing should be worn where employees are exposed to the liquid or high vapor concentration. Masks with fullface protection and organic vapor canisters should be worn. Gloves and goggles are advisable in any area where spill or splash might occur.

Bibliography

Arieff, A.J. 1965. Acute, toxic, polioencephalitis (creosote). *J. Am. Med. Assoc.* 193:745.

CRESOL

Description: $CH_3C_6H_4OH$, cresol, is a mixture of the three isomeric cresols, ortho-, meta-, and para-cresol, and is a colorless, yellowish, brownish-yellow, or pinkish liquid with a phenolic odor. Cresols are soluble in alcohol, glycol, and dilute alkalis. Also they may be combustible.

Synonyms: Cresylic acid, cresylol, hydroxytoluene, methyl phenol, oxytoluene, tricresol.

Potential Occupational Exposures: Cresol is used as a disinfectant, as an ore flotation agent, and as an intermediate in the manufacture of chemicals, dyes, plastics, and antioxidants. A mixture of isomers is generally used; the concentrations of the components are determined by the source of the cresol.

A partial list of occupations in which exposure may occur includes:

Antioxidant makers	Paint remover workers
Chemical disinfectant workers	Pitch workers
Dye makers	Plastic makers
Flotation agent makers	Resin makers
Foundry workers	Stain workers
Insulation enamel workers	Wool scourers

Permissible Exposure Limits: The Federal standard is 5 ppm (22 mg/m^3).

Routes of Entry: Inhalation or percutaneous absorption of liquid or vapor.

Harmful Effects

Local—Cresol is very corrosive to all tissues. It may cause burns if it is not removed promptly and completely and in case of extensive exposure, if it is not removed completely from contaminated areas of the body very quickly, death may result. When it contacts the skin, it may not produce any sensation immediately. After a few moments, prickling and intense burning occur. This is followed by loss of feeling. The affected skin shows wrinkling, white discoloration, and softening. Later gangrene may occur. If the chemical contacts the eyes, it may cause extensive damage and blindness. A skin rash may result from repeated or prolonged exposure of the skin to low concentrations of cresol. Discoloration of the skin may also occur from this type of exposure.

Systemic—When cresol is absorbed into the body either through the lungs, through the skin, or mucous membranes, or by swallowing, it may cause systemic poisoning. The signs and symptoms of systemic poisoning may develop in 20 or 30 minutes. These toxic effects include: weakness of the muscles, headache, dizziness, dimness of vision, ringing of the ears, rapid breathing, mental confusion, loss of consciousness, and sometimes death.

Prolonged or repeated absorption of low concentrations of cresol through the skin, mucous membranes, or respiratory tract may cause chronic systemic poisoning. Symptoms and signs of chronic poisoning include vomiting, difficulty in swallowing, salivation, diarrhea, loss of appetite, headache, fainting, dizziness, mental disturbances, and skin rash. Death may result if there has been severe damage to the liver and kidneys.

Medical Surveillance: Consider the skin, eyes, respiratory system, and liver and kidney function in placement or periodic examinations.

Special Tests: None.

Personal Protective Methods: Protective goggles and clothing should be worn to prevent direct contact with cresol. Masks with organic vapor canisters are advisable in areas of vapor concentration.

Bibliography

American Industrial Hygiene Association. 1969. Community air quality guides. Phenol and cresol. *Am. Ind. Hyg. Assoc. J.* 30:425.

CYCLOHEXANONE (See "Ketones")

D

DIACETONE ALCOHOL (See "Ketones")

2,4-DIAMINOANISOLE

Description: 2,4-Diaminoanisole has the structural formula:

Synonyms: 4-methoxy-m-phenylenediamine.

Potential Occupational Exposures: The principal use of 2,4-diaminoanisole is as a component of oxidation (permanent) hair and fur dye formulations. Approximately three-quarters of the current oxidation hair dye formulations contain 2,4-diaminoanisole in concentrations ranging from approximately 0.05% to approximately 2%. The concentration is determined by the shade of the dye. Oxidation hair dyes are very common among professional as well as over-the-counter products and account for approximately $200 million of the $280 million annual retail expenditure for hair dyes. NIOSH is unaware of any current domestic production of 2,4-diaminoanisole. Imports of 2,4-diaminoanisole are on the order of 25,000 pounds per year.

NIOSH estimates that approximately 400,000 workers have potential occupational exposure to 2,4-diaminoanisole. Hairdressers and cosmetologists comprise the largest portion of workers with potential exposure. (Gloves are usually worn by hairdressers when applying hair dyes). A relatively small number of fur dyers are probably exposed to higher levels of 2,4-diaminoanisole.

Permissible Exposure Limits: While the carcinogenicity of 2,4-diamino-anisole is being further evaluated, the National Institute for Occupational

Safety and Health recommends, as an interim and prudent measure, that occupational exposure to 2,4-diaminoanisole and its salts be minimized. Exposures should be limited to as few employees as possible, while minimizing workplace exposure levels with engineering and work practice controls. In particular, skin exposures should be avoided.

Routes of Entry: Skin contact.

Harmful Effects

Epidemiologic Studies—NIOSH has conducted two epidemiologic studies which suggest excess cancer among cosmetologists.

A report (1) based on data from a case-control study of 25,416 hospital admissions between 1956 and 1965 at Roswell Park Memorial Institute suggests an excess of cancer of specific genital sites (corpus uteri, ovaries) among hairdressers and cosmetologists.

Another study currently being conducted by NIOSH is also suggestive of excess cancer among cosmetologists. This study involves a sample of 53,183 records which are representative of the 417,795 Social Security disability awards made to female workers between 1969 and 1972. Age and race adjusted proportional morbidity ratios (PMbR's) have been constructed for 24 selected occupational groups. Among cosmetologists, elevated PMbR's were observed for cancer of the digestive organs, respiratory system, trachea, bronchus and lung, breast, and genital organs. Cosmetologists had a greater number of elevated PMbR's for specific primay malignant neoplasms than any other tabulated occupational group.

Thus, the preliminary analysis of the Social Security Administration disability data is consistent with the hypothesis that persons employed in occupations classified within the broad category of cosmetology may be at elevated risks of developing a neoplasm from exposures of occupational origin. Other relevant epidemiologic studies with conflicting results have been reported (2)-(6). These studies do not clearly demonstrate an association between hair dyes and cancer. NIOSH believes that its studies do suggest an association between cancer and employment as cosmetologists and hairdressers. However, it is recognized that cosmetologists and hairdressers are exposed to a large variety of substances, and it is difficult at this time to attribute any excess incidence of cancer to either hair dyes in general or 2,4-diaminoanisole in particular.

Laboratory Studies—Preliminary analyses of National Cancer Institute data indicate that male and female laboratory rats and mice fed 2,4-diaminoanisole sulfate in their diets for seventy-eight weeks experienced a statistically significant excess of site-specific malignant tumors as compared to controls.

Groups of fifty male and fifty female Fisher 344 rats and B6C3F1 mice were used in the test. Feed containing 0.05% or 0.12% technical grade 2,4-diamino-anisole sulfate was administered to each group of treated rats; each group of treated mice received feed containing 0.12% or 0.24% technical grade 2,4-diaminoanisole sulfate. Fifty animals of each sex of each species served as controls. After the seventy-eight week treatment period, observation of the mice continued for an additional thirteen weeks and observation of the rats continued for an additional twenty-six weeks.

Significant excess cancer was observed in the thyroid gland and integumentary system (skin) of high dose exposed rats of both sexes, as well as in the thyroid gland of high dose exposed mice, and in the lymphatic system of low dose exposed mice.

In other studies, 2,4-diaminoanisole was tested by skin application to laboratory rodents. Testing by skin application has considerable merit since this route of administration approximates that resulting from the use of hair dyes. Laboratory mice and rats painted with 2,4-diaminoanisole have been reported to experience no statistically significant excess of cancer (7)(8). Kinkle and Holzman, for example, reported applying a mixture containing 0.4% 2,4-diaminoanisole to the shaved backs of Sprague-Dawley rats twice weekly for two years, and then continuing to observe the surviving animals for an additional six months (8). However, the interpretation of the reported data is complicated by experimental design (9) and these experiments do not convincingly establish the safety of 2,4-diaminoanisole applied to skin.

NIOSH understands that recent and still unpublished data acquired by the Food and Drug Administration indicate that 2,4-diaminoanisole penetrates the skin and thereby enters the system of both man and rhesus monkey. This indicates that skin contact with 2,4-diaminoanisole must be avoided in the workplace.

There are reports indicating that 2,4-diaminoanisole is mutagenic in bacterial systems (9) and in *Drosophila* (10). Mutagenic activity per se should be considered an important liability. In addition, empirical correlations have suggested a relationship between mutagenicity, specially in bacterial strains and carcinogenicity in higher animals.

Medical Surveillance: No specific measures have been proposed by NIOSH.

Special Tests: None in common use.

Personal Protective Methods: As noted above, gloves are usually worn by hairdressers when applying hair dyes. Beyond that, NIOSH recommends minimization of exposure.

References

(1) *A Retrospective Survey of Cancer in Relation to Occupation.* DHEW (NIOSH) Publication No. 77-178, U.S. Department of Health, Education, and Welfare, Public

Health Service, Center for Disease Control, National Institute for Occupational Safety and Health. Cincinnati, Ohio, 1977.

(2) Garfinkel, J., Selvin, S. and Brown, S.M. Possible increased risk of lung cancer among beauticians. *J. National Cancer Institute.* 58, 141, 1977.

(3) Hammond, E.C., Some negative findings (polio, smallpox, tetanus and diptheria-vaccines; beauticians) and evaluation of risks. Presented at the American Cancer Society's Nineteenth Science Writers' Seminar. Sarasota, Florida, April 1977.

(4) Kinlen, L.J. et al. Use of hair dyes by patients with breast cancer: a case-control study. *British Medical Journal.* 2, 366, 1977.

(5) Menck, H.R. et al. Lung cancer risk among beauticians and other female workers. *J. National Cancer Institute.* 59, 1423, 1977.

(6) Shafer, N., and Shafer, R.W. Potential of carcinogenic effects of hair dyes. *New York State Journal of Medicine.* 76, 394, 1976.

(7) Burnett, C. et al. Long-term toxicity studies on oxidation hair dyes. *Food and Cosmetics Toxicology.* 13, 353, 1975.

(8) Kinkel, H.J., and Holzmann, S. Study of long-term percutaneous toxicity and carcinogenicity of hair dyes (oxidizing dyes) in rats. *Food and Cosmetics Toxicology.* 11, 641, 1973.

(9) Ames, B.N. et al. Hair dyes are mutagenic: identification of a variety of mutagenic ingredients. *Proceedings National Academy of Sciences USA* 72, 2423, (1975).

(10) Blijleven, W.G.H. Mutagenicity of four hair dyes in *Drosophila melanogaster. Mutation Research.* 48, 181, 1977.

Bibliography

National Institute for Occupational Safety and Health. *2,4-Diaminoanisole in Hair and Fur Dyes.* Current Intelligence Bulletin No. 19, Wash., DC (Jan. 13. 1978).

4,4'-DIAMINODIPHENYLMETHANE

Description: 4,4'-Diaminodiphenylmethane, $H_2NC_6H_4CH_2C_6H_4NH_2$, is a crystalline solid melting at 91.5°-92.0°C.

Synonyms: DDM, p,p'-methylenedianiline, MDA, bis(4-aminophenyl)methane, DAPM.

Potential Occupational Exposure: Approximately 99% of the DDM produced is consumed in its crude form (occasionally containing not more than 50% DDM and polyDDM) at its production site by reaction with phosgene in the preparation of isocyanates and polyisocyanates. These isocyanates and polyisocyanates are employed in the manufacture of rigid polyurethane foams which find application as thermal insulation. Polyisocyanates are also used in the preparation of the semiflexible polyurethane foams used for automotive safety cushioning.

DDM is also used as:

- an epoxy hardening agent

- a raw material in the production of polyurethane elastomers
- in the rubber industry
  a curative for neoprene (1)
  an antifrosting agent (antioxidant) in footwear
- a raw material in the production of Qiana nylon.
- a raw material in the preparation of poly(amide-imide) resins (used in magnet wire enamels)

It is estimated that 2,500 workers are exposed to DDM. Many of these exposures to DDM are in the preparation of isocyanates and polyisocyanates and, on construction sites, in the application of epoxy coatings.

Permissible Exposure Limits: ACGIH (1978) has proposed TWA values of 0.1 ppm (0.8 mg/m^3).

Routes of Entry: Inhalation, skin absorption.

Harmful Effects

In 1965, the hepatotoxic effects of DDM in humans were first seen in the so-called "Epping Jaundice" outbreak in Great Britain. In this incident 84 people who had eaten DDM-contaminated bread, experienced hepatocellular damage evidenced by elevated SGOT and SGPT levels (2)(3). DDM has also been shown to produce liver lesions in a group of intragastrically fed rats (4) and has caused liver degeneration and spleen lesions in another group of DDM fed rats (5).

DDM in the occupational environment has been implicated in a number of cases of toxic hepatitis. During an 18 month period beginning April 1972, six cases of hepatitis developed among about 300 men who used epoxy resins in the construction of a nuclear power plant in Alabama. Two chemicals, DDM and 2-nitropropane, were held suspect in this study (6)(7).

In another study, 13 cases of hepatitis developed between 1966 and 1972 among workers who added DDM to a mixture to produce a hard plastic insulating material. All of these men became ill within a few days of working intensively with DDM (8). One other case of hepatitis, possibly associated with DDM, was reported by a person who wrote to the Environmental Protection Agency describing an episode of acute hepatitis as well as CNS and pulmonary symptoms he experienced following exposure to a surfacing agent containing DDM.

The carcinogenic effects of DDM have also been studied. In one study, 16 rats were given 4 or 5 20-mg DDM doses by stomach tube over 8 months. A hepatoma and a haemangioma-like tumor of the kidney were found in one rat after 18 months and an adenocarcinoma of the uterus was found in another after 24 months (9). In another report, of 48 rats given DDM intragastrically 5 times weekly, all developed liver cirrhosis, four developed hepatomas (2 benign) and other, miscellaneous tumors (10). In a third report, 50% of 50 DDM injected rats developed tumors (4 hepatomas) compared with 26% of a control group (11). There have been no reported human cancers associated with DDM.

If, as hypothesized in the Center for Disease Control study of nuclear power plant construction workers, not all workers are susceptible to liver injury after exposure to DDM, and if the 1-2% incidence of liver disease seen in this study were applied to all workers with possible exposure to DDM, one would expect to see 25 to 50 cases of DDM associated toxic hepatitis a year.

Medical Surveillance: See 4,4-Methylenebis(2-chloroaniline) for a related material.

Special Tests: None commonly used.

Personal Protective Methods: Worker exposure by all routes should be carefully controlled to levels consistent with animal and human experience, according to ACGIH.

References

(1) Lloyd, J.W., DeCoufle, P. and Moore, R.M., Jr., Current intelligence—background information on chloroprene. *Journal of Occupational Medicine.* 17:263-265, 1975.
(2) Kopelman, H., Robertson, M.H., Sanders, P.G., and Ash, I. The Epping Jaundice. *Brit. Med. J.* 1:514, 1966.
(3) Kopelman, H., Scheuer, P.J., and Williams, R. The liver lesion of the Epping Jaundice. *Quart. J. Med.* 35:553, 1966.
(4) Gohlke, R. and Schmidt, P. 4,4'-Diaminodiphenyl methane: histological, enzyme histochemical and autoradiographic investigations in acute and subacute experiments in rats with and without the additional stress of heat. *Int. Arch. Arbeitsmed.* 32(3):217-31, 1974.
(5) Pludro, G., Karlowski, K., Makowska, M., Woggon, H., and Uhde, W.T. Toxicological and chemical studies of some epoxy resins and hardeners I. Determination of acute and subacute toxicity of phthalic acid anhydride, 4,4'-diaminodiphenylmethane and of the epoxy resin: Epilox EG-34. *Acta Poloniae Pharmaceutica.* 26(4):352-57, 1969.
(6) Unpublished data, Center for Disease Control, 1974.
(7) Williams, S.V., Bryan, J.A., and Wolf, F.S. Toxic hepatitis and methylenedianiline. *New Eng. J. of Med.* Vol. 291:1256, 1974.
(8) McGill, D.B. and Motto, J.D. An industrial outbreak of toxic hepatitis due to methylenedianiline. *New Eng. J. of Med.* 291(6):278-82, 1974.
(9) Schoental, R. Carcinogenic and chronic effects of 4,4'-diaminodiphenylmethane, an epoxy resin hardener. *Nature.* 219:1162-63, 1968.
(10) Munn, A. Occupational bladder tumors and carcinogens: recent developments in Britain. In: Deichmann, W. and Lampe, K.F., eds. *Bladder Cancer. A Symposium.* Birmingham, Alabama, Aesculapius, P. 187.
(11) Steinhoff, D. and Grundmann, E. Zur Cancerogenen Wirkung von 4,4'-diaminodiphenylmethan und 2,4'-diaminodiphenylmethan. *Naturwissenschaften.* 57:247-48, 1970.

DIBROMOCHLOROPROPANE

Description: 1,2-Dibromo-3-chloropropane (DBCP) has the chemical formula $CH_2BrCHBrCH_2Cl$. It has a gram molecular weight of 236.36. An amber to brown liquid with a pungent odor, DBCP has a specific gravity of 2.09 at 14°C referred to water. It has a vapor pressure of less than 1 mm of Hg at 21°C and boils at 199°C under a pressure of 760 mm of Hg.

Synonyms: DBCP.

Potential Occupational Exposures: DBCP has been used in agriculture as a nematocide since 1955, being supplied for such use in the forms of liquid concentrate, emulsifiable concentrate, powder, granules, and solid material. The estimated use of DBCP in the United States in recent years has ranged from 3.6 million pounds in 1971 to 12.3 million pounds in 1972 (USDA estimates).

Estimates of worker exposure to DBCP are not available but the number of workplaces affected are indicated by the fact that in August 1977, NIOSH alerted approximately 80 manufacturers and formulators to the potential hazard of worker exposure to DBCP.

Permissible Exposure Limits: The National Institute for Occupational Safety and Health (NIOSH) recommends that employee exposure to dibromochloropropane shall be controlled in the workplace so that no employee is exposed to airborne dibromochloropropane at a concentration greater than 10 parts per billion (approximately 0.1 mg/m^3) determined as a time-weighted average (TWA) concentration for up to a 10-hour work shift, 40-hour workweek.

Routes of Entry: Inhalation of vapors.

Harmful Effects

The possible effects on the health of employees chronically exposed to DBCP may include sterility, diminished renal function, and degeneration and cirrhosis of the liver. In addition, ingestion of daily doses of DBCP by mice and rats has been found to result in the appearance of gastric cancers in both sexes of both species and in mammary cancers in female rats. Although an increased risk for cancer has not been seen with inhalation exposures, these results are not definitive, therefore the risk of cancer due to occupational exposure to DBCP remains a continuing concern.

There are indications from in vitro experiments that mutagenic effects may occur also, but there has been no study yet of this possibility with mammalian subjects. Employees should be told of these possible effects and informed that some 20-25 years of experience in the manufacture and formulation of

DBCP has not yet called such effects in employees of the pesticide industry to the notice of physicians and epidemiologists.

Medical Surveillance: Medical surveillance shall be made available to employees as outlined below:

(a) Comprehensive preplacement or initial medical and work histories with emphasis on reproductive experience and menstrual history.

(b) Comprehensive physical examination with emphasis on the genito-urinary tract including testicle size and consistency in males.
1) Semen analysis to include sperm count, motility and morphology.
2) Other tests, such as serum testosterone, serum follicle stimulating hormone (FSH), and serum lutenizing hormone (LH) may be carried out if, in the opinion of the responsible physician, they are indicated. In addition, screening tests of the renal and hepatic systems may be considered.
3) A judgment of the worker's ability to use positive pressure respirators.

(c) Employees shall be counseled by the physician to ensure that each employee is aware that DBCP has been implicated in the production of effects on the reproductive system including sterility in male workers. In addition, they should be made aware that cancer was produced in some animals. While the relevancy of these findings is not yet clearly defined, they do indicate that both employees and employers should do everything possible to minimize exposure to DBCP.

(d) Periodic examinations containing the elements of the preplacement or initial examination shall be made available on at least an annual basis.

(e) Examinations of current employees shall be made available as soon as practicable after the promulgation of a standard for DBCP.

(f) Medical surveillance shall be made available to any worker suspected of having been exposed to DBCP.

(g) Pertinent medical records shall be maintained for all employees subject to exposure to DBCP in the workplace. Such records shall be maintained for 30 years and shall be available to medical representatives of the U.S. Government, the employer and the employee.

Special Tests: None commonly used.

Personal Protective Methods

Respiratory Protection—Engineering controls shall be used wherever needed to keep airborne dibromochloropropane concentrations below the recommended occupational exposure limit. Compliance with this limit may be achieved by the use of respirators under the following conditions only:

1) During the time necessary to install or test the required engineering controls.

2) For nonroutine operations, such as emergency maintenance or repair activities.

3) During emergencies when air concentrations of dibromochloropropane may exceed the recommended occupational exposure limit.

When a respirator is permitted, it shall be selected and used pursuant to detailed requirements set forth by NIOSH (1).

Eye Protection—Eye protection shall be provided by the employer and used by the employees where eye contact with liquid dibromochloropropane is likely. Selection, use, and maintenance of eye protective equipment shall be in accordance with the provisions of the American National Standard Practice for Occupational and Educational Eye and Face Protection, ANSI Z87.1-1968. Unless eye protection is afforded by a respirator hood or facepiece, protective goggles [splash-proof safety goggles (cup-cover type dust and splash safety goggles) which comply with 29 CFR 1910.133(a) (2)–(a)(6)] or a face shield (8-inch minimum) shall be worn at operations where there is danger of contact of the eyes with liquid dibromochloropropane because of spills or splashes. If there is danger of liquid dibromochloropropane striking the eyes from underneath or around the sides of the face shield, safety goggles shall be worn as added protection.

Protective Clothing—Protective clothing shall be resistant to the penetration and to the chemical action of dibromochloropropane. Additional protection, including gloves, bib-type aprons, boots, and overshoes, shall be provided for, and worn by, each employee during any operation that may cause direct contact with liquid dibromochloropropane. Supplied-air hoods or suits resistant to penetration by dibromochloropropane shall be worn when entering confined spaces, such as pits or storage tanks.

In situations where heat stress is likely to occur, supplied-air suits, preferably cooled, are recommended. The employer shall ensure that all personal protective clothing is inspected regularly for defects and is maintained in a clean and satisfactory condition by the employee.

Reference

(1) National Institute for Occupational Safety and Health. *Criteria for a Recommended Standard: Occupational Exposure to Dibromochloropropane.* NIOSH Doc. No. 78-115 (1978).

1,2-DIBROMOETHANE

Description: $BrCH_2CH_2Br$, 1,2-dibromoethane, is a colorless nonflammable liquid with a chloroform-like odor.

Synonyms: Ethylene dibromide, ethylene bromide, sym-dibromoethane.

Potential Occupational Exposures: 1,2-Dibromoethane is used principally as a fumigant for ground pest control and as a constituent of ethyl gasoline. It is also used in fire extinguishers, gauge fluids, and waterproofing preparations; and it is used as a solvent for celluloid, fats, oils, and waxes.

A partial list of occupations in which exposure may occur includes:

Antiknock compound makers	Gum processors
Cabbage growers	Lead scavenger makers
Drug makers	Resin makers
Farmers	Termite controllers
Fat processors	Tetraethyllead makers
Fire extinguisher makers	Wool reclaimers
Fumigant workers	

The concentration of ethylene dibromide in gasoline is variable, but is in the order of 0.025% (wt/vol) (1). (Ethylene dichloride is also used in admixture with DBE.) The chief sources of ethylene dibromide and dichloride emissions are from automotive sources via evaporation from the fuel tank and carburetor of cars operated on leaded fuel. Emissions from these sources have been estimated to range from 2 to 25 mg/day for 1972 through 1974 model-year cars in the U.S. (1).

Very limited and preliminary air monitoring data for ethylene dibromide, show air concentration values of 0.07-0.11 $\mu g/m^3$ (about 0.01 ppb) in the vicinity of gasoline stations along traffic arteries in 3 major cities; 0.2-1.7 $\mu g/m^3$ (about 0.1 ppb) at an oil refinery and 90-115 $\mu g/m^3$ (10-15 ppb) at DBE manufacturing sites in the U.S., suggesting that DBE is present in ambient air at very low concentrations (1).

It should be noted that the increased use of unleaded gasoline should result in lower ambient air levels of ethylene dibromide from its major sources of emissions (1)(2).

Ethylene dibromide has also been found in concentrations of 96 μg/m^3, up to a mile away from a U.S. Dept. of Agriculture fumigation center (2).

Concentrations of ethylene dibromide on the order of 1 ppm have been found in samples from streams of water on industrial sites. Limited information suggests that ethylene dibromide degrades at moderate rates in both water and soil. The use of ethylene dichloride and dibromide in fumigant mixtures of disinfecting fruits, vegetables, food grains, tobacco, seeds, seed beds, mills and warehouses, suggests the possibility that their residues per se or that of their respective hydrolytic products (e.g., ethylene chlorohydrin or bromohydrin) may be present in fumigated materials (3)-(6).

Although materials such as ethylene dichloride and dibromide are volatile, and their actual occurrence in processed or cooked foods can possibly be considered negligible, more significant exposure is considered more likely among agricultural workers or those fumigating grain and crops in storage facilities and the field, than among consumers of the food products (7).

Permissible Exposure Limits: The Federal standard is 20 ppm (145 mg/m^3) as an 8-hour TWA with an acceptable ceiling concentration of 300 ppm; acceptable maximum peaks above the ceiling of 50 ppm are allowed for 5 minutes duration. OSHA has recommended as of 1977 that the employer shall control workplace concentrations of ethylene dibromide so that no employee is exposed in his workplace to concentrations greater than 1.0 mg/m^3 (0.13 ppm) as a ceiling limit, as determined by a sampling period of 15 minutes. ACGIH (1978) has cited ethylene dibromide as a "substance with recognized carcinogenic potential awaiting reassignment of TLV pending further data acquisition." In November 1978, NCI called EDB "the most potent cancer-causing substance ever found in the animal test program of NCI." Industry sources rebutted this and a decision by EPA on use restriction was imminent.

Routes of Entry: Inhalation of the vapor and absorption through the skin.

Harmful Effects

Local—Prolonged contact of the liquid with the skin may cause erythema, blistering, and skin ulcers. These reactions may be delayed 24-48 hours. Dermal sensitization to the liquid may develop. The vapor is irritating to the eyes and to the mucous membranes of the respiratory tract.

Systemic—Inhalation of the vapor may result in severe acute respiratory injury, central nervous system depression, and severe vomiting. Animal experiments have produced injury to the liver and kidneys. Ethylene dibromide induced squamous cell carcinomas in the stomachs of both Osborne-Mendel rats and (C56B1XC3H)f1 mice when administered via chronic oral intubation at maximum tolerated doses (MTD) and at half MTD's (7).

Ethylene dibromide, without metabolic activation induces base-pair substitution reverse mutations in *S. typhimurium* TA 1530, TA 1535, TA 100, and G 46 plate assays (8)-(11). When tested in polymerase assays which are believed to be indicators of repairable DNA damage, ethylene dibromide was more toxic to *E. coli* p3478 (pol A⁻) than to *E. coli* W3110 (pol A⁺) hence suggesting that it can damage DNA (8).

Ethylene dibromide was not mutagenic in plate assays with *Serratia marcescens* 21 or in the host-mediated assay in mice (10). Ethylene dibromide induced recessive lethal mutations in the *ad*-3 region of a two-component heterokaryon of *Neurospora crassa* (12)(13) as well as X-chromosomal recessive lethals in *Drosophila* (14) and visible mutations in mutable clones in the *Tradescantia* stamen hair somatic test system (15)(16) where it exhibited good dose-response relationships with surface exposures as low as 3.6 ppm compared to 5 ppm with ethyl methane sulfonate (EMS) (15).

Ethylene dibromide did not cause dominant lethal mutations in mice when administered orally (5 doses totalling 50 or 100 mg/kg) or by i.p. injections (18 or 90 mg/kg) (17). It did not cause chromosome breaking effects in human lymphocytes or in *Alluim* roots (18).

The mutagenicity of the vicinal 1,2-dibromides was suggested to be a consequence of their ability to react to form highly unstable bromonium ions in solution,

$$\underset{\substack{| \quad | \\ Br \ Br}}{H_2C-CH_2} \longrightarrow Br^- + \underset{\substack{\backslash +/ \\ Br}}{H_2C-CH_2}$$

"biological alkylating agents" which can alkylate cellular nucleophiles including DNA. The initial product of alkylation of a hetero-atom such as O, N, or S would be the 2-bromoethyl derivatives, which would be a "half-mustard" type reagent capable of another alkylation reaction. Hence, 1,2-dibromoethane (as well as 1,2-dichloroethane) could be considered as a bifunctional alkylating agent, capable of introducing crosslinks into biological molecules (14).

Nauman et al (16) and Ehrenberg et al (19) also designated ethylene dibromide as an alkylating agent suggesting that it reacts via an S_N1 mechanism.

Antifertility effects of ethylene dibromide have been attributed by Edwards et al (20) to a direct alkylating effect of its primary metabolite, the glutathione conjugate of bromoethane, which is a more reactive alkylating agent than ethylene bromide.

Medical Surveillance: Preemployment and periodic examinations should evaluate the skin and eyes, respiratory tract, and liver and kidney functions.

Special Tests: None commonly used.

Personal Protective Methods: 1,2-Dibromoethane can penetrate most types of rubber and leather protective clothing, and, therefore, protective clothing made from other materials should be provided. Masks must be worn in areas with excessive vapor concentrations.

References

(1) Environmental Protection Agency. *Sampling and Analysis of Selected Toxic Substances, Task II—Ethylene Dibromide.* Final Report. Office of Toxic Substances, EPA, Wash., DC, Sept. (1975).

(2) Anon. Ethylene dibromide "ubiquitous" in air, EPA report says. *Toxic Materials News, 3* (1976) 132.

(3) Berck, B. Fumigant residues of carbon tetrachloride, ethylene dichloride, and ethylene dibromide in wheat, flour, bran, middlings and bread. *J. Agr. Food Chem., 22* (1974) 977-984.

(4) Wit, S.L., Besemer, A.F.H., Das, H.A., Goedkoop, W., Loostes, F.E., and Meppelink. Rept. No. 36/39, *Toxicology* (1969) National Institute of Public Health, Bilthoven, Netherlands.

(5) Fishbein, L. Potential hazards of fumigant residues. *Env. Hlth. Persp.* 14 (1976) 39-45.

(6) Olomucki, E., and Bondi, A. Ethylene dibromide fumigation of cereals I. Sorption of ethylene dibromide by grain. *J. Sci. Food Agr.* 6 (1955) 592.

(7) Olson, W.A., Haberman, R.T., Weisburger, E.K., Ward, J.M., and Weisburger, J.H. Induction of stomach cancer in rats and mice by halogenated aliphatic fumigants. *J. Natl. Cancer Inst.* 51 (1973).

(8) Brem, H., Stein, A.B. and Rosenkranz, H.S. The mutagenicity and DNA-modifying effect of haloalkanes. *Cancer Res.* 34 (1974) 2576-79.

(9) McCann, J., Choi, E., Yamasaki, E., and Ames, B.N. Detection of carcinogens as mutagens in the *Salmonella*/microsome test: Assay of 300 chemicals. *Proc. Natl. Acad. Sci.* 72 (1975) 5135-5139.

(10) Buselmaier, V.W., Rohrborn, G., and Propping, P. Mutagenitats-untersuchungen mit pestiziden im hos-mediated assay und mit dem dominanten-letal test an der maus. *Biol. Zbl.* 91 (1972) 311-325.

(11) Buselmaier, V.W., Rohrborn, G., and Propping, P. Comparative investigation of the mutagenicity of pesticides in mammalian test systems. *Mutation Res.* 21 (1973) 25-26.

(12) Malling, H.V. Ethylene dibromide: A potent pesticide with high mutagenic activity. *Genetics.* 61 (1969) s 39.

(13) DeSerres, F.J., and Malling, H.V. Genetic analysis of *ad*-3 mutants of *Neurospora crassa* induced by ethylene dibromide, a commonly used pesticide. *EMS Newsletter.* 3 (1970) 36-37.

(14) Vogel, E., and Chandler, J.L.R. Mutagenicity testing of cyclamate and some pesticides in *Drosophila melanogaster. Experientia.* 30 (1974) 621-623.

(15) Sparrow, A.H., Schairer, L.A., and Villalobos-Pietrini, R. Comparison of somatic mutation rates induced in *Tradescantia* by chemical and physical mutagens. *Mutation Res.* 26 (1974) 265-276.

(16) Nauman, C.H., Sparrow, A.H., and Schairer, L.A. Comparative effects of ionizing radiation and two gaseous chemical mutagens on somatic mutation induction in one mutable and two nonmutable clones of *Tradescantia. Mutation Res.* 38 (1976) 53-70.

(17) Epstein, S.S., Arnold, E., Andrea, J., Bass, W., and Bishop, Y. Detection of chemical

mutagens by the dominant lethal assay in the mouse. *Toxicol. Appl. Pharmacol.* 23 (1972) 288-325.

(18) Kristoffersson, U. Genetic effects of some gasoline additives. *Hereditas* 78 (1974) 319.

(19) Ehrenberg, L., Osterman-Holkar, S., Singh, D., Lundquist, U. On the reaction kinetics and mutagenic activity of methylating and β-halogenoethylating gasoline derivatives. *Radiat. Biol.* 15 (1974) 185-194.

(20) Edwards, K., Jackson, H., and Jones, A.R. Studies with alkylating esters, II. A chemical interpretation through metabolic studies of the infertility effects of ethylene dimethanesulfonate and ethylene dibromide. *Biochem. Pharmacol.* 19 (1970) 1783-1789.

Bibliography

Alumot, (Olomucki), E. and Hardof, Z. Impaired uptake of labelled protein by the ovarian follicles of hens treated with ethylene dibromide. *Comp. Biochem. Physiol.* 39B:6168 (1970).

Amir, D. Sites of spermicidal action of ethylene dibromide in bulls. *J. Reprod. Fert.* 35(3):519-525 (1973).

Memorandum of Alert: Ethylene Dibromide. National Cancer Institute (Oct. 4, 1974).

Occupational Health and Safety Administration. *Criteria for a Recommended Standard: Occupational Exposure to Ethylene Dibromide.* NIOSH Doc. No. 77-221 (1977).

Occupational Safety and Health Standards: Subpart Z—Toxic and Hazardous Substances. 20 CFR 1910.100, Table Z-2.

Olmstead, E.V. 1960. Pathological changes in ethylene dibromide poisoning. *AMA Arch. Indust. Health* 21:525.

Olson, et al. Induction of stomach cancer in rats and mice with halogenated aliphatic fumigants. *J. National Cancer Inst.* 51(6) 1993-1995 (1973).

Regulations for Fuels and Fuel Additives. 40 Federal Register 51995-52337 (Nov. 7, 1975).

Regulations for Rebuttable Presumption Against Registration. 40 CFR 162.11.

Review of Selected Literature on Ethylene Dibromide. EPA, Office of Toxic Substances (Nov. 1974).

Rowe, V.K., H.D. Spencer, D.D. McCollister, R.L. Hollingsworth, and E.M. Adams. 1952. Toxicity of ethylene dibromide determined on experimental animals. *AMA Arch. Indust. Health* p. 158.

Sampling and Analysis of Selected Toxic Substances, Task II — Ethylene Dibromide. EPA, Office of Toxic Substances (performed under Contract No. 68-01-2646) (Publication No. EPA-560/6-75-001, Sept. 1975).

Toxicological Studies of Selected Chemicals, Task II: Ethylene Dibromide. EPA, Office of Toxic Substances (performed under Contract No. 68-01-2646. Preliminary findings, personal communication by C.C. Lee, Project Director) (March 1976).

3,3'-DICHLOROBENZIDINE AND SALTS

Description: $C_6H_3ClNH_2C_6H_3ClNH_2$, 3,3'-dichlorobenzidine, is a gray or purple crystalline solid.

Synonyms: 4,4'-Diamino-3,3'-dichlorobiphenyl, 3,3'-dichlorobiphenyl-4,4'-diamine, 3,3'-dichloro-4,4'-biphenyldiamine.

Potential Occupational Exposures: The major uses of dichlorobenzidine are in the manufacture of pigments for printing ink, textiles, plastics, and crayons and as a curing agent for solid urethane plastics. There are no substitutes for many of its uses.

A partial list of occupations in which exposure may occur includes:

Pigment makers   Polyurethane workers

Permissible Exposure Limits: 3,3'-Dichlorobenzidine and its salts are included in a Federal standard for carcinogens; all contact with it should be avoided.

Routes of Entry: Inhalation and probably percutaneous absorption.

Harmful Effects

Local—May cause allergic skin reactions.

Systemic—3,3'-Dichlorobenzidine was shown to be a potent carcinogen in rats and mice in feeding and injection experiments, but no bladder tumors were produced. However, no cases of human tumors have been observed in epidemiologic studies of exposure to the pure compound.

Medical Surveillance: Preplacement and periodic examinations should include history of exposure to other carcinogens, smoking, alcohol, medication, and family history. The skin, lung, kidney, bladder, and liver should be evaluated; sputum or urinary cytology may be helpful.

Special Tests: None in common use.

Personal Protective Methods: These are designed to supplement engineering controls and to prevent all skin or respiratory contact. Full body protective clothing and gloves should be used by those employed in handling operations. Fullface supplied air respirators of continuous flow or pressure demand type should also be used. On exit from a regulated area, employees should shower and change into street clothes, leaving their protective clothing and equipment at the point of exit to be placed in impervious containers at the end of the work shift for decontamination or disposal. Effective methods should be used to clean and decontaminate gloves and clothing.

Bibliography

Glassman, J.M., and Meigs, J.W. 1951. Benzidine (4,4'-diaminobiphenyl) and substituted benzidines. A microchemical screening technique for estimating levels of industrial exposure from urine and air samples. *AMA Arch. Ind. Hyg. Occup. Med.* 4:519.

Sciarini, L.J., and Meigs, J.W. 1961. Biotransformation of the benzidines—III. Studies on diorthotolidine, dianisidine, and dichlorobenzidine: 3,3'-disubstituted congeners of benzidine (4,4'-diaminophenyl). *Arch. Environ. Health* 2:584.

1,2-DICHLOROETHANE

Description: $ClCH_2CH_2Cl$, 1,2-dichloroethane, is a colorless, flammable liquid which has a pleasant odor, sweetish taste.

Synonyms: Ethylene dichloride, sym-dichloroethane, ethylene chloride, glycol dichloride.

Potential Occupational Exposures: In recent years, 1,2-dichloroethane has found wide use in the manufacture of ethyl glycol, diaminoethylene, polyvinyl chloride, nylon, viscose rayon, styrene-butadiene rubber, and various plastics. It is a solvent for resins, asphalt, bitumen, rubber, cellulose acetate, cellulose ester, and paint; a degreaser in the engineering, textile and petroleum industries; and an extracting agent for soybean oil and caffeine. It is also used as an antiknock agent in gasoline, a pickling agent, a fumigant, and a dry-cleaning agent. It has found use in photography, xerography, water softening, and also in the production of adhesives, cosmetics, pharmaceuticals, and varnishes.

A partial list of occupations in which exposure may occur includes:

Adhesive makers	Insecticide makers
Bakelite processors	Metal degreasers
Camphor workers	Ore upgraders
Dry cleaners	Solvent workers
Exterminators	Textile cleaners
Gasoline blenders	Vinyl chloride makers

In an early document (1) issued in 1976 NIOSH estimated that 18,000 workers were potentially exposed to ethylene dichloride. In a subsequent document issued in 1978 (2) NIOSH estimates that as many as 2 million workers may have occupational exposure to ethylene dichloride. Of these workers, an estimated 34,000 are exposed to ethylene dichloride 4 hours or more per day. These projections are based on the NIOSH National Occupational Hazards Survey (NOHS) conducted between 1972 and 1974, which encompassed some 500,000 employees at approximately 4,775 facilities.

Permissible Exposure Limits: The Federal standard is 50 ppm (200 mg/m³) as an 8-hour TWA with an acceptable ceiling concentration of 100 ppm. Acceptable maximum peaks above the ceiling of 200 ppm are allowed for 5 minutes duration in any 3-hour period.

The current Department of Labor, Occupational Safety and Health Administration (OSHA) standard for occupational exposure to ethylene dichloride is 50 ppm (8-hour time-weighted average). In March 1976, NIOSH recommended an exposure limit of 5 ppm (time-weighted average for up to a 10-hour workday, 40-hour workweek) (2). Neither of these levels may provide adequate protection from potential carcinogenic effects because they were selected to prevent toxic effects other than cancer.

The ACGIH (1978) proposed a TWA value of 10 ppm (40 mg/m^3). In September 1978, NIOSH recommended that the TWA value be lowered to 1.0 ppm and that the ceiling exposure during any 15-minute period be lowered from 15 to 2 ppm. The new limits reflect the limits of analytical sampling and detection ability now available and are said to be justified by indications that EDC may be a carcinogen.

As an interim and prudent measure while the carcinogenicity of ethylene dichloride is being further evaluated, NIOSH recommends (2) that occupational exposure be minimized. Exposures should be limited to as few employees as possible, while minimizing workplace exposure levels with engineering and work practice controls.

Routes of Entry: Inhalation of vapor and skin absorption of liquid.

Harmful Effects

Local—Repeated contact with liquid can produce a dry, scaly, fissured dermatitis. Liquid and vapor may also cause eye damage.

Systemic—Inhalation of high concentrations may cause nausea, vomiting, mental confusion, dizziness, and pulmonary edema. Chronic exposure has been associated with liver and kidney damage.

Laboratory Animal Studies for Carcinogenicity: On March 6, 1978, the Clearinghouse on Environmental Carcinogens (NCI) reviewed and accepted the results of the bioassay of ethylene dichloride, which was performed under contract for the National Cancer Institute (3).

In the bioassay, technical grade 1,2-dichloroethane was tested for possible carcinogenicity in Osborne-Mendel rats and B6C3F1 mice. 1,2-Dichloroethane in corn oil was force-fed at either of two dosages, to groups of 50 male and 50 female animals of each species. Untreated and vehicle control animals were also used. The time-weighted average high and low doses of 1,2-dichloroethane in the chronic study were 95 and 47 mg/kg/day, respectively for rats of both sexes. The high and low time-weighted average doses for the male mice were 195 and 97 mg/kg/day, respectively, and 299 and 149 mg/kg/day, respectively for the female mice.

To relate some of the above information to the work environment, a 70 kg man breathing a typical 10 m^3/day (over an 8-hour work shift) of air contaminated with 50 ppm of ethylene dichloride (the current OSHA standard for exposure to ethylene dichloride) would have an inhalation exposure of about 30 mg/kg/day. Because respiration rate increases with exertion, jobs with higher exertion are likely to be associated with increased respiratory intake.

The National Cancer Institute has concluded that under the conditions of the study, 1,2-dichloroethane was carcinogenic to Osborne-Mendel rats and B6C3F1 mice. The following table summarizes the statistically significant tumors found in the NCI bioassay of 1,2-dichloroethane (3).

Statistically Significant Tumors Found in NCI Bioassay of 1,2-Dichloroethane

Species/Sex	Adverse Effect	Site
Rats/Male	Squamous-Cell Carcinomas	Forestomach
	Hemangiosarcomas	Circulatory System
	Fibromas	Subcutaneous Tissue
Rats/Female	Adenocarcinomas	Mammary Gland
Mice/Female	Adenocarcinomas	Mammary Gland
Mice/Female	Stromal Polyps	Endometrium
	Stromal Sarcomas	Endometrium
Mice/Male and Female	Adenomas	Alveoli and Bronchioli

Two additional studies to assess the carcinogenic potential of ethylene dichloride are currently underway. In an ethylene dichloride inhalation study being conducted in Italy by Dr. C. Maltoni, animals will be exposed to ethylene dichloride for 2 years and observed until the end of their natural lives. No evidence of any exceptional tumor in rats or mice has been found following 100 weeks of exposure (4). Also, Dr. B.M. Goldschmidt has informed NIOSH of bioassays of 1,2-dichloroethane being conducted at the New York University Intitute of Environmental Medicine. Groups of 30 female mice received skin applications of 1,2-dichloroethane for more than one year. None of the animals developed skin tumors, and autopsies did not reveal any unexpected internal lesions or tumors (5).

Other Laboratory Animal Studies: The toxic effects of ethylene dichloride exposure have been studied in a large number of animal species (2). Acute exposure to ethylene dichloride seemed to most frequently affect the cardiovascular system as evidenced by extreme lowering of blood pressure, and cardiac impairment. Other toxic effects include pulmonary edema, fatty degeneration of the liver and kidney (renal tubules) and degeneration of the adrenal cortex. Also, ethylene dichloride is reported to be a weak mutagen in bacteria and mutagenic in fruit flies (6). In the rat, ethylene dichloride has been reported to cross the placental barrier, accumulate in the placenta and fetal tissues, and cause abnormal development of the fetus (7)(8)(9).

Human Toxicity: The acute effects of ethylene dichloride are similar for all routes of entry: ingestion, inhalation, and skin absorption. Acute exposures result in nausea, vomiting, dizziness, internal bleeding, bluish-purple discoloration of the mucous membranes and skin (cyanosis), rapid but weak pulse and unconsciousness. Acute exposures can lead to death from respiratory and circulatory failure. Autopsies in such situations have revealed widespread bleeding and damage in most internal organs. Repeated long-term exposures to ethylene dichloride has resulted in neurologic changes, loss of appetite and other gastrointestinal problems, irritation of the mucous membranes, liver and kidney impairment, and death (1).

Medical Surveillance: Preplacement and periodic examinations should include an evaluation of the skin and liver and kidney functions.

Special Tests: None commonly used; can be determined in expired air.

Personal Protective Methods: There are four basic methods of limiting employee exposure to ethylene dichloride. None of these is a simple industrial hygiene or management decision and careful planning and thought should be used prior to implementation of any of these.

Product Substitution—The substitution of an alternative material with a lower potential health risk is one method. However, extreme care must be used when selecting possible substitutes. Alternatives to ethylene dichloride should be fully evaluated with regard to possible human effects. Unless the toxic effects of the alternative have been thoroughly evaluated, a seemingly safe replacement, possibly only after years of use, may be found to induce serious health effects.

Contaminant Controls—The most effective control of ethylene dichloride, where feasible, is at the source of contamination by enclosure of the operation and/or local exhaust ventilation.

If feasible, the process or operation should be enclosed with a slight vacuum so that any leakage will result in the flow of external air into the enclosure.

The next most effective means of control would be a well designed local exhaust ventilation system that physically encloses the process as much as possible, with sufficient capture velocity to keep the contaminant from entering the work atmosphere.

To ensure that ventilation equipment is working properly, effectiveness (e.g., air velocity, static pressure, or air volume) should be checked at least every three months. System effectiveness should be checked soon after any change in production, process, or control which might result in significant increases in airborne exposures to ethylene dichloride.

Employee Isolation—A third alternative is the isolation of employees. It frequently involves the use of automated equipment operated by personnel observing from a closed control booth or room. The control room is maintained at a greater air pressure than that surrounding the process equipment so that air flow is out of, rather than into, the room. This type of control will not protect those employees who must do process checks, adjustments, maintenance, and related operations.

Personal Protective Equipment—The least preferred method specified in the 1978 NIOSH publication (2), although the only method specified in the 1976 NIOSH publication (1) is the use of personal protective equipment. This equipment, which may include respirators, goggles, gloves, etc., should not be used as the only means to prevent or minimize exposure during routine operations.

Exposure to ethylene dichloride should not be controlled with the use of respirators except:

- During the time period necessary to install or implement engineering or work practice controls; or
- In work situations in which engineering and work practice controls are technically not feasible; or
- For maintenance; or
- For operations which require entry into tanks or closed vessels; or
- In emergencies.

Only respirators approved by the National Institute for Occupational Safety and Health (NIOSH) should be used. Refer to *NIOSH Certified Equipment, December 15, 1975,* NIOSH publication #76-145 and *Cumulative Supplement June 1977, NIOSH Certified Equipment,* NIOSH publication #77-195. The use of faceseal coverlets or socks with any respirator voids NIOSH approvals.

Quantitative faceseal fit test equipment (such as sodium chloride, dioctyl phthalate, or equivalent) should be used. Refer to *A Guide to Industrial Respiratory Protection,* NIOSH publication #76-189 for guidelines on appropriate respiratory protection programs.

In addition, proper maintenance procedures, good housekeeping in the work area and education of employees concerning the nature of the hazard, its control and personal hygiene are all aspects of a good control program.

References

(1) National Institute for Occupational Safety and Health. *Criteria for a Recommended Standard: Occupational Exposure to Ethylene Dichloride.* NIOSH Doc. No. 76-139 (1976).
(2) National Institute for Occupational Safety and Health. *Ethylene Dichloride.* NIOSH Current Intelligence Bulletin No. 25; Wash., DC (April 19, 1978).

(3) *Bioassay of 1,2-Dichloroethane for Possible Carcinogenicity.* DHEW Publication No. (NIH) 78-1305. U.S. Department of Health, Education, and Welfare, Public Health Service, National Institutes of Health, National Cancer Institute, (Camera-Ready Copy dated 1/10/78).
(4) Letter to Director, National Institute for Occupational Safety and Health (NIOSH) from Mr. Albert C. Clark, Manufacturing Chemists Association (MCA), Dec. 22, 1977. Letter to Dr. Arthur Gregory (NIOSH) from Ms. Lucille C. Henschel (MCA), February 28, 1978.
(5) Letter to Dr. Arthur Gregory, National Institute for Occupational Safety and Health (NIOSH) from B.M. Goldschmidt, Ph.D., New York University Medical Center, Institute of Environmental Medicine, February 28, 1978.
(6) Fishbein, L. Industrial mutagens and potential mutagens I. Halogenated aliphatic derivatives. *Mutation Research.* 32:267-308, 1976.
(7) Vozovaya, M.A. The effect of dichloroethane on the sexual cycle and embryogenesis of experimental animals. *Akusk. Ginekol.* (Moskow). 2:57-59, 1977.
(8) Vozovaya, M.A. The effect of small concentrations of benzene and dichloroethane separately and combined on the reproductive function of animals. *Gig. Sanit.* (Moskow). 6:100-102, 1976.
(9) Vozovaya, M.A. The effect of small concentrations of benzene, dichloroethane alone and combined, on the reproductive function of animals and the development of the progeny. *Gig. Tr. Prof. Zabol.* (Moskow). 7:20-23, 1975.

Bibliography

Heppel, L.A., Neal, P.A., Perrin, T.L., Endicott, K.M. and Porterfield, V.T. 1945. The toxicity of 1,2-dichloroethane (ethylene) [sic] III. Its acute toxicity and the effect of protective agents. *J. Pharmcol. Exp. Ther.* 84:53.
McLaughlin, R.S. 1946. Chemical burns of human cornea. *Am. J. Ophthalmol.* 29:1355.
Menshick, H. 1957. Acute inhalation poisoning from symmetrical dichloroethane. *Arch. Gewerbepathol. Gewerbehyg.* 15:241.
Spencer, H.F., Rowe, V.L., Adams, E.M., McCollister, D.D. and Irish, D.D. 1951. Vapor of ethylene dichloride determined by experiments on laboratory animals. *AMA Arch. Indust. Hyg. Occup. Med.* 4:482.
Yodiken, R.E. and Babcock, J.R., Jr., 1973. 1,2-Dichloroethane poisoning. *Arch Environ. Health* 26:281.

1,2-DICHLOROETHYLENE

Description: ClCH=CHCl, 1,2-dichloroethylene, exists in two isomers, cis 60% and trans 40%. There are variations in toxicity between these two forms. At room temperature, it is a liquid with a slight acrid, ethereal odor. Gradual decomposition results in hydrochloric acid formation in the presence of ultraviolet light or upon contact with hot metal.

Synonyms: Acetylene dichloride, sym-dichloroethylene, 1,2-dichloroethene.

Potential Occupational Exposures: 1,2-Dichloroethylene is used as a solvent for waxes, resins, and acetylcellulose. It is also used in the extraction of rubber, as a refrigerant, in the manufacture of pharmaceuticals and artificial pearls, and in the extraction of oils and fats from fish and meat.

A partial list of occupations in which exposure may occur includes:

Carbolic acid processors	Solvent workers
Drug makers	Wax makers
Dry cleaners	

Permissible Exposure Limits: The Federal standard is 200 ppm (790 mg/m^3).

Route of Entry: Inhalation of the vapor.

Harmful Effects

Local—This liquid can act as a primary irritant producing dermatitis and irritation of mucous membranes.

Systemic—1,2-Dichloroethylene acts principally as a narcotic, causing central nervous system depression. Symptoms of acute exposure include dizziness, nausea and frequent vomiting, and central nervous system intoxication similar to that caused by alcohol. Renal effects, when they do occur, are transient.

Medical Surveillance: Consider possible irritant effects on skin or respiratory tract as well as liver and renal function in preplacement or periodic examinations.

Special Tests: None commonly used; expired air analyses may be useful in detecting exposure.

Personal Protective Methods: Barrier creams and gloves are needed to protect the skin. In high vapor concentrations masks and protective clothing are required.

Bibliography

McBirney, R.S. 1954. Trichloroethylene and dichloroethylene poisoning. *AMA Arch. Ind. Hyg. Occup. Med.* 10:130.

DICHLOROETHYL ETHER

Description: ClCH$_2$CH$_2$OCH$_2$CH$_2$Cl, dichloroethyl ether, is a clear, colorless liquid with a pungent, fruity odor.

Synonyms: Dichloroether, dichloroethyl oxide, sym-dichloroethyl ether, bis-(2-chloroethyl) ether, 2,2'-dichloroethyl ether.

Potential Occupational Exposures: Dichloroethyl ether is used in the manufacture of paint, varnish, lacquer, soap, and finish remover. It is also used as a solvent for cellulose esters, naphthalenes, oils, fats, greases, pectin, tar, and gum; in dry cleaning; in textile scouring; and in soil fumigation.

A partial list of occupations in which exposure may occur includes:

Cellulose ester makers	Oil processors
Degreasers	Paint makers
Dry cleaners	Soap makers
Ethylcellulose processors	Tar processors
Fat processors	Textile scourers
Gum processors	Varnish workers
Lacquer makers	

Permissible Exposure Limits: The Federal standard for dichloroethyl ether is 15 ppm (90 mg/m^3); however, the ACGIH recommended TLV in 1978 was 5 ppm (30 mg/m^3).

Route of Entry: Inhalation of vapor, percutaneous absorption.

Harmful Effects

Local—Irritation of the conjunctiva of the eyes with profuse lacrimation, irritation to mucous membranes of upper respiratory tract, coughing, and nausea may result from exposure to vapor. The liquid when placed in animal eyes has produced damage. Vapors in minimal concentrations (3 ppm) are distinctly irritating and serve as a warning property.

Systemic—In animal experiments dichloroethyl ether has caused severe irritation of the respiratory tract and pulmonary edema. Animal experiments have also shown dichloroethyl ether to be capable of causing drowsiness, dizziness, and unconsciousness at high concentrations. Except for accidental inhalation of high concentrations, the chief hazard in industrial practice is a mild bronchitis which may be caused by repeated exposure to low concentrations.

Medical Surveillance: Consideration should be given to the skin, eyes, and respiratory tract, and to the central nervous system in placement or periodic examinations.

Special Tests: None have been proposed.

Personal Protective Methods: In cases of vapor concentrations, protective clothing with fullface respirator with air supply should be worn. Skin protection (gloves, protective clothing) is needed to prevent skin absorption. Goggles should be used to prevent eye burns.

Bibliography

Schrenk, H.H. 1933. Acute response of guinea pigs to vapors of some new commercial organic compounds. VII. Dichloroethyl ether. Public Health Reports 48:1389.

DIISOBUTYL KETONE (See "Ketones")

N,N-DIMETHYLACETAMIDE

Description: $CH_3CON(CH_3)_2$, dimethylacetamide, is a colorless, nonvolatile liquid.

Synonyms: Acetic acid dimethylamide, DMA, DMAC, acetyl dimethylamide.

Potential Occuptional Exposures: Dimethylacetamide is used commercially as a solvent in various industries.

A partial list of occupations in which exposure may occur includes:

Solvent workers

Permissible Exposure Limits: The Federal standard is 10 ppm (35 mg/m^3).

Routes of Entry: Inhalation of vapor and absorption through intact skin.

Harmful Effects

Local—None known.

Systemic—Jaundice has been noted in workers exposed chronically to dimethylacetamide vapor although skin absorption may also have occurred. Liver injury consists of cord-cell degeneration, but recovery is usually rapid. Other symptoms from large oral doses as an anticancer drug include depression, lethargy, and visual and auditory hallucinations.

Medical Surveillance: Preplacement and periodic medical examinations should give special attention to skin, central nervous system, and liver function or disease.

Special Tests: None commonly used.

Personal Protective Methods: Organic vapor masks or air supplied respirators may be required in elevated vapor concentrations. Percutaneous absorption should be prevented by gloves and other protective clothing. Goggles should be used to prevent eye splashes. In cases of spills or splashes, the wet clothing should be immediately removed and the involved skin area thoroughly cleaned. Clean clothing should be issued to workers on a daily basis and showers taken before changing to street clothes.

Bibliography

Horn, H.J. 1961. Toxicology of dimethylacetamide. *Toxicol. Appl. Pharmacol.* 3:12.

4-DIMETHYLAMINOAZOBENZENE

Description: $C_6H_5NNC_6H_4N(CH_3)_2$, 4-dimethylaminoazobenzene, is a flaky yellow crystal.

Synonyms: Aniline-N,N-dimethyl-p-(phenylazo), benzeneazo dimethylaniline, fat yellow, oil yellow, butter yellow, methyl yellow.

Potential Occupational Exposures: 4-Dimethylaminoazobenzene is only used for research purposes. It was formerly used as a dye, but has been substituted by diethylaminoazobenzene. It was also formerly used for coloring margarine and butter.

A partial list of occupations in which exposure may occur includes:

Research workers

Permissible Exposure Limits: 4-Dimethylaminoazobenzene is included in the Federal standard for carcinogens; all contact with it should be avoided.

Routes of Entry: Probably inhalation and percutaneous absorption.

Harmful Effects

Local—Unknown.

Systemic—Cancer of the liver has been produced in rats and mice in feeding experiments. No human effects have been reported.

Medical Surveillance: Preplacement and periodic examinations should include a history of exposure to other carcinogens; use of alcohol, smoking, and medications; and family history. Special attention should be given to liver size and liver function tests.

Special Tests: None commonly used.

Personal Protective Methods: These are designed to supplement engineering controls and to prevent all contact with skin and the respiratory tract. Protective clothing and gloves should be provided, and also appropriate type dust or supplied air respirators. On exit from a regulated area, employees should shower and change into street clothes, leaving their clothes at the point of exit, to be placed in impervious containers at the end of the work shift for decontamination or disposal.

Bibliography

Miller, J.A., and E.C. Miller. 1953. The carcinogenic aminoazo dyes. *Adv. Cancer Res.* 1:339.

N,N-DIMETHYLFORMAMIDE

Description: $HCON(CH_3)_2$, dimethylformamide, is a colorless liquid which at 25°C is soluble in water and organic solvents. It has a fishy, unpleasant odor at relatively low concentrations, but the odor has no warning property.

Synonyms: DMF, the "universal organic solvent," DMFA.

Potential Occupational Exposures: Dimethylformamide has powerful solvent properties for a wide range of organic compounds. Because of dimethyl formamide's physical properties, it has been used when solvents with a slow rate of evaporation are required.

It finds particular usage in the manufacture of polyacrylic fibers, butadiene, purified acetylene, pharmaceuticals, dyes, petroleum products, and other organic chemicals.

A partial list of occupations in which exposure may occur includes:

Acetylene purifiers	Organic chemical synthesizers
Butadiene makers	Petroleum refinery workers
Drug makers	Resin makers
Dye makers	Solvent workers
	Synthetic fiber makers

Permissible Exposure Limits: The Federal standard is 10 ppm (30 mg/m³).

Routes of Entry: Inhalation of vapor, and it is readily absorbed through intact skin.

Harmful Effects

Local—Dimethylformamide exposure may cause dermatitis.

Systemic—Inhalation of dimethylformamide or skin contact with this chemical may cause colicky abdominal pain, anorexia, nausea, vomiting, constipation, diarrhea, facial flushing (especially after drinking alcohol), elevated blood pressures, hepatomegaly, and other signs of liver damage. This chemical has produced kidney damage in animals.

Medical Surveillance: Preplacement and periodic examinations should be concerned particularly with liver and kidney function and with possible effects on the skin.

Special Tests: None in common use.

Personal Protective Methods: Organic vapor masks or air supplied respirators may be required in elevated vapor concentrations. Percutaneous absorption should be prevented by gloves and other protective clothing. Goggles should be used to prevent eye splashes. In cases of spills or splashes, the wet clothing should be immediately removed and the skin area thoroughly cleaned. Clean clothing should be issued to workers on a daily basis and showers taken before changing to street clothes.

Bibliography

Clayton, J.W., Jr., J.R. Barnes, D.B. Hood, and G.W.H. Schepers. 1963. The inhalation toxicity of dimethylformamide (DMF). *Am. Ind. Hyg. Assoc. J.* 24:144.
Martelli, D. 1960. Toxicology of dimethylformamide. *Med. Lav.* 51:123.
Massmann, W. 1956. Toxcological investigations on dimethylformamide. *Br. J. Ind. Med.* 13:51.
Potter, H.P. 1973. Dimethylformamide-induced abdominal pain and liver injury. *Arch. Environ. Health* 27:340.

DIMETHYL SULFATE

Description: $(CH_3)_2SO_4$, dimethyl sulfate, is an oily, colorless liquid slightly soluble in water, but more soluble in organic solvents.

Synonyms: Sulfuric acid dimethyl ester.

Potential Occupational Exposures: Industrial use of dimethyl sulfate is based upon its methylating properties. It is used in the manufacture of methyl esters, ethers and amines, in dyes, drugs, perfume, phenol derivatives, and other organic chemicals (1)-(3). It is also used as a solvent in the separation of mineral oils.

A partial list of occupations in which exposure may occur includes:

Amine makers	Organic chemical synthesizers
Drug makers	Perfume makers
Dye makers	Phenol derivative makers

Permissible Exposure Limits: The Federal standard is 1 ppm (5 mg/m^3). The ACGIH (1978) set TWA values of 0.1 ppm (0.5 mg/m^3) with the note that dimethyl sulfate is an "Industrial Substance Suspect of Carcinogenic Potential for Man."

Routes of Entry: Inhalation of vapor; percutaneous absorption of liquid.

Harmful Effects

Local—Liquid is highly irritating and causes skin vesiculation and analgesia. Lesions are typically slow-healing and may result in scar tissue while analgesia may last several months. Liquid and vapor are irritating to the mucous membranes, and exposure produces lacrimation, rhinitis, edema of the mucosa of the mouth and throat, dysphagia, sore throat, and hoarseness. Irritation of the skin and mucous membranes may be delayed in appearance. Eye irritation may result in conjunctivitis, keratitis, and photophobia. In severe cases corneal opacities, perforation of the nasal septum and permanent or persistent visual disorders have been reported.

Systemic—The toxicity of dimethyl sulfate is based upon its alkylating properties and its hydrolysis to sulfuric acid and methyl alcohol. Acute exposure may cause respiratory dysfunctions such as pulmonary edema, bronchitis, and pneumonitis following a latent period of 6 to 24 hours. Cerebral edema and other central nervous system effects such as drowsiness, temporary blindness, tachycardia or bradycardia may be linked to dimethyl sulfate's effect on nerve endings. Secondary pulmonary effects such as susceptibility to infection, as well as more pronounced effects in those persons with pre-existing respiratory disorders, are also noteworthy. Chronic poisoning occurs only rarely and is usually limited to ocular and respiratory disabilities. It has been reported to be carcinogenic in rats, but this has not been verified in man.

Dimethyl sulfate has been shown to be carcinogenic in the rat (the only species tested) by inhalation (1)(4), subcutaneous injection (4) and following prenatal exposure (4). It is carcinogenic to the rat in a single-dose exposure (1)(4). The possibility of carcinogenicity of dimethyl sulfate in man occupationally exposed for 11 years has been raised (5), however good epidemiological evidence is unavailable to confirm this (1)(5). Dimethyl sulfate is mutagenic in *Drosophila* (6)(7), *E. coli* (6) and *Neurospora* (8)(9) and induces chromosome breakage in plant material (10).

Medical Surveillance: Preplacement and periodic medical examinations should give special consideration to the skin, eyes, central nervous system, lungs. Chest x-rays should be taken and lung, liver, and kidney functions evaluated. Sputum and urinary cytology may be useful in detecting the presence or absence of carcinogenic effects.

Special Tests: None in common use.

Personal Protective Methods: These are designed to supplement engineering controls and to reduce skin, eye, or respiratory contact to a negligible level. The liquid and the vapor of dimethyl sulfate are extremely irritating so that the skin, eyes, as well as the respiratory tract should be protected at all times. Protective clothing, gloves, goggles, face shields, aprons, and boots should be used in areas where there is danger of splash or spill. Fullface vapor masks or supplied air respirators may be necessary in areas of vapor build-up or leaks. Attention should be given to personal hygiene with a change of work clothes daily and shower before change to street clothes.

References

(1) *IARC*. Dimethyl Sulfates, in Vol. 4, International Agency for Research on Cancer, Lyon, France (1974) pp. 271-276.
(2) *IARC*. Diethyl Sulfate, in Vol. 4, International Agency for Research on Cancer, Lyon, France (1974), 277-281.
(3) Fishbein, L., Flamm, W.G., and Falk, H.L. *Chemical Mutagens*. Academic Press, New York, (1970) pp. 216-217.
(4) Druckrey, H., Kruse, H., Preussmann, R., Ivankovic, S., and Landschutz, C. Cancerogene alkylierende substanzen. III. Alkylhalogenide, -sulfate, -sulfonate und ringgespannte heterocyclen. *Z. Krebsforsch.* 74 (1970) 241.
(5) Druckrey, H., Preussmann, R., Nashed, N., and Ivankovic, S. Carcinogene alkylierende substanzen I. Dimethylsulfate, carcinogene wirkung an ratten und wahrscheinliche ursache von berufskrebs. *Z. Krebsforsch.* 68 (1966) 103.
(6) Alderson, T. Ethylation versus methylation in mutation of *E. coli and Drosophila. Nature* 203 (1964) 1404.
(7) Pelecanos, M. and Alderson, T. The mutagenic action of diethylsulfate in *Drosophila melanogaster* I. The dose-mutagenic response to larval and adult feeding. *Mutation Res.* 1 (1964) 173.
(8) Malling, H.V. Identification of the genetic alterations in nitrous acid-induced AD-3 mutants of *Neurospora crassa. Mutation Res.* 2 (1965) 320.
(9) Kolmark, G. Mutagenic properties of esters of inorganic acids investigated by the *Neurospora* back mutation test. *Ser. Physiol.* 26 (1956) 205-220.
(10) Loveless, A., and Ross, W.C.J. Chromosome alteration and tumour inhibition by nitrogen mustards: The hypothesis of cross-linking alkylation. *Nature* 166 (1950) 1113.

Bibliography

Haswell, R.W. 1960. Dimethyl sulphate poisoning by inhalation. *J. Occup. Med.* 2:454.
Littler, T.R., and R.B. McConnell. 1955. Dimethyl sulphate poisoning. *Br. J. Ind. Med.* 12:54.

Thiess, A.M., and P.J. Goldman. 1968. Arbeitsmedizinische fragen im zusammenhang mit der dimethylsufat-intoxikation. Beobachtungen aus 30 jahren in der BASF. *Zentralbl. Arbeitsmed.* 18:195.

DINITROBENZENE

Description: $C_6H_4(NO_2)_2$, dinitrobenzene, may exist in three isomers; the meta form is the most widely used.

Synonyms: Dinitrobenzol.

Potential Occupational Exposures: Dinitrobenzene is used in the synthesis of dyestuffs, dyestuff intermediates, and explosives and in celluloid production.

A partial list of occupations in which exposure may occur includes:

Celluloid makers   Explosives workers
Dye makers   Organic chemical synthesizers

Permissible Exposure Limits: The Federal standard for all isomers of dinitrobenzene is 1 mg/m^3 (0.15 ppm).

Routes of Entry: Inhalation and percutaneous absorption of liquid.

Harmful Effects

Local—Exposure to dinitrobenzene may produce yellowish coloration of the skin, eyes, and hair.

Systemic—Exposure to any isomer of dinitrobenzene may produce methemoglobinemia, symptoms of which are headache, irritability, dizziness, weakness, nausea, vomiting, dyspnea, drowsiness, and unconsciousness. If treatment is not given promptly, death may occur. Consuming alcohol, exposure to sunlight, or hot baths may make symptoms worse. Dinitrobenzene may also cause a bitter almond taste or burning sensation in the mouth, dry throat, and thirst. Reduced vision may occur. In addition liver damage, hearing loss, and ringing of the ears may be produced. Repeated or prolonged exposure may cause anemia.

Medical Surveillance: Preemployment and periodic examinations should be concerned particularly with a history of blood dyscrasias, reactions to medications, alcohol intake, eye disease, and skin and cardiovascular status. Liver and renal functions should be evaluated periodically as well as blood and general health.

Special Tests: Methemoglobin levels should be followed until normal in all cases of suspected cyanosis. Dinitrobenzene can be determined in the urine; levels greater than 25 mg/liter may indicate significant absorption.

Personal Protective Methods: Dinitrobenzene is readily absorbed through intact skin and its vapors are highly toxic. Protective clothing impervious to the liquid should be worn in areas where the likelihood of splash or spill exists. When splash or spill occurs on ordinary work clothes, they should be removed immediately and the area washed thoroughly. In areas of elevated vapor concentrations fullface masks with organic vapor canisters or air supplied respirators with fullface piece should be used. Daily changes of work clothing and mandatory showering at the end of each shift before changing to street clothes should be enforced.

Bibliography

Beritic, T. 1956. Two cases of meta-dinitrobenzene poisoning with unequal response. *Brit. J. Ind. Med.* 13:114.

DINITRO-o-CRESOL

Description: $CH_3C_6H_2(NO_2)_2OH$, dinitro-o-cresol, exists in 9 isomeric forms of which 3,5-dinitro-o-cresol is the most important commercially. It is a yellow crystalline solid.

Synonyms: DNOC; 4,6-dinitro-o-cresol is also known as 3,5-dinitro-o-cresol, 2-methyl-4,6-dinitrophenol, 3,5-dinitro-2-hydroxytoluene.

Potential Occupational Exposures: DNOC is widely used in agriculture as a herbicide and pesticide; it is also used in the dyestuff industry.

A partial list of occupations in which exposure may occur includes:

Dye makers   Pesticide workers
Herbicide workers

Permissible Exposure Limits: The Federal standard for all isomers of DNOC is 0.2 mg/m^3.

Routes of Entry: Inhalation and percutaneous absorption.

Harmful Effects

Local—None reported except for staining of skin and hair.

Systemic—DNOC blocks the formation of high energy phosphate compounds, and the energy from oxidative metabolism is liberated as heat. Early symptoms of intoxication by inhalation or skin absorption are elevation of the basal metabolic rate and rise in temperature accompanied by fatigue, excessive sweating, unusual thirst, and loss of weight. The clinical picture resembles in part a thyroid crisis. Weakness, fatigue, increased respiratory rate, tachycardia, and fever may lead to rapid deterioration and death. Bilateral cataracts have been seen following oral ingestion for therapeutic purposes.

These have not been seen during industrial or agricultural use.

Medical Surveillance: Consider eyes, thyroid, and cardiovascular system, as well as general health.

Special Tests: None commonly used.

Personal Protective Methods: Since dinitro-o-cresol is used extensively in agriculture as well as industry, worker education to the toxic properties of the chemical are necessary. Where there is a possibility of skin contamination or vapor inhalation, full protection should be provided. Impervious protective clothing and fullface masks with organic vapor canisters or air supplied respirators are advised. A clean set of work clothes daily and showers following each shift before change to street clothes are essential.

Bibliography

Bistrup, P.L., and D.J.H. Payne. 1951. Poisoning by dinitro-ortho-cresol; report of eight fatal cases occuring in Great Britain. *Br. Med. J.* 2:16.

Harvey, D.G., P.L. Bistrup, and J.A.L. Bonnell. 1951. Poisoning by dinitro-ortho-cresol; some observations on the effects of dinitro-ortho-cresol administered by mouth to human volunteers. *Brit. Med. J.* 2:13.

Hayes, W.J., Jr. *Clinical Handbook on Economic Poisons.* Pub. 476, p. 109. U.S. Government Printing Office, Washington.

Markicevic, A., D. Prpic-Majic, and N. Bosnar-Turk. 1972. Rezultati ciljanih pregleda radnika eksponiranih dinitroorth krezolu (DNOC). *Ark. Hig. Rad. Toksikol.* 23:1.

DINITROPHENOL

Description: There are six isomers of dinitrophenol, $C_6H_3(NO_2)_2OH$, of which 2,4-dinitrophenol is the most important industrially. It is an explosive, yellow crystalline solid.

Synonyms: DNP.

Potential Occupational Exposures: 2,4-DNP is used in the manufacturing of dyestuff intermediates, wood preservatives, pesticides, herbicides, explosives, chemical indicators, photograph developers, and also in chemical synthesis.

A partial list of occupations in which exposure may occur includes:

Chemical indicator makers   Organic chemical synthesizers
Dye makers   Photographic developer makers
Explosive workers   Wood preservative workers
Herbicide workers

Permissible Exposure Limits: There is no Federal standard for DNP. A useful guideline of 0.2 mg/m^3 is based on data for dinitro-o-cresol.

Routes of Entry: Percutaneous absorption and inhalation of dust and vapors.

Harmful Effects

Local—DNP causes yellow staining of exposed skin. Dermatitis may be due to either primary irritation or allergic sensitivity.

Systemic—The isomers differ in their toxic effects. In general, DNP disrupts oxidative phosphorylation (as in the case of DNOC) which results in increased metabolism, oxygen consumption, and heat production. Acute intoxication is characterized by sudden onset of fatigue, thirst, sweating, and oppression of the chest. There is rapid respiration, tachycardia, and a rise in body temperature. In less severe poisoning, the symptoms are nausea, vomiting, anorexia, weakness, dizziness, vertigo, headache, and sweating. The liver may be sensitive to pressure, and there may also be jaundice. DNP poisoning is more severe in warm environments. If not fatal, the effects are rapidly and completely reversible. Chronic exposure results in kidney and liver damage and cataract formation. Occasional hypersensitivity reactions, e.g., neutropenia, skin rashes, peripheral neuritis, have been seen after oral use.

Medical Surveillance: Consider skin, eyes, thyroid, blood, central nervous system, liver and kidney function, as well as general health in preplacement and periodic examinations.

Special Tests: Can be measured in urine as such or as an aminophenol derivative.

Personal Protective Methods: Because of its wide use in agriculture, lumbering, photography, as well as in the petrochemical industry, worker education to the toxic properties of dinitrophenol are important. Impervious protective clothing, fullface masks with organic vapor canisters or air supplied respirators are necessary in areas of high concentration of dust or vapor. Spills and splashes that contaminate clothing require the worker to immediately change clothes and wash the area thoroughly. Workers should have clean work clothes on every shift and should be required to shower prior to changing to street clothing.

Bibliography

Gisclard, J.B., and M.M. Woodward. 1946. 2,4-Dinitrophenol poisoning: a case report. *J. Ind. Hyg. Toxicol.* 28:47.
Gosselin, R.E., H.C. Hodge, R. P. Smith, and M.N. Gleason. 1976. *Clinical Toxicology of Commercial Products,* 4th ed. Williams and Wilkins Co.. Baltimore.

DINITROTOLUENE

Description: Six isomers of dinitrotoluene, $C_6H_3(NO_2)_2CH_3$, exist, the most important being 2,4-dinitrotoluene.

Synonyms: Dinitrotoluol, DNT.

Potential Occupational Exposures: DNT is used in the manufacture of explosives and dyes in organic synthesis, e.g., trinitrotoluene.

A partial list of occupations in which exposure may occur includes:

Dye makers   Organic chemical synthesizers
Explosive workers

Permissible Exposure Limits: The Federal standard is 1.5 mg/m^3.

Routes of Entry: Inhalation of vapor and percutaneous absorption of liquid.

Harmful Effects

Local—None.

Systemic—The effects from exposure to dinitrotoluene are caused by its capacity to produce anoxia due to the formation of methemoglobin. Cyanosis may occur with headache, irritability, dizziness, weakness, nausea, vomiting, dyspnea, drowsiness, and unconsciousness. If treatment is not given promptly, death may occur. The onset of symptoms may be delayed. The ingestion of alcohol may cause increased susceptibility. Repeated or prolonged exposure may cause anemia.

Medical Surveillance: Preemployment and periodic examinations should be concerned particularly with a history of blood dyscrasias, reactions to medications, alcohol intake, eye disease, skin, and cardiovascular status. Liver and renal functions should be evaluated periodically as well as blood and general health.

Special Tests: None commonly used. Forms a blue color with alcoholic NaOH.

Personal Protective Methods: Liquid soaked clothing should be immediately removed and the skin area washed thoroughly. Impervious protective clothing should be provided if skin exposure to liquid is anticipated. In areas of elevated vapor concentration, fullface masks with organic vapor canisters or air-supplied respirators should be required.

Bibliography

Norwood, W.D. 1943. Trinitrotoluene (TNT), its effective removal from the skin by a special liquid soap. *Ind. Med.* 12:206.

DIOXANE

Description: $OCH_2CH_2OCH_2CH_2$, dioxane, is a volatile, colorless liquid that may form explosive peroxides during storage.

Synonyms: 1,4-Diethylene dioxide, diethylethene ether, 1,4-dioxane, para-dioxane.

Potential Occupational Exposures: Dioxane finds its primary use as a solvent for cellulose acetate, dyes, fats, greases, lacquers, mineral oil, paints, polyvinyl polymers, resins, varnishes, and waxes. It finds particular usage in paint and varnish strippers, as a wetting agent and dispersing agent in textile processing, dye baths, stain and printing compositions, and in the preparation of histological slides.

A partial list of occupations in which exposure may occur includes:

Adhesive workers	Histology technicians
Cellulose acetate workers	Lacquer makers
Cement workers	Metal cleaners
Degreasers	Oil processors
Deodorant makers	Paint makers
Detergent workers	Polish makers
Emulsion makers	Shoe cream makers
Fat processors	Varnish remover makers
Glue makers	Wax makers

Permissible Exposure Limits: The Federal standard is 100 ppm (360 mg/m³); however, the ACGIH 1978 recommended TLV was 50 ppm (180 mg/m³) of technical grade. NIOSH has promulgated a recommended standard as of 1977 such that occupational exposure to dioxane shall be controlled so that employees are not exposed at airborne concentrations greater than 1 ppm (3.6 mg/m³) based on a 30-minute sampling period.

Route of Entry: Inhalation of vapor as well as percutaneous absorption.

Harmful Effects

Local—Liquid and vapor may be irritating to eyes, nose, and throat.

Systemic—Exposure to dioxane vapor may cause drowsiness, dizziness, loss of appetite, headache, nausea, vomiting, stomach pain, and liver and kidney damage. Prolonged skin exposure to the liquid may cause drying and cracking.

The principal toxic effects of dioxane have long been known to be centrilobular, hepatocellular and renal tubular, epithelial degeneration and necrosis (4)-(7).

More recent reports by Argus et al (8) and Hoch-Ligeti et al (9) described nasal and hepatic carcinomas in rats ingesting water containing large doses of dioxane (up to 1.8% of dioxane in the drinking water for over 13 months). For example, Argus et al (8) reported hepatomas in Wistar rats maintained on drinking water containing 1% dioxane for 63 weeks while Hoch-Ligeti et al (9) described the induction of nasal cavity carcinomas in Wistar rats maintained on drinking water containing from 0.75 to 1.8% dioxane for over 13 months.

Studies by Kociba et al (10) in 1974 indicated a dose response for the toxicity of dioxane in Sherman strain rats. Daily administration of 1% dioxane (in drinking water to male and female rats of 1015 and 1599 mg/kg/day respectively) for up to 2 years caused pronounced toxic effects including the occurrence of hepatic and nasal tumors. There was an induction of untoward effects, liver and kidney damage, but not tumor induction in male and female rats receiving 0.1% dioxane (equivalent to approximately 94 and 148 mg/kg/day respectively) in the drinking water, and female rats receiving 0.01% dioxane in the drinking water (equivalent to approximately 9.6 and 19.0 mg/kg/day, respectively) showed no evidence of tumor formation or other toxic effects considered to be related to treatment.

Gehring et al (11) postulated that the toxicity and carcinogenicity of 1,4-dioxane (as well as vinyl chloride) are expressed only when doses are sufficient to overwhelm their detoxification mechanisms. When such doses are given, there is a disproportionate retention of the compound per se and/or its metabolites in the body. Also observed are changes in the biochemical status of the animals consistent with accepted mechanisms for cancer induction (11).

Although dioxane is considered to be a weak to moderate hepatic carcinogen (8)(12), the mechanism of its carcinogenic action is not understood. Earlier suggestions were advanced that by virtue of the potent hydrogen bond breaking (13) and protein denaturing action (14) of dioxane, the molecular basis for carcinogenic action lies in the inactivation of key cellular macromolecules involved in metabolic control. Although acute toxicity studies suggested involvement of microsomal mixed function oxidases, pretreatment with enzyme inducers had little or no effect on covalent binding (12). No microsome-catalyzed dioxane binding to exogenous DNA was observed under conditions that allowed significant binding of benzo[a]pyrene. Incubation of isolated microsomes or nuclei also showed no enzyme-catalyzed binding of dioxane (12).

It had been earlier postulated by Hoch-Ligeti et al (9) that a reactive free radical or a carbonium ion may arise in the metabolism of dioxane and may represent a proximate carcinogen. Another possibility was that a peroxide of dioxane may account for its carcinogenicity (9).

Medical Surveillance: Preplacement and periodic examinations should be directed to symptoms of headache and dizziness, as well as nausea and other gastrointestinal disturbances. The condition of the skin and of renal and liver function should be considered.

Special Tests: No specific bio-monitoring tests are available. However, the pharmacokinetic and metabolic fate of 1,4-dioxane has been shown to be dose dependent in rats due to a limited capacity to metabolize dioxane to β-hydroxyethoxyacetic acid (HEAA) (1)(2), the major urinary metabolite.

1,4-Dioxane and HEAA were found in the urine of dioxane plant personnel exposed to a time-weighted average concentration of 1.6 ppm dioxane for 7.5 hours. The average concentrations of dioxane and HEAA in samples of urine collected at the end of each workday were 3.5 and 414 μmol/liter respectively (3). The high ratio of HEAA to dioxane (118 to 1) suggests that at low-exposure concentrations, dioxane is rapidly metabolized to HEAA.

Personal Protective Methods: In areas of vapor concentration, protective clothing, barrier creams, gloves, and masks should be used.

References

(1) Young, J.D., and Gehring, P.J. The dose-dependent fate of 1,4-dioxane in male rats. *Toxicol. Appl. Pharmcol. 33* (1975) 183.

(2) Young, J.D., Braun, W.H. LeBeau, J.E., and Gehring, P.J. Saturated metabolism as the mechanism for the dose-dependent fate of 1,4-dioxane in rats. *Toxicol. Appl. Pharmacol. 37* (1976) 138.

(3) Young, J.D., Braun, W.H., Gehring, P.J., Horvath, B.S., and Daniel, R.L. 1,4-Dioxane and β-hydroxyethoxyacetic acid excretion in urine of humans exposed to dioxane vapors. *Toxicol. Appl. Pharmacol. 38* (1976) 643-646.

(4) De Navasquex, A. Experimental tubular necrosis of the kidney accompanied by liver changes due to dioxan poisoning. *J. Hyg. 35* (1935) 540-548.

(5) Fairley, A., Linton, E.C., and Ford-Moore, A.H. The toxicity to animals of 1,4-dioxane. *J. Hyg. 34* (1934) 486-501.

(6) Schrenk, H.H., and Yant, W.P. Toxicity of dioxane. *J. Ind. Hyg. Toxicol. 18* (1936) 448-460.

(7) Kesten, H.D., Mulinos, M.G., and Pomerantz, L. Pathologic effects of certain glycols and related compounds. *Arch. Pathol.* 27 (1939) 447-465.

(8) Argus, M.F., Sohal, R.S., Bryant, G.M., Hoch-Ligeti, C., and Arcos, J.C. Dose-response and ultrastructural alterations in dioxane carcinogenesis. *Eur. J. Cancer,* 9 (1973) 237-243.

(9) Hoch-Ligeti, C., Argus, M.F., and Arcos, J.C. Induction of carcinomas in the nasal cavity of rats by dioxane. *Brit. J. Cancer,* 24 (1970) 164-170.

(10) Kociba, R.J., McCossister, S.B., Park, C., Torkelson, T.R., and Gehring, P.J. 1,4-Dioxane. I. Results of a 2-year ingestion study in rats. *Toxicol. Appl. Pharmacol.* 30 (1974) 275-286.

(11) Gehring, P.J., Watanabe, P.G. and Young, J.D. The relevance of dose-dependent pharmacokinetics and biochemical alterations in the assessment of carcinogenic hazard of chemicals. Presented at Meeting of Origins of Human Cancer, Cold Spring Harbor, NY, Sept. 7-14 (1976).

(12) Woo, Y.T., Argus, M.F., and Arcos, J.C. Dioxane carcinogenesis—apparent lack of enzyme-catalyzed covalent binding to macromolecules. *The Pharmacologist* (1976) 158.

(13) Argus, M.F., Arcos, J.C., Alam, A., and Mathison, J.H. A viscometric study of hydrogen-bonding properties of carcinogenic nitrosamines and related compounds. *J. Med. Pharm. Chem.,* 7 (1964) 460-465.

(14) Bemis, J.A., Argus, M.F., and Arcos, J.C. Studies on the denaturation of biological macromolecules by chemical carcinogens. III. Optical rotary dispersion and light scattering changes in ovalbumin during denaturation and aggregation by water soluble carcinogens. *Biochim. Biophys. Acta,* 126 (1966) 275-285.

Bibliography

Barber, H. 1934. Haemorrhagic nephritis and necrosis of the liver from dioxane poisoning. *Guy's Hosp. Rep.* 84:267.

Johnstone, R.T. 1959. Death due to dioxane? *Arch. Ind. Health* 20:445.

National Institute for Occupational Safety and Health, *Criteria for a Recommended Standard: Occuptional Exposure to Dioxane,* NIOSH Doc. No. 77-226 (1977).

DIPHENYL

Description: $C_6H_5C_6H_5$, diphenyl, is a colorless to light yellow, leaflet solid with a potent characteristic odor.

Synonyms: Biphenyl, phenylbenzene.

Potential Occupational Exposures: Diphenyl is a fungistat for oranges which is applied to the inside of shipping containers and wrappers. It is also used as a heat transfer agent and as an intermediate in organic synthesis. Diphenyl is produced by thermal dehydrogenation of benzene.

A partial list of occupations in which exposure may occur includes:

Orange packers   Fungicide workers
Organic chemical synthesizers

Permissible Exposure Limits: The Federal standard is 0.2 ppm (1.5 mg/m^3).

Routes of Entry: Inhalation of vapor or dust; percutaneous absorption.

Harmful Effects

Local—Repeated exposure to dust may result in irritation of skin and res-piratory tract. The vapor may cause moderate eye irritation. Repeated skin con-tact may produce a sensitization dermatitis.

Systemic—In acute exposure, diphenyl exerts a toxic action on the central nervous system, on the peripheral nervous system, and on the liver. Symp-toms of poisoning are headache, diffuse gastrointestinal pain, nausea, indi-gestion, numbness and aching of limbs, and general fatigue. Liver function tests may show abnormalities. Chronic exposure is characterized mostly by central nervous system symptoms, fatigue, headache, tremor, insomnia, sen-sory impairment, and mood changes. Such symptoms are rare, however.

Medical Surveillance: Consider skin, eye, liver function and respiratory tract irritation in any preplacement or periodic examination.

Special Tests: None in common use.

Personal Protective Methods: Because of its low vapor pressure and low order of toxicity, it does not usually present a major problem in industry. Pro-tective creams, gloves, and masks with organic vapor canisters for use in areas of elevated vapor concentrations should suffice. Elevated temperature may increase the requirement for protective methods or ventilation.

Bibliography

Hakkinen, I., E. Siltanen, S. Herberg, A.M. Seppalainen, P. Karli, and E. Vikkula. 1973. Di-phenyl poisoning in fruit paper production. *Arch. Environ. Health* 26:70.

E

EPICHLOROHYDRIN

Description: CH$_2$OCHCH$_2$Cl, epichlorohydrin, is a colorless liquid with a chloroform-like odor.

Synonyms: Epi, chloropropylene oxide, 1-chloro-2,3-epoxypropane, chloromethyloxirane, 2-epichlorohydrin.

Potential Occupational Exposures: Epichlorohydrin is used in the manufacture of many glycerol and glycidol derivatives and epoxy resins, as a stabilizer in chlorine-containing materials, as an intermediate in the preparation of cellulose esters and ethers, paints, varnishes, nail enamels, and lacquers, and as a cement for celluloid.

A partial list of occupations in which exposure may occur includes:

Cellulose ether workers
Epoxy resin makers
Glycerol derivative makers
Glycerophosphoric acid makers
Glycidol derivative makers
Gum processors

Lacquer makers
Nail enamel makers
Organic chemical synthesizers
Paint makers
Resin makers
Solvent workers

NIOSH estimates that approximately 50,000 workers in the United States are potentially exposed to epichlorohydrin.

Permissible Exposure Limits: The Federal standard is 5 ppm (19 mg/m^3). NIOSH has recommended a TWA limit of 2 mg/m^3 with ceiling concentration of 19 mg/m^3 based on a 15-minute sampling period. NIOSH recommended in late 1978 that epichlorohydrin be handled as a human carcinogen.

Route of Entry: Inhalation of vapor, percutaneous absorption of liquid.

Harmful Effects

Local—Epichlorohydrin is highly irritating to eyes, skin, and respiratory tract. Skin contact may result in delayed blistering and deep-seated pain. Allergic eczematous contact dermatitis occurs occasionally.

192

Systemic—The earliest symptoms of intoxication may be referable to the gastrointestinal tract (nausea, vomiting, abdominal discomfort) or pain in the region of the liver. Labored breathing, cough, and cyanosis may be evident and the onset of chemical pneumonitis may occur several hours after exposure. Animals exposed repeatedly to this chemical have developed lung, kidney, and liver injury.

Epichlorohydrin has recently been reported to produce squamous cell carcinomas of the nasal epithelium in rats following inhalation at levels of 100 ppm for 6 hours/day (1). Epichlorohydrin has been previously shown to induce local sarcomas in mice following subcutaneous injection (2).

Epichlorohydrin (without metabolic activation) at concentrations of 1-50 mmol per 1 hr, induced reverse mutations in *S. typhimurium* G46 and TA 100 tester strains (3). The mutagenic activity with TA 1535 tester strain was markedly reduced in the presence of liver homogenates (4). Epichlorohydrin produced reverse mutations in *E. coli* (5) and in *Neurospora crassa* (6), recessive lethal mutations in *Drosophila melanogaster* (7), and was mutagenic in *Klebsiella pneumoniae* (8).

Doses of 50 and 100 mg/kg of epichlorohydrin after 3 hours increased the frequency of reverse mutations using *S. typhimurium* strains G46, TA 100 and TA 1950 in ICR female mice in a host-mediated assay (3).

Mutagenic activity (as determined with the TA 1535 strain of *S. typhimurium*) was detected in the urine of mice after oral administration of 200-400 mg/kg epichlorohydrin (4). Although an initial evaluation of mutagenic activity (utilizing the above system) in the urine of 2 industrial workers exposed to a concentration in excess of 25 ppm was regarded as borderline, additional mutagenic testing revealed more definitive evidence of activity, with the active compound appearing as a conjugate (4).

Epichlorohydrin induced dose-dependent chromosome abnormalities in bone marrow of ICR mice injected i.p. with a single dose of 1-50 mg/kg or repeated doses of 5 × 5-20 mg/kg, or given p.o. in a single dose of 5-100 mg/kg or repeated doses of 5 × 20 mg/kg (3). Epichlorohydrin did not induce dominant lethal mutation in ICR mice when given i.p. in a single dose of 5-40 mg/kg (3), 150 mg/kg (9), repeated doses of 5 × 1-10 mg/kg, p.o. in a single dose of 20 or 40 mg/kg or by repeated doses at 5 × 4-20 mg/kg (3). It was 4 to 5 times less mutagenic than the polyfunctional mutagenic agent TEPA when tested analogously (10); the epichlorohydrin induced changes were mainly classified as chromatid and isochromatid breaks and exchanges (3). These results demonstrated the ability of epichlorohydrin to induce gene and chromosome mutations in somatic cells. The finding of no changes in gametic cells was suggested to be the result of biotransformation changes of epichlorohydrin into forms which then cannot reach gametic cells in a concentration capable of inducing dominant lethal effects (11).

Medical Surveillance: Consider possible effects on the skin, eyes, lungs, liver, and kidney in preplacement or periodic examinations.

Special Tests: None currently used.

Personal Protective Methods: Goggles and rubber, protective clothing should be worn. Epichlorohydrin slowly penetrates rubber, so all contaminated clothing should be thoroughly washed. Respirators are required in areas of vapor concentrations.

References

(1) Anon. Epichlorohydrin causes nose cancers in rats. NYU's Nelson Reports. *Pesticide & Toxic News* 5 (19) (1977) 27-28.
(2) Van Duuren, B.L., Goldschmidt, B.N., Katz, C., Siedman, I., and Paul, J.S. Carcinogenic activity of alkylating agents. *J. Natl. Cancer Inst.* 53 (1974) 695-700.
(3) Sram, R., Cerna, M., and Kucerova, M. The genetic risk of epichlorohydrin as related to the occupational exposure. *Biol. Zbl.* 95 (1976) 451-462.
(4) Kilian, D.J., Pullin, T.G., Connor, T.H., Legator, M.S., and Edwards, H.N. Mutagenicity of epichlorohydrin in the bacterial assay system: Evaluation by direct in vitro activity and in vivo activity of urine from exposed humans and mice. Presented at 8th Annual Meeting of Environmental Mutagen Society, Colorado Springs, Colo. Feb. 13-17 (1977) p. 35.
(5) Strauss, B., and Okubo, S. Protein synthesis and the induction of mutations in *E. coli* by alkylating agents. *J. Bact.* 79 (1960) 464-473.
(6) Kolmark, G. and Giles, N.H. Comparative studies on monoepoxides as inducers of reverse mutations in *Neurospora. Genetics* 40 (1955) 890-902.
(7) Rapoport, L.A. Dejstvie okisi etilena, glitsida i glikoles na gennye mutatsii (Action of ethylene oxide glycides and glycols on genetic mutations). *Dokl. Akad. Nouk. SSSR* 60 (1948) 469-72.
(8) Voogd, C.E. Mutagenic action of epoxy compounds and several alcohols. *Mutation Res.* 21 (1973) 52-53.
(9) Epstein, S.S., Arnold, E., Andrea, J., Bass, W., and Bishop, Y. Detection of chemical mutagens by the dominant lethal assay in the mouse. *Toxicol. Appl. Pharmacol.* 23 (1972) 288-325.
(10) Kucerova, M., Polivakova, Z., Sram, R., and Matousek, V. Mutagenic effect of epichlorohydrin. I. Testing on human lymphocytes in vitro in comparison with TEPA, *Mutation Res.* 34 (1976) 271-278.
(11) Schalet, A. *Drosophila Info. Service* 28 (2) (1954) 155.

Bibliography

Hahn, J.D. 1970. Post-testicular antifertility effects of epichlorohydrin and 2,3-epoxypropanol. *Nature* 226:87.
Lawrence, W.H., Malik, M., Turner, J.E., and Austin, J. 1972. Toxicity profile of epichlorohydrin. *J. Pharm. Sci.* 61:1712.
National Institute for Occupational Safety and Health. *Criteria for a Recommended Standard: Occupational Exposure to Epichlorohydrin.* NIOSH Doc. No. 76-206, Wash., D.C. (1976).
Shell Chemical Company. 1972. *Epichlorohydrin*—Industrial Hygiene Bulletin. Shell Chemical Company, New York.

ETHANOLAMINES

Description: $H_2NCH_2CH_2OH$—monoethanolamine, $HN(CH_2CH_2OH)_2$—diethanolamine, $N(CH_2CH_2OH)_3$—triethanolamine. All three compounds are water soluble liquids. Monoethanolamine has a low vapor pressure while the vapor pressures of the other ethanolamines are very low. Monoethanolamine and diethanolamine have ammonia odors while triethanolamine has only a faint nonammonia odor. The acid salts have less odor and are of low volatility. Ethanolamines can be detected by odor as low as 2 to 3 ppm.

Synonyms

Monoethanolamine: Ethanolamine, 2-aminoethanol, colamine
Diethanolamine: 2,2'-Iminodiethanol
Triethanolamine: 2,2',2''-Nitrilotriethanol

Potential Occupational Exposures: Monoethanolamine is widely used in industry to remove carbon dioxide and hydrogen from natural gas, to remove hydrogen sulfide and carbonyl sulfide, as an alkaline conditioning agent, and as an intermediate for soaps, detergents, dyes, and textile agents.

Diethanolamine is an absorbent for gases, a solubilizer for 2,4-dichlorophenoxyacetic acid (2,4-D), and a softener and emulsifier intermediate for detergents. It also finds use in the dye and textile industry.

Triethanolamine is used as a plasticizer, neutralizer for alkaline dispersions, lubricant additive, corrosion inhibitor, and in the manufacture of soaps, detergents, shampoos, shaving preparations, face and hand creams, cements, cutting oils, insecticides, surface active agents, waxes, polishes, and herbicides.

A partial list of occupations in which exposure may occur includes:

Cement makers
Detergent makers
Dye makers
Emulsifier makers
Herbicide makers
Insecticide makers
Natural gas workers

Plastic workers
Polish makers
Soap makers
Surfactant makers
Textile workers
2,4-D makers

Permissible Exposure Limits: The Federal standard for monoethanolamine is 3 ppm (6 mg/m^3). There are no standards for the other compounds.

Routes of Entry: Inhalation of vapor and percutaneous absorption.

Harmful Effects

Local—Ethanolamine has had wide use in industry, yet reports of injury in man are lacking. Ethanolamine in animal experiments was highly irritating to the skin, eyes, and respiratory tract. Diethanolamine and triethanolamine produced much less irritation. In human experiments, ethanolamine produced only redness of the skin.

Systemic—No specific published data on human exposure are available. Animal experiments indicate that it is a central nervous system depressant. Acute high level exposures produced pulmonary damage and nonspecific hepatic and renal lesions in animals.

Medical Surveillance: Evaluate possible irritant effects on skin and eyes.

Special Tests: None in common use.

Personal Protective Methods: Protective clothing should be worn, and in areas of elevated vapor concentrations, full face masks should be supplied.

Bibliography

Weeks, M.H., Downing, T.O., Musselman, N.P., Carson, T.R., and Groff, W.A. 1960. The effects of continuous exposure of animals to ethanolamine vapor. *Am. Ind. Hyg. Assoc.* J. 21:374.

ETHYL ALCOHOL

Description: CH_3CH_2OH, ethyl alcohol, is a colorless, volatile, flammable liquid. Ethyl alcohol is produced by fermentation and distillation or by synthesis.

Synonyms: Ethanol, grain alcohol, spirit of wine, cologne spirit, ethyl hydroxide, ethyl hydrate.

Potential Occupational Exposures: Ethyl alcohol is used in the chemical synthesis of a wide variety of compounds such as acetaldehyde, ethyl ether, ethyl chloride, and butadiene. It is a solvent or processing agent in the manufacture of pharmaceuticals, plastics, lacquers, polishes, plasticizers, perfumes, cosmetics, rubber accelerators, explosives, synthetic resins, nitrocellulose, adhesives, inks, and preservatives. It is also used as an antifreeze and as a fuel.

A partial list of occupations in which exposure may occur includes:

Acetaldehyde makers	Detergent makers
Acetic anhydride makers	Distillers
Adhesive makers	Dye makers
Beverage makers	Ethyl ether makers

Histology technicians
Ink makers
Lacquer makers
Motor fuel blenders
Organic chemical synthesizers

Rubber makers
Shellac processors
Solvent workers
Stainers

Permissible Exposure Limits: The Federal standard is 1,000 ppm (1,900 mg/m^3).

Route of Entry: Inhalation of vapor and percutaneous absorption.

Harmful Effects

Local—Mild irritation of eye and nose occurs at very high concentrations. The liquid can defat the skin, producing a dermatitis characterized by drying and fissuring.

Systemic—Prolonged inhalation of high concentrations, besides the local effect on the eyes and upper respiratory tract, may produce headache, drowsiness, tremors, and fatigue. Tolerance may be a factor in individual response to a given air concentration.

Bizarre symptoms (other than typical manifestations of intoxication) may result from the denaturants often present in industrial ethyl alcohol. Ethyl alcohol may act as an adjuvant, increasing the toxicity of other inhaled, absorbed, or ingested chemical agents. An exception is methanol where ethyl alcohol counteracts methanol toxicity.

Medical Surveillance: No special considerations needed.

Special Tests: Ethyl alcohol can readily be determined in blood, urine, and expired air.

Personal Protective Methods: Personal protective equipment is recommended where skin contact may occur.

Bibliography

Gonzales, T.A., Vance, M., Halpern, M., and Umberger, C.H. 1954. *Legal Medicine: Pathology and Toxicology,* 2nd ed. Appleton-Century-Crofts, New York. p. 1083.
Henson, E.V. 1960. The toxicity of some aliphatic alcohols, part II. *J. Occup. Med.* 2:497.
Jain, N., and Cravey, R.H. 1972. Analysis of alcohol. I. A review of chemical and infrared methods. *J. Chromatogr. Sci.* 10:257.
Jain, N.C., and Cravey, R.H. 1972. Analysis of alcohol. II. A review of gas chromatographic methods. *J. Chromatogr. Sci.* 10:263.

ETHYLBENZENE (See "Styrene/Ethyl Benzene")

ETHYL BROMIDE [See "Methyl (and Ethyl) Bromide"]

ETHYL CHLORIDE

Description: CH_3CH_2Cl, ethyl chloride, is a flammable gas with an ethereal odor and a burning taste. It is flammable, and the products of combustion include phosgene and hydrogen chloride.

Synonyms: Monochloroethane, hydrochloric ether, chloroethane, chlorethyl.

Potential Occupational Exposures: Ethyl chloride is used as an ethylating agent in the manufacture of tetraethyl lead, dyes, drugs, and ethyl cellulose. It can be used as a refrigerant and as a local anesthetic (freezing).

A partial list of occupations in which exposure may occur includes:

Anesthetists	Phosphorus and sulfur processors
Dentists	Physicians
Drug makers	Refrigeration workers
Ethylation workers	Resin makers
Fat and oil processors	Sulfur processors
Nurses	Tetraethyl lead makers
Perfume makers	Wax makers

Permissible Exposure Limits: The Federal standard is 1,000 ppm (2,600 mg/m^3).

Routes of Entry: Inhalation of gas and slight percutaneous absorption.

Harmful Effects

Local—The liquid form of ethyl chloride is mildly irritating to skin and eyes. Frostbite can occur due to rapid liquid evaporation.

Systemic—Ethyl chloride exposure may produce headache, dizziness, incoordination, stomach cramps, and eventual loss of consciousness. In high concentrations, it is a respiratory tract irritant, and death due to cardiac arrest has been recorded. Renal damage has been reported in animals. Effects from chronic exposure have not been reported.

Medical Surveillance: Consider possible acute cardiac effects in any preplacement or periodic examination.

Special Tests: None commonly used. Ethyl chloride is excreted in expired air.

Personal Protective Methods: In high vapor concentrations, use mask with full face protection, gloves, and protective clothing.

Bibliography

Von Oettingen, W.F. 1955. *The Halogenated Aliphatic, Olefinic, Cyclic, Aromatic, and Aliphatic-Aromatic Hydrocarbons Including the Halogenated Insecticides, Their Toxicity and Potential Dangers.* Public Health Service Publication No. 414. U.S. Public Health Service.

ETHYLENE CHLOROHYDRIN

Description: CH_2ClCH_2OH, ethylene chlorohydrin, is a colorless liquid with an ether-like odor.

Synonyms: Glycol chlorohydrin, 2-chloroethanol, β-chloroethyl alcohol.

Potential Occupational Exposures: Ethylene chlorohydrin is used in the synthesis of ethylene glycol, ethylene oxide, amines, carbitols, indigo, malonic acid, novocaine, and in other reactions where the hydroxyethyl group is introduced into organic compounds, for the separation of butadiene from hydrocarbon mixtures, in dewaxing and removing cycloalkanes from mineral oil, in the refining of rosin, in the manufacture of certain pesticides, and in the extraction of pine lignin. In the lacquer industry, it is used as a solvent for cellulose acetate, cellulose esters, resins and waxes, and in the dyeing and cleaning industry, it is used to remove tar spots, as a cleaning agent for machines, and as a solvent in fabric dyeing. It has also found use in agriculture in speeding up sprouting of potatoes and in treating seeds to inhibit biological activity.

A partial list of occupations in which exposure may occur includes:

Cellulose acetate makers	Novocaine makers
Drug makers	Organic chemical synthesizers
Dye makers	Potato growers
Ethyl cellulose workers	Procaine makers
Indigo makers	Resin workers
Lacquer makers	Textile dyers and printers

Permissible Exposure Limits: The Federal standard is 5 ppm (16 mg/m^3). The ACGIH (1978) has set TWA values of 1 ppm (3 mg/m^3).

Routes of Entry: Inhalation of vapor; percutaneous absorption of liquid.

Harmful Effects

Local—High vapor concentrations are irritating to the eyes, nose, throat, and skin.

Systemic—Ethylene chlorohydrin is extremely toxic and in addition to local irritation of eyes, respiratory tract, and skin, inhalation of the vapor may produce nausea, vomiting, dizziness, headache, thirst, delirium, low blood pressure, collapse, and unconsciousness. The urine may show red cells, albumin, and casts. Death may occur in high concentrations with damage to the lung and brain. There is little margin of safety between early reversible symptoms and fatal intoxication. The toxic effects may be related to its metabolites, chloroacetaldehyde and chloroacetic acid.

Medical Surveillance: Preplacement examination, including a complete history and physical should be performed. Examination of the respiratory system, liver, kidneys, and central nervous system should be stressed. The skin should be examined. A chest x-ray should be taken and pulmonary function tests performed (FVC-FEV).

The above procedures should be repeated on an annual basis, except that the x-ray is needed only when indicated by pulmonary function testing.

Special Tests: None commonly used. Presence in blood can probably be determined by appropriate gas chromatographic methods.

Personal Protective Methods: The liquid readily penetrates rubber. Protective clothing should be discarded at first sign of deterioration. Barrier creams may be used and scrupulous personal hygiene should be practiced.

Bibliography

Bush, A.F., Abrams, H.K., and Brown, H.V. 1949. Fatality and illness caused by ethylene chlorohydrin in an agricultural occupation. *Ind. Hyg. and Toxicol.* 31:352.
Vallotta, F., Bertagni, P., and Troisi, F.M. 1953. Acute poisoning caused by ingestion of ethylene chlorohydrin. *J. Ind. Med.* 10:161.

ETHYLENEDIAMINE

Description: $H_2NCH_2CH_2NH_2$, ethylenediamine, is a strongly alkaline, colorless, clear, thick liquid with an ammonia odor.

Synonyms: Ethanediamine, 1,2-diaminoethane.

Potential Occupational Exposures: Ethylenediamine is used as a solvent, an emulsifier for casein and shellac solutions, a stabilizer in rubber latex, a chemical intermediate in the manufacture of dyes, corrosion inhibitors, synthetic waxes, fungicides, resins, insecticides, asphalt wetting agents, and pharmaceuticals, and also in controlling acidity or alkalinity.

A partial list of occupations in which exposure may occur includes:

Albumin processors	Fungicide makers
Casein processors	Insecticide makers
Drug makers	Oil neutralizers
Dye makers	Resin makers
Emulsion workers	Rubber makers
Ethylenediamine tetraacetic acid (EDTA) makers	Shellac processors
	Surfactant makers

Permissible Exposure Limits: The Federal standard is 10 ppm (25 mg/m^3).

Routes of Entry: Inhalation of vapor and percutaneous absorption.

Harmful Effects

Local—Ethylenediamine vapor may cause irritation of the nose and tingling of the face. Cutaneous sensitivity has been reported.

In animal experiments, the liquid has produced severe irritation of the eyes and corneal damage. It has also produced severe irritation and necrosis.

Systemic—In animal experiments, high concentrations of ethylenediamine vapor have produced damage to liver, lungs, and kidneys.

Medical Surveillance: Consider possible irritant effects on skin, eyes and respiratory system. History of allergic redness of skin or asthmatic symptoms may be important in placement and periodic examinations.

Special Tests: None have been developed.

Personal Protective Methods: Protective clothing, gloves, and goggles should be worn to protect the skin and eyes. Full face masks with organic vapor canisters must be used in areas of high vapor concentrations. Recent reports indicate that a nonoccupational allergic contact dermatitis may develop after use of pharmaceuticals containing ethylenediamine.

Bibliography

Dernehl, C.U. 1951. Clinical experiences with exposures to ethylene amines. *Ind. Med. Surg.* 20:541.
Epstein, E., and Maibach, H.I. 1968. Ethylenediamine, allergic contact dermatitis. *Arch. Dermatol.* 98:476.

ETHYLENE DIBROMIDE (See "Dibromoethane")

ETHYLENE DICHLORIDE (See "1,2-Dichloroethane")

ETHYLENE GLYCOL

Description: $HOCH_2CH_2OH$, ethylene glycol, is a colorless, odorless, viscous liquid with a sweetish taste.

Synonyms: 1,2-Ethanediol, glycol alcohol, glycol, EG.

Potential Occupational Exposures: Because of ethylene glycol's physical properties, it is used in antifreeze, hydraulic fluids, electrolytic condensors, and heat exchangers. It is also used as a solvent and as a chemical intermediate for ethylene glycol dinitrate, glycol esters, and resins.

A partial list of occupations in which exposure may occur includes:

Antifreeze makers	Metal polishers
Brake fluid makers	Paint makers
Explosive makers	Resin makers
Glue makers	Textile makers
Hydraulic fluid makers	Tobacco workers
Ink makers	Wax makers
Metal cleaners	

Permissible Exposure Limits: There is no Federal standard; however, ACGIH in 1978 recommended a TLV of 10 mg/m^3 for particulate ethylene glycol and 100 ppm (250 mg/m^3) for the vapor form.

Route of Entry: Inhalation of particulate or vapor. Percutaneous absorption may also contribute to intoxication.

Harmful Effects

Local—None.

Systemic—Ethylene glycol's vapor pressure is such that at room temperature toxic concentrations are unlikely to occur. Poisoning resulting from vapor usually occurs only if ethylene glycol liquid is heated; therefore, occupational exposure is rare. Chronic symptoms and signs include: anorexia, oliguria, nystagmus, lymphocytosis, and loss of consciousness. Inhalation seems to primarily result in central nervous system depression and hematopoietic dysfunction, whereas, ingestion may result in depression followed by respiratory and cardiac failure, renal and brain damage.

Medical Surveillance: No special considerations are needed.

Special Tests: Urinalysis for oxalic acid, an ethylene glycol metabolite, may be useful in diagnosis of poisoning by oral ingestion.

Personal Protective Methods: Masks should be worn in areas of vapor concentration.

Bibliography

Ahmed, M.D. 1971. Ocular effects of antifreeze poisoning. *Br. J. Ophthalmol.* 55:845.
Aquino, H.C., and Leonard, C.D. 1972. Ethylene glycol poisoning: report of three cases. *J.K. Med. Assoc.* 70:463.
Gallyas, F., Jaray, J., and Csata, S. 1971. Acute renal failure following ethyleneglycol poisoning. *Acta. Chir. Acad. Sci. Hung.* 12:225.
Troisi, F.M. 1950. Chronic intoxication by ethylene glycol vapor. *Br. J. Ind. Med.* 7:65.

ETHYLENE GLYCOL DINITRATE
(See "Nitroglycerin and Ethylene Glycol Dinitrate")

ETHYLENE GLYCOL ETHERS AND DERIVATIVES

Description

Ethylene glycol monoethyl ether: $CH_3CH_2OCH_2CH_2OH$.
Ethylene glycol monoethyl ether acetate: $CH_3CH_2OCH_2CH_2OOCCH_3$.
Ethylene glycol monomethyl ether: $CH_3OCH_2CH_2OH$.
Ethylene glycol monomethyl ether acetate: $CH_3OCH_2CH_2OOCCH_3$.
Ethylene glycol monobutyl ether: $CH_3CH_2CH_2CH_2OCH_2CH_2OH$.

These substances are colorless liquids with a slight odor.

Synonyms

Ethylene glycol monoethyl ether: cellosolve, 2-ethoxyethanol.
Ethylene glycol monoethyl ether acetate: cellosolve acetate, 2-ethoxyethyl acetate.
Ethylene glycol monomethyl ether: methyl cellosolve, 2-methoxyethanol.
Ethylene glycol monomethyl ether acetate: methyl cellosolve acetate, 2-methoxyethyl acetate.

Ethylene glycol monobutyl ether: butyl cellosolve, 2-butoxyethanol.

Potential Occupational Exposures: Ethylene glycol ethers are used as solvents for resins, lacquers, paints, varnishes, gum, perfume, dyes and inks, and as a constituent of painting pastes, cleaning compounds, liquid soaps, cosmetics, nitrocellulose, and hydraulic fluids. Acetate derivatives are used as solvents for oils, greases and ink, in the preparation of lacquers, enamels, and adhesives, and to dissolve resins and plastics.

A partial list of occupations in which exposure may occur includes:

Cellophane sealers	Nail polish makers
Cleaning solution makers	Oil processors
Dry cleaners	Plastic makers
Film makers	Printers
Hydraulic fluid makers	Stainers
Ink makers	Textile dyers
Lacquer makers	Wax processors

Permissible Exposure Limits: The Federal standards for these compounds are:

Ethylene glycol monoethyl ether	200 ppm	740 mg/m^3
Ethylene glycol monoethyl ether acetate	100 ppm	540 mg/m^3
Ethylene glycol monomethyl ether	25 ppm	80 mg/m^3
Ethylene glycol monomethyl ether acetate	25 ppm	120 mg/m^3
Ethylene glycol monobutyl ether	50 ppm	240 mg/m^3

ACGIH in 1978 recommended a TLV of 25 ppm (120 mg/m^3) for ethylene glycol monomethyl ether acetate.

Route of Entry: Inhalation of vapor and percutaneous absorption of liquid.

Harmful Effects

Local—Ethylene glycol ethers are only mildly irritating to the skin. Vapor may cause conjunctivitis and upper respiratory tract irritation. Temporary corneal clouding may also result and may last several hours. Acetate derivatives cause greater eye irritation than the parent compounds. The butyl and methyl ethers may penetrate skin readily.

Systemic—Acute exposure to these compounds results in narcosis, pulmonary edema, and severe kidney and liver damage. Symptoms from repeated overexposure to vapors are fatigue and lethargy, headache, nausea, anorexia, and tremor. Anemia and encephalopathy have been reported with ethylene glycol monomethyl ether. Rats show increased hemolysis of erythrocytes from inhalation of ethylene glycol monobutyl ether. This has not been shown in man. Acute poisoning by ingestion resembles ethylene glycol toxicity, with death from renal failure.

Medical Surveillance: Preplacement and periodic examinations should evaluate blood, central nervous system, renal and liver functions, as well as the skin and respiratory tract.

Special Tests: None currently used.

Personal Protective Methods: Use glasses and protective clothing to prevent skin absorption. Respiratory protection may be needed if ventilation is poor or if compounds are heated or atomized.

Bibliography

Carpenter, C.P., Pozzani, U.C., Weil, C.S., Nair, J.H. III, Keck, G.A., and Smyth, H.F. 1956. The toxicity of butyl cellosolve solvent. *AMA Arch. Ind. Health* 14:114.
Nitter-Hauge, S. 1971. Poisoning with ethylene glycol monomethyl ether: report of two cases. *Acta. Med. Scand.* 188:277.
Zavon, M.R. 1963. Methyl cellosolve intoxication. *Am. Ind. Hyg. Assoc. J.* 24:36.

ETHYLENEIMINE

Description: H_2CNHCH_2, ethyleneimine, is a colorless volatile liquid with an ammoniacal odor.

Synonyms: Azacyclopropane, aziridine, dimethyleneimine, ethylenimine, vinylamine, azirane, dihydroazirine, EI.

Potential Occupational Exposures: Ethyleneimine is a highly reactive compound and is used in many organic syntheses. The polymerization products, polyethyleneimines, are used as auxiliaries in the paper industry and as flocculation aids in the clarification of effluents. It is also used in the textile industry for increasing wet strength, flameproofing, shrinkproofing, stiffening, and waterproofing (1)(2).

A partial list of occupations in which exposure may occur includes:

Effluent treaters   Organic chemical synthesizers
Paper makers   Textile makers
Polyethyleneimine makers

Permissible Exposure Limits: Ethyleneimine was included in the Federal standard for carcinogens; all contact with it should be avoided. ACGIH (1978) set a TWA of 0.5 ppm (1.0 mg/m^3) with no caution on carcinogenic nature.

Routes of Entry: Inhalation and percutaneous absorption.

Harmful Effects

Local—The vapor is strongly irritating to the conjunctiva and cornea, the mucous membranes of the nose, throat, and upper respiratory tract, and the skin. The liquid is a severe irritant and vesicant in humans, and severe eye burns have followed contact with the cornea. Skin sensitization has occurred.

Systemic—Acute exposures in humans have caused nausea, vomiting, headaches, dizziness, and pulmonary edema. In mice acute lethal exposures to vapor produced pulmonary edema, renal damage, and hematuria. Compounds with the aziridine structure have some of the properties of alkylating agents. Ethyleneimine has been reported to induce mutagenic effects in in vitro cultures, microorganisms, plants, and animals.

In repeated exposures rodents have developed pancytopenia and gonadal effects. Rats given twice weekly subcutaneous injections of ethyleneimine in oil for about 33 weeks developed sarcoma at the injection site and one case of transitional cell carcinoma in the kidney was observed. Feeding experiments with mice at 13 ppm in the diet for 74 weeks produced hepatomas and pulmonary tumors. These effects have not been reported in humans.

Aziridine is carcinogenic in two strains of mice following its oral administration producing an increased incidence of liver-cell pulmonary tumors (1)(3).

Aziridine induces both transmissible translocations and sex-linked recessive lethal mutations in *Drosophila melanogaster* (4)(5); specific locus mutations in silkworms (*Bombyx mori*) (6). Aziridine also produces leaky mutants, mutants with polarized and nonpolarized complementation patterns, and noncomplementing mutants and multilocus deletions in *Neurospora crassa* (7) and induces mitotic recombination (8) and gene conversion in *Saccharomyces cerevisiae* (9).

Aziridine induces chromosome aberrations in cultured human cells (10), mouse embryonic skin cultures (11) and Crocker mouse Sarcoma 188 (11). When rabbits were inseminated with spermatozoa which had been treated with aziridine in vitro, only 40% of embryos were found to be viable relative to the number of corpora lutea in comparison to 78% in controls (12). Aziridine has been reported to possess teratogenic activity (13).

Medical Surveillance: Based partly on animal experimental data, examinations should include history of exposure to other carcinogens, smoking, alcohol, medications, and family history. The skin, eye, lung, liver, and kidney should be evaluated. Sputum or urine cytology may be helpful.

Special Tests: None in common use. Chromosomal studies have been made, but are probably not useful for routine surveillance.

Personal Protective Methods: These are designed to supplement engineering control and prevent all skin or respiratory exposure. Full body protective clothing and gloves should be used. Full face supplied air respirators with continuous flow or pressure demand type should also be used. Eyes should be protected at all times. On exit from a regulated area employees should shower and change to street clothes, leaving their protective clothing and equipment at the point of exit, to be placed in impervious containers at the end of the work shift for decontamination or disposal.

References

(1) *IARC.* Aziridine. Vol. 9, International Agency for Research on Cancer. Lyon (1975).

(2) Fishbein, L., Flamm, W.G., and Falk, H.L. *Chemical Mutagens.* Academic Press, New York (1970).

(3) Innes, J.R.M., Ulland, B.M., Valerio, M.G., Petrucelli, L., Fishbein, L., Hart, E.R., Pallotta, A.J., Bates, R.R., Falk, H.L., Gart, J.J., Klein, M., Mitchell, I., and Peters, J. Bioassay of pesticides and industrial chemicals for tumorigenicity in mice: A preliminary note. *J. Natl. Cancer Inst.* 42 (1969) 1101.

(4) Alexander, M.L., and Glanges, E. Genetic damage induced by ethylenimine. *Proc. Nat. Acad. Sci.* 53 (1965) 282.

(5) Sram, R.J. The effect of storage on the frequency of translocations in *Drosophila melanogaster. Mutation Res.* 9 (1970) 243.

(6) Inagaki, E., and Oster, I.I. Changes in the mutational response of silkworm spermatozoa exposed to mono- and polyfunctional alkylating agents following storage. *Mutation Res.* 7. (1969) 425.

(7) Ong, T.M., and deSerres, F.J. Mutagenic activity of ethylenimine in *Neurospora crassa. Mutation Res.* 18 (1973) 251.

(8) Zimmermann, F.K., and VonLaer, U. Induction of mitotic recombination with ethylenimine in *Saccharomyces cerevisiae. Mutation Res.* 4 (1967) 377.

(9) Zimmermann, F.K. Induction of mitotic gene conversion by mutagens. *Mutation Res.* 11 (1971) 327.

(10) Chang, T.H., and Elequin, F.T. Induction of chromosome aberrations in cultured human cells by ethylenimine and its relation to cell cycle. *Mutation Res.* 4 (1967) 83.

(11) Biesele, J.J., Philips, F.S., Thiersch, J.B., Burchenal, J.H., Buckley, S.M., and Stock, C.C. *Nature* 166 (1950) 1112.

(12) Nuzhdin, N.I., and Nizhnik, G.V. Fertilization and embryonic development of rabbits after treatment of spermatozoa in vitro with chemical mutagens. *Dokl. Akad. Nauk. SSSR Otd. Biol.* 181 (1968) 419.

(13) Murphy, M.L., Del Moro, A., and Lacon, C. The comparative effects of five polyfunctional alkylating agents on the rat fetus with additional notes on the chick embryo. *Ann. N.Y. Acad. Sci.* 68 (1958) 762.

Bibliography

Dermer, O.C., and Ham, G.E. 1969. *Ethyleneimine and Other Aziridines.* Academic Press, New York.

Walpole, A.L., Roberts, D.C., Rose, F.L., Hendry, J.A., and Homer, R.E. 1954. Cytotoxic agents: IV, the carcinogenic actions of some monofunctional ethyleneimine derivatives. *Br. J. Pharmacol. Chemother.* 9:306.

ETHYLENE OXIDE

Description: $H_2\overline{COCH}_2$, ethylene oxide, is a colorless gas with a sweetish odor.

Synonyms: 1,2-Epoxyethane, oxirane, dimethylene oxide, anprolene.

Potential Occupational Exposures: Ethylene oxide is used as an intermediate in organic synthesis for ethylene glycol, polyglycols, glycol ethers, esters, ethanolamines, acrylonitrile, plastics, and surface-active agents. It is also used as a fumigant for foodstuffs and textiles, an agricultural fungicide, and for sterilization, especially for surgical instruments (1).

A partial list of occupations in which exposure may occur includes:

Acrylonitrile makers	Fungicide workers
Butyl cellosolve makers	Gasoline sweeteners
Detergent makers	Grain elevator workers
Disinfectant makers	Organic chemical synthesizers
Ethanolamine makers	Polyglycol makers
Ethylene glycol makers	Polyoxirane makers
Exterminators	Rocket fuel handlers
Foodstuff fumigators	Surfactant makers
Fumigant makers	Textile fumigators

Permissible Exposure Limits: The Federal standard is 50 ppm (90 mg/m^3).

Route of Entry: Inhalation of gas.

Harmful Effects

Local—Aqueous solutions of ethylene oxide or solutions formed when the anhydrous compound comes in contact with moist skin are irritating and may lead to a severe dermatitis with blisters, blebs, and burns. It is also absorbed by leather and rubber and may produce burns or irritation. Allergic eczematous dermatitis has also been reported. Exposure to the vapor in high concentrations leads to irritation of the eyes. Severe eye damage may result if the liquid is splashed in the eyes. Large amounts of ethylene oxide evaporating from the skin may cause frostbite.

Systemic—Breathing high concentrations of ethylene oxide may cause nausea, vomiting, irritation of the nose, throat, and lungs. Pulmonary edema may occur. In addition, drowsiness and unconsciousness may occur. Ethylene oxide has been found to cause cancer in female mice exposed to it for prolonged periods.

Ethylene oxide is a well-known mutagen in commercial use in plants. No mutagenic effect has been demonstrated in man.

In limited studies, no carcinogenic effect was found when ethylene oxide was tested in ICR/Ha Swiss mice by skin application and in rats by subcutaneous injection (2).

Ethylene oxide reacts with DNA, primarily at the N-7 position of guanosine, forming N-7-hydroxyethylguanine (3). Ethylene (1,2-^{3}H) oxide alkylated protein fractions taken from different organs of mice exposed to air containing 1.15 ppm of the labelled agent (4). The highest activity was found in the lung followed by the liver, kidney, spleen and testis (4).

Ethylene oxide produces reverse mutations in *S. typhimurium* TA 1535 strains (without activation) (5) and in *Neurospora crassa* at the adenine locus (6), induces recessive lethals (7)(8)(9), translocations (10)(11), and minute mutations (12) in *Drosophila melanogaster.*

Exposure of male Long-Evans rats for 4 hours to 1.83 g/m^3 (1,000 ppm) ethylene oxide produced dominant lethal mutations (5), while chromosome aberrations were found in bone-marrow cells of male rats of the same strain exposed to 0.45 g/m^3 (250 ppm) ethylene oxide for 7 hours/day for 3 days (5). Ethylene oxide induces chromosome aberrations in mammalian somatic cells (13)(14).

Medical Surveillance: Preplacement and periodic examinations should consider the skin and eyes, allergic history, the respiratory tract, blood, liver, and kidney function.

Special Tests: None in common use.

Personal Protective Methods: Eyes and skin should be protected and protective clothing changed when it is contaminated. In areas of high vapor concentration, respirators should be supplied to cover the face, including eyes. Shoes contaminated by this chemical should be discarded.

References

(1) Gutsche, C.D., and Pasto, D.J. *Fundamentals of Organic Chemistry.* Prentice-Hall, Englewood Cliffs, NJ (1975) p. 296.
(2) *IARC,* Vol. 11. Cadmium, nickel, some epoxides, miscellaneous industrial chemicals and general considerations on volatile anaesthetics. International Agency for Research on Cancer. Lyon, 1976, pp. 157-167, 191-196.
(3) Brookes, P., and Lawley, P.D. The alkylation of guanosine and guanylic acid, *J. Chem. Soc.* (1961) 3923-3928.
(4) Ehrenberg, L., Hiesche, K.D., Osterman-Golkar, S., and Wennberg, I. Effects of genetic risks of alkylating agents: Tissue doses in the mouse from air contaminated with ethylene oxide. *Mutation Res.* 24 (1974) 83-103.
(5) Embree, J.W., and Hine, C.H. Mutagenicity of ethylene oxide. *Toxicol. Appl. Pharmacol.* 33 (1975) 172-173.
(6) Kolmark, G., and Westergaard, M. Further studies on chemically induced reversions at the adenine locus of Neurospora. *Hereditas* 39 (1953) 209-224.

(7) Bird, M.J. Chemical production of mutations in *Drosophila:* Comparison of techniques. *J. Genet.* 50 (1952) 480–485.

(8) Rapoport, I.A. Dejstvie okisi etilena, glitsida i glikoles na gennye mutatsii (Action of ethylene oxide glycides and glycols on genetic mutations). *Dokl. Acad. Nauk. SSSR* 60 (1948) 469–472.

(9) Rapoport, I.A. Alkylation of gene molecule. *Dokl. Acad. Nauk. SSSR* 59 (1948) 1183–1186.

(10) Watson, W.A.F. Further evidence on an essential difference between the genetic effects of mono- and bifunctional alkylating agents. *Mutation Res.* 3 (1966) 455–457.

(11) Nakao, Y., and Auerbach, C. Test of possible correlation between cross-linking and chromosome breaking abilities of chemical mutagens. *Z. Vererbungsl.* 92 (1961) 457–461.

(12) Fahmy, O.G., and Fahmy, M.J. Gene elimination in carcinogenesis: Reinterpretation of the somatic mutation theory. *Cancer Res.* 30 (1970) 195–205.

(13) Strekalova, E.E. Mutagenic action of ethylene oxide on mammals. *Toksikol. Nov. Prom. Khimschesk. veshchestv.* 12 (1971) 72–78.

(14) Fomenko, V.N., and Strekalova, E.E. Mutagenic action of some industrial poisons as a function of concentration and exposure time. *Toksikol. Nov. Promysclen. Khimschesk. Veshchestv.* 13 (1973) 51–57.

Bibliography

Biro, L., Fisher, A.A., and Price, E. 1974. Ethylene oxide burns. *Arch. Derm.* 110:924.

Jacobson, K.H., Hackley, E.B., and Feinsilver, L. 1956. The toxicity of inhaled ethylene oxide and propylene oxide vapors. *AMA Arch. Ind. Health* 13:237.

Sexton, R.J., and Henson, E.V. 1949. Dermatological injuries by ethylene oxide. *J. Ind. Hyg. Toxicol.* 31:297.

Sulovska, K., Lindgren, D., Eriksson, G., and Ehrenberg, L. 1969. The mutagenic effect of low concentrations of ethylene oxide in air. *Hereditas* 62:264.

ETHYLENE THIOUREA

Description: Ethylene thiourea,

$$NH-\underset{\|}{\overset{S}{C}}-NH-CH_2-CH_2$$

is a white needle-like crystalline solid which melts at 203° to 204°C. Commercial ethylene thiourea is available as a solid powder, as a dispersion in oil (which retards the formation of fine dust dispersions in workplace air), and "encapsulated" in a matrix of compatible elastomers. In this latter form, ethylene thiourea may be least likely to escape into the workplace air.

Synonyms: 2-Imidazolidinethione, ETU.

Potential Occupational Exposures: Ethylene thiourea is used extensively as an accelerator in the curing of polychloroprene (Neoprene) and other elasto-

mers. NIOSH estimates that approximately 3,500 workers in the rubber industry have potential occupational exposure to ethylene thiourea. This estimate
is based on the NIOSH National Occupational Hazard Survey which was conducted between 1972 and 1974, and included over 500,000 employees at 4,775
facilities. In addition, exposure to ethylene thiourea also results from the very
widely used ethylene bisdithiocarbamate fungicides. Ethylene thiourea may
be present as a contaminant in the ethylene bisdithiocarbamate fungicides
and can also be formed when food containing the fungicides is cooked.

Permissible Exposure Limits: Pending completion of a NIOSH Special Occupational Hazard Review, NIOSH believes it would be prudent to minimize occupational exposure to ethylene thiourea. There is no current Occupational
Safety and Health Administration (OSHA) exposure standard for ethylene
thiourea.

Route of Entry: Inhalation of dust.

Harmful Effects

Ethylene thiourea has been shown to be carcinogenic and teratogenic (causing malformations in offspring) in laboratory animals. In addition, ethylene
thiourea can cause myxedema (the drying and thickening of skin, together
with a slowing down of physical and mental activity), goiter, and other effects
related to decreased output of thyroid hormone.

In a recent publication from E.I. du Pont de Nemours and Company's Haskell
Laboratory, for example, Stula and Krauss report "marked teratogenic effects
were demonstrated" when ethylene thiourea was applied to the skin of laboratory rats (1). In this study, 50 milligrams ethylene thiourea per kilogram
body weight was applied as a 20% solution in dimethylsulfoxide (DMSO) to
the skin of pregnant, female Sprague-Dawley rats on the twelfth and also on
the thirteenth day of gestation. Malformations were observed in all of seventy-
three fetuses upon sacrifice of the pregnant females on the twentieth day of
gestation. Appropriate control animals showed no fetal abnormality. A number of other reports in the literature further document the teratogenic potential
of ethylene thiourea [e.g., (2) and references cited therein].

The Stula and Krauss study (1) further demonstrates that dose as well as day
of gestation on which exposures occur are critical factors. When 50 mg/kg
ethylene thiourea applications were made as above, but on the tenth and
eleventh day of gestation, only five of eighty-three fetuses exhibited abnormalities; DMSO controls in this case exhibited one abnormality in forty fetuses. However, no fetal abnormalities were found among pregnant rats similarly exposed to 25 mg/kg ethylene thiourea on the tenth and also on the
eleventh day of gestation.

Ethylene thiourea has been shown to cause cancer in laboratory animals. In
one study, for example, Charles River CD rats were fed ethylene thiourea at

175 or 350 ppm in their diets for eighteen months and then observed while on a control diet for an additional six months (3). Thyroid carcinomas were observed in seventeen of twenty-six males (two with pulmonary metastases) and eight of twenty-six females at the high dose level, as well as three of twenty-six males and three of twenty-six females at the low dose level. In addition, solid-cell adenomas of the thyroid were observed in one female at the high dose level as well as two females at the low dose level. No thyroid carcinoma was found among the thirty-two male or thirty-two female controls. The International Agency for Research in Cancer (IARC) has concurred that oral administration of ethylene thiourea produces thyroid carcinoma in rats (4). Other authors (5)(6) have also reported carcinogenic effects of ethylene thiourea in laboratory animals.

Medical Surveillance: Initial and routine employee exposure surveys should be made by competent industrial hygiene and engineering personnel. These surveys are necessary to determine the extent of employee exposure and to ensure that controls are effective.

The *NIOSH Occupational Exposure Sampling Strategy Manual,* NIOSH Publication #77-173, may be helpful in developing efficient programs to monitor employee exposures to ethylene thiourea. The manual discusses determination of the need for exposure measurements, selection of appropriate employees for exposure evaluation, and selection of sampling times.

Employee exposure measurements should consist of 8-hour TWA (time-weighted average) exposure estimates calculated from personal or breathing zone samples (air that would most nearly represent that inhaled by the employees). Area and source measurements may be useful to determine problem areas, processes, and operations.

Special Tests: None in use at present.

Personal Protective Methods: There are four basic methods of limiting employee exposure to ethylene thiourea. None of these is a simple industrial hygiene or management decision and careful planning and thought should be used prior to implementation of any of these.

Product Substitution—The substitution of an alternative material with a lower potential health and safety risk is one method. However, extreme care must be used when selecting possible substitutes. Alternatives to ethylene thiourea should be fully evaluated with regard to possible human effects. Unless the toxic effects of the alternative have been thoroughly evaluated, a seemingly safe replacement, possibly only after years of use, may be found to induce serious health effects.

Contaminant Controls—The most effective control of ethylene thiourea, where feasible, is at the source of contamination by enclosure of the operation and/or local exhaust ventilation.

If feasible, the process or operation should be enclosed with a slight vacuum so that any leakage will result in the flow of air into the enclosure.

The next most effective means of control would be a well designed local exhaust ventilation system that physically encloses the process as much as possible, with sufficient capture velocity to keep the contaminant from entering the work atmosphere.

To ensure that ventilation equipment is working properly, effectiveness (e.g., air velocity, static pressure, or air volume) should be checked at least every three months. System effectiveness should be checked soon after any change in production, process or control which might result in significant increases in airborne exposures to ethylene thiourea.

Employee Isolation—A third alternative is the isolation of employees. It frequently involves the use of automated equipment operated by personnel observing from a closed control booth or room. The control room is maintained at a greater air pressure than that surrounding the process equipment so that air flow is out of, rather than into, the room. This type of control will not protect those employees that must do process checks, adjustments, maintenance, and related operations.

Personal Protective Equipment—The least preferred method is the use of personal protective equipment. This equipment, which may include respirators, goggles, gloves, and related items, should not be used as the only means to prevent or minimize exposure during routine operations.

Exposure to ethylene thiourea should not be controlled with the use of respirators except:

— During the time necessary to install or implement engineering or work practice controls; or

— In work situations in which engineering and work practice controls are technically not feasible; or

— For maintenance; or

— For operations which require entry into tanks or closed vessels; or

— In emergencies.

Only respirators approved by the National Institute for Occupational Safety and Health (NIOSH) should be used. Refer to *NIOSH Certified Equipment, December 15, 1975,* NIOSH Publication #76-145 and *Cumulative Supplement June 1977, NIOSH Certified Equipment,* NIOSH Publication #77-195. The use of faceseal coverlets or socks with any respirator voids NIOSH approvals.

Quantitative faceseal fit test equipment (such as sodium chloride, dioctyl phthalate, or equivalent) should be used. Refer to *A Guide to Industrial Respi-*

ratory Protection, NIOSH Publication #76-189 for guidelines on appropriate respiratory protection programs.

References

(1) Stula, E.F., and Krauss, W.C. Embryotoxicity in rats and rabbits from cutaneous application of amide-type solvents and substituted ureas. *Toxicology and Applied Pharmacology* 41, 35-55, 1977.
(2) Khera, K.S., and Tryphonas, L. Ethylenethiourea-induced hydrocephalus: pre- and postnatal pathogenesis in offspring from rats given a single oral dose during pregnancy. *Toxicology and Applied Pharmacology* 42, 85-97, 1977.
(3) Ulland, B.M. et al. Thyroid cancer in rats from ethylene thiourea intake. *J. Nat. Cancer Inst.* 49, 583-584, 1972.
(4) *IARC Monographs on the Evaluation of Carcinogenic Risk of Chemicals to Man.* 7, 45-52, 1974.
(5) Innes, J.R.M. et al. Bioassay of pesticides and industrial chemicals for tumorigenicity in mice: A preliminary note. *J. Nat. Cancer Inst.* 42, 1101-1114, 1969.
(6) Graham, S.L. et al. Effects of prolonged ethylene thiourea ingestion on the thyroid of the rat. *Food and Cosmetic Toxicology* 13, 493-499, 1975.

Bibliography

National Institute for Occupational Safety and Health. *Ethylene Thiourea.* Current Intelligence Bulletin 22, Wash., D.C. (April 11, 1978).

ETHYL ETHER

Description: $CH_3CH_2OCH_2CH_3$, ethyl ether, is a colorless, mobile, highly flammable, volatile liquid with a characteristic pungent odor.

Synonyms: Anesthetic ether, diethyl ether, diethyl oxide, ether, ethoxyethane, ethyl oxide, sulfuric ether.

Potential Occupational Exposures: Ethyl ether is used as a solvent for waxes, fats, oils, perfumes, alkaloids, dyes, gums, resins, nitrocellulose, hydrocarbons, raw rubber, and smokeless powder. It is also used as an inhalation anesthetic, a refrigerant, in diesel fuels, in dry cleaning, as an extractant, and as a chemical reagent for various organic reactions.

A partial list of occupations in which exposure may occur includes:

Acetic acid makers	Explosive makers
Alcohol denaturers	Fat processors
Collodion makers	Gasoline engine primers
Diesel fuel blenders	Dye makers
Drug makers	Gum processors

Motor fuel makers	Rayon makers
Nitrocellulose makers	Rubber workers
Oil processors	Smokeless powder makers
Perfume makers	Wax makers

Permissible Exposure Limits: The Federal standard is 400 ppm (1,200 mg/m^3).

Route of Entry: Inhalation of vapor.

Harmful Effects

Local—Ethyl ether vapor is mildly irritating to the eyes, nose, and throat. Contact with liquid may produce a dry, scaly, fissured dermatitis.

Systemic—Ethyl ether has predominantly narcotic properties. Overexposed individuals may experience drowsiness, vomiting, and unconsciousness. Death may result from severe overexposure. Chronic exposure results in some persons in anorexia, exhaustion, headache, drowsiness, dizziness, excitation, and psychic disturbances. Albuminuria has been reported. Chronic exposure may cause an increased susceptibility to alcohol.

Medical Surveillance: Preplacement or periodic examinations should evaluate the skin and respiratory tract, liver, and kidney function. Persons with a past history of alcoholism may be at some increased risk due to possibility of ethyl ether addiction (known as "ether habit").

Special Tests: Tests for exposure may include expired breath for unmetabolized ethyl ether and blood for ethyl ether content by oxidation with chromate solution or by gas chromatographic methods.

Personal Protective Methods: Barrier creams, gloves, protective clothing, and, in areas of vapor concentration, full face respirator should be used.

Bibliography

Hamilton, A., and Minot., G.R. 1920. Ether poisoning in the manufacture of smokeless powder. *J. Ind. Hyg.* 2:41.
Von Oettingen, W.F. 1958. *Poisoning, A Guide to Clinical Diagnosis and Treatment,* 2nd ed. W.B. Saunders Co., Philadelphia.

ETHYL SILICATE

Description: $(C_2H_5O)_4Si$, ethyl silicate, is a colorless, flammable liquid with a sharp odor detectable at 85 ppm.

Synonyms: Tetraethyl orthosilicate, tetraethoxy silane.

Potential Occupational Exposures: Ethyl silicate is used in production of cases and molds for casting of metals and as a hardener for water and weather-resistant concrete.

A partial list of occupations in which exposure may occur includes:

Acidproof cement makers	Heat resistant paint makers
Adhesive makers	Lacquer makers
Brick preserver makers	Metal casters
Building preserver makers	Silicate paint makers

Permissible Exposure Limits: There is no Federal standard; however, the ACGIH recommended TLV is 100 ppm (approximately 850 mg/m^3) determined as a time-weighted average. This TLV has not been confirmed in human exposure. At 3,000 ppm, ethyl silicate vapors are intolerable. ACGIH (1978) has proposed a TWA of 10 ppm (85 mg/m^3).

Route of Entry: Inhalation of vapor.

Harmful Effects

Local—Ethyl silicate is a primary irritant to the eyes and the nose.

Systemic—Damage to the lungs, liver, and kidneys, and anemia have been observed in animal experiments but have not been reported for human exposure.

Medical Surveillance: Placement or periodic examinations should include the skin, eyes, respiratory tract, as well as liver and kidney functions.

Special Tests: None currently used.

Personal Protective Methods: Full face masks in areas of vapor concentration.

Bibliography

Pozzini, U.C., and Carpenter, C.P. 1951. Response of rodents to repeated inhalation of vapors of tetraethyl orthosilicate. *Arch. Ind. Hyg. Occup. Med.* 4:465.

F

FIBROUS GLASS

Description: Fibrous glass is the name for a manufactured fiber in which the fiber-forming substance is glass. Glasses are a class of materials made from silicon dioxide with oxides of various metals and other elements, that solidify from the molten state without crystallization. A fiber is considered to be a particle with a length-to-diameter ratio of 3 to 1 or greater.

Most fibrous glass that is manufactured consists of fibers with diameters 3.5 μm or larger. The volume of small diameter fiber production has not been determined. Fibers with diameters less than 1 μm are estimated to comprise less than 1% of the fibrous glass market.

Synonyms: Glass fiber, glass wool, Fiberglas.

Potential Occupational Exposures: It is estimated that fibrous glass is used in over 30,000 product applications. The major uses of fibrous glass are in thermal, electrical, and acoustical insulation, weather proofing, plastic reinforcement, filtration media, and in structural and textile materials. NIOSH estimates that 200,000 workers are potentially exposed to fibrous glass.

Permissible Exposure Limits: NIOSH recommends that occupational exposure to fibrous glass be controlled so that no worker is exposed at an airborne concentration greater than 3,000,000 fibers/m^3 of air (3 fibers/cc of air) having a diameter equal to or less than 3.5 μm and a length equal to or greater than 10 μm determined as a time weighted average (TWA) concentration for up to a 10-hour work shift in a 40-hour workweek; airborne concentrations determined as total fibrous glass shall be limited to a TWA concentration of 5 mg/m^3 of air. This differs from the present Federal standard which classifies fibrous glass as an inert or nuisance dust with the limits of exposure being 15 million particles per cubic foot (mppcf) or 5 mg/m^3 for the respirable fraction and 50 mppcf or 15 mg/m^3 total dust, both as 8-hour TWA concentrations.

Routes of Entry: Inhalation and skin contact.

Harmful Effects

Different dimensions of fibrous glass will produce different biologic effects.

Large diameter (greater than 3.5 μm) glass fibers have been found to cause skin, eye, and upper respiratory tract irritation; a relatively low frequency of fibrotic changes; and a very slight indication of an excess mortality due to nonmalignant respiratory disease. Smaller diameter (less than 3.5 μm) fibrous glass has not been conclusively related to health effects in humans but glass fibers of this dimension have only been regularly produced since the 1960s. Smaller diameter fibers have the ability to penetrate to the alveoli and this potential is cause for concern and the primary reason that fibers 3.5 μm or smaller are subject to special controls. Experimental studies in animals have demonstrated carcinogenic effects with the long (greater than 10 μm) and thin fibers (usually less than 1 μm in diameter). However, these studies were performed by implanting fibrous glass in the pleural or peritoneal cavities.

The data from studies with these routes of exposure cannot be directly extrapolated to conditions of human exposure. On the basis of available information, NIOSH does not consider fibrous glass to be a substance that produces cancer as a result of occupational exposure. The data on which to base this conclusion are limited. Fibrous glass does not appear to possess the same potential as asbestos for causing health hazard. Glass fibers are not usually of the fine submicron diameters as are asbestos fibrils and the concentrations of glass fibers in workplace air are generally orders of magnitude less than for asbestos. In one study, glass fibers were found to be cleared from the lungs more readily than asbestos.

Medical Surveillance: NIOSH recommends that workers subject to fibrous glass exposure have comprehensive preplacement medical examinations with emphasis on skin susceptibility and prior exposure in dusty trades. Subsequent annual examinations should give attention to the skin and respiratory system with attention to pulmonary function.

Special Tests: None in common use.

Personal Protective Methods

> (a) Skin Protection
> Protective clothing shall be worn to prevent fibrous glass contact with skin especially hands, arms, neck, and underarms.
>
> (b) Eye Protection
> Safety goggles or face shields and goggles shall be worn during tear-out or blowing operations or when applying fibrous glass materials overhead. They should be used in all areas where there is a likelihood that airborne glass fibers may contact the eyes.
>
> (c) Respiratory Protection
> Engineering controls should be used wherever feasible to maintain fibrous glass concentrations at or be-

low the prescribed limits. Respirators should only be used when engineering controls are not feasible; for example, in certain nonstationary operations where permanent controls are not feasible.

Bibliography

National Institute for Occupational Safety and Health. *Occupational Exposure to Fibrous Glass: A Symposium.* NIOSH Doc. No. 76-151 (1976).
National Institute for Occupational Safety and Health. *Criteria for a Recommended Standard: Occupational Exposure to Fibrous Glass.* NIOSH Doc. No. 77-152 (1977).

FLUORINE AND COMPOUNDS

Description: F_2, molecular fluorine, is a yellow gas. Sulfuric acid reacts with fluorspar producing hydrofluoric acid (HF) which is starting material for synthesis of most fluorine compounds. Fluorine forms fluorides but not fluorates or perfluorates.

Synonyms

Fluorine: None.

Hydrogen fluoride: Hydrofluoric acid gas, fluorohydric acid gas, anhydrous hydrofluoric acid.

Fluorides: None.

Potential Occupational Exposures: Elemental fluorine is used in the conversion of uranium tetrafluoride to uranium hexafluoride, in the synthesis of organic and inorganic fluorine compounds, and as an oxidizer in rocket fuel.

Hydrogen fluoride, its aqueous solution hydrofluoric acid, and its salts are used in production of organic and inorganic fluorine compounds such as fluorides and plastics; as a catalyst, particularly in paraffin alkylation in the petroleum industry; as an insecticide; and to arrest the fermentation in brewing. It is utilized in the fluorination processes, especially in the aluminum industry, in separating uranium isotopes, in cleaning cast iron, copper, and brass, in removing efflorescence from brick and stone, in removing sand from metallic castings, in frosting and etching glass and enamel, in polishing crystal, in decomposing cellulose, in enameling and galvanizing iron, in working silk, in dye and analytical chemistry, and to increase the porosity of ceramics.

Fluorides are used as an electrolyte in aluminum manufacture, a flux in smelting nickel, copper, gold, and silver, as a catalyst for organic reactions, a wood preservative, fluoridation agent for drinking water, a bleaching agent for cane seats, in pesticides, rodenticides, and as a fermentation inhibitor. They are utilized in the manufacture of steel, iron, glass, ceramics, pottery, enamels, in the coagulation of latex, in coatings for welding rods, and in cleaning graphite, metals, windows, and glassware. Exposure to fluorides may also occur during preparation of fertilizer from phosphate rock by addition of sulfuric acid.

A partial list of occupations in which exposure may occur includes:

Aluminum fluoride makers	Fluorochemical workers
Aluminum makers	Glass etchers
Bleachers	Incandescent lamp frosters
Brass cleaners	Insecticide makers
Casting cleaners	Ore dissolvers
Ceramic workers	Stone cleaners
Copper cleaners	Uranium refiners
Crystal glass polishers	Yeast makers
Fermentation workers	

Major users of fluoride compounds are the aluminum, chemical, and steel industries, and NIOSH estimates that 350,000 workers are potentially exposed to fluorides.

Permissible Exposure Limits: The applicable Federal standards are: fluorine 1.0 ppm (2.0 mg/m^3), fluoride as dust (2.5 mg/m^3), hydrogen fluoride 3 ppm, ceiling 5 ppm and peak 10 ppm for 30 minutes. For hydrogen fluoride NIOSH has recommended 2.5 mg/m^3 (fluoride ion) TWA with a ceiling of 5 mg/m^3 (fluoride ion) for a 15-minute sampling period.

Routes of Entry: Inhalation of gas, mist, dust, or fume; ingestion of dust.

Harmful Effects

Local—Fluorine and some of its compounds are primary irritants of skin, eyes, mucous membranes, and lungs. Thermal or chemical burns may result from contact; the chemical burns cause deep tissue destruction and may not become symptomatic until several hours after contact, depending on dilution. Nosebleeds and sinus trouble may develop on chronic exposure to low concentration of fluoride or fluorine in air. Accidental fluoride burns, even when they involve small body areas (less than 3%), can cause systemic effects of fluoride poisoning by absorption of the fluoride through the skin.

Systemic—Inhalation of excessive concentration of elemental fluorine or of hydrogen fluoride can produce bronchospasm, laryngospasm, and pulmonary edema. Gastrointestinal symptoms may be present. A brief exposure to 25 ppm has caused sore throat and chest pain, irreparable damage to the lungs, and death.

Most cases of acute fluoride intoxication result from ingestion of fluoride compounds. The severity of systemic effects is directly proportional to the irritating properties and the amount of the compound that has been ingested. Gastrointestinal symptoms of nausea, vomiting, diffuse abdominal cramps, and diarrhea can be expected. Large doses produce central nervous system involvement with twitching of muscle groups, tonic and clonic convulsions, and coma.

The systemic effects of prolonged absorption of fluorides from either dusts or vapors have long been a source of some uncertainty. Fluorides are retained preferentially in bone, and excessive intake may result in an osteosclerosis that is recognizable by x-ray. The first signs of changes in density appear in the lumbar spine and pelvis. Usually some ossification of ligaments occurs. Recent investigations suggest that rather severe skeletal fluorosis can exist in workers without any untoward physiological effects, detrimental effects on their general health, or physical impairment.

Fluorides occur in nature and enter the human body through inhalation or ingestion (natural dusts and water). In children, mottling of the dental enamel may occur from increased water concentrations. These exposures are usually minimal and occur over extended periods. Residential districts which adjoin manufacturing areas can be subjected to continual exposures at minimal levels, or to heavy exposure in the event of accident or plant failure, as in the case of the Meuse Valley disaster.

Medical Surveillance: Preemployment and periodic examinations should consider possible effects on the skin, eyes, teeth, respiratory tract, and kidneys. Chest x-rays and pulmonary function should be followed. Kidney function should be evaluated. If exposures have been heavy and skeletal fluorosis is suspected, pelvic x-rays may be helpful. Intake of fluoride from natural sources in food or water should be known.

Special Tests: In the case of exposure to fluoride dusts, periodic urinary fluoride excretion levels have been very useful in evaluating industrial exposures and environmental dietary sources.

Personal Protective Methods: In areas with excessive gas or dust levels for any type of fluorine, worker protection should be provided. Respiratory protection by dust masks or gas masks with an appropriate canister or supplied air respirator should be provided. Goggles or fullface masks should be used. In areas where there is a likelihood of splash or spill, acid resistant clothing including gloves, gauntlets, aprons, boots, and goggles or face shields should be provided to the worker. Personal hygiene should be encouraged, with showering following each shift and before change to street clothes. Work clothes should be changed following each shift, especially in dusty areas. Attention should be given promptly to any burns from fluorine compounds due to absorption of the fluorine at the burn site and the possibility of developing systemic symptoms from absorption from burn sites.

Bibliography

Biologic Assay Committee, American Industrial Hygiene Association. 1971. Biologic monitoring guides: fluorides. *Am. Ind. Hyg. Assoc. J.* 32:274.

Burke, W.J., Hoegg, U.R., and Phillips, R.E. 1973. Systemic fluoride poisoning resulting from a fluoride skin burn. *J. Occup. Med.* 15:39.

Dinman, B.D., Bovard, W.J., Bonney, T.B., Cohen, J.M., and Colwell, M.O. 1976. Absorption and excretion of fluoride immediately after exposure—Pt. 1. *J. Occup. Med.* 18:7.

Hodge, H.C., and Smith, F.A. 1970. Air quality criteria for the effects of fluorides on man. *J. Air. Pollut. Control Assoc.* 20:232.

Kaltreider, N.L., Elder, M.J., Cralley, L.V., and Colwell, M.O. 1972. Health survey of aluminum workers with special reference to fluoride exposure. *J. Occup. Med.* 17:531.

National Institute for Occupational Safety and Health. *Criteria for a Recommended Standard: Occupational Exposure to Inorganic Fluoride.* NIOSH Doc. No. 76-103. (1976).

National Institute for Occupational Safety and Health. *Criteria for a Recommended Standard: Occupational Exposure to Hydrogen Fluoride.* NIOSH Doc. No. 76-143 (1976).

Neefus, J.D., Cholak, J., and Saltzman, B.E. 1970. The determination of fluoride in urine using a fluoride-specific ion electrode. *Am. Ind. Hyg. Assoc. J.* 31:96.

Princi, F. 1960. Fluorides: a critical review. III. The effects on man of the absorption of fluoride. *J. Occup. Med.* 2:92.

FLUOROCARBONS

Description: Chemically, fluorocarbons are hydrocarbons containing fluorine and include compounds that may contain other halogens in addition to fluorine. Generally, these compounds are colorless, nonflammable gases, though a few are liquids at room temperature. Decomposition of chlorine-containing fluoromethanes caused by contact with an open flame or hot metal produces hydrogen chloride, hydrogen fluoride, phosgene, carbon dioxide, and chlorine.

Synonyms: See listing of economically important fluorocarbons under "Permissible Exposure Limits." Also, Freon is a trademark for a number of fluorocarbons used particularly in refrigeration products and equipment.

Potential Occupational Exposures: The fluorocarbons are used primarily as refrigerants and polymer intermediates. They are also used as aerosol propellants, anesthetics, fire extinguishers, foam blowing agents, in drycleaning, and in degreasing of electronic equipment. The fluorocarbons have found wide use due to their relatively low toxicity.

A partial list of occupations in which exposure may occur includes:

Aerosol bomb workers   Drug makers
Ceramic mold makers   Fire extinguisher workers
Drycleaners   Heat transfer workers

Metal conditioners	Refrigeration makers
Plastic makers	Rocket fuel makers
Pressurized food makers	Solvent workers

Permissible Exposure Limits: Federal standards for selected fluorocarbons of economic importance are listed as follows:

| Compound | Federal Standards | |
	ppm	mg/m^3
Bromotrifluoromethane Fluorocarbon 13B1	1,000	6,100
Dibromodifluoromethane Fluorocarbon 12B2	100	860
Dichlorodifluoromethane Fluorocarbon 12	1,000	4,950
Dichloromonofluoromethane Fluorocarbon 21	1,000	4,200
Dichlorotetrafluoroethane Fluorocarbon 114	1,000	7,000
Fluorotrichloromethane Fluorocarbon 11	1,000	5,600
1,1,1,2-Tetrachloro-2,2-difluoroethane Fluorocarbon 112	500	4,170
1,1,2,2-Tetrachloro-1,2-difluoroethane Fluorocarbon 112	500	4,170
1,1,2-Trichloro-1,2,2-trifluoroethane Fluorocarbon 113	1,000	7,600
Bromochlorotrifluoroethane Fluorocarbon 123B1	No Standard	
Chlorodifluoromethane Fluorocarbon 22 (ACGIH TLV of 1,000 ppm)	No Standard	
Chloropentafluoroethane Fluorocarbon 115	No Standard	
Chlorotrifluoroethylene Fluorocarbon 1113	No Standard	
Chlorotrifluoromethane Fluorocarbon 13	No Standard	
Difluoroethylene Fluorocarbon 1131	No Standard	
Fluoroethylene Fluorocarbon 1141	No Standard	
Hexafluoropropylene Fluorocarbon 1216	No Standard	
Octafluorocyclobutane Fluorocarbon C-318	No Standard	
Tetrafluoroethylene Fluorocarbon 1114	No Standard	

Route of Entry: Inhalation of vapor or gas (from volatile fluorocarbon products or from decomposition of fluorocarbon polymers).

Harmful Effects

Local—These compounds may produce mild irritation to the upper respiratory tract. Dermatitis occurs only rarely. Decomposition products may also be the cause of these effects.

Systemic—Mild central nervous system depression may occur in cases of exposure to very high concentrations of fluorocarbons. Symptoms from acute exposure may manifest themselves in occasional tremor and incoordination. It has been reported that dizziness has resulted from an exposure to 5% dichlorodifluoromethane and unconsciousness from exposure to 15%. Cardiac arrhythmias, with sudden death, have occurred from breathing some of these chemicals. Typically, fluorocarbons have very low levels of toxicity, and their predominant hazard is from simple asphyxia.

Fluoroalkenes are more toxic than fluoroalkanes. Liver and kidney damage has been reported to occur from chronic exposure to fluoroalkenes, whereas no chronic effects have been reported from fluoroalkanes.

Medical Surveillance: Though these compounds are of a low level of toxicity, they should not be considered inert. There are no specific diagnostic tests for the toxic effects occurring at very high concentrations. Preplacement and periodic examinations should consider possible cardiac effects from acute exposure.

Special Tests: None commonly used. Compounds are usually excreted rapidly in expired air.

Personal Protective Methods: Simple ventilation can avert acute poisoning. Masks are rarely needed.

Bibliography

Imbus, H.R., and Adkins, C. 1972. Physical examinations of workers exposed to trichlorotrifluoroethane. *Arch. Environ. Health* 24:257.
National Institute for Occupational Safety and Health. *Criteria for a Recommended Standard: Occupational Exposure to Decomposition Products of Fluorocarbon Polymers.* NIOSH Doc. No. 77-193 (1977).
Pattison, F.L.M. 1959. *Toxic Aliphatic Fluorine Compounds.* Elsevier Publishing Co., Amsterdam and Princeton.
Smith, P.E. Jr. 1971. Human exposure to fluorocarbon 113 (1,1,2-trichloro-1,2,2-trifluoroethane). *Am. Ind. Hyg. Assoc. J.* 32:143.
Thyrum, P.T. 1972. Editorial views. Fluorinated hydrocarbons and the heart. *Anesthesiology* 36:103.

FORMALDEHYDE

Description: HCHO, formaldehyde, is a colorless, pungent gas. It is produced commercially by the catalytic oxidation of methyl alcohol and sold in aqueous solution containing 30% to 50% formaldehyde and from 0% to 15% methanol, which is added to prevent polymerization. Formaldehyde solution is called formalin, Formol, or Morbicid.

Synonyms: Oxomethane, oxymethylene, methylene oxide, formic aldehyde, methyl aldehyde.

Potential Occupational Exposures: Formaldehyde has found wide industrial usage as a fungicide, germicide, and in disinfectants and embalming fluids. It is also used in the manufacture of artificial silk and textiles, latex, phenol, urea, thiourea and melamine resins, dyes, and inks, cellulose esters and other organic molecules, mirrors, and explosives. It is also used in the paper, photographic, and furniture industries.

A partial list of occupations in which exposure may occur includes:

Anatomists	Hide preservers
Biologists	Ink makers
Deodorant makers	Latex makers
Disinfectant makers	Photographic film makers
Embalming fluid makers	Textile printers
Formaldehyde resin makers	Wood preservers

The demand for formaldehyde, calculated as a 37% solution, has been estimated to be 6,300 million pounds for 1976 (1). NIOSH estimates that some 8,000 employees are potentially exposed.

Permissible Exposure Limits: The Federal standard is 3 ppm determined as a TWA. The acceptable ceiling concentration is 5 ppm with an acceptable maximum peak above this value of 10 ppm for a maximum duration of 30 minutes. ACGIH in 1978 recommended a TLV of 2 ppm (3 mg/m^3) as a ceiling value. NIOSH has recommended (2) a ceiling of 1 ppm (1.2 mg/m^3) for any 30-minute sampling period.

Route of Entry: Inhalation of gas.

Harmful Effects

Local—Formaldehyde gas may cause severe irritation to the mucous membranes of the respiratory tract and eyes. The aqueous solution splashed in the eyes may cause eye burns. Urticaria has been reported following inhalation of gas. Repeated exposure to formaldehyde may cause dermatitis either from irritation or allergy.

Systemic—Systemic intoxication is unlikely to occur since intense irritation of upper respiratory passages compels workers to leave areas of exposure. If workers do inhale high concentrations of formaldehyde, coughing, difficulty in breathing, and pulmonary edema may occur. Ingestion, though usually not occurring in industrial experience, may cause severe irritation of the mouth, throat, and stomach.

The mutagenicity of formaldehyde has been described most extensively for *Drosophila* (3)-(11) (with hydrogen peroxide) (12) and established for *Neurospora cassida* (also with hydrogen peroxide) (13)-(15) and *E. coli* (15)-(18).

Formaldehyde effects on *E. coli* B/r in a special mutant lacking a DNA polymerase (pol A$^-$) and, therefore, a repair deficient strain were elaborated by Rosenkranz (17). Formaldehyde treatment of pol A$^+$ and pol A$^-$ strains showed differential toxicity, determined by the "zone of inhibition" surrounding a formaldehyde-soaked disc placed on the surface of the growth agar. There was a preferential inhibition of growth in the pol A$^-$ strain, indicating that some repair capability may affect the survival of formaldehyde treated bacteria.

In the above studies, Rosenkranz (17) also described the interaction of known carcinogens (e.g., methyl methanesulfonate and N-hydroxylaminofluorene) with both the pol A$^+$ and pol A$^-$ strains of E. coli and concluded that "in view of the present findings and because the procedure used seems to be quite reliable for detecting carcinogens, it would seem that continued use of formaldehyde requires reevaluation and monitoring as exposure to even low levels of this substance might be deleterious especially if it occurs over prolonged periods of time, a situation which probably increases the chance of carcinogenesis."

Formaldehyde is also mutagenic in E. coli B/r strains which were altered in another repair function, Hcr$^-$. (This strain lacks the ability to reactivate phage containing UV-induced thymine dimers because it lacks an excision function.) Strains of E. coli B/r which were Hcr$^-$ showed more mutation to streptomycin resistance or to tryptophan independence than did the repair competent Hcr$^+$ strain. Ultraviolet inactivation of Hcr$^-$ strains was enhanced by treatment with formaldehyde, possibly indicating some effect of formaldehyde on the repair function (18).

Formaldehyde has been found to combine with RNA or its constituent nucleotides (19)-(21) with the formation of hydroxymethylamide (HOCH$_2$CONH$_2$) and hydroxymethylamine (NH$_2$CH$_2$OH) by hydroxymethylation of amido ($-$CONH$_2$) and amino groups respectively.

Formaldehyde has been found to combine more readily with single stranded polynucleotides such as replicating DNA (22) or synthetic poly A (23). The reaction products may also include condensation products of adenosine such as methylene bis AMP. The possibility of formation of these compounds in vivo has led to the postulation that adenine dimers may be found in polynucleotides in situ or may be erroneously incorporated into polynucleotides (21)(22).

An alternative mechanism of action for formaldehyde involving the formation of peroxidation products by autooxidation of formaldehyde or by its reaction with other molecules to form free radicals has been proposed (15). The synergism between hydrogen peroxide and formaldehyde in producing mutations in Neurospora has been described (15). The combination of formaldehyde and H$_2$O$_2$ was found to be differentially mutagenic at two loci, adenine and inositol utilization. These two loci showed divergent dose response curves when similarly treated with formaldehyde and H$_2$O$_2$. This was taken as evidence for a mutagenic peroxidation product.

A number of carcinogenic studies of formaldehyde per se (24)-(26) as well as that of hexamethylene tetramine (27)(28) (an agent known to release formaldehyde) have been reported. To date the assessment of the carcinogenicity based on these studies would appear to be equivocal.

Medical Surveillance: Consider the skin, eyes, and respiratory tract in any preplacement or periodic examination, especially if the patient has a history of allergies.

Special Tests: None in common use.

Personal Protective Methods: Prevention of intoxication may be easily accomplished by supplying adequate ventilation and protective clothing. Barrier creams may also be helpful. In areas of high vapor concentrations, full protective face masks with air supply is needed, as well as protective clothing.

References

(1) Environmental Protection Agency. *Investigation of Selected Potential Environmental Contaminants—Formaldehyde, Final Report.* Office of Toxic Substances, Environmental Protection Agency, August, 1976.
(2) National Institute for Occupational Safety and Health. *Criteria for a Recommended Standard: Occupational Exposure to Formaldehyde.* NIOSH Doc. No. 77-126 (1977).
(3) Alderson, T. Chemically induced delayed germinal mutation in *Drosophila. Nature* 207 (1965) 164.
(4) Auerbach, C. Mutation tests on *Drosophila melanogaster* with aqueous solutions of formaldehyde. *Am. Naturalist* 86 (1952) 330-332.
(5) Khishin, A.F.E. The requirement of adenylic acid for formaldehyde mutagenesis *Mutation Res.* 1 (1964) 202.
(6) Burdette, W.J. Tumor incidence and lethal mutation rate in a tumor strain of *Drosophila* treated with formaldehyde. *Cancer Res.* 11 (1951) 555.
(7) Auerbach, C. Analysis of the mutagenic action of formaldehyde food. III. Conditions influencing the effectiveness of the treatment. *Z. Vererbungslehre* 81 (1956) 627-647.
(8) Auerbach, C. *Drosophila* tests in pharmacology. *Nature* 210 (1966) 104.
(9) Rapoport, I.A. Carbonyl compounds and the chemical mechanisms of mutations. *C.R. Acad. Sci. USSR* 54 (1946) 65.
(10) Kaplan, W.D. Formaldehyde as a mutagen in *Drosophila. Science* 108 (1948).
(11) Auerbach, C. The mutagenic mode of action of formalin. *Science* 110 (1949) 119.
(12) Sobels, F.H. Mutation tests with a formaldehyde-hydrogen peroxide mixture in *Drosophila. Am. Naturalist* 88 (1954) 109-112.
(13) Dickey, F.H., Cleland, G.H., and Lotz, C. The role of organic peroxides in the induction of mutations. *Proc. Natl. Acad. Sci.* 35 (1949) 581.
(14) Jensen, K.A., Kirk, I., Kolmark, G., and Westergaard, M. Chemically induced mutations in *Neurospora.* Cold Spring Symp. *Quant. Biol.* 16 (1951) 245.
(15) Auerbach, C., and Ramsey, D. Analysis of a case of mutagen specificity in *Neurospora crassa. Mol. Gen. Genetics* 103 (1968) 72.
(16) Demerec, M., Bertani, G., and Flint, J. Chemicals for mutagenic action on *E. coli. Am. Naturalist* 85 (1951) 119.
(17) Rosenkranz, H.S. Formaldehyde as a possible carcinogen. *Bull. Env. Contam. Toxicol.* 8 (1972) 242.

(18) Nishioka, H. Lethal and mutagenic action of formaldehyde, Hcr+ and Hcr⁻ strains, *E. coli. Mutation Res.* 17 (1973) 261.

(19) Hoard, D.E. The applicability of formol titration to the problem of end-group determinations in polynucleotides. A Preliminary Investigation. *Biochem. Biophys. Acta.* 40 (1960) 62.

(20) Haselkorn, R., and Doty, P. The reaction of formaldehyde with polynucleotides. *J. Biol. Chem.* 236 (1961) 2730.

(21) Alderson, T. Significance of ribonucleic acid in the mechanism of formaldehyde induced mutagenesis. *Nature* 185 (1960) 904.

(22) Voronina, E.N. Study of the spectrum of mutations caused by formaldehyde in *E. coli* K-12 in different periods of a synchronized lag period. *Sov. Genet.* 7 (1971) 788.

(23) Fillippova, L.M. Pan'shin, O.A. and Koslyankovskii, F.K. Chemical mutagens. *Genetika* 3 (1967) 135.

(24) Watanabe, F., Matsunaga, T., Soejima, T., and Iwata, Y. Study on the carcinogenicity of aldehyde, 1st report. Experimentally produced rat sarcomas by repeated injections of aqueous solution of formaldehyde. *Gann.* 45 (1954) 451.

(25) Watanabe, F., and Sugimoto, S. Study on the carcinogenicity of aldehyde, 2nd report. Seven cases of transplantable sarcomas of rats appearing in the area of repeated subcutaneous injections of urotropin. *Gann* 46 (1955) 365.

(26) Horton, A.W., Tye, R., and Stemmer, K.L. Experimental carcinogenesis of the lung, inhalation of gaseous formaldehyde on an aerosol tar by C3H mice. *J. Natl. Cancer Inst.* 30 (1963) 31.

(27) Della Porta, G., Colnagi, M.I., and Parmiani, G. Non-carcinogenicity of hexamethylenetetramine in mice and rats. *Food Cosmet. Toxicol.* 6 (1968) 707.

(28) Brendel, R. Untersuchungen an ratten zur vertraglickeit von hexamethylene tetramin. *Arznei-Forsch* 14 (1964) 51.

Bibliography

Bartone, N.F., Grieco, R.V., and Herr, B.S. Jr. 1968. Corrosive gastritis due to ingestion of formaldehyde without esophageal impairment. *J. Am. Med. Assoc.* 103:104.

Hendrick, D.J., and Lane, D.J. 1975. Formalin asthma in hospital staff. *Brit. Med. J.* 1:607.

Henson, E.V. 1959. The toxicology of some aliphatic aldehydes. *J. Occup. Med.* 1:457.

O'Quinn, S.E., and Kennedy, C.B. 1965. Contact dermatitis due to formaldehyde in clothing textiles. *J. Am. Med. Assoc.* 194:593.

Porter, A.H. 1975. Acute respiratory distress following formaldehyde inhalation. *Lancet* 2:603.

FORMATES

Description: Methyl formate, $HCOOCH_3$; ethyl formate, $HCOOC_2H_5$. These are colorless, mobile, flammable liquids with agreeable odors.

Synonyms: Methyl formate, methyl methanoate; ethyl formate, ethyl methanoate.

Potential Occupational Exposures: Formates are solvents for cellulose nitrate, oils, greases, fats, cellulose acetate, fatty acids, acetylcellulose, collodion, and celluloid. They are also used as larvicides, fumigants, flavoring agents in the production of lemonade, rum, arrack, and essences, and they are used in chemical synthesis.

A partial list of occupations in which exposure may occur includes:

Cellulose acetate workers   Nitrocellulose workers
Flavoring makers   Organic chemical synthesizers
Fumigant makers   Pesticide workers
Fumigators   Tobacco fumigators
Grain fumigators

Permissible Exposure Limits: The Federal standards are:

Methyl formate   100 ppm   250 mg/m^3
Ethyl formate   100 ppm   300 mg/m^3

Routes of Entry: Inhalation, ingestion, and skin absorption.

Harmful Effects

Local—Methyl formate is a mild irritant to mucous membranes, especially eyes and respiratory system. Ethyl formate may be irritating to skin and mucous membranes in high concentrations.

Systemic—Methyl formate has an irritant and narcotic effect, and in high concentrations may cause drowsiness and unconsciousness. Systemic intoxication in industry is rare.

Medical Surveillance: Consider eye and respiratory disease or symptoms in any placement or follow-up examinations.

Special Tests: None in common use.

Personal Protective Methods: Barrier creams should be used to protect the skin, and masks should be used in areas of vapor concentration.

Bibliography

Von Oettingen, W.F. 1959. The aliphatic acids and their esters—toxicity and potential dangers. The saturated monobasic aliphatic acids and their esters. *AMA Arch. Ind. Health* 20:517.

FORMIC ACID

Description: HCOOH, formic acid, is a colorless, flammable, fuming liquid, with a pungent odor.

Synonyms: Methanoic acid, formylic acid, hydrogen carboxylic acid.

Potential Occupational Exposures: Formic acid is a strong reducing agent and is used as a decalcifier. It is used in dyeing color fast wool, electroplating, coagulating latex rubber, regenerating old rubber, and dehairing, plumping, and tanning leather. It is also used in the manufacture of acetic acid, airplane dope, allyl alcohol, cellulose formate, phenolic resins, and oxalate; and it is used in the laundry, textile, insecticide, refrigeration, and paper industries.

A partial list of occupations in which exposure may occur includes:

Airplane dope makers	Leather makers
Allyl alcohol makers	Paper makers
Dyers	Perfume makers
Electroplaters	Rubber workers
Insecticide makers	Textile makers
Lacquer makers	Wine makers
Laundry workers	

Permissible Exposure Limits: The Federal standard is 5 ppm (9 mg/m^3).

Routes of Entry: Inhalation of vapor and percutaneous absorption.

Harmful Effects

Local—The primary hazard of formic acid results from severe irritation of the skin, eyes, and mucous membranes. Lacrimation, increased nasal discharge, cough, throat discomfort, erythema, and blistering may occur depending upon solution concentrations.

Systemic—These have not been reported from inhalation exposure and are unlikely due to its good warning properties.

Swallowing formic acid has caused a number of cases of severe poisoning and death. The symptoms found in this type of poisoning include salivation, vomiting, burning sensation in the mouth, bloody vomiting, diarrhea, and pain. In severe poisoning, shock may occur. Later, breathing difficulties may develop. Kidney damage may also be present.

Medical Surveillance: Consideration should be given to possible irritant effects on the skin, eyes, and lungs in any placement or periodic examinations.

Special Tests: None currently used.

Personal Protective Methods: Workers should be supplied with protective clothing, gloves, and goggles. Respiratory protection will be needed in areas of high vapor exposure.

Bibliography

Henson, E.V. 1959. Toxicology of the fatty acids. *J. Occup. Med.* 1:339.

Malorny, G. 1969. Die akute und chronische toxizitat der ameisenaure und ihrer formiate. *Ernaechrungswiss* 9:332.

Von Oettingen, W.F. 1959. The aliphatic acids and their esters—toxicity and potential dangers. The saturated monobasic aliphatic acids and their esters. *AMA Arch. Ind. Health* 20:517.

FURFURAL

Description: $C_5H_4O_2$, furfural, is an aromatic heterocyclic aldehyde with an amber color and aromatic odor. This liquid is obtained from cereal straws and brans containing pentosans by hydrolysis and dehydration with sulfuric acid.

Synonyms: Furfurol (a misnomer), furfuraldehyde, artificial ant oil, pyromucic aldehyde, furol, 2-furaldehyde.

Potential Occupational Exposures: Furfural is used as a solvent for wood resin, nitrated cotton, cellulose acetate, and gums. It is used in the production of phenolic plastics, thermosetting resins, refined petroleum oils, dyes, and varnishes. It is also utilized in the manufacture of pyromucic acid, vulcanized rubber, insecticides, fungicides, herbicides, germicides, furan derivatives, polymers, and other organic chemicals.

A partial list of occupations in which exposure may occur includes:

Adipic acid makers	Lysine makers
Butadiene refiners	Metal refiners
Cellulose acetate makers	Nylon makers
Disinfectant workers	Organic chemical synthesizers
Dye makers	Pyromucic acid makers
Herbicide makers	Road builders

Permissible Exposure Limits: The Federal standard is 5 ppm (20 mg/m^3).

Routes of Entry: Inhalation of vapor, percutaneous absorption.

Harmful Effects

Local—Liquid and concentrated vapor are irritating to the eyes, skin, and mucous membranes of the upper respiratory tract. Eczematous dermatitis as well as skin sensitization, resulting in allergic contact dermatitis and photo-sensitivity, may develop following repeated exposure.

Systemic—Workers chronically exposed to the vapor have had complaints of headache, fatigue, itching of the throat, lacrimation, loss of the sense of taste, numbness of the tongue, and tremor. Occupational overexposure is relatively rare due to the liquid's low vapor pressure, and symptoms usually disappear rapidly after removal from exposure.

Medical Surveillance: Consider skin irritation and skin allergies (especially to aldehydes) in preplacement or periodic examinations. Also consider possible respiratory irritant effects.

Special Tests: None commonly used.

Personal Protective Methods: Protective clothing and adequate ventilation should be provided in areas where toxic exposure may occur. In areas of vapor concentrations, fullface masks may be required.

Bibliography

Castellino, N., Elmino, O., and Rozera, G. 1963. Experimental research on toxicity of furfural. *Arch. Environ. Health* 7:574.

Dunlop, A.P., and Peters, F.N. 1953. *The Furans.* American Chemical Society Monograph Series No. 119. Reinhold Publishing Corp., New York p. 171.

Ubaydullayev, R. 1970. Biological effect of low concentrations of furfural under experimental conditions. *J. Hyg. Epidemiol. Microbiol. Immunol.* (Praha). 14:240.

G

GASOLINE

Description: Gasoline is a highly flammable, mobile liquid with a characteristic odor.

Synonyms: Petrol, motor spirits, benzin.

Potential Occupational Exposures: Gasoline is used as a fuel, diluent, and solvent throughout industry.

A partial list of occupations in which exposure may occur includes:

Filling station attendants
Garage mechanics
Gasoline engine operators
Motor transport drivers

Pipeline workers
Refinery workers
Tank car cleaning crews

Permissible Exposure Limits: Presently, the composition of gasoline is so varied that a single Federal standard for all types of gasoline is not applicable. It is recommended, however, that atmospheric concentrations should be limited by the aromatic hydrocarbon content.

Route of Entry: Most cases of poisoning reported have resulted from inhalation of vapor and ingestion. It is not known whether gasoline poisoning may be compounded by percutaneous absorption.

Harmful Effects

Local—Gasoline is irritating to skin, conjunctiva, and mucous membranes. Dermatitis may result from repeated and prolonged contact with the liquid, which may defat the skin. Certain individuals may develop hypersensitivity.

Systemic—Gasoline vapor acts as a central nervous system depressant. Exposure to low concentrations may produce flushing of the face, staggering gait, slurred speech, and mental confusion. In high concentrations, gasoline vapor may cause unconsciousness, coma, and possibly death resulting from respiratory failure.

Other signs also may develop following acute exposure. These signs are early acute hemorrhage of the pancreas, centrilobular cloudy swelling and fatty degeneration of the liver, fatty degeneration of the proximal convoluted tubules and glomeruli of the kidneys, and passive congestion of the spleen.

Ingestion and aspiration of the liquid gasoline usually occurs during siphoning.

Chemical pneumonitis, pulmonary edema, and hemorrhage may follow. Aromatic hydrocarbon content may also cause hematopoietic changes. Absorption of alkyl lead antiknock agents contained in many gasolines poses an additional problem especially where there is prolonged skin contact. The existence of chronic poisoning has not been established.

Medical Surveillance: No special considerations are necessary.

Special Tests: None in common use.

Personal Protective Methods: Barrier creams and impervious gloves, protective clothing. Masks in heavy exposure to vapors.

Bibliography

Hunter, G.A. 1968. Chemical burns on the skin after contact with petrol. *Br. J. Plast. Surg.* 21:337.

Machle, W. 1941. Gasoline intoxication. *J. Am. Med. Assoc.* 117:1965.

Nagata, T., and S. Fujiwara. 1968. Gas chromatographic detection of gasoline in the blood: a case report. *Jpn. J. Med.* 22:274.

Wang, C.C., and G.V. Irons. 1961. Acute gasoline intoxication. *Arch. Environ. Health* 2:114.

Zuckler, R., E.D. Kilbourne, and J.B. Evans. 1950. Pulmonary manifestations of gasoline intoxication. A review with report of a case. *Arch. Ind. Hyg. Occup. Med.* 2:17.

GERMANIUM

Description: Ge, germanium, is a greyish-white, lustrous, brittle metalloid. It is never found free and occurs most commonly in argyrodite and germanite. It is generally produced from germanium containing minerals or as a by-product in zinc production or coal processing. Germanium is insoluble in water.

Synonyms: None.

Potential Occupational Exposures: Because of its semiconductor properties, germanium is widely used in the electronic industry in rectifiers, diodes, and transistors. It is alloyed with aluminum, aluminum-magnesium, antimony, bronze, and tin to increase strength, hardness, or corrosion resistance. In the process of alloying germanium and arsenic, arsine may be released; stibine is released from the alloying of germanium and antimony. Germanium is also used in the manufacture of optical glass, lenses for infrared applications, red-fluorescing phosphors, and cathodes for electronic valves, and in electroplating, in the hydrogenation of coal, and as a catalyst, particularly at low temperatures. Certain compounds are used medically.

Industrial exposures to the dust and fumes of the metal or oxide generally occur during separation and purification of germanium, welding, multiple-zone melting operations, or cutting and grinding of crystals. Germanium tetrahydride (germanium hydride, Germane, monogermane) and other hydrides are produced by the action of a reducing acid on a germanium alloy.

A partial list of occupations in which exposure may occur includes:

Alloy makers	Rectifier makers
Dental alloy makers	Semiconductor makers
Electroplaters	Transistor makers
Glass makers	Vacuum tube makers
Phosphor makers	Residue workers

Permissible Exposure Limits: There is no Federal standard for germanium or its compounds; however, the ACGIH recently added a TLV for germanium tetrahydride of 0.2 ppm (0.6 mg/m^3).

Route of Entry: Inhalation of gas, vapor, fume, or dust.

Harmful Effects

Local—The dust of germanium dioxide is irritating to the eyes. Germanium tetrachloride causes irritation of the skin.

Systemic—Germanium tetrachloride is an upper respiratory irritant and may cause bronchitis and pneumonitis. Prolonged exposure to high level concentrations may result in damage to the liver, kidney, and other organs. Germanium tetrahydride is a toxic hemolytic gas capable of producing kidney damage.

Medical Surveillance: Consider respiratory, liver, and kidney disease in any placement or periodic examinations.

Special Tests: None commonly used, but can be determined in urine.

Personal Protective Methods: In dust areas, protective clothing and gloves may be necessary to protect the skin, and goggles to protect the eyes. In areas where germanium tetrachloride is in high concentrations, dust-fume masks or supplied air respirators with full facepiece should be supplied to all workers. Personal hygiene is to be encouraged, with change of clothes following each shift and showering prior to change to street clothes.

Bibliography

Dudley, H.C., and E.J. Wallace. 1952. Pharmacological studies of radiogermanium (G371). *AMA Arch. Ind. Hyg. Occup. Med.* 6:263.

Hueper, W.C. 1947. Germanium. *Occup. Med.* 4:208.

Rosenfeld, G., and E.J. Wallace. 1953. Studies of the acute and chronic toxicity of germanium. *AMA Arch. Ind. Hyg. Occup. Med.* 8:466.

GRAPHITE

Description: Graphite is crystallized carbon and usually appears as soft, black scales. There are two types of graphite, natural and artificial.

Synonyms: Plumbago, black lead, mineral carbon.

Potential Occupational Exposures: Natural graphite is used in foundry facings, steelmaking, lubricants, refractories, crucibles, pencil "lead," paints, pigments, and stove polish. Artificial graphite may be substituted for these uses with the exception of clay crucibles; other types of crucibles may be produced from artificial graphite. Additionally, it may be used as a high temperature lubricant or for electrodes. It is utilized in the electrical industry in electrodes, brushes, contacts, and electronic tube rectifier elements; as a constituent in lubricating oils and greases; to treat friction elements, such as brake linings; to prevent molds from sticking together; and in moderators in nuclear reactors.

A partial list of occupations in which exposure may occur includes:

Brake lining makers	Match makers
Cathode ray tube makers	Nuclear reactor workers
Commutator brush makers	Paint makers
Crucible makers	Pencil lead makers
Electrode makers	Pigment makers
Explosive makers	Refractory material makers
Foundry workers	Steel makers
Lubricant makers	Stove polish makers

Permissible Exposure Limits: The Federal standard for natural graphite is 15 mppcf. Synthetic graphite is designated a "nuisance particulate" by ACGIH (1978) with a TLV of 30 mppcf or 10 mg/m^3.

Route of Entry: Inhalation of dust.

Harmful Effects

Local—None.

Systemic—Exposure to natural graphite may produce a progressive and disabling pneumoconiosis similar to anthracosilicosis. Symptoms include headache, coughing, depression, decreased appetite, dyspnea, and the production of black sputum. Some individuals may be asymptomatic for many years then suddenly become disabled. It has not yet been determined whether the free crystalline silica in graphite is solely responsible for development of the disease. There is evidence that artificial graphite may be capable of producing a pneumoconiosis.

Medical Surveillance: Preemployment and periodic examinations should be directed toward detecting significant respiratory disease, through chest x-rays and pulmonary function tests.

Special Tests: None.

Personal Protective Methods: Workers in exposed areas should be provided with dust masks with proper cartridges and should be instructed in their maintenance.

Bibliography

Harding, H.E., and G.B. Oliver. 1949. Changes in the lungs produced by natural graphite. *Br. J. Ind. Med.* 6:91.

Pendergrass, E.P., A.J. Vorwald, N.M. Mishkin, J.G. Whildin, and C.W. Werley. 1967/1968. Observations on workers in the graphite industry. *Med. Radiogr. Photogr.* 43:71; 44:2.

Ranasinha, K.W., and C.G. Uragoda. 1972. Graphite pneumoconiosis. *Br. J. Ind. Med.* 29:178.

Zenz, C. 1975. *Occupational Medicine.* Yearbook Medical Publications, Chicago.

H

n-HEPTANE

Description: $CH_3(CH_2)_5CH_3$, n-heptane, is a clear liquid which is highly flammable and volatile.

Synonyms: None.

Potential Occupational Exposures: n-Heptane is used as a solvent and as a standard in testing knock of gasoline engines.

A partial list of occupations in which exposure may occur includes:

> Process workers (where heptane is used as the solvent)
> Refinery laboratory workers
> Refinery workers

Permissible Exposure Limits: The Federal standard for n-heptane is 500 ppm $(2,000 \, mg/m^3)$. (Note: The 1976 ACGIH lists a TLV of 400 ppm $(1,600 \, mg/m^3)$.

Route of Entry: Inhalation of the vapor.

Harmful Effects

Local—n-Heptane can cause dermatitis and mucous membrane irritation. Aspiration of the liquid may result in chemical pneumonitis, pulmonary edema, and hemorrhage.

Systemic—Systemic effects may arise without complaints of mucous membrane irritation. Exposure to high concentrations causes narcosis producing vertigo, incoordination, intoxication characterized by hilarity, slight nausea, loss of appetite, and a persisting gasoline taste in the mouth. These effects may be first noticed on entering a contaminated area. n-Heptane may cause low order sensitization of the myocardium to epinephrine.

Medical Surveillance: Preplacement examinations should evaluate the skin and general health, including respiratory, liver, and kidney function.

Special Tests: None have been commonly used.

Personal Protective Methods: Barrier creams and gloves. Masks where exposed to vapor. (Note: The reader is also referred to the section on "Alkanes.")

HEXACHLOROBENZENE

Description: Hexachlorobenzene, C_6Cl_6, is a solid, crystallizing in needles, MP 231°C and boiling at 323°-326°C.

Synonyms: Perchlorobenzene, HCB.

Potential Occupational Exposures: About 90% of the estimated 8 million pounds of HCB produced annually in the United States is as a by-product at 10 perchloroethylene, 5 trichloroethylene, and 11 carbon tetrachloride manufacturing plants. HCB is commonly detected in solid wastes and liquid effluents. Most of the remaining production is as a by-product at more than 70 other sites producing chlorine and certain pesticides. About 45,000 pounds per year are released into the environment during pesticide use. HCB has also been found in the waste tars from vinyl chloride and other chlorine-product plants.

Permissible Exposure Limits: In the wake of widespread HCB contamination of cattle in Louisiana in 1973, and concern over possible contamination of sheep in California, EPA established an interim tolerance of 0.5 ppm. Concurrently, the State of Louisiana and several companies took immediate steps to tighten up solid waste practices from manufacturing through disposal. Also, supplies of Dacthal containing 10% HCB as an inert ingredient were voluntarily withdrawn from the California market. Several toxicological, monitoring, economics, and related projects were initiated by EPA to provide a better basis for further actions, including the establishment of a tolerance. Also, additional toxicological efforts were undertaken by USDA.

As soon as the needed toxicological data are available, food tolerance will be established. Also, all pesticidal uses of HCB, including pesticides which contain HCB as a contaminant, will be reviewed. Studies of land and other disposal methods have been completed. Ocean dumping of HCB-laden tars is prohibited. Although not directly addressed by the NPDES permit program, provisions relating to suspended solids, and oil and grease may provide some degree of control if HCB enters the effluent stream.

Route of Entry: In 1975, 46% of the soil samples collected at 26 locations along a 150-mile transect in Louisiana were contaminated with HCB at levels from 20 to 440 ppb. Parallel sampling of aquatic sediments revealed concentrations of 40 to 850 ppb. Although water samples were generally below 3 ppb, one sample below an industrial discharge contained 90 ppb. Air immediately adjacent to production facilities has shown concentrations from 1.0 to 23.6 $\mu g/m^3$. Most of the HCB appeared to be associated with particulate, but low levels were found in the gaseous phase as well, which might result from volatization from solid wastes. Samples collected from pastureland near a known HCB production site revealed concentrations in the vegetation from 0.01-630 ppm and in the soil from 0.01-300 ppb.

HCB residues have been found in soil, wildlife, fish, and food samples collected from all over the world. In the United States, HCB residues have been reported in birds and bird eggs collected from Maine to Florida, duck tissue collected from across the country, and fish and fish eggs from the East Coast and Oregon. Animal foods, including chicken feed, fish food, and general laboratory feeds, have been found to contain HCB residues. The frequency of detection of HCB residues in domestic meats has been steadily increasing since 1972, in part because of closer scrutiny. HCB has been detected in trace amounts in only two drinking water supplies.

EPA's monitoring of human adipose tissues collected from across the United States reveals that about 95% of the population has trace HCB residues.

Harmful Effects

The death of breast-fed infants and an epidemic of skin sores and skin discoloration were associated with accidental consumption of HCB-contaminated seed grain in Turkey in the mid-1950s. Doses were estimated at 50 to 200 mg/day for several months to two years. Clinical manifestations included weight loss, enlargement of the thyroid and lymph nodes, skin photosensitization, and abnormal growth of body hair.

HCB levels of up to 23 ppb in blood are believed to have contributed to enzyme disruptions in the population of a small community in southern Louisiana in 1973.

Long-term (up to 3 years) animal ingestion studies show a detectable increase in deaths at 32 ppm, cellular alteration at 1 ppm, biochemical effects at 0.5 ppm, and behavioral alteration between 0.5 and 5 ppm.

Apparently, the effective dosage to offspring is increased by exposure to the parent. A 12% reduction in offspring survival resulted when exposure to very low levels had been continuous for three generations. Teratogenic effects appear minimal.

While HCB appears to have little effect on aquatic organisms, a bioaccumulation factor of 15,000 has been demonstrated in catfish. HCB is toxic to some birds. Eighty ppm caused death, and 5 ppm caused liver enlargement and other effects in quail. The half life of HCB in cattle and sheep is almost 90 days.

HCB is very stable. It readily vaporizes from soil into the air; emissions to air in turn contaminate the soil.

Medical Suveillance: Preplacement and regular medical examinations.

Special Tests: None in common use.

Personal Protective Methods: Chemical safety goggles are recommended for eye protection. Respirators are required to prevent inhalation.

Bibliography

Assessing Potential Ocean Pollutants. (p 188-208) National Academy of Sciences (1975).

Burns, J.E. and Miller, F.M. Hexachlorobenzene contamination: Its effects on a Louisiana population. *Arch. Environ. Health* 30:44-48 (1975).

Cam, C. and Nygogosyan, G. Acquired porphyria cutanea tarda due to hexachlorobenzene. *J. Am. Med. Assoc.* 183:88-91 (1963).

An Ecological Study of Hexachlorobenzene and Hexachlorobutadiene. Final report under EPA Contract No. 68-01-2689.

Environmental Contamination from Hexachlorobenzene. EPA, Office of Toxic Substances (July 1973).

Industrial Pollution of the Lower Mississippi River in Louisiana. EPA, Office of Water and Hazardous Materials (April 1972).

Sampling and Analysis for Selected Toxic Substances—Task 1, HCB and HCBD. EPA, Office of Toxic Substances (performed under Contract No. 68-01-2646).

Survey of Industrial Processing Data, Task 1—Hexachlorobenzene and Hexachlorobutadiene Pollution from Chlorocarbon Processes. EPA, Office of Toxic Substances, Report No. 560/3-75-003 (June 1975).

Survey of Methods Used to Control Wastes Containing Hexachlorobenzene. EPA, Office of Solid Waste Management (Nov. 1975).

Vos, J.G., et al. Toxicity of hexachlorobenzene in Japanese quail with special reference to porphyria, liver damage, reproduction and tissue residues. *Toxicol. Appl. Pharmacol.* 18:944-957 (1971).

Zitko, V., and Choi, P.M.K. HCB and p,p'-DDE in eggs of cormorants, gulls and ducks from the Bay of Fundy, Canada. *Bull. Env. Contam. and Toxicology,* 7(51):63-64 (1972).

HEXAMETHYLENETETRAMINE

Description: $(CH_2)_6N_4$, hexamethylenetetramine, is an odorless, crystalline solid.

Synonyms: Methenamine, hexamine, formamine, ammonioformaldehyde.

Potential Occupational Exposures: Hexamethylenetetramine is used as an accelerator in the rubber industry, as a curing agent in thermosetting plastics, as a fuel pellet for camp stoves, and in the manufacture of resins, pharmaceuticals, and explosives.

A partial list of occupations in which exposure may occur includes:

Drug makers	Resin makers
Explosive makers	Rubber makers
Fuel tablet makers	Textile makers
Phenol-formaldehyde resin workers	Urea-formaldehyde resin workers

Permissible Exposure Limits: There is no Federal standard for hexamethyl-enetetramine.

Routes of Entry: Ingestion and skin contact.

Harmful Effects

Local—Very mild skin irritant.

Systemic—None. Side effects from ingestion are urinary tract irritation, skin rash, and digestive disturbances. Large oral doses can cause severe nephritis which may be fatal.

Medical Surveillance: No specific considerations are necessary.

Special Tests: None.

Personal Protective Methods: If repeated or prolonged skin exposure is likely, gloves or protective clothing may be needed.

Bibliography

The reader is referred to references (27) and (28) in the section on "Formaldehyde."

HEXAMETHYLPHOSPHORIC TRIAMIDE

Description: Hexamethylphosphoric triamide, $[(CH_3)_2N]_3PO$, is a colorless liquid with a density of 1.03 g/ml and a boiling point of 232°C.

Synonyms: Synonyms for hexamethylphosphoric triamide include ENT 50882, hempa, hexametapol, hexamethylphosphamide, hexamethylphos-phoramide, hexamethylphosphoric acid triamide, hexamethylphosphorotri-amide, hexamethylphosphotriamide, HMPA, HMPT, HPT, phosphoric tris-(dimethylamide), phosphoryl hexamethyltriamide, tris(dimethylamino) phosphine oxide, and tris(dimethylamino) phosphorus oxide (1).

Potential Occupational Exposure: Hexamethylphosphoric triamide is a material possessing unique solvent properties and is widely used as a solvent, in small quantities, in organic and organometallic reactions in laboratories (2)(3). This is the major source of occupational exposure to HMPA in the United States.

Du Pont, the major manufacturer of hexamethylphosphoric triamide in the United States, periodically produces HMPA at its Chamber Works, Deepwater, New Jersey. Other producers of HMPA in the United States include Chemical Samples Company and Fike Chemical Company. None of Du Pont's HMPA is marketed; all is used internally at its Spruance Plant in Richmond, Virginia, as a processing solvent in the production of Kevlar aramid fiber. Du Pont reports that Kevlar contains less than 1 ppm (w/w) of the HMPA which is so firmly held by the fiber that Du Pont believes there is no hazard to customers or employees handling the final fiber product.

Hexamethylphosphoric triamide had been manufactured and distributed in the past by Dow Chemical Company (as Dorcol) and Eastman Chemical Products, Inc. (as Inhibitor HPT). Both firms have advised NIOSH that they discontinued these products several years ago (4). HMPA has been evaluated for use as an ultraviolet light inhibitor in polyvinyl chloride formulations, as an additive for antistatic effects, as a flame retardant, and as a deicing additive for jet fuels (4)(5)(6).

Hexamethylphosphoric triamide has also been extensively investigated as an insect chemosterilant (7)(8).

It is estimated that 5,000 people are occupationally exposed to hexamethylphosphoric triamide. More than 90% of these exposures are in research laboratories.

Permissible Exposure Limits: There is no current OSHA standard for hexamethylphosphoric triamide exposure. ACGIH (1978) classifies HMPA as an "Industrial Substance Suspect of Carcinogenic Potential for Man."

Routes of Entry: Inhalation of vapors.

Harmful Effects

Animal—HMPA is known to have a variety of toxic effects on laboratory animals. Acute toxic effects seen in rats fed HMPA include kidney disease, severe bronchiectasis and bronchopneumonia with squamous metaplasia and fibrosis in lungs (9)(10). In rabbits, repeated application of HMPA to the skin caused dose related weight loss, altered gastrointestinal function and apparent nervous-system dysfunction (11). Testicular atrophy and aspermia have been observed in rats following oral treatment with HMPA (9)(12). Oral treatment with HMPA has also been highly inhibitory to testicular development in cockerels (13).

HMPA is known to produce mutagenic effects in fruit flies *(Drosophila melanogaster)* (14). However, studies of the effects of HMPA on human (15) and mice chromosomes (16) showed no greater frequency of HMPA induced chromosomal aberrations when compared with controls.

The E.I. du Pont de Nemours and Company (Du Pont) reported to the National Institute for Occupational Safety and Health in a letter dated Sept. 24, 1975, that nasal tumors (squamous cell carcinoma) have been observed in rats exposed to hexamethylphosphoric triamide. NIOSH has also been advised that Du Pont has notified its customers and employees of these findings.

Preliminary results of the inhalation toxicity study of HMPA, recently released by Du Pont, show nasal tumors in rats exposed daily to 400 and 4,000 parts per billion (ppb) HMPA after 8 months of exposure. In some cases, the tumors originating from the epithelial lining of the nasal turbinate bones filled the nasal cavity and penetrated into the brain. No nasal tumors were reported among rats exposed to 50 ppb HMPA and controls.

Prior to the Du Pont observations, the only other known report of tumors associated with exposure to HMPA was a long-term feeding study by Kimbrough. While lung tumors were observed, the results of this study were inconclusive because the tumor incidence among HMPA exposed rats was not greater than among the control rats (17).

Human—There are no data available on the toxic effects of hexamethylphosphoric triamide in humans.

Medical Surveillance: In light of the potential risk of human exposure to this chemical in the work environment, the National Institute for Occupational Safety and Health is advising the occupational health community of the above findings. Preplacement and regular physical examinations are indicated.

Special Tests: None in use.

Personal Protective Methods: Conventional laboratory precautions of goggles, gloves and good hood ventilation are indicated.

References

(1) *Chemline.* National Library of Medicine, Bethesda, Maryland, Oct. 1975.
(2) Fieser, M. and Fieser, L.F. *Reagents for Organic Synthesis.* John Wiley and Sons, Inc., New York. 4:244, 1974; 3:149, 1972; 2:208, 1969; 1:430, 1967.
(3) *Chemical Abstracts, Chemical Substances Index.* Chemical Abstracts Service, Columbus, Ohio, Vol. 81, 1974.
(4) Personal communication with representatives of Dow Chemical Company, Midland, Michigan and Eastman Chemical Products, Inc., Kingsport, Tenn.
(5) Eastman Technical Data Sheet No. X-203, Eastman Chemical Products, Inc., Kingsport, Tenn.
(6) *The Merck Index,* 8th Ed. Merck and Company, Inc., Rahway, NJ, p. 528, 1968.
(7) Bull, D.L. and Borkovec, A.B. Metabolism of carbon-14-labeled hempa by adult boll weevils. *Arch. Environ. Contam. Toxicol.* 1(2):148-58, 1973.

(8) Landa, V. Action of chemosterilants in the reproductive organs and tissues of insects. *Proc. Int. Cong. Entomol.* 13th, 3:423-24, 1972.

(9) Kimbrough, R.D. and Gaines, T.B. Toxicity of hexamethylphosphoramide in rats. *Nature.* 211:146-47, 1966.

(10) Kimbrough, R.D. and Sedlak, V.A. Lung morphology in rats treated with hexamethylphosphoramide. *Toxicol. Appl. Pharmacol.* 12(1):60-7, 1968.

(11) Shott, L.D., Borkovec, A.B., and Knapp, W.A., Jr. Toxicology of hexamethylphosphoric triamide in rats and rabbits. *Toxicol. Appl. Pharmacol.* 18(13):499-506, 1971.

(12) Jackson, H., Jones, A.R., and Cooper, E.R.A. Effects of hexamethylphosphoramide on rat spermatogenesis and fertility. *J. Repro. Fertil.* 20:263-69, 1969.

(13) Sherman, M. and Herrick, R.B. Acute toxicity of five insect chemosterilants, hemel, hempa, tepa, metepa, and methotrexate, for cockerels. *Toxicol. Appl. Pharmacol.* 16:100-07, 1970.

(14) Benes, V. and Sram, R.J. Mutagenic activity of some pesticides in *Drosophila melanogaster. Indus. Med. Surg.* 38(12):442-44, 1969.

(15) Chang, T.H. and Klassen, W. Comparative effects of tretamine, tepa, apholate, and their structural analogs on human chromosomes in vitro. *Chromasoma.* 24(3):314-23, 1968.

(16) Manna, G.K. and Das, P.K. Effect of two chemosterilants apholate and hempa on the bone-marrow chromosomes of mice. *Can. J. Genet. Cytol.* 15(3):451-59, 1973.

(17) Kimbrough, R.D. and Gaines, T.B. The chronic toxicity of hexamethylphosphoramide in rats. *Bull. Environ. Contam. and Toxicol.* 10(4):225-26, 1973.

n-HEXANE

Description: $CH_3(CH_2)_4CH_3$, n-hexane, is a colorless, volatile liquid and is highly flammable.

Synonyms: None.

Potential Occupational Exposures: n-Hexane is used as a solvent, particularly in the extraction of edible fats and oils, as a laboratory reagent, and as the liquid in low temperature thermometers. Technical and commercial grades consist of 45 to 85% n-hexane, as well as cyclopentanes, isohexane, and from 1 to 6% benzene.

A partial list of occupations in which exposure may occur includes:

> Fat processors
> Oil processors
> Thermometer makers

Permissible Exposure Limits: The Federal standard is 500 ppm (1800 mg/m^3) for workroom exposure to n-hexane. [Note: In 1978 ACGIH lists a TLV of 100 ppm (360 mg/m^3)].

Route of Entry: Inhalation of vapor.

Harmful Effects

Local—Dermatitis and irritation of mucous membranes of the upper respiratory tract.

Systemic—Asphyxia may be produced by high concentrations. Acute exposure may cause narcosis resulting in slight nausea, headache, and dizziness. Myocardial sensitization to epinephrine may occur but is of low order. Peripheral neuropathy has been reported resulting from exposure to n-hexane.

Medical Surveillance: Consider the skin, respiratory system, central and peripheral nervous system, and general health in preplacement and periodic examinations.

Special Tests: None in use.

Personal Protective Methods: Barrier creams and gloves are recommended, as are masks where workers are exposed to vapors.

Bibliography

Paulson, G.W., and G.W. Waylonis. 1976. Polyneuropathy due to n-hexane. *Arch. Intern. Med.* 136:880.
(Note: The reader is also referred to the section on "Alkanes.")

HYDRAZINE AND DERIVATIVES

Description: Hydrazine (H_2N-NH_2) is a colorless, oily liquid with an ammoniacal odor. Phenylhydrazine ($C_6H_5NHNH_2$) is an oily, colorless liquid or a crystalline solid. Dimethylhydrazine, UDMH, $(CH_3)_2NNH_2$, is a hygroscopic mobile liquid. Hydrazine and UDMH are soluble in water and alcohol. Phenylhydrazine is slightly soluble in water.

Synonyms

       Hydrazine: Hydrazine base, diamine.
       Phenylhydrazine: Hydrazinobenzene.
       Dimethylhydrazine: UDMH, 1,1-dimethylhydrazine,
          asymmetrical dimethylhydrazine.

Potential Occupational Exposures: Both UDMH and hydrazine are used in liquid rocket fuels. Because of its strong reducing capabilities, hydrazine is used as an intermediate in chemical synthesis and in photography and metallurgy. It is also used in the preparation of anticorrosives, textile agents, and pesticides, and as a scavenging agent for oxygen in boiler water. Hydrazine salts find use as fluxes in soft soldering and aluminum soldering. Phenylhydrazine is very reactive with carbonyl compounds and is a widely used reagent in conjunction with sugars, aldehydes, and ketones, in addition to its use in the synthesis of dyes, pharmaceuticals such as antipyrin, cryogenin, and pyramidone, and other organic chemicals. The hydrochloride salt is used in the treatment of polycythemia vera.

A partial list of occupations in which exposure may occur includes:

Acrylic and vinyl textile dyers	Insecticide makers
Agricultural chemical makers	Jet fuel workers
Anticorrosion additive makers	Oxygen scavenger makers
Antioxidant workers	Photographic developer makers
Boiler operators	Rocket fuel workers
Chemists	Solder flux makers
Drug makers	Water treaters

NIOSH has estimated the number of workers exposed to the hydrazines as follows:

Hydrazine	9,000
Hydrazine dihydrochloride	89,000
Hydrazine sulfate	2,500
Hydrazine hydrobromide	1,500
Hydrazine hydrate	1,700
1,1-dimethylhydrazine	1,500
1,2-dimethylhydrazine	Unknown but small
Phenylhydrazine	5,000

Permissible Exposure Limits: The Federal standards for the compounds are:

Hydrazine 1 ppm (1.3 mg/m^3)
Phenylhydrazine 5 ppm (22 mg/m^3)
Dimethylhydrazine 0.5 ppm (1 mg/m^3)

More recently, NIOSH (23) has promulgated a recommended standard for hydrazine as follows: Occupational exposure to hydrazines shall be controlled so that employees are not exposed at concentrations greater than those specified below, expressed as milligrams of the free base per cubic meter of air (mg/m^3), determined as ceiling concentrations in any 2-hour period:

hydrazine	- 0.04 mg/m^3 (0.03 ppm)*
methylhydrazine	- 0.08 mg/m^3 (0.04 ppm)
1,1-dimethylhydrazine	- 0.15 mg/m^3 (0.06 ppm)
phenylhydrazine	- 0.6 mg/m^3 (0.14 ppm)

*Approximate equivalents in parts of free base per
  million parts of air (ppm).

These recommended limits are the lowest concentrations measured by the recommended method of analysis. No limit is recommended for 1,2-dimethylhydrazine, since an acceptable method of sampling and analysis is presently unavailable.

Routes of Entry: Inhalation and percutaneous absorption.

Harmful Effects

Local—All three compounds have similar toxic local effects due to their irritant properties. The vapor is highly irritating to the eyes, upper respiratory tract, and skin, and causes delayed eye irritation. Severe exposure may produce temporary blindness. The liquid is corrosive, producing penetrating burns and severe dermatitis. Permanent corneal lesions may occur if the liquid is splashed in the eyes. A sensitization dermatitis may be produced.

Systemic—Inhalation of hydrazine may cause dizziness and nausea. In animals hydrazine has caused liver and kidney damage and pulmonary edema. It has also been reported to cause adenocarcinoma of the lung and liver in animals.

Hydrazine or hydrazine salts (e.g., sulfate) have been shown to be carcinogenic in mice after oral and intraperitoneal administration and in rats following oral administration (1)-(6).

Hydrazine has been shown to be an effective agent in producing mutations in phage (7), bacteria (8)-(11), higher plants (12)(13), and *Drosophila* (14)(15). Hydrazine did not produce detectable levels of dominant lethals in mice (16)(17) though it did produce mutations in *S. typhimurium* used in the host-mediated assay in mice (18). It should be noted that the dominant lethal assay tests primarily for chromosomal aberrations, whereas most of the successful tests with hydrazine have been for mutations more likely to be single locus events (10).

UDMH has been reported to induce abnormalities in the morphology of sperm in the Cauda epididymis of mice which reached maximum levels ≤3 weeks after exposure of the animals to their agent (14). UDMH is carcinogenic in mice after oral administration (15)(16).

Substituted hydrazine derivatives are receiving considerable attention in pharmacology and toxicology because of their widespread use as herbicides and rocket fuels, as intermediates in chemical synthesis, and as therapeutic agents for the treatment of tuberculosis, depression, and cancer. Besides the liver necrosis found in therapy with isoniazid and iproniazid, hydrazines are known to produce many other toxic responses including methemoglobinemia, hemolysis, fatty liver, mutagenesis, and carcinogenesis (17)-(19).

The mutagenicity of hydrazine and some of its derivatives has recently been reviewed by Kimball (20). Hydrazine can react with the pyrimidine in DNA to saturate the 5,6 double bond (especially of thymine) to form N^4-amino-cytosine. It can also open up the pyrimidine ring with consequent loss of pyrimidines from DNA. Hydrazine can react either directly with DNA or through intermediate radical reactions including the formation of H_2O_2. A number of substituted hydrazines can also act in much the same way. Other hydrazines, especially the methyl derivatives, can act as alkylating agents to alkylate purines, primarily (20).

Although hydrazine per se has not been reported to produce chromosomal aberrations, several of its derivatives including isoniazid (which is believed to produce hydrazine in vivo) have been reported to produce chromosomal aberrations and other nuclear anomalies. It should be noted that most of the compounds producing chromosomal effects are methylated derivatives of hydrazine which might be acting as alkylating agents (20).

The metabolic fate of hydrazines (21)(22) and hydrazides (21) have been recently reviewed. In humans and most other mammalian species, one of the most important biotransformation reactions affecting hydrazine-hydrazide compounds is acetylation of the terminal nitrogen group. Biotransformation appears to be essential to the capacity of many hydrazine derivatives to inhibit monoamine oxidase. Other reactions of the hydrazines, e.g., hydrolysis, oxidation, and reduction frequently results in the formation of metabolites with potent biological effects (21)(22).

The reader is also referred to an extensive text and bibliography concerning the harmful effects of exposure to the various hydrazines in a recent NIOSH document (23).

Medical Surveillance: Based partly on experimental animal data, placement should include a history of exposure to other carcinogens, smoking, alcohol, medications, and family history. The skin, eye, lungs, liver, kidney, blood, and central nervous system should be evaluated. Sputum or urine cytology may give useful information.

Special Tests: Hydrazine may be detected in the blood; UDMH has been measured in blood and urine. Some phenylhydrazine metabolites are known. None of these are in common use, however.

Personal Protective Methods: Protective clothing, gloves, and goggles should be worn to reduce any skin or eye contact. Fullface supplied air masks and full protective clothing may be required if vapor concentrations are significant. Clean work clothes should be supplied on a daily basis, and workers should shower prior to change to street clothes.

References

(1) *IARC,* Vol. 4. International Agency for Research on Cancer, Lyon, France, pp. 127-136; 137-143 (1974).

(2) Biancifiori, C., and Severi, L. The relation of isoniazid (INH) and allied compounds to carcinogenesis in some species of small laboratory animals: a review. *Brit. J. Cancer* 20 (1966) 528.

(3) Kelly, M.G., O'Gara, R.W., Yancey, S.T., Gadekar, K., Botkin, C., and Oliverio, V.T. Comparative carcinogenicity of N-isopropyl-α-(2-methylhydrazino)-p-tolua-mide·HCl (procarbazine hydrochloride), its degradation products, other hydra-zines, and isonicotinic acid hydrazide. *J. Natl. Cancer Inst.* 42 (1969) 337.

(4) Toth, B. Investigation on the relationship between chemical structure and carcino-genic activity of substituted hydrazines. *Proc. Am. Ass. Cancer Res.* 12 (1971) 55.

(5) Toth, B. Lung tumor induction and inhibition of breast adenocarcinomas by hy-drazine sulfate in mice. *J. Natl. Cancer Inst.* 42 (1969) 469.

(6) Severi, L., and Biancifiori, C. Hepatic carcinogenesis in CBA/Cb/Se mice and Cb/Se rats by isonicotinic acid hydrazide and hydrazine sulfate. *J. Natl. Cancer Inst.* 41 (1968) 331.

(7) Chandra, S.V.S.G., and Reddy, G.M. Specific locus mutations in maize by chemical mutagens. *Curr. Sci.* 40 (1971) 136-137.

(8) Lingens, F. Mutagene wirkung von hydrazin auf *E. coli*-zellen. *Naturwiss.* 48 (1961) 480.

(9) Lingens, F. Erzeugung biochemischer mangel mutanten von *E. coli* mit hilfe von hy-drazin und hydrazin derivaten. *Z. Naturforsch.* 19B (1964) 151-156.

(10) Kimball, R.F., and Hirsch, B.F. Tests for the mutagenic action of a number of chemi-cals on *Haemophilus influenzae* with special emphasis on hydrazine. *Mutation Res.* 30 (1975) 9-20.

(11) Kimball, R.F. Reversions of proline-requiring auxotrophs of *Haemophilus influenzae* by N-methyl-N'-nitro-nitrosoguanidine and hydrazine. *Mutation Res.* 36 (1976) 29-38.

(12) Jain, H.K., Raut, R.N., and Khamanker, Y.G. Base-specific chemicals and mutation analysis in Lycopersicon. *Heredity* 23 (1968) 247.

(13) Chu, B.C.F., Brown, D.M., and Burdon, M.G. Effect of nitrogen and catalase on hy-droxylamine and hydrazine mutagenesis. *Mutation Res.* 20 (1973) 265-270.

(14) Wyrobek, A.J., and London, S.A. Effect of hydrazines on mouse sperm cells. *Proc. Ann. Conf. Environ. Toxicol.* 4th, (1973), AD-781031, pp. 417-432, *Chem.* 82 (1975) 150111U.

(15) Toth, B. Comparative studies with hydrazine derivatives. Carcinogenicity of 1,1-dimethylhydrazine, unsymmetrical (1,1-DMH) in the blood vessels, lung, kidneys, and liver of Swiss mice. *Proc. Am. Assoc. Cancer Res.* 13 (1972) 34.

(16) Toth, B. 1,1-Dimethylhydrazine (unsymmetrical) carcinogenesis in mice. Light microscopic and ultrastructural studies on neoplastic blood vessels. *J. Natl. Cancer Inst.* 50 (1973) 181.

(17) Back, K.C. and Thomas, A.A. Aerospace problems in pharmacology and toxicology. *Annual Rev. Pharmacol.* 10 (1970) 395.

(18) Juchau, M.R. and Herita, A. Metabolism of hydrazine derivatives of interest. *Drug Metab. Rev.* 1 (1972) 71.

(19) Toth, B. Synthetic and naturally occurring hydrazines as possible causative agents. *Cancer Res.* 35 (1975) 3693.

(20) Kimball, R.F. The mutagenicity of hydrazine and some of its derivatives. *Mutation Res.* 39 (1977) 111-126.

(21) Colvin, L.B. Metabolic fate of hydrazines and hydrazides. *J. Pharm. Sci.* 58 (1969) 1433-1443.

(22) Juchau, M.R. and Horita, A. Metabolism of hydrazine derivatives of pharmacologic interest. *Drug Metabolism Reviews* 1 (1973) 71-100.

(23) National Institute for Occupational Safety and Health. *Criteria for a Recommended Standard: Occuptional Exposure to Ketones.* NIOSH Doc. No. not yet assigned. Wash., DC (In press - 1978).

Bibliography

Jacobson, K.H., J.H. Clem, W.E. Rinehart, and N. Mayes. 1955. The acute toxicity of the vapors of some methylated hydrazine derivatives. *AMA Arch. Ind. Health.* 12:609.
Krop, S. 1954. Toxicology of hydrazine. *AMA Arch. Ind. Hyg. Occup. Med.* 9:199.
Shook, B.S., Sr., and O.H. Cowart. 1957. Health hazards associated with unsymmetrical dimethylhydrazine. *Ind. Med. Surg.* 26:333.
Von Oettingen, W.F. 1941. *The Aromatic Amino and Nitro Compounds, Their Toxicity and Potential Dangers.* Public Health Bulletin No. 271. U.S. Public Health Service p. 158.

HYDROGEN BROMIDE (See "Bromine/Hydrogen Bromide")

HYDROGEN CHLORIDE

Description: HCl, hydrogen chloride, is a colorless, nonflammable gas, soluble in water. The aqueous solution is known as hydrochloric acid or muriatic acid and may contain as much as 38% HCl.

Synonyms: Anhydrous hydrochloric acid, chlorohydric acid.

Potential Occupational Exposures: Hydrogen chloride itself is used in the manufacture of pharmaceutical hydrochlorides, chlorine, vinyl chloride from acetylene, alkyl chlorides from olefins, arsenic trichloride from arsenic trioxide; in the chlorination of rubber; as a gaseous flux for babbitting operations; and in organic synthesis involving isomerization, polymerization, alkylation, and nitration reactions.

The acid is used in the production of fertilizers, dyes, dyestuffs, artificial silk, and paint pigments; in refining edible oils and fats; in electroplating, leather tanning, ore refining, soap refining, petroleum extraction, pickling of metals, and in the photographic, textile, and rubber industries.

A partial list of occupations in which exposure may occur includes:

Battery makers	Metal cleaners
Bleachers	Oil well workers
Chemical synthesizers	Organic chemical synthesizers
Dye makers	Photoengravers
Electroplaters	Plastic workers
Fertilizer makers	Rubber makers
Food processors	Soap makers
Galvanizers	Tannery workers
Glue makers	Textile workers

Tantalum ore refiners   Wire annealers
Tin ore refiners

Permissible Exposure Limits: The Federal standard for hydrogen chloride is 5 ppm (7 mg/m^3) as a ceiling value.

Route of Entry: Inhalation of gas or mist.

Harmful Effects

Local—Hydrochloric acid and high concentrations of hydrogen chloride gas are highly corrosive to eyes, skin, and mucous membranes. The acid may produce burns, ulceration, and scarring on skin and mucous membranes, and it may produce dermatitis on repeated exposure. Eye contact may result in reduced vision or blindness. Dental discoloration and erosion of exposed incisors occur on prolonged exposure to low concentrations. Ingestion may produce fatal effects from esophageal or gastric necrosis.

Systemic—The irritant effect of vapors on the respiratory tract may produce laryngitis, glottal edema, bronchitis, pulmonary edema, and death.

Medical Surveillance: Special consideration should be given to the skin, eyes, teeth, and respiratory system. Pulmonary function studies and chest x-rays may be helpful in following recovery from acute overexposure.

Special Tests: None in common use.

Personal Protective Methods: Appropriate gas masks with canister or supplied air respirators should be provided when vapor concentrations are excessive. Acid resistant clothing including gloves, gauntlets, aprons, boots, and goggles or face shield should be provided in all areas where there is likelihood of splash or spill of liquid. Personal hygiene and showering after each shift should be encouraged.

Bibliography

American Medical Association. 1946. Effects of hydrochloric acid fumes. *J. Amer. Med. Assoc.* 131:1182.
Thiele, E. 1953. Fatal poisoning from use of hydrochloric acid in a confined space. *Zentralbl. Arbeitsmed. Arbeitsschutz* 3:146 (*Indust. Hyg. Digest Abst.* No. 387, 1954.)

HYDROGEN CYANIDE

Description: Hydrogen cyanide, a colorless gas or liquid, intensely poisonous with the odor of bitter almonds, is highly flammable and explosive and is a very weak acid. Hydrogen cyanide, HCN (together with its soluble salts), owes its toxicity to the −CN moiety and not to its acid properties. HCN vapor is released when cyanide salts come in contact with any acid.

Synonyms: Hydrocyanic acid, prussic acid.

Potential Occupational Exposures: Hydrogen cyanide is used as a fumigant, in electroplating, and in chemical synthesis of acrylates and nitriles, particularly acrylonitrile. It may be generated in blast furnaces, gas works, and coke ovens. Cyanide salts have a wide variety of uses, including electroplating, steel hardening, fumigating, gold and silver extraction from ores, and chemical synthesis.

A partial list of occupations in which exposure may occur includes:

Acid dippers	Fumigant workers
Acrylate makers	Gas workers
Ammonium salt makers	Gold extractors
Blast furnace workers	Jewelers
Cellulose product treaters	Organic chemical synthesizers
Coke oven workers	Polish makers
Cyanogen makers	Silver extractors
Electroplaters	Steel workers

The number of workers with potential exposure to HCN has been estimated by NIOSH to be approximately 1,000.

Permissible Exposure Limits: The Federal standard for hydrogen cyanide is 10 ppm (11 mg/m^3). NIOSH has recommended 5 mg/m^3 expressed as cyanide and determined as a ceiling concentration based on a 10-minute sampling period.

Routes of Entry: Inhalation of vapor and percutaneous absorption of liquid and concentrated vapor.

Harmful Effects

Local—Hydrogen cyanide is a mild upper respiratory irritant and may cause slight irritation of the nose and throat. There may also be irritation from skin and eye contact with the liquid. Hydrogen cyanide liquid may cause eye irritation.

Systemic—Hydrogen cyanide is an asphyxiant. It inactivates certain enzyme systems, the most important being cytochrome oxidase, which occupies a fundamental position in the respiratory process and is involved in the ultimate electron transfer to molecular oxygen. Inhalation, ingestion, or skin absorption of hydrogen cyanide may be rapidly fatal. Larger doses may cause loss of consciousness, cessation of respiration, and death. Lower levels of exposure may cause weakness, headache, confusion, nausea, and vomiting. These symptoms may be followed by unconsciousness and death.

Medical Surveillance: Preplacement and periodic examinations should include the cardiovascular and central nervous systems, liver and kidney function, blood, history of fainting or dizzy spells.

Special Tests: Blood CN levels may be useful during acute intoxication. Urinary thiocyanate levels have been used but are nonspecific and are elevated in smokers.

Personal Protective Methods: If personal protective equipment is necessary, air supplied or self-contained gas masks specific for hydrogen cyanide, and clothing impervious to HCN vapor should be worn. Eye protection can be provided by fullface respirators or goggles. All personnel working with processes involving cyanides should be specially trained so that they fully understand the hazard, and so they will faithfully follow all rules laid down for safe handling.

Bibliography

Amdur, M.L. 1959. Accidental group exposure to acetonitrile. *J. Occup Med.* 1:627.
National Institute for Occupational Safety and Health. *Criteria for a Recommended Standard: Occupational Exposure to Hydrogen Cyanide.* NIOSH Doc. No. 77-108 (1977).
Wolfsie, J.H. and C.B. Shaffer. 1959. Hydrogen cyanide—hazards, toxicology prevention and management of poisoning. *J. Occup. Med.* 1:281.

HYDROGEN FLUORIDE (See "Fluorine and Compounds")

HYDROGEN PEROXIDE

Description: H_2O_2, anhydrous hydrogen peroxide, is a colorless rather unstable liquid with a bitter taste. Hydrogen peroxide is completely miscible with water and is commercially sold in concentrations of 3, 35, 50, 70, and 90% solutions.

Synonyms: Peroxide, hydrogen dioxide, hydroperoxide.

Potential Occupational Exposures: Hydrogen peroxide is used in the manufacture of acetone, antichlor, antiseptics, benzol peroxide, buttons, disinfectants, pharmaceuticals, felt hats, plastic foam, rocket fuel, and sponge rubber. It is also used in bleaching bone, feathers, flour, fruit, fur, gelatin, glue, hair, ivory, silk, soap, straw, textiles, wax, and wood pulp, and as an oxygen source in respiratory protective equipment. Other specific occupations with potential exposure include liquor and wine agers, dyers, electroplaters, fat refiners, photographic film developers, wool printers, veterinarians, and water treaters.

A partial list of occupations in which exposure may occur includes:

Acetone makers	Fat refiners
Alcoholic beverage agers	Hide disinfectors
Antichlor makers	Metal cleaners
Antiseptic makers	Photographic film developers
Benzol peroxide makers	Plastic foam makers
Button makers	Rocket fuel workers
Disinfectant makers	Sponge rubber makers
Drug makers	Veterinarians
Dyers	Water treaters
Electroplaters	Wool printers

Permissible Exposure Limits: The Federal standard for hydrogen peroxide (90%) is 1 ppm (1.5 mg/m^3).

Route of Entry: Inhalation of vapor or mist.

Harmful Effects

Local—The skin, eyes, and mucous membranes may be irritated by concentrated vapor or mist. Bleaching and a burning sensation may occur at lower levels, while high concentrations may result in blistering and severe eye injury, which may be delayed in appearance.

Systemic—Inhalation of vapor or mist may produce pulmonary irritation ranging from mild bronchitis to pulmonary edema. No chronic systemic effects have been observed.

Hydrogen peroxide has been found mutagenic in *E. coli* (1)(2), *Staphylococcus aureus* (3)-(5), and *Neurospora* (6)-(8) (including mixtures of hydrogen peroxide and acetone, and hydrogen peroxide and formaldehyde) (7). The inactivation of transforming DNA by hydrogen peroxide (9)-(12), as well as by peroxide-producing agents (e.g., compounds which contain a free NOH group as hydroxylamine, N-methylhydroxylamine, hydroxyurea, hydroxyurethane and hydrazines on exposure to oxygen) has been described (12)(13). Hydrogen peroxide has induced chromosome aberrations in strains of ascites tumors in mice (14) and in *Vicia faba* (15).

Hydrogen peroxide has also been found to be nonmutagenic in bacteria

(16)(17) and *Drosophila* (18). Recent studies by Thacker and Parker (19) on the induction of mutation in yeast by hydrogen peroxide suggest that it is ineffective in the induction of nuclear gene mutation. This could be because radical action produces certain types of lesions, leading to inactivation and mitochondrial genome mutation, but not to point mutational changes.

There is general agreement that the effects of hyrogen peroxide, (as well as hydrogen peroxide-producing agents) on DNA are caused by the free radicals they generate.

H_2O_2 decomposes into two $\cdot$OH radicals in response to UV irradiation or spontaneously at elevated temperatures. It also gives rise to HOO$\cdot$ radicals in the presence of ions of transition metals in their lower valence state (e.g., Fe^{++}, Cu^+). These radicals can then react with organic molecules to produce relatively more stable organic peroxy radicals and organic peroxides which may decompose again into free radicals. This process of "radical-exchange" sustains the effectiveness of short-lived radicals such as $\cdot$OH, HOO$\cdot$, and H$\cdot$ and gives them an opportunity to reach the genome where they can exert their effect (13).

Medical Surveillance: Preplacement and periodic examinations should be directed to evaluation of the general health with particular reference to the skin, eyes, mucous membranes, and respiratory tract.

Special Tests: None.

Personal Protective Methods: In areas where concentrated hydrogen peroxide is being used, if there is danger of spill or splash, skin protection should be provided by protective clothing, gloves, goggles, and boots. Where fumes or vapor are excessive, workers should be provided with gas masks with full-face pieces and proper canisters or supplied air respirators. Additional health hazards may occur from the decomposition of hydrogen peroxide. Oxygen, possibly at high pressure, may form, which may create an explosion hazard. Hydrogen peroxide is generally handled in a closed system to prevent contamination.

References

(1) Demerec, M., Bertani, G., and Flint, J. Chemicals for mutagenic action on *E. coli*. *Am. Naturalist* 85 (1951) 119.

(2) Iyer, V.N., and Szybalski, W. Two simple methods for the detection of chemical mutagens. *Appl. Microbiol.* 6 (1958) 23.

(3) Wyss, O., Stone, W.S. and Clark, J.B. The production of mutations in *Staphylococcus aureus* by chemical treatment of the substrate. *J. Bacteriol.* 54 (1947) 767.

(4) Wyss, O., Clark, J.B., Haas, F., and Stone, W.S. The role of peroxide in the biological effects of irradiated broth. *J. Bacteriol.* 56 (1948) 51.

(5) Haas, F.L., Clark, J.B., Wyss, O., and Stone, W.S. Mutations and mutagenic agents in bacteria. *Am. Naturalist.* 74 (1950) 261.

(6) Jensen, K.A., Kirk, I., Kolmark, G., and Westergaard, M. Chemically induced mutations in *Neurospora. Cold Springs Harbor Symp. Quant. Biol.* 16 (1951) 245-261.

(7) Dickey, F.H., Cleland, G.H., and Lotz, C. The role of peroxides in the induction of mutations. *Proc. Natl. Acad. Sci. US* 35 (1949) 581.

(8) Wagner, R.P., Haddox, C.R., Fuerst, R., and Stone, W.S. The effect of irradiated medium, cyanide and peroxide on the mutation rate in *Neurospora. Genetics* 35 (1950) 237.

(9) Latarjet, R., Rebeyrotte, N., and Demerseman, P. In *Organic Peroxides in Radiobiology.* (ed. Haissinsky, M.), Pergamon Press, Oxford (1958) p. 61.

(10) Luzzati, D., Schweitz, H., Bach, M.L., and Chevallier, M.R. Action of succinic peroxide on deoxyribonucleic acid (DNA). *J. Chim Phys.* 58 (1961) 1021.

(11) Zamenhof, A., Alexander, H.E., and Leidy, G. Studies on the chemistry of the transforming activity. I. Resistance to physical and chemical agents. *J. Exptl. Med.* 98 (1953) 373.

(12) Freese, E.B., Gerson, J., Taber, H., Rhaese, H.J., and Freese, E.E. Inactivating DNA alterations by peroxides and peroxide inducing agents. *Mutation Res.* 4 (1967) 517.

(13) Fishbein, L., Flamm, W.G., and Falk, H.L. *Chemical Mutagens.* Academic Press, New York (1970) pp. 269-273.

(14) Schöneich, J. The induction of chromosomal aberrations by hydrogen peroxide in strains of ascites tumors in mice. *Mutation Res.* 4 (1967) 385.

(15) Lilly, L.J. and Thoday, J.M. Effects of cyanide on the roots of *Vicia faba. Nature,* 177 (1956) 338.

(16) Doudney, C.O. Peroxide effects on survival and mutation induction in UV light exposed and photo-reactivated bacteria. *Mutation Res.* 6 (1968) 345-353.

(17) Kimball, R.F., Hearon, J.Z., and Gaither, N. Tests for a role of H_2O_2 in x-ray mutagenesis. II. Attempts to induce mutation by peroxide. *Radiation Res.* 3 (1955) 435-443.

(18) Sobels, F.H. Peroxides and the induction of mutations by x-rays, ultraviolet, and formaldehyde. *Radiation Res. Suppl. 3* (1963) 171-183.

(19) Thacker, J., Parker, W.F. The induction of mutation in yeast by hydrogen peroxide. *Mutation Res.* 38 (1976) 43-52.

Bibliography

Oberst, F.W., C. Comstock, and E.B. Hackley. 1954. Inhalation toxicity of 90% hydrogen peroxide vapor. Acute, subacute, and chronic exposures of laboratory animals. *AMA Arch. Ind. Hyg. Occup. Med.* 10:319.

HYDROGEN SULFIDE

Description: H_2S, hydrogen sulfide, is a flammable, colorless gas with a characteristic rotten-egg odor and is soluble in water.

Synonyms: Sulfuretted hydrogen, hydrosulfuric acid, stink damp.

Potential Occupational Exposures: Hydrogen sulfide is used in the synthesis of inorganic sulfides, sulfuric acid, and organic sulfur compounds, as an analytical reagent, as a disinfectant in agriculture, and in metallurgy. It is generated in many industrial processes as a by-product and also during the decomposition of sulfur-containing organic matter, so potential for exposure exists in a variety of situations. Hydrogen sulfide is found in natural gas, volcanic gas, and in certain natural spring waters.

It may also be encountered in the manufacture of barium carbonate, barium salt, cellophane, depilatories, dyes, and pigments, felt, fertilizer, adhesives, viscose rayon, lithopone, synthetic petroleum products; in the processing of sugar beets; in mining, particularly where sulfide ores are present; in sewers and sewage treatment plants; during excavation of swampy or filled ground for tunnels, wells, and caissons; during drilling of oil and gas wells; in purification of hydrochloric acid and phosphates; during the low temperature carbonization of coal; in tanneries, breweries, slaughterhouses; in fat rendering; and in lithography and photoengraving.

A partial list of occupations in which exposure may occur includes:

Barium carbonate makers	Rayon makers
Brewery workers	Sewage treatment plant workers
Caisson workers	Sewer workers
Cellophane makers	Silk makers
Coke oven workers	Slaughterhouse workers
Depilatory makers	Soap makers
Dye makers	Sugar beet processors
Fat renderers	Sulfuric acid purifiers
Felt makers	Sulfur makers
Lithographers	Synthetic fiber makers
Miners	Tannery workers
Natural gas makers	Tunnel workers
Paper pulp makers	Well diggers

NIOSH estimates that 125,000 workers in the United States are exposed or potentially exposed to hydrogen sulfide.

Permissible Exposure Limits: The Federal standard is a ceiling value of 20 ppm (30 mg/m^3) with a maximum peak above this value for an 8-hour shift of 50 ppm (75 mg/m^3) for a maximum duration of 10 minutes once only if no other measurable exposure occurs. ACGIH (1978) gives a TWA value of 10 ppm (15 mg/m^3).

NIOSH recommends that exposure to hydrogen sulfide be limited to a ceiling concentration of 15 mg/m^3 (10 ppm) for 10 minutes. A requirement for continuous monitoring when there is a potential for exposure to hydrogen sulfide at a concentration of 70 mg/m^3 (50 ppm) or higher is also recommended. Occupational exposure to hydrogen sulfide has been defined as exposure at or above the ceiling concentration of 15 mg/m^3.

Route of Entry: Inhalation of gas.

Harmful Effects

Local—Palpebral edema, bulbar conjunctivitis, keratoconjunctivitis, and ocular lesions may occur when hydrogen sulfide comes in contact with the eyes. Photophobia and lacrimation may also develop. Direct irritation of the respiratory tract may cause rhinitis, pharyngitis, bronchitis, and pneumonia. Hydrogen sulfide may penetrate deep into the lungs and cause hemorrhagic pulmonary edema. Hydrogen sulfide's irritative effects are due to the formation of alkali sulfide when the gas comes in contact with moist tissues.

Systemic—Acute exposure may cause immediate coma which may occur with or without convulsions. Death may result with extreme rapidity from respiratory failure. Postmortem signs include a typical greenish cyanosis of the chest and face with green casts found in viscera and blood. The toxic action of hydrogen sulfide is thought to be due to inhibition of cytochrome oxidase by binding iron which is essential for cellular respiration. Subacute exposure results in headache, dizziness, staggering gait, and excitement suggestive of neurological damage, and nausea and diarrhea suggestive of gastritis. Recovery is usually complete although rarely polyneuritis may develop as a result of vestibular and extrapyramidal tract damage. Tremors, weakness, and numbness of extremities may also occur. Physicians may observe a "rotten-egg" breath and abnormal electrocardiograms in victims. Systemic effects from chronic exposure to hyrogen sulfide have not been established.

Medical Surveillance: Preplacement medical examinations should evaluate any preexisting neurological, eye, and respiratory conditions and any history of fainting seizures.

It is recommended by NIOSH that preplacement and periodic examinations (once every 3 years) be made available to all workers occupationally exposed to hydrogen sulfide. These are to include physical examinations which give particular attention to the eyes and the nervous and respiratory systems.

Special Tests: NIOSH has recommended sampling and analytical methods using a methylene blue technique adapted from the NIOSH/OSHA Standards Completion Program. The methylene blue method for sampling and analysis was selected because of its sensitivity and because it has already demonstrated a wide applicability in industry. The use of NIOSH validated detector tubes is also recommended for this purpose. For continuous monitoring, no specific instrument is recommended. However, minimum capabilities such as response time, sensitivity, and range are specified.

Special Tests: None in common use for surveillance purposes.

Personal Protective Methods: Hydrogen sulfide's strong odor, noticeable at low concentrations, is a poor warning sign as it may cause olfactory paralysis, and some persons are congenitally unable to smell H_2S.

Accidental exposure may occur when workers enter sewage tanks and other confined areas in which hydrogen sulfide is formed by decomposition. In a number of cases workers enter unsuspectingly and collapse almost immediately. Workers, therefore, should not enter enclosed spaces without proper precautions.

All Federal standard and other safety precautions must be observed when tanks or other confined spaces are to be entered. In areas where the exposure to hydrogen sulfide exceeds the standards, workers should be provided with fullface canister gas masks or preferably supplied air respirators.

Bibliography

Adelson, L., and I. Sushine. 1966. Fatal hydrogen sulfide intoxication. Report of three cases in a sewer. *Arch. Pathol.* 81:375.

Kleinfeld, M., C. Giel, and A. Rosso. 1964. Acute hydrogen sulfide intoxication; an unusual source of exposure. *Ind. Med. Sur.* 33:656.

McCabe, L.C., and G.D. Clayton. 1952. Air pollution by hydrogen sulfide in Poza Rica, Mexico. An evaluation of the incident of Nov. 24, 1950. *AMA Arch. Ind. Hyg. Occup. Med.* 6:199.

Milby, T.H. 1962. Hydrogen sulfide intoxication. Review of the literature and report of unusual accident resulting in two cases of nonfatal poisoning. *J. Occup. Med.* 4:431.

National Institute for Occupational Safety and Health. *Criteria for a Recommended Standard: Occupational Exposure to Hydrogen Sulfide.* NIOSH Doc. No. 77-158 (1977).

Simson, R.E., and G.R. Simpson. 1971. Fatal hydrogen sulfide poisoning associated with industrial waste exposure. *Med. J. Aust.* 1:331.

Winek, C.L., W.D. Collom, and C.H. Wecht. 1968. Death from hydrogen sulfide fumes. *Lancet* 1:1096.

HYDROQUINONE

Description: $C_6H_4(OH)_2$, hydroquinone, exists as colorless, hexagonal prisms.

Synonyms: Quinol, hydroquinol, p-diphenol, hydrochinone, dihydroxybenzene, p-dihydroxybenzene, p-hydroxyphenol, 1,4-benzenediol.

Potential Occupational Exposures: Hydroquinone is a reducing agent and is used as a photographic developer and as an antioxidant or stabilizer for certain materials which polymerize in the presence of oxidizing agents. Many of its derivatives are used as bacteriostatic agents, and others, particularly 2,5-bis(ethyleneimino) hydroquinone, have been reported to be good antimitotic and tumor-inhibiting agents.

A partial list of occupations in which exposure may occur includes:

Antioxidant makers	Drug makers
Bacteriostatic agent makers	Fur processors

Motor fuel blenders	Plastic stabilizer workers
Organic chemical synthesizers	Stone coating workers
Paint makers	Styrene monomer workers
Photographic developer makers	

Permissible Exposure Limits: The Federal standard is 2 mg/m^3.

Route of Entry: Inhalation of dust.

Harmful Effects

Local—The dust is a mild primary irritant. Skin sensitization to the dry solid is very rare but does occur on occasion from contact with its alkaline solutions. The skin may be depigmented by repeated applications of ointments of hydroquinone, but this virtually never occurs from contact with dust or dilute water solutions. Following prolonged exposure to elevated dust levels, brownish conjunctiva stains may appear. These may be followed by corneal opacities and structural changes in the cornea which may lead to loss of visual acuity. The early pigmentary stains are reversible, while the corneal changes tend to be progressive.

Systemic—Oral ingestion of large quantities of hydroquinone may produce blurred speech, tinnitus, tremors, sense of suffocation, vomiting, muscular twitching, headache, convulsions, dyspnea and cyanosis from methemo-globinemia, and coma and collapse from respiratory failure. The urine is usually green or brownish green. No systemic symptoms have been found following inhalation of hydroquinone dust.

Medical Surveillance: Careful examination of the eyes, including visual acuity and slit lamp examinations, should be carried out in preplacement and periodic examinations. Also the skin should be examined.

Special Tests: Hydroquinone is excreted in the urine as a sulfate ester. This has not been helpful in following worker exposure to dust.

Personal Protective Methods: The eyes should be protected by goggles or dust masks with fullface shield. Protective clothing is recommended along with good hygiene practice, clothes changing after each shift, and showering prior to dressing in street clothes. Oxidation of hydroquinone may produce quinone vapor which is highly irritating.

Bibliography

Anderson, B., and F. Oglesby. 1958. Corneal changes from quinone-hydroquinone exposure. *Arch. Ophthalmol.* 59:495.
Seutter, E., and A.H.M. Sutorius. 1972. Quantitative analysis of hydroquinone in urine. *Clin. Chim. Acta.* 38:231.

I

INDIUM AND COMPOUNDS

Description: Indium metal is malleable, ductile and softer than lead. The two most important indium compounds are the oxide and sulfate. Physical properties are presented below:

Chemical	Specific Gravity	Melting Point °C	Boiling Point °C	Solubility in Water
Indium	7.30^{20}	156.61	2000 ± 10	i (hot & cold)*
Indium sesquioxide	7.179		volat. 850	i cold
Indium sulfate	3.438			s cold, vs hot

*Finely divided indium forms hydroxide on contact with water.

Synonyms: None.

Potential Occupational Exposures: The uses of indium and its compounds are shown in the following table:

Compound	Use	Purpose
Indium	Dental alloy	
	Solder and alloy industries	Produces high quality solders and braze-bonded connectors
	Automotive bearings (Europe only)	
	Low pressure sodium lamps (Europe)	
	Other	Nuclear reactor control rod alloys
		Catalysts
		Indium oxide fuel cells
		Cryogenic gasket material

(continued)

Compound	Use	Purpose
Indium oxide	Glass	Used for coloring. A light to dark brown can be obtained, depending on the amount used
Indium sulfate	Electroplating	Used to prepare sulfate electrolytes
Radioisotopes of indium and indium compounds	Medical	Treatment of cancer and diagnostic organ scanning

Permissible Exposure Limits: The threshold limit value for indium and its compounds is 0.1 mg/m^3. This value is recommended in view of the character and severity of injury from indium salts, especially the pulmonary effects.

Route of Entry: Ingestion or inhalation.

Harmful Effects

Indium and its compounds are considered moderately toxic and as a local irritant on contact with the skin. However, they are rated extremely toxic on ingestion. The American Conference of Governmental Hygienists has recommended a threshold limiting value of 0.1 mg/m^3 based almost entirely on experiments with animals and the severity of injury from indium salts. The data involving humans is very limited. As a result possibly too much weight is given to a Russian report that individuals exposed to indium compounds during production complained of pains in joints and bones, tooth decay, nervous and gastrointestinal disorders, heart pains and general debility. This has not been reported in comparable United States activities.

Indium is poorly absorbed through the intestine, and oral levels of toxicity are quite high. The cells of the reticuloendothelial system phagocytize indium compound particles, and indium toxicity is apparently due to the concentration of heavy metals by these cells. Radioisotopes of indium are used for diagnostic x-ray scanning. Carrier-free isotopes can be used with relatively few toxic effects, although several cases of anaphylactic shock have occurred following injection of indium compounds.

Medical Surveillance: Preplacement and regular physical examinations are indicated.

Personal Protective Methods: Personnel who work with or are exposed to dust, salts or mists of indium and its compounds should, in the absence of exact data, wear safety equipment such as chemical safety goggles for the eyes and a respirator to avoid inhalation. It may be necessary to wear protective clothing to avoid skin contact.

Bibliography

Environmental Protection Agency. *Preliminary Investigation of Effects on the Environment of Boron, Indium, Nickel, Selenium, Tin, Vanadium and Their Compounds, Vol. II, Indium.* Report EPA-560/2-75-005B, Wash., D.C., Office of Toxic Substances (Aug. 1975).

IRON COMPOUNDS

Description: Fe, iron, is a malleable, silver-grey metal. Ferric oxide is a dense, dark red powder or lumps. Hematite is the most important iron ore and is generally found as red hematite (red iron ore, mainly Fe_2O_3) and brown hematite (brown iron ore, mainly limonite, a hydrated sesquioxide of iron). Magnetic iron oxide, Fe_3O_4, is black. Iron is insoluble in water. Iron oxide is soluble in hydrochloric acid.

Synonyms: None.

Potential Occupational Exposures: Iron is alloyed with carbon to produce steel. The addition of other elements (e.g., manganese, silicon, chromium, vanadium, tungsten, molybdenum, titanium, niobium, phosphorus, zirconium, aluminum, copper, cobalt, and nickel) imparts special characteristics to the steel.

Occupational exposures occur during mining, transporting, and preparing of ores and during the production and refining of the metal and alloys. In addition, certain workers may be exposed while using certain iron-containing materials: welders, grinders, polishers, silver finishers, metal workers, and boiler scalers.

A partial list of occupations in which exposure may occur includes:

Arc cutters	Metalizers
Bessemer operators	Seam welders
Electric arc welders	Stainless steel makers
Flame cutters	Steel foundry workers
Friction saw operators	

Permissible Exposure Limits: The Federal standard for iron oxide fume is 10 mg/m³. There are no standards for other iron compounds. ACGIH (1978) set a TWA of 5 mg/m³ for iron oxide fume. They also set 1.0 mg/m³ for soluble iron compounds and a TWA value of 0.01 ppm (0.08 mg/m³) for iron pentacarbonyl.

Route of Entry: Inhalation of dust.

Harmful Effects

Local—Soluble iron salts, especially ferric chloride and ferric sulfate, are cutaneous irritants and their aerosols are irritating to the respiratory tract. Iron compounds as a class are not associated with any particular industrial risk.

Systemic—The inhalation of iron oxide fumes or dust may cause a benign pneumoconiosis (siderosis). It is probable that the inhalation of pure iron oxide does not cause fibrotic pulmonary changes, whereas the inhalation of iron oxide plus certain other substances may cause injury.

On the basis of epidemiological evidence, exposure to hematite dust increases the risk of lung cancer for workers working underground, but not for surface workers. It may be, however, that hematite dust becomes carcinogenic only in combination with radioactive material, ferric oxide, or silica. There is no evidence that hematite dust or ferric oxide causes cancer in any part of the body other than the lungs.

Iron compounds derive their dangerous properties from the radical with which the iron is associated. Iron pentacarbonyl is one of the more dangerous metal carbonyls. It is highly flammable and toxic. Symptoms of overexposure closely resemble those caused by $Ni(CO)_4$ and consist of giddiness and headache, occasionally accompanied by fever, cyanosis, and cough due to pulmonary edema. Death may occur within 4 to 11 days due to pneumonia, liver damage, vascular injury, and central nervous system degeneration.

Medical Surveillance: Special consideration should be given to respiratory disease and lung function in placement and periodic examinations. Smoking history should be known. Chest x-rays and pulmonary function should be evaluated periodically especially if symptoms are present.

Personal Protective Methods: Dust masks are recommended for all workers exposed to areas of elevated dust concentrations and especially those workers in underground mines. In areas where iron oxide fumes are excessive, vapor canister masks or supplied air masks are recommended. Generally speaking, protective clothing is not necessary, but attention to personal hygiene, showering, and clothes changing should be encouraged.

Bibliography

International Agency for Research on Cancer. 1971. *IARC Monographs on the Evaluation of the Carcinogenic Risk of Chemicals to Man.* Volume 1.
Jones, J.G., and Warner, C.G. 1972. Chronic exposure to iron oxide, chromium oxide, and nickel oxide fumes of metal dressers in a steelworks. *Br. J. Ind. Med.* 29:169.
Sunderman, F.W., West, B., and Kincaid, J.F. 1954. A toxicity study of iron pentacarbonyl. *AMA Arch. Ind. Health* 19:11.

ISOCYANATES

Description: Both toluene diisocyanate (TDI) and methylene bisphenyl iso-cyanate (MDI) are liquids and may exist in different isomers: 2,4-toluene di-isocyanate and methylene bisphenyl 4,4'-diisocyanate. Other less commonly used isocyanates are hexamethylene diisocyanate (HDI) and 1,5-naphthalene diisocyanate (NDI).

Synonyms: Toluene diisocyanate: TDI, tolylene diisocyanate, diisocyanatotol-uene. Methylene bisphenyl isocyanate: MDI, diphenylmethane diisocyanate, methane diisocyanate.

Potential Occupational Exposures: TDI is more widely used than MDI. Poly-urethanes are formed by the reaction of isocyanates with polyhydroxy com-pounds. Since the reaction proceeds rapidly at room temperature, the re-actants must be mixed in pots or spray guns just before use. These resins can be produced with various physical properties, e.g., hard, flexible, semirigid foams, and have found many uses, e.g., upholstery padding, thermal insula-tion, molds, surface coatings, shoe innersoles, and in rubbers, adhesives, paints, and textile finishes. Because of TDI's high volatility, exposure can occur in all phases of its manufacture and use. MDI has a much lower volatil-ity, and problems generally arise only in spray applications.

A partial list of occupations in which exposure may occur includes:

Adhesive workers	Polyurethane makers
Insulation workers	Rubber workers
Isocyanate resin workers	Ship burners
Lacquer workers	Textile processors
Organic chemical synthesizers	Wire coating workers
Paint sprayers	

It is estimated that approximately 40,000 workers are potentially exposed to TDI and that many of these exposures are in small workplaces.

Permissible Exposure Limits: The Federal standard for MDI is 0.02 ppm (0.2 mg/m^3) as a ceiling value. The Federal standard for the 2,4 isomer of TDI is also 0.02 ppm (0.14 mg/m^3) as a ceiling value. However, the standard recom-mended in the NIOSH Criteria Document for TDI is 0.005 ppm (0.036 mg/m^3) as a TWA and 0.02 ppm for any 20-minute period. ACGIH (1978) proposes 0.002 ppm (0.015 mg/m^3) as a TWA.

Route of Entry: Inhalation of vapor.

Harmful Effects

Local—TDI and MDI may cause irritation of the eyes, respiratory tract, and skin. The irritation may be severe enough to produce bronchitis and pulmonary edema. Nausea, vomiting, and abdominal pain may occur. If liquid TDI is allowed to remain in contact with the skin, it may produce redness, swelling, and blistering. Contact of liquid TDI with the eyes may cause severe irritation, which may result in permanent damage if untreated. Swallowing TDI may cause burns of the mouth and stomach.

Systemic—Sensitization to TDI and MDI may occur, which may cause an asthmatic reaction with wheezing, dyspnea, and cough. These symptoms may first occur during the night following exposure to these chemicals. Some decrease in lung function in the absence of symptoms has been observed in some workers exposed to TDI for long periods of time.

Medical Surveillance: Preplacement and periodic medical examinations should include chest roentgenograph, pulmonary function tests, and an evaluation of any respiratory disease or history of allergy. Periodic pulmonary function tests may be useful in detecting the onset of pulmonary sensitization.

Special Tests: None in common use.

Personal Protective Methods: Protective clothing and goggles should be worn if there is a possibility of contact with the liquids. In areas of vapor concentration, fullface masks with organic vapor canisters or respirators with supplied air and full facepieces should be worn.

Bibliography

Konzen, R.B., Craft, B.F., Scheel, L.D., and Gorski, C.H. 1966. Human response to low concentrations of p,p-diphenylmethanediisocyanate (MDI). *Amer. Ind. Hyg. Assoc. J.* 27:121.
Longley, E.O. 1964. Methane diisocyanate: A respiratory hazard? *Arch. Environ. Health* 8:898.
National Institute for Occupational Safety and Health. *Criteria for a Recommended Standard: Occupational Exposure to Toluene Diisocyanate.* NIOSH Doc. No. 73-11022 (1973).

ISOPHORONE (See "Ketones")

ISOPROPYL ALCOHOL

Description: Isopropyl alcohol, $CH_3CHOHCH_3$, is a flammable liquid, boiling at 82.5°C with a slight odor resembling that of a mixture of ethanol and acetone.

Synonyms: 2-Propanol, isopropanol, secondary propyl alcohol, dimethyl carbinol.

Potential Occupational Exposures: Isopropyl alcohol is a chemical widely used in liniments, skin lotions, cosmetics, permanent wave preparations, pharmaceuticals, and hair tonics. It is also employed as a solvent in perfumes, in extraction processes, as a preservative, in lacquer formulations, and in many dye solutions. In addition, it has been employed as an ingredient of antifreezes, soaps, and window cleaners. It may also be used as a raw material for the manufacture of acetone and various isopropyl derivatives. NIOSH estimates that approximately 141,000 employees are potentially exposed to isopropyl alcohol in the United States.

Permissible Exposure Limits: NIOSH recommends adherence to environmental workplace limits of 400 ppm as a time-weighted average concentration and 800 ppm as a ceiling concentration for up to a 10-hour workday, 40-hour workweek.

Route of Entry: Inhalation of vapor.

Harmful Effects

Isopropyl alcohol is a relatively innocuous chemical when compared with several other compounds currently under study by NIOSH. At low airborne concentrations of isopropyl alcohol, the only effect observed is mild irritation of eyes and skin. At much higher concentrations, it is believed, largely on the basis of animal studies, that narcosis can occur.

On the basis of several studies conducted by Weil and his colleagues, NIOSH suspects that a carcinogen may have been present in the old "strong acid" isopropyl alcohol manufacturing process. Animal studies subsequently did not support the contention that isopropyl alcohol is a carcinogen. Although the data obtained from the animal studies were frequently inconsistent, evidence points to isopropyl oil obtained from the old process as a possible carcinogen. The unequivocal identification of the carcinogen has not been made. Epidemiologic and other studies are recommended by NIOSH to determine whether a carcinogen exists in the new "weak-acid" isopropyl alcohol manufacturing process. Until such information becomes available, NIOSH is recommending that special procedures be followed during the manufacturing process.

Medical Surveillance: NIOSH recommends that workers subject to isopropyl alcohol exposure have comprehensive preplacement medical examinations. Periodic medical examinations shall be made available on an annual basis. Particular attention will be given in these medical examinations to the skin, sinuses, and respiratory system.

Special Tests: None in use.

Personal Protective Methods: Engineering controls should be used wherever feasible to maintain isopropyl alcohol concentrations at or below the indicated limit, and respirators should only be used in certain prescribed circumstances.

Bibliography

National Institute for Occupational Safety and Health. *Criteria for a Recommended Standard: Occupational Exposure to Isopropyl Alcohol.* NIOSH Doc. No. 76-142 (1976).

K

KEPONE

Description: Kepone, $C_{10}Cl_{10}O$, is a crystalline solid which decomposes at 350°C.

Synonyms: Chlordecone

Potential Occupational Exposures: NIOSH has identified fewer than 50 establishments processing or formulating pesticides using kepone and has estimated that 600 workers are potentially exposed to kepone. (NIOSH is unaware of any plant in the United States which is currently manufacturing kepone; the only known plant manufacturing it was closed in July 1975.)

Permissible Exposure Limits: NIOSH recommends that the workplace environmental level for kepone be limited to 1 $\mu g/m^3$ as a time-weighted average concentration for up to a 10-hour workday, 40-hour workweek, as an emergency standard.

Route of Entry: Inhalation of dust.

Harmful Effects

In July 1975, a private physician submitted a blood sample to the Center for Disease Control (CDC) to be analyzed for kepone, a chlorinated hydrocarbon pesticide. The sample had been obtained from a kepone production worker who suffered from weight loss, nystagmus, and tremors. CDC notified the State epidemiologist that high levels of kepone were present in blood sample, and he initiated an epidemiologic investigation which revealed other employees suffering with similar symptoms. It was evident to the State official after visiting the plant that the employees had been exposed to kepone at extremely high concentrations through inhalation, ingestion, and skin absorption. He recommended that the plant be closed, and company management complied.

Of the 113 current and former employees of this kepone-manufacturing plant examined, more than half exhibited clinical symptoms of kepone poisoning.

270

Medical histories of tremors (called "kepone shakes" by employees), visual disturbances, loss of weight, nervousness, insomnia, pain in the chest and abdomen and, in some cases, infertility and loss of libido were reported. The employees also complained of vertigo and lack of muscular coordination. The intervals between exposure and onset of the signs and symptoms varied between patients but appeared to be dose-related.

NIOSH has recently received a report on a carcinogenesis bioassay of technical grade kepone which was conducted by the National Cancer Institute using Osborne-Mendel rats and B6C3F1 mice. Kepone was administered in the diet at two tolerated dosages. In addition to the clinical signs of toxicity, which were seen in both species, a significant increase ($P<0.05$) of hepatocellular carcinoma in rats given large dosages of kepone and in mice at both dosages was found. Rats and mice also had extensive hyperplasia of the liver.

In view of these findings, NIOSH must assume that kepone is a potential human carcinogen.

Medical Surveillance: Employers shall make medical surveillance available to all workers occupationally exposed to kepone, including personnel periodically exposed during routine maintenance or emergency operations. Periodic examinations shall be made available at least on an annual basis.

Special Tests: None in use.

Personal Protective Methods

(a) Protective Clothing
 1) Coveralls or other full-body protective clothing shall be worn in areas where there is occupational exposure to kepone. Protective clothing shall be changed at least daily at the end of the shift and more frequently if it should become grossly contaminated.
 2) Impervious gloves, aprons, and footwear shall be worn at operations where solutions of kepone may contact the skin. Protective gloves shall be worn at operations where dry kepone or materials containing kepone are handled and may contact the skin.
 3) Eye protective devices shall be provided by the employer and used by the employees where contact of kepone with eyes is likely. Selection, use, and maintenance of eye protective equipment shall be in accordance with the provisions of the American National Standard Practice for Occupational and Educational Eye and Face Protection, ANSI Z87.1-1968. Unless eye protection is afforded by a respirator hood or facepiece, protective goggles or a face shield shall be worn at operations where there is danger of contact of the eyes with dry or wet materials containing kepone because of spills, splashes, or excessive dust or mists in the air.
 4) The employer shall ensure that all personal protective devices are inspected regularly and maintained in clean and satisfactory working condition.

5) Work clothing may not be taken home by employees. The employer shall provide for maintenance and laundering of protective clothing.
6) The employer shall ensure that precautions necessary to protect laundry personnel are taken while soiled protective clothing is being laundered.
7) The employer shall ensure that kepone is not discharged into municipal waste treatment systems or the community air.

(b) Respiratory Protection from Kepone

Engineering controls shall be used wherever feasible to maintain airborne kepone concentrations at or below that recommended above. Compliance with the environmental exposure limit by the use of respirators is allowed only when airborne kepone concentrations are in excess of the workplace environmental limit because required engineering controls are being installed or tested, when nonroutine maintenance or repair is being accomplished, or during emergencies. When a respirator is thus permitted, it shall be selected and used in accordance with NIOSH requirements.

Bibliography

National Institute for Occupational Safety and Health. *Recommended Standard for Occupational Exposure to Kepone.* Wash., D.C. (Jan. 27, 1976).

KEROSENE

Description: Kerosene is a pale yellow or clear, mobile liquid, composed of a mixture of petroleum distillates, having a characteristic odor. Chemically, it is composed of aliphatic hydrocarbons with 10 to 16 carbons per molecule and benzene and naphthalene derivatives.

Synonyms: Kerosine, coal-oil, range-oil.

Potential Occupational Exposures: Kerosene is used as a fuel for lamps, stoves, jets, and rockets. It is also used for degreasing and cleaning metals and as a vehicle for insecticides.

A partial list of occupations in which exposure may occur includes:

Farmers	Insecticide workers
Garage workers	Jet fuel handlers
Grease removers	Metal cleaners
Heating fuel handlers	Petroleum refinery workers

Permissible Exposure Limits: Presently there is no Federal standard for kerosene vapor in workroom air.

Route of Entry: Inhalation of vapor.

Harmful Effects

Local—The liquid may produce primary skin irritation as a result of defatting. Aspiration of liquid may cause extensive pulmonary injury. Because of its low surface tension, kerosene may spread over a large area, causing pulmonary hemorrhage and chemical pneumonitis. Kerosene mist may also cause mucous membrane irritation.

Systemic—Inhalation of high concentrations may cause headache, nausea, confusion, drowsiness, convulsions, and coma. When kerosene is ingested, it may cause nausea, vomiting, and, in severe cases, drowsiness progressing to coma, and death by hemorrhagic pulmonary edema and renal involvement.

Medical Surveillance: No specific considerations are needed.

Special Tests: None in use.

Personal Protective Methods: Barrier creams, gloves, and protective clothing are recommended. Where workers are exposed to vapors, masks are recommended.

Bibliography

El-Habashi, A., Fahim, A., and Kamel, A. 1969. Toxi-pathological studies on kerosene. *J. Egypt. Med. Assoc.* 52:421.
Richardson, J.A., and Pratt-Thomas, H.R. 1951. Toxic effects of varying doses of kerosene administered by different routes. *Am. J. Med. Sci.* 221:531.

KETONES

Description

The ketone family includes:

$$\text{Acetone: } CH_3COCH_3$$
$$\text{Diacetone: } (CH_3)_2COHCH_2COCH_3$$
$$\text{Methyl ethyl ketone: } CH_3COCH_2CH_3$$
$$\text{Methyl n-propyl ketone: } CH_3CH_2CH_2COCH_3$$
$$\text{Methyl n-butyl ketone: } CH_3COCH_2CH_2CH_2CH_3$$
$$\text{Methyl isobutyl ketone: } (CH_3)_2CHCH_2COCH_3$$

Synonyms

Acetone: 2-Propanone, dimethyl ketone, β-ketopropane, pyro-acetic ether.

Diacetone: Diacetone alcohol, diacetonyl alcohol, dimethylace-
tonyl carbinol, 4-hydroxy-4-methyl-2-pentanone, 2-methyl-2-pen-
tanol-4-one.
Methyl ethyl ketone: Butanone, 2-butanone, MEK, ethyl methyl ketone.
Methyl n-propyl ketone: Ethyl acetone, 2-pentanone, MPK.
Methyl n-butyl ketone: n-Butyl methyl ketone, propyl acetone, 2-
hexanone, MBK.
Methyl isobutyl ketone: Hexone, isopropyl acetone, 4-methyl-2-
pentanone, MIBK.

Potential Occupational Exposures: This group of ketone solvents has many uses in common. They are low-cost solvents for resins, lacquers, oils, fats, collodion, cotton, cellulose acetate, nitrocellulose, cellulose esters, epoxy resins, gums, pigments, dyes, vinyl polymers, and copolymers. They are used as chemical intermediates, in the manufacture of smokeless powder and explosives, and in the paint, lacquer, varnish, plastics, dyeing, celluloid, photographic, cement, rubber, artificial silk and leather, synthetic rubber, and lubricating oil industries. They are also used in hydraulic fluids, metal cleaning compounds, quick drying inks, airplane dopes, compositions for paper and textiles, pharmaceuticals, cosmetics, and as paint removers and dewaxers.

A partial list of occupations in which exposure may occur includes:

Adhesive makers	Lacquer and oil processors
Celluloid makers	Shoemakers
Dope workers	Solvent workers
Dye makers	Varnish and stain makers
Explosive makers	Wax makers
Garage mechanics	

NIOSH (1) has estimated the following worker exposures to various ketones:

Acetone	2,816,000
Methyl ethyl ketone	3,031,000
Methyl n-propyl ketone	<500
Methyl n-butyl ketone	222,000
Methyl n-amyl ketone	67,000
Methyl isobutyl ketone	1,853,000
Methyl isoamyl ketone	19,000
Diisobutyl ketone	NA*
Cyclohexanone	1,190,000
Mesityl oxide	<500
Diacetone alcohol	1,350,000
Isophorone	1,507,000

*Not Available

Permissible Exposure Limits

The Federal standards are:

Acetone	1,000 ppm	2,400 mg/m^3
Diacetone alcohol	50 ppm	240 mg/m^3
Methyl ethyl ketone	200 ppm	590 mg/m^3
Methyl n-propyl ketone	200 ppm	700 mg/m^3
Methyl n-butyl ketone	25 ppm	100 mg/m^3
Methyl isobutyl ketone	100 ppm	410 mg/m^3

NIOSH (1) has specified that occupational exposure to ketones shall be controlled so that employees are not exposed at concentrations greater than the limits, in milligrams per cubic meter (mg/m^3) of air, in the following table as TWA concentrations for up to a 10-hour workshift, 40-hour workweek.

Recommended Exposure Limits for the Ketones

Ketone	TWA Concentration Limits	
	mg/m^3	Approximate ppm equivalents
Acetone	590	250
Methyl ethyl ketone	590	200
Methyl n-propyl ketone	530	150
Methyl n-butyl ketone	4	1
Methyl n-amyl ketone	465	100
Methyl isobutyl ketone	200	50
Methyl isoamyl ketone	230	50
Diisobutyl ketone	140	25
Cyclohexanone	100	25
Mesityl oxide	40	10
Diacetone alcohol	240	50
Isophorone	23	4

Routes of Entry: Inhalation of vapor, percutaneous absorption.

Harmful Effects

Local—These solvents may produce a dry, scaly, and fissured dermatitis after repeated exposure. High vapor concentrations may irritate the conjunctiva and mucous membranes of the nose and throat, producing eye and throat symptoms.

Systemic—In high concentrations, narcosis is produced, with symptoms of headache, nausea, lightheadedness, vomiting, dizziness, incoordination, and unconsciousness.

Recent reports indicate that exposure of workers to methyl n-butyl ketone has been associated with the development of peripheral neuropathy. Rat experiments have also shown nerve changes characteristic of peripheral neuropathy after exposure to 1,300 ppm.

Extensive data and a large bibliography (186 references) are available in a NIOSH report (1).

Medical Surveillance: Preplacement examinations should evaluate skin and respiratory conditions. In the case of methyl n-butyl ketone, special attention should be given to the central and peripheral nervous systems.

Special Tests: Acetone can be determined in the blood, urine, and expired air, and has been used as an index of exposure. A neurotoxic metabolite of methyl n-butyl ketone has been found in the rat. This is not present with methyl isobutyl ketone.

Personal Protective Methods: Contact with skin and eyes should be avoided by the use of protective clothing. In areas of high vapor concentration, masks should be used.

The employer shall provide respirators in accordance with the following table and shall ensure that the employees use the respirators properly when they are required. Respiratory protective devices shall be those approved by NIOSH and the Mine Safety and Health Administration.

Respirator Selection Guide for Ketones*

| | Multiples of TWA Concentration Limit | | | | | |
| | . Less Than or Equal to . | | | Greater Than | Emergency | Fire- |
Ketone	10X	50X	100X	100X	Entry	Fighting
Acetone	CDFHIJ	KL	KL	KL	KL	CK

(continued)

Multiples of TWA
..... Concentration Limit .....

| Ketone | . Less Than or Equal to . | | | Greater Than | Emergency | Fire- |
	10X	50X	100X	100X	Entry	Fighting
Methyl n-butyl ketone	ABEG	CDFJ	HI	KL	KL	K
Methyl ethyl and methyl n-propyl ketone	B	CDFJHI	KL	KL	KL	K
Other eight ketones**	B	CDFJ	HI	KL	KL	K

*See Respirator Code below.

**Methyl n-amyl ketone, methyl isobutyl ketone, methyl isoamyl ketone, diisobutyl ketone, cyclohexanone, mesityl oxide, diacetone alcohol, or isophorone.

Respirator Code

Code	Respirator Type Approved under Provisions of 30 CFR 11
A	Chemical cartridge respirator with half-mask facepiece and organic vapor cartridge
B	Chemical cartridge respirator with full facepiece and organic vapor cartridge
C	Gas mask with full facepiece and chin-type organic vapor canister
D	Gas mask with full facepiece and back- or front-mounted organic vapor canister
E	Type C supplied-air respirator, with half-mask facepiece, operated in demand (negative pressure) mode
F	Type C supplied-air respirator, with full facepiece, operated in demand (negative pressure) mode
G	Type C supplied-air respirator, with half-mask facepiece, operated in continuous-flow or pressure-demand mode
H	Type C supplied-air respirator, with full facepiece, operated in continuous-flow or pressure-demand mode
I	Type C supplied-air respirator with hood, helmet, or suit
J	Self-contained breathing apparatus, with full facepiece, operated in demand (negative pressure) mode
K	Self-contained breathing apparatus, with full facepiece, operated in pressure-demand or other positive pressure mode
L	Combination Type C supplied-air respirator, with full facepiece, operated in pressure-demand mode and equipped with auxiliary positive pressure self-contained air supply

References

(1) National Institute for Occupational Safety and Health. *Criteria for a Recommended Standard: Occupational Exposure to Ketones.* NIOSH Doc. No. not yet assigned, Wash., D.C. (In press—1978.)

Bibliography

Allen, N., Mendell, J.R., Billmaier, D.J., Fontaine, R.E., and O'Neill, J. 1975. Toxic poly-neuropathy due to methyl n-butyl ketone: An industrial outbreak. *Arch. Neurol.* 32:209.

Henson, E.V. 1959. Toxicology of some aliphatic ketones. *J. Occup. Med.* 1:607.

Mallor, J.S. 1976. MBK neuropathy among spray painters. *J. Amer. Med. Assoc.* 235:1455.

Mendell, J.R., Saida, K., Ganansia, M.F., Jackson, D.B., Weiss, H., Gardier, R.W., Chris-man, C., Allen, N., Couri, D., O'Neill, J., Marks, B., and Hetland, L. 1974. Toxic poly-neuropathy produced by methyl n-butyl ketone. *Science* 185:787.

L

LEAD—ALKYL

Description: Both tetraethyl and tetramethyl lead are colorless liquids; however, they are generally mixed with dyes to identify them. TEL is insoluble in water, but soluble in organic solvents. TML is only slightly soluble in organic solvents. Tetraethyllead will decompose in bright sunlight yielding needle-like crystals of tri-, di-, and monoethyllead compounds, which have a garlic odor.

Synonyms: Tetraethyllead: TEL. Tetramethyllead: TML.

Potential Occupational Exposures: TEL and TML are used singly or together as "antiknock" ingredients in gasoline. Exposure may occur during synthesis, handling, transport, or mixing with gasoline.

A partial list of occupations in which exposure may occur includes:

Gasoline additive workers   Storage tank cleaners

Permissible Exposure Limits: The Federal standard for tetraethyllead is 0.075 mg Pb/m^3 and for tetramethyllead 0.07 mg Pb/m^3. ACGIH (1978) sets a TWA value of 0.100 mg/m^3.

Routes of Entry: Inhalation of vapor and percutaneous absorption of liquid. TML is more volatile than TEL and therefore may present more of an inhalation hazard. If the tri-, di-, and monoethyllead compounds are dried, the dust may be inhaled producing the same symptomatology as TEL.

Harmful Effects

Local—Liquid alkyllead may penetrate the skin without producing appreciable local injury. However, the decomposition products of TEL (i.e., mono-, di-, triethyllead compounds) in dust form may be inhaled and result in irritation of the upper respiratory tract and possibly paroxysmal sneezing. This dust, when in contact with moist skin or ocular membranes, may cause itching, burning, and transient redness. TEL itself may be irritating to the eyes.

Systemic—The absorption of a sufficient quantity of tetraethyllead, whether

279

briefly at a high rate, or for prolonged periods at a lower rate, may cause acute intoxication of the central nervous system. Mild degrees of intoxication cause headache, anxiety, insomnia, nervous excitation, and minor gastrointestinal symptoms with a metallic taste in the mouth.

The most noticeable clinical sign of tetraethyllead poisoning is encephalopathy which may give rise to a variety of symptoms, which include mild anxiety, toxic delirium with hallucinations, delusions, convulsions, and acute toxic psychosis. Physical signs are not prominent; but bradycardia, hypotension, increased reflexes, tremor, and slight weight loss have been reported. No peripheral neuropathy has been observed. When the interval between the termination of (either brief or prolonged) exposure and the onset of symptoms is delayed (up to 8 days) the prognosis is guardedly hopeful, but when the interval is short (few hours), an early fatal outcome may result. Recovered patients show no residual damage to the nervous system, although recovery may be prolonged.

Diagnosis depends on developing a history of exposure to organic lead compounds, followed by the onset of encephalopathy. Biochemical measurements are helpful but not diagnostic. Blood lead is usually not elevated in proportion to the degree of intoxication. Urine aminolevulinic acid, and coproporphyrin excretion will show values close to normal with no correlation with the severity of intoxication. Erythrocyte protoporphyrin also remains within normal range.

No cases of poisoning from absorption of tetramethyllead have been found. The compound responsible for almost all cases of organic lead poisoning is tetraethyllead. Animal experimentation, however, indicates that a similar intoxication can be caused by tetramethyllead.

Medical Surveillance: In both preemployment and periodic physical examinations, the worker's general health should be evaluated, and special attention should be given to neurologic and emotional disorders.

Special Tests: None seem to be useful.

Personal Protective Methods: A training program should stress the importance of personal hygiene and encourage the proper use of personal protective equipment. Showers, lavatories, and locker rooms are necessary.

Workers should be required to make a complete change of clothing at the beginning and end of each shift and to shower prior to changing to street clothes. Eating should not be permitted in work areas. In areas where vapor concentrations of TEL exceed the standard, dust masks, organic vapor canister masks, or supplied air respirators should be furnished and required to be worn. In areas of spills or splash, impervious clothing should be worn and goggles furnished.

Bibliography

Beattie, A.D., M.R. Moore, and A. Goldberg. 1972. Tetraethyllead poisoning. *Lancet* 2:12.

Commission of the European Communities, Directorate General for Dissemination of Knowledge, Centre for Information and Documentation. 1973. *Proceedings of the International Symposium on the Environmental Health Aspects of Lead, Amsterdam.* Centre for Information and Documentation, Luxembourg.

de Treville, R.T.P., H.W. Wheeler, and T. Sterling. 1962. Occupational exposure to organic lead compounds—The relative degree of hazard in occupational exposure to airborne tetraethyllead and tetramethyllead. *Arch. Environ. Health* 5:532.

Kehoe, R.A., J. Cholak, J.A. Spence, and W. Hancock. 1963. Potential hazard of exposure to lead—I. Handling and use of gasoline containing tetramethyllead. *Arch. Environ. Health* 6:239.

Kehoe, R.A., J. Cholak, J.G. McIlhinney, G.A. Lofquist, and T.D. Sterling. 1963. Potential hazard of exposure to lead—II. Further investigations in the preparation, handling, and use of gasoline containing tetramethyllead. *Arch. Environ. Health* 6:255.

Olsen, E.D. and P.I. Jatlow. 1972. An improved delves cup atomic absorption procedure for determination of lead in blood and urine. *Clin. Chem.* 18:1312.

LEAD—INORGANIC

Description: Pb, inorganic lead, includes lead oxides, metallic lead, lead salts, and organic salts such as lead soaps, but excludes lead arsenate and organic lead compounds. Lead is a blue-grey metal which is very soft and malleable. Commercially important lead ores are galena, cerussite, anglesite, crocoisite, wulfenite, pyromorphite, matlockite, and vanadinite. Lead is slightly soluble in water in presence of nitrates, ammonium salts, and carbon dioxide.

Synonyms: None.

Potential Occupational Exposures: Metallic lead is used for lining tanks, piping, and other equipment where pliability and corrosion resistance are required such as in the chemical industry in handling corrosive gases and liquids used in the manufacture of sulfuric acid; in petroleum refining; and in halogenation, sulfonation, extraction, and condensation processes; and in the building industry.

It is also used as an ingredient in solder, a filler in the automobile industry, and a shielding material for x-rays and atomic radiation; in manufacture of tetraethyllead and organic and inorganic lead compounds, pigments for paints and varnishes, storage batteries, flint glass, vitreous enameling, ceramics as a glaze, litharge rubber, plastics, and electronic devices. Lead is utilized in metallurgy and may be added to bronze, brass, steel, and other alloys to improve their characteristics. It forms alloys with antimony, tin, copper, etc. It is also used in metallizing to provide protective coatings and as a heat treatment bath in wire drawing.

Exposures to lead dust may occur during mining, smelting, and refining, and to fume, during high temperature (above 500°C) operations such as welding or spray coating of metals with molten lead. There are numerous applications for lead compounds, some of the more common being in the plates of electric batteries and accumulators, as compounding agents in rubber manufacture, as ingredients in paints, glazes, enamels, glass, pigments, and in the chemical industry.

A partial list of occupations in which exposure may occur includes:

Battery makers	Insecticide workers
Brass founders	Lubricant makers
Ceramic makers	Match makers
Enamel workers	Painters
Glass makers	Plumbers
Imitation pearl makers	Solderers

It is estimated that approximately 783,000 industrial workers are potentially exposed to lead products.

Permissible Exposure Limits: The Federal standard for lead and its inorganic compounds was 0.2 mg/m^3 as a time-weighted average. The NIOSH Criteria Document recommends a time-weighted average value of 0.15 mg Pb/m^3. On November 14, 1978, OSHA set a final standard in which industries will be given 1 to 3 years to reach a 0.1 mg (100 μg)/m^3 level and from 1 to 10 years to reach a final standard of 0.05 mg (50μg)/m^3.

Routes of Entry: Ingestion of dust; inhalation of dust or fume.

Harmful Effects

Local—None.

Systemic—The early effects of lead poisoning are nonspecific and, except by laboratory testing, are difficult to distinguish from the symptoms of minor seasonal illnesses. The symptoms are decreased physical fitness, fatigue, sleep disturbance, headache, aching bones and muscles, digestive symptoms (particularly constipation), abdominal pains, and decreased appetite. These symptoms are reversible and complete recovery is possible.

Later findings include anemia, pallor, a "lead line" on the gums, and decreased hand-grip strength. Lead colic produces an intense periodic abdominal cramping associated with severe constipation and, occasionally, nausea and vomiting. Alcohol ingestion and physical exertion may precipitate these symptoms. The peripheral nerve affected most frequently is the radial nerve.

This will occur only with exposure over an extended period of time and causes "wrist drop." Recovery is slow and not always complete. When the central nervous system is affected, it is usually due to the ingestion or inhalation of large amounts of lead. This results in severe headache, convulsions, coma, delirium, and possibly death. The kidneys can also be damaged after long periods of exposure to lead, with loss of kidney function and progressive azotemia.

Because of more efficient material handling methods and biological monitoring, serious cases of lead poisoning are rare in industry today.

Medical Surveillance: In preemployment physical examinations, special attention is given to neurologic and renal disease and baseline blood lead levels. Periodic physical examinations should include hemoglobin determinations, tests for blood lead levels, and evaluation of any gastrointestinal or neurologic symptoms. Renal function should be evaluated.

Special Tests: Periodic evaluation of blood lead levels are widely used as an indicator of increased or excessive lead absorption. Other indicators are blood and urine coproporphyric III and delta amino low valence acid dehydrase (ALAD). Erythrocytic protoporphyrin determinations may also be helpful.

Personal Protective Methods: Workers should be supplied with full body work clothing and caps (hard hats). The dust should be removed (vacuumed) before leaving after the shift. Showering after each shift prior to changing to street clothes should be encouraged. Dust and fume masks or supplied air respirators should be supplied to all employees exposed to concentrations above the TWA standard and in all emergencies. Food should not be eaten in contaminated areas.

Bibliography

National Institute for Occupational Safety and Health. *Criteria for a Recommended Standard: Occupational Exposure to Inorganic Lead.* NIOSH Doc. No. 73-11010, Wash., DC (1973).
OSHA. Occupational exposure to lead: Final standard. *Federal Register* 43 No. 220, 52952-53014 (Nov. 14, 1978).

M

MAGNESIUM AND COMPOUNDS

Description: Magnesium is a light, silvery-white metal and is a fire hazard. It is found in dolomite, magnesite, brucite, periclase, carnallite, kieserite and as a silicate in asbestos, talc, olivine, and serpentine. It is also found in sea water, brine wells, and salt deposits. It is insoluble in water and ordinary solvents.

Synonyms: None.

Potential Occupational Exposures: Magnesium alloyed with manganese, aluminum, thorium, zinc, cerium, and zirconium is used in aircraft, ships, automobiles, hand tools, etc., because of its lightness. Dow metal is the general name for a large group of alloys containing over 85% magnesium. Magnesium wire and ribbon are used for degassing valves in the radio industry and in various heating appliances; as a deoxidizer and desulfurizer in copper, brass, and nickel alloys; in chemical reagents; as the powder in the manufacture of flares, incendiary bombs, tracer bullets, and flashlight powders; in the nuclear energy process; and in a cement of magnesium oxide and magnesium chloride for floors.

A partial list of occupations in which exposure may occur includes:

Alloy makers	Organic chemical synthesizers
Antiseptic makers	Pigment makers
Battery makers	Steel makers
Drug makers	Textile workers
Flare makers	Welders
Fungicide makers	

Permissible Exposure Limits: The Federal standard for magnesium oxide fume is 15 mg/m^3, according to NIOSH. ACGIH (1978) sets 10 mg/m^3 as a TWA.

Route of Entry: Inhalation of fume.

Harmful Effects

Local—Magnesium and magnesium compounds are mild irritants to the conjunctiva and nasal mucosa, but are not specifically toxic. Magnesium in finely

284

divided form is readily ignited by a spark or flame, and splatters and burns at above 2300°F. On the skin, these hot particles are capable of producing second and third degree burns, but they respond to treatment as other thermal burns do. Metallic magnesium foreign bodies in the skin cause no unusual problems in man. In animal experiments, however, they have caused "gas gangrene"—massive localized gaseous tumors with extensive necrosis.

Systemic—Magnesium in the form of nascent magnesium oxide can cause metal fume fever if inhaled in sufficient quantity. Symptoms are analogous to those caused by zinc oxide: cough, oppression in the chest, fever, and leukocytosis. There is no evidence that inhalation of magnesium dust has led to lung injury. It has been noted that magnesium workers show a rise in serum magnesium—although no significant symptoms of ill health have been identified. Some investigators have reported higher incidence of digestive disorders and have related this to magnesium absorption, but the evidence is scant. In foundry casting operations, hazards exist from the use of fluoride fluxes and sulfur-containing inhibitors which produce fumes of fluorides and sulfur dioxide.

Medical Surveillance: No specific recommendations.

Special Tests: None.

Personal Protective Methods: Employees should receive training in the use of personal protective equipment, proper methods of ventilation, and fire suppression. Protective clothing should be designed to prevent burns from splatters. Masks to prevent inhalation of fumes may be necessary under certain conditions, but generally this can be controlled by proper ventilation. Dust masks may be necessary in areas of dust concentration as in transfer and storage areas, but adequate ventilation generally provides sufficient protection.

Bibliography

Drinker, K., and Thomson, R.M. 1927. Metal fume fever—II. The effects of inhaling magnesium oxide fume. *J. Ind. Hyg. Toxicol.* 9:187.
Drinker, K.R., and Drinker, P. 1928. Metal fume fever—V. Results of inhalation by animal of zinc and magnesium oxide fume. *J. Ind. Hyg. Toxicol.* 10:56.

MALATHION

Description: Malathion, O,O-dimethyl-S-(1,2-dicarboethoxyethyl) dithiophosphate, $C_{10}H_{19}O_6PS_2$,

$$(CH_3O)_2P-S-CHCOOC_2H_5$$
$$\underset{S}{\overset{\parallel}{}} \quad \underset{CH_2COOC_2H_5}{\overset{\mid}{}}$$

It is a colorless, light to amber liquid.

Synonyms: None.

Potential Occupational Exposures: In 1971, the annual production of malathion in the United States totaled about 35 million pounds.

Malathion is marketed as 99.6% technical grade liquid. Available formulations include wettable powders (25% and 50%), emulsifiable concentrates, dusts, and aerosols.

Malathion is used in the control of certain insect pests on fruits, vegetables, and ornamental plants. It has been used in the control of houseflies, mosquitoes, and lice, and on various farm and livestock animals.

NIOSH estimates that approximately 75,000 workers in the United States are occupationally exposed to malathion, in the course of manufacture, formulation and application.

Permissible Exposure Limits: When skin exposure is prevented, exposure to malathion in the workplace shall be controlled so that employees are not exposed to malathion at a TWA concentration greater than 15 mg/m^3 of air for up to a 10-hour work shift, 40-hour workweek, according to NIOSH (1).

A skin exposure limit of 10 mg/m^3 has been set by the American Conference of Governmental Industrial Hygienists as of 1978.

Routes of Entry: Inhalation of vapor or skin exposure.

Harmful Effects

As described and documented in detail in an NIOSH document (1), the main signs and symptoms of malathion intoxication are increased bronchial secretion and excessive salivation, nausea, vomiting, excessive sweating, miosis, and muscular weakness and fasciculations. These signs and symptoms are induced by the inhibition of functional acetylcholine esterase (AChE) in the nervous system.

Medical Surveillance

1) Preplacement and periodic medical examinations shall include:
   (a) Comprehensive initial or interim medical and work histories.
   (b) A physical examination which shall be directed toward, but not limited to, evidence of frequent headache, dizziness, nausea, tightness of the chest, dimness of vision, and difficulty in focusing the eyes.
   (c) Determination, at the time of the preplacement examination, of a baseline or working baseline erythrocyte ChE activity.
   (d) A judgement of the worker's physical ability to use negative or positive pressure regulators as defined in 29 CFR 1910.134.

2) Periodic examinations shall be made available on an annual basis or at some other interval determined by the responsible physician.

3) Medical records shall be maintained for all workers engaged in the manufacture or formulation of malathion and such records shall be kept for at least one year after termination of employment.

4) Pertinent medical information shall be available to medical representatives of the U.S. Government, the employer and the employee.

Special Tests: Erythrocyte cholinesterase levels as noted above and as described in detail by NIOSH (1).

Personal Protective Methods

(a) *Protective Clothing*
Any employee whose work involves likely exposure of the skin to malathion or malathion formulations, e.g., mixing or formulating, shall wear full-body coveralls or the equivalent, impervious gloves, and impervious footwear and, when there is danger of malathion coming in contact with the eyes, safety goggles shall be provided and worn. Any employee who applies malathion shall be provided with and required to wear the following protective clothing and equipment: goggles, whole-body coveralls, and impervious footwear.

(b) *Respiratory Protection*
Engineering controls shall be used wherever feasible to maintain airborne malathion concentrations below the recommended workplace environmental limit. Compliance with the workplace environmental limit by the use of respirators is allowed only when airborne malathion concentrations are in excess of the workplace environmental limit because required engineering controls are being installed or tested, when nonroutine maintenance or repair is being accomplished, or during emergencies. When a respirator is thus permitted, it shall be selected and used in accordance with NIOSH requirements (1).

References

(1) National Institute for Occupational Safety and Health. *Criteria for a Recommended Standard: Occupational Exposure to Malathion.* NIOSH Doc. No. 76-205, Wash., D.C. (June 1976).

MANGANESE AND COMPOUNDS

Description: Mn, manganese, is a reddish-grey or silvery, soft metal. The most important ore containing manganese is pyrolusite. Manganese may also be produced from ferrous scrap used in the production of electric and open-hearth steel. Manganese decomposes in water and is soluble in dilute acid.

Synonyms: None.

Potential Occupational Exposures: Most of the manganese produced is used in the iron and steel industry in steel alloys, e.g., ferromanganese, silicomanganese, Manganin, spiegeleisen, and as an agent to reduce oxygen and sulfur content of molten steel. Other alloys may be formed with copper, zinc, and aluminum. Manganese and its compounds are utilized in the manufacture of dry cell batteries (MnO_2), paints, varnishes, inks, dyes, matches and fireworks, as a fertilizer, disinfectant, bleaching agent, laboratory reagent, drier for oils, an oxidizing agent in the chemical industry particularly in the synthesis of potassium permanganate, and as a decolorizer and coloring agent in the glass and ceramics industry.

Organomanganese compounds such as methylcyclopentadienyl manganese tricarbonyl (MMT) have been proposed as supplements and/or replacements for tetraethyllead (TEL) as an antiknock in gasoline.

Exposure may occur during the mining, smelting and refining of manganese, in the production of various materials, and in welding operations with manganese coated rods.

A partial list of occupations in which exposure may occur includes:

Battery makers	Glass makers
Ceramic makers	Ink makers
Drug makers	Paint makers
Electric arc welders	Varnish makers
Foundry workers	Water treaters

Permissible Exposure Limits: The Federal standard for manganese is 5 mg/m^3 as a ceiling value. The Illinois Environmental Health Resource Center recommends an environmental standard for particulate manganese of 0.006 $\mu g/m^3$. ACGIH (1978) has set 0.2 mg/m^3 as a TWA for MMT and 1.0 for manganese tetroxide.

The Federal government in Sept. 1978 affirmed a ban on MMT use in the United States, not because of any toxicity of MMT but because MMT hinders catalytic converter operation and causes increased air pollution.

Routes of Entry: Inhalation of dust or fume; limited percutaneous absorption of liquids.

Harmful Effects

Local—Manganese dust and fumes are only minor irritants to the eyes and mucous membranes of the respiratory tract, and apparently are completely innocuous to the intact skin.

Systemic—Chronic manganese poisoning has long been recognized as a clinical entity. The dust or fumes (manganous compounds) enter the respiratory tract and are absorbed into the blood stream. Manganese is then deposited in major body organs with a special predilection for the liver, spleen, and certain nerve cells of the brain and spinal cord. Among workers there is a very marked variation in individual susceptibility to manganese. Some workers have worked in heavy exposure for a lifetime and have shown no signs of the disease; others have developed manganese intoxication with as little as 49 days of exposure.

The early phase of chronic manganese poisoning is most difficult to recognize, but it is also important to recognize since early removal from the exposure may arrest the course of the disease. The onset is insidious, with apathy, anorexia, and asthenia. Headache, hypersomnia, spasms, weakness of the legs, arthralgias, and irritability are frequently noted. Manganese psychosis follows with certain definitive features: unaccountable laughter, euphoria, impulsive acts, absentmindedness, mental confusion, aggressiveness, and hallucinations. These symptoms usually disappear with the onset of true neurological disturbances, or may resolve completely with removal from manganese exposure.

Progression of the disease presents a range of neurological manifestations that can vary widely among individuals affected. Speech disturbances are common: monotonous tone, inability to speak above a whisper, difficult articulation, incoherence, even complete muteness. The face may take on masklike quality, and handwriting may be affected by micrographia. Disturbances in gait and balance occur, and frequently propulsion, retropropulsion, and lateropropulsion are affected, with no movement for protection when falling. Tremors are frequent, particularly of the tongue, arms, and legs. These will increase with intentional movements and are more frequent at night. Absolute detachment, broken by sporadic or spasmodic laughter, ensues, and as in extrapyramidal affections, there may be excessive salivation and excessive sweating. At this point the disease is indistinguishable from classical Parkinson's disease.

Chronic manganese poisoning is not a fatal disease although it is extremely disabling.

Manganese dust is no longer believed to be a causative factor in pneumonia. If there is any relationship at all, it appears to be as an aggravating factor to a preexisting condition. Freshly formed fumes have been reported to cause fever and chills similar to metal fume fever.

Medical Surveillance: Preemployment physical exams should be directed toward the individual's general health with special attention to neurologic and personality abnormalities. Periodic physical examinations may be required as often as every two months. Special emphasis should be given to behavioral and neurological changes: speech defects, emotional disturbances, hypertonia, tremor, equilibrium, difficulty in walking or squatting, adiadochokinesis, and handwriting.

Special Tests: There are no laboratory tests which can be used to diagnose manganese poisoning.

Personal Protective Methods: In areas where the ceiling value standards are exceeded, dust masks or respirators are necessary. Education in the use and necessity of these devices is important.

Bibliography

Cook, D.G., Fahn, S., and Brait, K.A. 1974. Chronic manganese intoxication. *Arch. Neurol.* 30:59.
Illinois Institute for Environmental Quality. *Airborne Manganese Health Effects and Recommended Standard.* Doc. No. 75-18, Chicago, Ill. (Sept. 1975).
National Academy of Sciences. *Manganese.* (In a series on medical and biologic effects of environmental pollutants), Wash., D.C. (1973).
Rodier, J. 1955. Manganese poisoning in Moroccan miners. *Br. J. Ind. Med.* 12:21.
Smyth, L.T., Ruhf, R.C., Whitman, N.E., and Dugan, T. 1973. Clinical manganism and exposure to manganese in the production and processing of ferromanganese alloy. *J. Occup. Med.* 15:101.

MERCAPTANS

Description: Methyl mercaptan: CH_3SH; ethyl mercaptan: CH_3CH_2SH; n-butyl mercaptan: $CH_3(CH_2)_2CH_2SH$; and perchloromethyl mercaptan: CCl_3SCl.

These compounds are typically flammable liquids except methyl mercaptan which is a gas. Perchloromethyl mercaptan is yellow; the rest are colorless. A strong unpleasant odor is the most characteristic property of mercaptans and may be detected at very low levels, i.e., less than 0.5 ppm. Perchloromethyl mercaptan is insoluble in water, but others are slightly soluble.

Synonyms

Methyl mercaptans: Methanethiol, mercaptomethane, thiomethyl alcohol, methyl sulfhydrate.

Ethyl mercaptan: Ethanethiol, mercaptoethane, ethyl sulfhydrate, thioethyl alochol.

n-Butyl mercaptan: 1-Butanethiol, n-butyl thioalcohol, thiobutyl alcohol.

Potential Occupational Exposures: In general, mercaptans find use as intermediates in the manufacture of pesticides, fumigants, dyes, pharmaceuticals, and other chemicals, and as gas odorants, i.e, to serve as a warning property for hazardous odorless gases. Particular usages for methyl mercaptan include the synthesis of methionine and the manufacture of fungicides and jet fuels. Ethyl mercaptan is used as an adhesive stabilizer and butyl mercaptan may be used as a solvent.

A partial list of occupations in which exposure may occur includes:

Drug makers	Methionine makers
Dye makers	Organic chemical synthesizers
Fumigant makers	Pesticide makers
Fumigators	Warning agent workers
Jet fuel blenders	

Permissible Exposure Limits: The Federal standard for each mercaptan is:

Methyl mercaptan	10 ppm	20 mg/m^3
Ethyl mercaptan	10 ppm	25 mg/m^3
Butyl mercaptan	10 ppm	35 mg/m^3
Perchloromethyl		
mercaptan	0.1 ppm	0.8 mg/m^3

The above standards are determined as TWAs, except ethyl and methyl mercaptan which are ceiling values. ACGIH has lowered the TLVs of all but perchloromethyl mercaptan: methyl mercaptan, 0.5 ppm (1 mg/m^3); ethyl mercaptan, 0.5 ppm (1.0 mg/m^3); butyl mercaptan, 0.5 ppm (1.5 mg/m^3); all TWAs.

Route of Entry: Inhalation of gas or vapor.

Harmful Effects

Local—Mercaptans have an intensely disagreeable odor and are irritating to skin, eyes, and mucous membranes of the upper respiratory tract. Liquid may cause contact dermatitis and vapor may cause irritation to nose and throat. Perchloromethyl mercaptan is stronger in its irritant ability than the other mercaptans which cause only slight to moderate irritation.

Systemic—Methyl mercaptan acts toxicologically like hydrogen sulfide and may depress the central nervous system resulting in respiratory paralysis and death. Victims who survive severe exposures may suffer from headache, dizziness, staggering gait, nausea, and vomiting. Respiratory tract irritation may lead to pulmonary edema and possibly renal and hepatic damage. The above effects are based primarily on animal experimentation. In a recent case of acute methyl mercaptan exposure, a worker developed acute anemia and methemoglobinemia 24 hours following coma.

Medical Surveillance: Preplacement and periodic medical examinations should consider skin, eyes, lung and central nervous system as well as liver and kidney functions. Blood studies may be helpful in following acute intoxication from methyl mercaptan.

Special Tests: None commonly used.

Personal Protective Methods: In areas where liquid mercaptan is likely to be spilled or splashed on the skin, impervious clothing, gloves, gauntlets, aprons, and boots should be supplied. Otherwise protective methods are as for sulfur dioxide. (See Sulfur Dioxide.)

Bibliography

Blinova, E.A. 1965. O normirovanii konstentratsii veshchestvs s sil'nym zapakhom v vozdukhe proizvodstvennykh pom eshchenit. *Gig. Sanit.* 30:18.

Fairchild, E.J., Stokinger, H.E. 1958. Toxicologic studies on organic sulfur compounds. 1. Acute toxicity of some aliphatic and aromatic thiols (mercaptans). *Am. Ind. Hyg. Assoc. J.* 19:171.

Gobbato, F., and Terribile, P.M. 1968. Toxicologic properties of mercaptans. *Folia Medica (Napoli)* 51:329.

Schults, W.T., Fountain, E.N., and Lynch, E.C. 1970. Methanethiol poisoning. *J. Am. Med. Assoc.* 211:2153.

MERCURY—ALKYL

Description: Methyl mercury compounds: methyl mercury dicyandiamide—$CH_3HgNHC(NH)NHCN$. Soluble in water.

Ethyl mercury compounds: ethylmercuric chloride: C_2H_5HgCl. Insoluble in water. Ethylmercuric phosphate: $(C_2H_5HgO)_3PO$. Soluble in water. N-(Ethylmercuric)-p-toluenesulfonanilide: $C_6H_5N(HgC_2H_5)SO_2C_6H_4CH_3$. Practically insoluble in water.

Synonyms

Methyl mercury compounds
   Methyl mercury dicyandiamide: cyano (methyl mercury) guanidine, Panogen.

Ethyl Mercury compounds
   Ethylmercuric chloride: Ceresan.
   Ethylmercuric phosphate: New Ceresan.
   N-(Ethylmercuric)-p-toluenesulfonanilide: Ceresan M.

Potential Occupational Exposures: These compounds are used in treating seeds for fungi and seedborne diseases, as timber preservatives, and disinfectants.

A partial list of occupations in which exposure may occur includes:

Disinfectant makers   Seed handlers
Fungicide makers   Wood preservers

Permissible Exposure Limits: The Federal standard is 0.01 mg/m^3 as an 8-hour TWA with an acceptable ceiling of 0.04 mg/m^3.

Routes of Entry: Inhalation of dust, percutaneous absorption.

Harmful Effects

Local—Alkyl mercury compounds are primary skin irritants and may cause dermatitis. When deposited on the skin, they give no warning, and if contact is maintained, can cause second-degree burns. Sensitization may occur.

Systemic—The central nervous system, including the brain, is the principal target tissue for this group of toxic compounds. Severe poisoning may produce irreversible brain damage resulting in loss of higher functions.

The effects of chronic poisoning with alkyl mercury compounds are progressive. In the early stages, there are fine tremors of the hands, and in some cases, of the face and arms. With continued exposure, tremors may become coarse and convulsive; scanning speech with moderate slurring and difficulty in pronunciation may also occur. The worker may then develop an unsteady gait of a spastic nature which can progress to severe ataxia of the arms and legs. Sensory disturbances including tunnel vision, blindness, and deafness are also common.

A later symptom, constriction of the visual fields, is rarely reversible and may be associated with loss of understanding and reason which makes the victim completely out of touch with his environment. Severe cerebral effects have been seen in infants born to mothers who had eaten large amounts of methyl mercury contamined fish.

Medical Surveillance: Preplacement and periodic physical examinations should be concerned particularly with the skin, vision, central nervous system, and kidneys. Consideration should be given to the possible effects on the fetus of alkyl mercury exposure in the mother. Constriction of visual fields may be a useful diagnostic sign. (See Mercury—Inorganic.)

Special Tests: Blood and urine levels of mercury have been studied, especially in the case of methyl mercury. A precise correlation has not been found between exposure levels and concentrations. They may be of some value in indicating that exposure has occurred, however.

Personal Protective Methods: (See Mercury—Inorganic.)

Bibliography

Ahlmark, A. 1948. Poisoning by methyl mercury compounds. *Br. J. Ind. Med.* 5:119.
Kark, R.A.P., Poskanzer, D.C., Bullock, J.D., and Boylan, G. 1971. Mercury poisoning and its treatment with N-acetyl-d,1-penicillamine. *N. Engl. J. Med.* 285:10.
Lundgren, K.D., and Swensson, A. 1949. Occupational poisoning by alkyl mercury compounds. *J. Ind. Hyg. Toxicol.* 31:190.
Report of an International Committee. 1969. Maximum allowable concentrations of mercury compounds. *Arch. Environ. Health* 19:891.

MERCURY—INORGANIC

Description: Hg, inorganic mercury, is here taken to include elemental mercury, inorganic mercury compounds, and organic mercury compounds, excluding alkyl mercury compounds. Metallic mercury is a silver-white liquid at room temperature. It occurs as the free metal or as cinnabar (HgS). Mercury is produced from the ore by roasting or reduction.

Synonyms: Quicksilver, hydrargyrum.

Potential Occupational Exposures: Elemental and inorganic mercury compounds are used in the manufacture of scientific instruments (barometers, thermometers, etc.), electric equipment (meters, switches, batteries, rectifiers, etc.), mercury vapor lamps, incandescent electric lamps, x-ray tubes, artificial silk, radio valves, amalgams with copper, tin, silver, or gold, and solders with lead and tin. In the chemical industry, it is used as a fluid cathode for the electrolytic production of caustic soda (sodium hydroxide), chlorine, and acetic acid. It is utilized in gold, silver, bronze, and tin plating, tanning and dyeing, feltmaking, taxidermy, textile manufacture, photography and photoengraving, in extracting gold and silver from ores, in paints and pigments, in the preparation of drugs and disinfectants in the pharmaceutical industry, and as a chemical reagent.

The aryl mercury compounds such as phenylmercury are primarily used as disinfectants, fungicides for treating seeds, antiseptics, herbicides, preservatives, mildew-proofing agents, denaturants for ethyl alcohol, germicides, and bactericides.

Hazardous exposure may occur during mining and extraction of mercury and in the use of mercury and its compounds. Elemental mercury readily volatilizes at room temperature.

A partial list of occupations in which exposure may occur includes:

Amalgam makers	Caustic soda makers
Bactericide makers	Dental amalgam makers
Battery makers	Fungicide makers

Gold extractors	Photographers
Jewelers	Taxidermists
Paper makers	

It is estimated that approximately 150,000 workers are exposed to mercury.

Permissible Exposure Limits: The Federal standard for mercury is 0.1 mg/m^3 as a ceiling value. The NIOSH recommended standard is 0.05 mg Hg/m^3 as a TWA. ACGIH (1978) concurs in this value.

Routes of Entry: Inhalation of dust or vapor; percutaneous absorption of elemental mercury.

Harmful Effects

Local—Mercury is a primary irritant of skin and mucous membranes. It may occasionally be a skin sensitizer.

Systemic— Acute poisoning due to mercury vapors affects the lungs primarily, in the form of acute interstitial pneumonitis, bronchitis, and bronchiolitis.

Exposure to lower levels over prolonged periods produces symptom complexes that can vary widely from individual to individual. These may include weakness, fatigability, loss of appetite, loss of weight, insomnia, indigestion, diarrhea, metallic taste in the mouth, increased salivation, soreness of mouth or throat, inflammation of gums, black line on the gums, loosening of teeth, irritability, loss of memory, and tremors of fingers, eyelids, lips, or tongue. More extensive exposures, either by daily exposures or one-time, can produce extreme irritability, excitability, anxiety, delirium with hallucinations, melancholia, or manic depressive psychosis. In general, chronic exposure produces four classical signs: gingivitis, sialorrhea, increased irritability, and muscular tremors. Rarely are all four seen together in an individual case.

Either acute or chronic exposure may produce permanent changes to affected organs and organ systems.

Medical Surveillance: Preemployment and periodic examinations should be concerned especially with the skin, respiratory tract, central nervous system, and kidneys. The urine should be examined and urinary mercury levels determined periodically. Signs of weight loss, gingivitis, tremors, personality changes, and insomnia would be suggestions of possible mercury intoxication.

Special Tests: Urine mercury determination may be helpful as an index of amount of absorption. Opinions vary as to the significance of a given level. Generally, 0.1 to 0.5 mg Hg/l of urine is considered significant.

Personal Protective Methods: In areas where the exposures are excessive, respiratory protection shall be provided either by full face canister type mask or supplied air respirator, depending on the concentration of mercury fumes. Above 50 mg Hg/m^3 requires supplied air positive pressure full face respirators. Full body work clothes including shoes or shoe covers and hats should be supplied, and clean work clothes should be supplied daily. Showers should be available and all employees encouraged to shower prior to change to street clothes. Work clothes should not be stored with street clothes in the same locker. Food should not be eaten in the work area.

Bibliography

National Institute for Occupational Safety and Health. *Criteria for a Recommended Standard: Occupational Exposure to Inorganic Mercury.* NIOSH Doc. No. 73-11024 (1973).

MESITYL OXIDE (See "Ketones")

METHYL ALCOHOL

Description: CH$_3$OH, methyl alcohol, is a colorless, volatile liquid with a mild odor.

Synonyms: Methanol, carbinol, wood alcohol, wood spirit.

Potential Occupational Exposures: Methyl alcohol is used as a starting material in organic synthesis of chemicals such as formaldehyde, methacrylates, methyl amines, methyl halides, and ethylene glycol, and as an industrial solvent for inks, resins, adhesives, and dyes for straw hats. It is an ingredient in paint and varnish removers, cleaning and dewaxing preparations, spirit duplicating fluids, embalming fluids, antifreeze mixtures, and enamels and is used in the manufacture of photographic film, plastics, celluloid, textile soaps, wood stains, coated fabrics, shatterproof glass, paper coating, waterproofing formulations, artificial leather, and synthetic indigo and other dyes. It has also found use as an extractant in many processes, an antidetonant fuel-injection fluid for aircraft, a rubber accelerator, and a denaturant for ethyl alcohol.

A partial list of occupations in which exposures may occur includes:

Acetic acid makers	Ester makers
Art glass workers	Feather workers
Bookbinders	Felt hat makers
Bronzers	Foundry workers
Dyers	Gilders
Enamel makers	Ink makers

Lasters   Painters
Leather workers   Photoengravers
Millinery workers

NIOSH estimates that approximately 175,000 workers in the United States are potentially exposed to methyl alcohol.

Permissible Exposure Limits: The Federal standard is 200 ppm (260 mg/m^3). NIOSH recommends adherence to the present Federal standard of 200 ppm; 262 milligrams of methyl alcohol per cubic meter of air as a time-weighted average for up to a 10-hour workday, 40-hour workweek. In addition NIOSH recommends a ceiling of 800 ppm; 1,048 milligrams of methyl alcohol per cubic meter of air as determined by a sampling time of 15 minutes.

Routes of Entry: Inhalation of vapor; percutaneous absorption of liquid.

Harmful Effects

Local—Contact with liquid can produce defatting and a mild dermatitis. Methyl alcohol is virtually nonirritating to the eyes or upper respiratory tract below 2,000 ppm, and it is difficult to detect by odor at less than this level.

Systemic—Methyl alcohol may cause optic nerve damage and blindness. Its toxic effect is thought to be mediated through metabolic oxidation products, such as formaldehyde or formic acid, and may result in blurring of vision, pain in eyes, loss of central vision, or blindness. Other central nervous system effects result from narcosis and include headache, nausea, giddiness, and loss of consciousness. Formic acid may produce acidosis. These symptoms occur principally after oral ingestion and are very rare after inhalation.

Medical Surveillance: Consider eye disease and visual acuity in any periodic or placement examinations, as well as skin and liver and kidney functions.

Special Tests: Determination of methyl alcohol in blood, and methyl alcohol and formic acid in urine. Estimation of alkali reserve which may be impaired because of acidosis following accidental ingestion.

Personal Protective Methods: Barrier creams and protective clothing.

Bibliography

Crook, J.E., and McLaughlin, J.S. 1966. Methyl alcohol poisoning. *J. Occup. Med.* 8:467.
Kane, R.L., Talbert, W., Harlan, J., Sizemore, G., and Cataland, S. 1968. A methanol poisoning outbreak in Kentucky. *Arch. Environ. Health* 17:119.
Keeney, A.H., and Mellingkoff, S.M. 1951. Methyl alcohol poisoning. *Ann. Intern. Med.* 34:331.
National Institute for Occupational Safety and Health. *Criteria for a Recommended Standard: Occupational Exposure to Methyl Alcohol.* NIOSH Doc. No. 76-148, Wash., D.C. (1976).

METHYL i-AMYL KETONE (See "Ketones")

METHYL n-AMYL KETONE (See "Ketones")

METHYL (AND ETHYL) BROMIDE

Description: CH_3Br, methyl bromide, is a colorless, nearly odorless gas. It is synthesized from sodium bromide, methyl alcohol, and sulfuric acid.

C_2H_5Br, ethyl bromide, is a colorless, volatile, flammable liquid possessing an etherlike odor, burning taste. It becomes yellowish on exposure to air. It is produced from potassium bromide, ethyl alcohol and sulfuric acid.

Synonyms

Methyl bromide: bromomethane, monobromomethane.
Ethyl bromide: bromoethane, monobromoethane.

Potential Occupational Exposures: Methyl bromide: The primary use of methyl bromide is as an insect fumigant for soil, grain, warehouses, mills, ships, etc. It is also used as a chemical intermediate and a methylating agent, a refrigerant, a herbicide, a fire extinguishing agent, a low-boiling solvent in aniline dye manufacture, for degreasing wool, for extracting oils from nuts, seeds, and flowers, and in ionization chambers.

Ethyl bromide: This chemical is used as an ethylating agent in organic synthesis and gasoline, as a refrigerant, and as an extraction solvent. It has limited use as a local anesthetic.

A partial list of occupations in which exposure may occur includes:

Anesthetists	Grain fumigators
Color makers	Refrigerant makers
Dye makers	Soil fumigators
Fire extinguisher workers	Solvent workers
Fruit fumigators	Wool degreasers

Permissible Exposure Limits: The Federal standard for methyl bromide is 20 ppm (80 mg/m^3) as a ceiling value. The Federal standard for ethyl bromide is 200 ppm (890 mg/m^3) as a TWA. The ACGIH recommended TLV for methyl bromide is 15 ppm (60 mg/m^3) as a TWA.

Routes of Entry: Inhalation and percutaneous absorption.

Harmful Effects

Local—Methyl bromide is irritating to the eyes, skin, and mucous membranes of the upper respiratory tract. In cases of moderate skin exposure, there may be an itching dermatitis, and in severe cases, vesicles and second-degree burns. Methyl bromide may be absorbed by leather, resulting in prolonged skin contact. Repeated or prolonged skin contact with ethyl bromide may cause irritation.

Systemic—High concentrations of either methyl or ethyl bromide may cause lung irritation which may result in pulmonary edema and death. Acute exposure to methyl bromide may produce delayed effects. Onset of symptoms is usually delayed from 30 minutes to 6 hours; the first to appear are malaise, visual disturbances, headaches, nausea, vomiting, somnolence, vertigo, and tremor in the hands. The tremor may become more severe and widespread, developing into epileptiform type convulsions followed by coma and death due to pulmonary or circulatory failure or both. A period of delirium and mania may precede convulsions, but convulsions have been reported without any other warning symptoms. Kidney damage may occur; permanent brain damage may result.

In chronic poisoning, the effects of methyl bromide are usually limited to the central nervous system: lethargy; muscular pains; visual, speech, and sensory disturbances; and mental confusion being the most prominent complaints. Ethyl bromide exposure has not been associated with chronic effects other than skin irritation.

Medical Surveillance

Methyl bromide: Evaluate the central nervous system, respiratory tract, and skin in preplacement and periodic examinations.
Ethyl bromide: No specific considerations needed.

Special Tests: None in common use. Blood bromide levels have been measured in cases of methyl bromide intoxication, but their value in routine monitoring of exposure has not been established.

Personal Protective Methods: Rubber, not leather, protective clothing should be utilized. When masks are worn, they should provide fullface protection.

Bibliography

Hine, C.H. 1969. Methyl bromide poisoning. A review of ten cases. *J. Occup. Med.* 11:1.
Rathus, E.M., and Landy, P.J. 1961. Methyl bromide poisoning. *Br. J. Ind. Med.* 18:53.
Von Oettingen, W.F. 1946. *The Toxicity and Potential Dangers of Methyl Bromide with Special Reference to Its Use in the Chemical Industry, in Fire Extinguishers, and in Fumigation.* National Institute of Health Bulletin No. 185. U.S. Public Health Service.

METHYL i-BUTYL KETONE (See "Ketones")

METHYL n-BUTYL KETONE (See "Ketones")

METHYL CHLORIDE

Description: CH_3Cl, methyl chloride, is a colorless gas possessing a faint, sweet odor.

Synonyms: Monochloromethane, chloromethane.

Potential Occupational Exposures: Methyl chloride is used as a methylating and chlorinating agent in organic chemistry. In petroleum refineries it is used as an extractant for greases, oils, and resins. Methyl chloride is also used as a solvent in the synthetic rubber industry, as a refrigerant, and as a propellant in polystyrene foam production. In the past it has been used as a local anesthetic (freezing).

A partial list of occupations in which exposure may occur includes:

Aerosol packagers
Drug makers
Flavor extractors
Low temperature solvent workers
Methylation workers

Methylcellulose makers
Polystyrene foam makers
Refrigeration workers
Rubber makers
Vapor pressure thermometer makers

Permissible Exposure Limits: The Federal standard is 100 ppm (210 mg/m^3) as an 8-hour TWA, with an acceptable ceiling concentration of 200 ppm; acceptable maximum peaks above the ceiling of 300 ppm are allowed for 5 minutes duration in a 3-hour period.

Route of Entry: Percutaneous absorption.

Harmful Effects

Local—Skin contact with the discharge from the pressurized gas may cause frostbite. The liquid may damage eyes.

Systemic—Signs and symptoms of chronic exposure include staggering gait, difficulty in speech, nausea, headache, dizziness, and blurred vision. Vomiting has also occurred in some cases. These effects may be observed following a latency period of several hours.

Acute exposure is much like chronic except that the latency period is shorter and the effects more severe. Coma or convulsive seizures may occur. Acute poisoning predominantly depresses the central nervous system, but renal and hepatic damage may also occur. Recently noted in these cases is the depression of bone marrow activity. Recovery from severe exposure may take as long as 2 weeks.

Methyl chloride has recently been reported to be highly mutagenic in *S. typhimurium* tester strain TA 1535 (1) (which can detect mutagens causing base pair substitutions). It should be noted that metabolic activation was not required to detect mutagenesis.

Medical Surveillance: Preplacement and periodic examinations should give careful consideration to a previous history of the central nervous system, and to renal or hepatic disorders.

Special Tests: None in common use.

Personal Protective Methods: Masks should be used in areas of high vapor concentrations.

References

(1) Andrews, A.W., Zawistowski, E.S., and Valentine, C.R. A comparison of the mutagenic properties of vinyl chloride and methyl chloride. *Mutation Res.*, 40 (1976) 273.

Bibliography

Hansen, H., Weaver, N.K., and Venable, F.S. 1953. Methyl chloride intoxication. Report of fifteen cases. *AMA Arch. Ind. Hyg. Occup. Med.* 8:328.
MacDonald, J.D.C. 1964. Methyl chloride intoxication. Report of cases. *J. Occup. Med.* 6:81.
Mackie, I.J. 1961. Methyl chloride intoxication. *Med. J. Aust.* 1:203.
Scharnweber, H.C., Spears, G.N., and Cowles, S.R. 1974. Chronic methyl chloride intoxication in six industrial workers. *J. Occup. Med.* 16:112.

4,4'-METHYLENEBIS(2-CHLOROANILINE)

Description: $CH_2(C_6H_3ClNH_2)_2$, 4,4'-methylenebis(2-chloroaniline) or Moca, is a yellow to light grey-tan pellet and is also available in liquid form.

Synonyms: Moca, 4,4'-diamino-3,3'-dichlorodiphenylmethane, 4,4'-methylene-2,2'-dichloroaniline.

Potential Occupational Exposures: Moca is primarily used in the production of solid elastomeric parts. Other uses are as a curing agent for epoxy resins and in the manufacture of crosslinked urethane foams used in automobile seats and safety padded dashboards; it is also used in the manufacture of gun mounts, jet engine turbine blades, radar systems, and components in home appliances.

A partial list of occupations in which exposure may occur includes:

Elastomer makers   Polyurethane foam workers
Epoxy resin workers

Permissible Exposure Limits: Moca is included in the Federal standard for carcinogens; all contact with it should be avoided.

Routes of Entry: Inhalation; percutaneous absorption.

Harmful Effects

Local—None reported.

Systemic—Feeding experiments with rats produced liver and lung cancer. No tumors were found in experiments with dogs. No tumors or other illness have been reported from chronic exposure in man except a mild cystitis which subsided within a week.

Moca when administered at levels of 0.2% and 0.1% in the diet of mice produced vascular tumors at the higher dose level and hepatomas at both levels (1) and hepatomas and lung tumors in rats maintained throughout their lifespan on a low protein diet containing 0.1% Moca (2). Moca is mutagenic in the Salmonella/microsome test (3).

Medical Surveillance: Preplacement and periodic examinations should include a history of exposure to other carcinogens, alcohol and smoking habits, use of medications, and family history. Special attention should be given to liver size and function and to any changes in lung symptoms or x-rays.

Special Tests: None commonly used.

Personal Protective Methods: These are designed to supplement engineering controls and to prevent all contact with skin and the respiratory tract. Protective clothing and gloves should be provided, and also appropriate type dust masks or supplied air respirators. On exit from a regulated area, employees should shower and change into street clothes, leaving the protective clothing and equipment at the point of exit, to be placed in impervious containers at the end of the work shift for decontamination or disposal.

References

(1) Russfield, A.B., Homberger, F., Boger, E., Weisburger, E.K., and Weisburger, J.H.

The carcinogenic effect of 4,4'-methylene-bis-(2-chloroaniline) in mice and rats. *Toxicol. Appl. Pharmacol.* (1973).
(2) Grundmann, E., and Steinhoff, D. Leber und lungentumoren nach, 3,3'-dichlor-4,4'-diaminodiphenylmethan bei ratten. *Z. Krebsforsch.* 74 (1970) 28.
(3) McCann, J., Choi, E., Yamasaki, E., and Ames, B.N. Detection of carcinogens as mutagens in the Salmonella/microsome test: Assay of 300 chemicals. *Proc. Nat. Acad. Sci.* 72 (1975) 5135.

Bibliography

Linch, A.L., O'Connor, G.B., Barnes, J.R., Killian, A.S. Jr., and Neeld, W.E. Jr. 1971. Methylene-bis-ortho-chloroaniline (MOCA): Evaluations of hazards and exposure control. *Am. Ind. Hyg. Assoc. J.* 32:802.
Mastromatteo, E. 1965. Recent health experiences in Ontario. *J. Occup. Med.* 7:502.

METHYLENE CHLORIDE

Description: CH_2Cl_2, methylene chloride, is a nonflammable, colorless liquid with a pleasant aromatic odor noticeable at 300 ppm (this, however, should not be relied upon as an adequate warning of unsafe concentrations).

Synonyms: Dichloromethane, methylene dichloride, methylene bichloride.

Potential Occupational Exposures: Methylene chloride is used mainly as a low temperature extractant of substances which are adversely affected by high temperature. It can be used as a solvent for oil, fats, waxes, bitumen, cellulose acetate, and esters. It is also used as a paint remover and as a degreaser.

A partial list of occupations in which exposure may occur includes:

Aerosol packagers	Leather finish workers
Anesthetic makers	Oil processors
Bitumen makers	Paint remover makers
Degreasers	Resin makers
Fat extractors	Solvent workers
Flavoring makers	Stain removers

NIOSH estimates that 70,000 workers are exposed to methylene chloride.

Permissible Exposure Limits: The Federal standard is 500 ppm (1,740 mg/m³) as an 8-hour TWA with an acceptable ceiling concentration of 1,000 ppm; acceptable maximum peaks above the ceiling of 2,000 ppm are allowed for 5 minutes duration in a 2-hour period. The 1976 TWA value set by ACGIH was 200 ppm (720 mg/m³). In 1978 TWA value proposed by ACGIH is 100 ppm (360 mg/m³). NIOSH has recommended a TWA value of 75 ppm.

A major consideration in the recommended NIOSH standard is concomitant exposure to carbon monoxide since a major metabolic product of methylene chloride is carbon monoxide. NIOSH recommends that occupational exposure to methylene chloride be controlled so that workers are not exposed in excess of 75 ppm (261 mg/m^3) determined as a time-weighted average concentration for up to a 10-hour workday, 40-hour workweek, or to peak concentrations in excess of 500 ppm (1,740 mg/m^3) as determined by any 15-minute sampling period in the absence of CO exposure >9 ppm.

Where the time-weighted average occupational exposure to carbon monoxide exceeds 9 ppm (the current Federal air pollution standard, 40 CFR 50.8), NIOSH recommends that occupational exposure to methylene chloride be controlled so that workers are not exposed to methylene chloride and carbon monoxide in combinations which exceed unity according to the following equation:

$$\frac{C(CO)}{L(CO)} + \frac{C(CH_2Cl_2)}{L(CH_2Cl_2)} \leqslant 1$$

C(CO) = TWA exposure concentration of carbon monoxide, ppm
L(CO) = recommended TWA exposure limit of carbon monoxide = 35 ppm
C(CH$_2$Cl$_2$) = TWA exposure concentration of methylene chloride, ppm
L(CH$_2$Cl$_2$) = recommended TWA exposure limit of methylene chloride = 75 ppm

Routes of Entry: Inhalation of vapors and percutaneous absorption of liquid.

Harmful Effects

Local—Repeated contact with methylene chloride may cause a dry, scaly, and fissured dermatitis. The liquid and vapor are irritating to the eyes and upper respiratory tract at higher concentrations. If the liquid is held in contact with the skin, it may cause skin burns.

Systemic—Methylene chloride is a mild narcotic. Effects from intoxication include headache, giddiness, stupor, irritability, numbness, and tingling in the limbs. Irritation to the eyes and upper respiratory passages occurs at higher dosages. In severe cases, observers have noted toxic encephalopathy with hallucinations, pulmonary edema, coma, and death. Cardiac arrythmias have been produced in animals but have not been common in human experiences. Exposure to this agent may cause elevated carboxyhemoglobin levels which may be significant in smokers, or workers with anemia or heart disease, and those exposed to CO.

Medical Surveillance: Changes in liver, respiratory tract, and central nervous system should be considered during preplacement or periodic medical examinations. Smoking history should be known; anemias or cardiovascular disease may increase the hazard.

Special Tests: The metabolism and excretion of methylene chloride has been thoroughly studied. Blood and expired air analyses are useful indicators of exposure. Carboxyhemoglobin levels may be useful indicators of excessive exposure, especially in nonsmokers.

Personal Protective Methods: Protective clothing, gloves to prevent skin contact, and, in areas of high concentration, fullface masks.

Bibliography

Deplace, Y., Covigneaux, A., and Cobasson, G. 1962. Affections professionelles dues au chlorure de methylene et au dichloroethane. *Arch. Mal. Prof.* 23:816.

Golubovskii, I.E., and Kamchatnova. V.P. 1964. *Hyg. Sanit. (USSR).* 29:145.

Kuzelova, M., and Vlasak, R. 1966. Vliv metylenchloridu na zdravi pracujicich pri vyrobe filmove folie a sledovani kyseliny mravenci jako metabolitu methylenchloridu. *Prac. Lek.* 18:167.

Moskowitz, S., and Shapiro, H. 1952. Fatal exposure to methyl chloride vapor. *Arch. Ind. Hyg. Occup. Med.* 6:116.

National Institute for Occupational Safety and Health. *Criteria for a Recommended Standard: Occupational Exposure to Methylene Chloride.* NIOSH Doc. No. 76-138 (1976).

Stewart, R.D., Fisher, T.N., Hosko, M.J., Peterson, J.E., Baretta, E.D., and Dodd., H.C. 1972. Experimental human exposure to methylene chloride. *Arch. Environ. Health* 25:342.

Weiss, G. 1967. Toxische enzephalose beim beruflichen umgang mit methylenchlorid. *Zentralbl. Arbeitsmed.* 17:282.

METHYL ETHYL KETONE (See "Ketones")

METHYL PARATHION

Description: Methyl parathion,

$$(CH_3O)_2\overset{\displaystyle S}{\overset{\displaystyle \|}{P}}-O-C_6H_4-NO_2$$

is a crystalline solid melting at 37° to 38°C. It is used as an insecticide.

Synonyms: Dimethyl parathion, O-p-nitrophenyl thiophosphate, Metaphos.

Potential Occupational Exposures: In recent years, methyl parathion has been produced in increasingly greater quantities in the United States and

NIOSH estimates that approximately 150,000 United States workers are potentially exposed to methyl parathion in occupational settings.

Permissible Exposure Limits: NIOSH recommends promulgation of an environmental limit of 0.2 milligram of methyl parathion per cubic meter of air as a time-weighted average for up to a 10-hour workday, 40-hour workweek.

Routes of Entry: Adherence to all provisions of the standard is required in workplaces using methyl parathion regardless of the airborne methyl parathion concentration because of serious effects produced by contact with the skin, mucous membranes, and eyes. Since methyl parathion does not irritate or burn the skin, no warning of skin exposure is likely to occur. However, methyl parathion is readily absorbed through the skin, mucous membranes, and eyes and presents a potentially great danger from these avenues of absorption. It is extremely important to emphasize that available evidence indicates that the greatest danger to employees exposed to methyl parathion is from skin contact.

Harmful Effects

Methyl parathion, an organophosphorus insecticide, is converted in the environment and in the body to methyl paraoxon, a potent inactivator of acetylcholinesterase, an enzyme responsible for terminating the transmitter action of acetylcholine at the junction of cholinergic nerve endings with their effector organs or postsynaptic sites.

The scientific basis for the recommended environmental limit is the prevention of medically significant inhibition of acetylcholinesterase. If functional acetylcholinesterase is greatly inhibited, a cholinergic crisis may ensue resulting in signs and symptoms of methyl parathion poisoning, including nausea, vomiting, abdominal cramps, diarrhea, involuntary defecation and urination, blurring of vision, muscular twitching, and difficulty in breathing.

Medical Surveillance: NIOSH recommends that medical surveillance, including preemployment and periodic examinations, shall be made available to workers who may be occupationally exposed to methyl parathion. Biologic monitoring is also recommended as an additional safety measure.

Special Tests: None in common use.

Personal Protective Methods: Personal protective equipment and protective clothing are recommended for those workers occupationally exposed to methyl parathion to further reduce exposure. In certain instances, such as emergency situations, NIOSH recommends that respirators be worn.

NIOSH also recommends that stringent work practices and engineering controls be adhered to in order to further reduce the likelihood of poisoning that may result from skin contact, ingestion, or from inhalation of methyl parathion.

Bibliography

National Institute for Occupational Safety and Health. *Criteria for a Recommended Standard: Occupational Exposure to Methyl Parathion.* NIOSH Doc. No. 77-106 (1977).

METHYL n-PROPYL KETONE (See "Ketones")

MOLYBDENUM AND COMPOUNDS

Description: Mo, molybdenum, is a silver-white metal or a greyish-black powder. Molybdenite is the only important commercial source. This ore is often associated with copper ore. Molybdenum is insoluble in water and soluble in hot concentrated nitric and sulfuric acid.

Synonyms: None.

Potential Occupational Exposures: Most of the molybdenum produced is used in alloys: steel, stainless steel, tool steel, cast iron, steel mill rolls, manganese, nickel, chromium, and tungsten. The metal is used in electronic parts (contacts, spark plugs, x-ray tubes, filaments, screens, and grids for radios), induction heating elements, electrodes for glass melting, and metal spraying applications. Molybdenum compounds are utilized as lubricants; as pigments for printing inks, lacquers, paints, for coloring rubber, animal fibers, and leather, and as a mordant; as catalysts for hydrogenation cracking, alkylation, and reforming in the petroleum industry, in Fischer-Tropsch synthesis, in ammonia production, and in various oxidation-reduction and organic cracking reactions; as a coating for quartz glass; in vitreous enamels to increase adherence to steel; in fertilizers, particularly for legumes; in electroplating to form protective coatings; and in the production of tungsten.

Hazardous exposures may occur during high-temperature treatment in the fabrication and production of molybdenum products, spraying applications, or through loss of catalyst. MoO_3 sublimes above 800°C.

A partial list of occupations in which exposure may occur includes:

Ceramic makers	Metal platers
Drug makers	Petroleum refinery workers
Electroplaters	Steel alloy makers
Fertilizer makers	Tannery workers
Glass makers	Vacuum tube makers

Permissible Exposure Limits: The Federal standards are: Molybdenum: soluble compounds, 5 mg/m³. Molybdenum: insoluble compounds, 15 mg/m³.

Route of Entry: Inhalation of dust or fume.

Harmful Effects

Local—Molybdenum trioxide may produce irritation of the eyes and mucous membranes of the nose and throat. Dermatitis from contact with molybdenum is unknown.

Systemic—No reports of toxic effects of molybdenum in the industrial setting have appeared. It is considered to be an essential trace element in many species, including man. Animal studies indicate that insoluble molybdenum compounds are of a low order of toxicity (e.g, disulfide, oxides, and halides). Soluble compounds (e.g., sodium molybdate) and freshly generated molybdenum fumes, however, are considerably more toxic. Inhalation of high concentrations of molybdenum trioxide dust is very irritating to animals and has caused weight loss, diarrhea, loss of muscular coordination, and a high mortality rate. Molybdenum trioxide dust is more toxic than the fumes. Large oral doses of ammonium molybdate in rabbits caused some fetal deformities. Excessive intake of molybdenum may produce signs of a copper deficiency.

Medical Surveillance: Preemployment and periodic physical examinations should evaluate any irritant effects to the eyes or respiratory tract and the general health of the worker. Although molybdenum compounds are of a low order of toxicity, animal experimentation indicates protective measures should be employed against the more soluble compounds and molybdenum trioxide dust and fumes. The normal intake of copper in the diet appears to be sufficient to prevent systemic toxic effects due to molybdenum poisoning.

Special Tests: None in common use.

Personal Protective Methods: Where dust and fumes exceed the standard, molybdenum workers should be supplied with dust masks or supplied air respirators. Full body work clothes are advisable with daily change of clothes and showering before changing to street clothes.

Bibliography

Environmental Protection Agency. *Molybdenum—A Toxicological Appraisal.* Report EPA-600/1-75-004, Research Triangle Park, N.C. Health Effects Research Laboratory (Nov. 1975).
Fairhall, L.T., Dunn, R.C., Sharpless, N.E. and Pritchard, E.A. 1945. *The Toxicity of Molybdenum.* Public Health Bulletin No. 293. U.S. Government Printing Office, Wash., D.C.

N

NAPHTHA

Description: Naphthas derived from both petroleum and coal tar are included in this group. Petroleum naphthas composed principally of aliphatic hydrocarbons are termed "close-cut" fractions. "Medium-range" and "wide-range" fractions are made up of 40 to 80% aliphatic hydrocarbons, 25 to 50% naphthenic hydrocarbons, 0 to 10% benzene, and 0 to 20% other aromatic hydrocarbons.

Coal tar naphtha is a mixture of aromatic hydrocarbons, principally toluene, xylene, and cumene. Benzene, however, is present in appreciable amounts in those coal tar naphthas with low boiling points.

Synonyms: Petroleum naphtha—ligroin, benzine, petroleum ether, petroleum benzine.

Potential Occupational Exposures: Naphthas are used as organic solvents for dissolving or softening rubber, oils, greases, bituminous paints, varnishes, and plastics. The less flammable fractions are used in dry cleaning, the heavy naphthas serving as bases for insecticides.

A partial list of occupations in which exposure may occur includes:

Chemical laboratory workers	Petroleum refinery workers
Detergent makers	Rubber coaters
Dry cleaners	Solvent workers
Fat processors	Stainers
Insecticide workers	Varnish makers
Metal degreasers	Wax makers
Oil processors	Wool processors
Painters	Xylene makers

Permissible Exposure Limits: The Federal standard for petroleum naphtha is 500 ppm (2,000 mg/m^3); for coal tar naphtha it is 100 ppm (400 mg/m^3).

Route of Entry: Inhalation of vapor. Percutaneous absorption of liquid is probably not important in development of systemic effects unless benzene is present.

Harmful Effects

Local—The naphthas are irritating to the skin, conjunctiva, and the mucous membranes of the upper respiratory tract. Skin "chapping" and photosensitivity may develop after repeated contact with the liquid. If confined against skin by clothing, the naphthas may cause skin burn.

Systemic—Petroleum naphtha has a lower order of toxicity than that derived from coal tar, where the major hazard is brought about by the aromatic hydrocarbon content. Sufficient quantities of both naphthas cause central nervous system depression. Symptoms include inebriation, followed by headache and nausea. In severe cases, dizziness, convulsions, and unconsciousness occasionally result. Symptoms of anorexia and nervousness have been reported to persist for several months following an acute overexposure, but this appears to be rare. One fraction, hexane, has been reported to have been associated with peripheral neuropathy. (See Hexane.) If benzene is present, coal tar naphthas may produce blood changes such as leukopenia, aplastic anemia, or leukemia. The kidneys and spleen have also been affected in animal experiments. (See Benzene.)

Medical Surveillance: Preplacement and periodic medical examinations should include the central nervous system. If benzene exposure is present, workers should have a periodic complete blood count (CBC) including hematocrit, hemoglobin, white blood cell count and differential count, mean corpuscular volume and platelet count, reticulocyte count, serum bilirubin determination, and urinary phenol in the preplacement examination and at 3-month intervals. There are no specific diagnostic tests for naphtha exposure but urinary phenols may indicate exposure to aromatic hydrocarbons. It should be noted that benzene content of vapor may be higher than predicted by content in the liquid.

Special Tests: None in common use.

Personal Protective Methods: Workers should use barrier creams, protective clothing, gloves and masks where exposure to the vapor is likely.

Bibliography

Pagnotto, L.D., Elkins, H.B., Brugsch, H.G., and Walkley, J.E. 1961. Industrial benzene exposure from petroleum naphtha. 1. Rubber coating industry. *Am. Ind. Hyg. Assoc. J.* 22:417.

NAPHTHALENE

Description: $C_{10}H_8$, naphthalene, is a white crystalline solid with a characteristic "moth ball" odor.

Synonyms: Naphthalin, moth flake, tar camphor, white tar.

Potential Occupational Exposures: Naphthalene is used as a chemical intermediate or feedstock for synthesis of phthalic, anthranilic, hydroxyl (naphthols), amino (naphthylamines), and sulfonic compounds which are used in the manufacture of various dyes. Naphthalene is also used in the manufacture of hydronaphthalenes, synthetic resins, lampblack, smokeless powder, and celluloid. Naphthalene has been used as a moth repellent.

A partial list of occupations in which exposure may occur includes:

Beta naphthol makers	Lampblack makers
Celluloid makers	Moth repellent workers
Coal tar workers	Phthalic anhydride makers
Dye chemical makers	Smokeless powder makers
Fungicide makers	Tannery workers
Hydronaphthalene makers	Textile chemical makers

Permissible Exposure Limits: The Federal standard is 10 ppm (50 mg/m^3).

Route of Entry: Inhalation of vapor or dust.

Harmful Effects

Local—Naphthalene is a primary irritant and causes erythema and dermatitis upon repeated contact. It is also an allergen and may produce dermatitis in hypersensitive individuals. Direct eye contact with the dust has produced irritation and cataracts.

Systemic—Inhaling high concentrations of naphthalene vapor or ingesting may cause intravascular hemolysis and its consequences. Initial symptoms include eye irritation, headache, confusion, excitement, malaise, profuse sweating, nausea, vomiting, abdominal pain, and irritation of the bladder.

There may be progressive jaundice, hematuria, hemoglobinuria, renal tubular blockage, and acute renal shutdown. Hematologic features include red cell fragmentation, icterus, severe anemia with nucleated red cells, leukocytosis, and dramatic decreases in hemoglobin, hematocrit, and red cell count. Individuals with a deficiency of glucose-6-phosphate dehydrogenase in erythrocytes are more susceptible to hemolysis by naphthalene.

Medical Surveillance: Consider eyes, skin, blood, liver, and renal function in placement and follow-up examinations. Low erythrocyte glucose 6-phosphate dehydrogenase increases risk.

Special Tests: None in common use.

Personal Protective Methods: As used in industry, they are rarely necessary.

In dusty areas and areas of high vapor concentration, dust type or organic vapor canister masks should be supplied. Skin protection with gloves, barrier creams, or protective clothing may be useful.

ALPHA-NAPHTHYLAMINE

Description: $C_{10}H_7NH_2$, alpha-naphthylamine, exists as white needlelike crystals which turn red on exposure to air.

Synonyms: 1-Aminonaphthalene, naphthalidine.

Potential Occupational Exposures: α-Naphthylamine is used in the manufacture of dyes, condensation colors, and rubber, and in the synthesis of many chemicals such as α-naphthol, sodium naphthionate, o-naphthionic acid, Neville and Winther's acid, sulfonated naphthylamines, α-naphthylthiourea (a rodenticide), and N-phenyl-α-naphthylamine.

A partial list of occupations in which exposure may occur includes:

Dye makers   Rubber workers
Chemical synthesizers

Permissible Exposure Limits: α-Naphthylamine is included in the Federal standard for carcinogens; all contact with it should be avoided.

Routes of Entry: Inhalation and percutaneous absorption.

Harmful Effects

Local—None reported.

Systemic—It has not been established whether α-naphthylamine is a human carcinogen per se or is associated with an excess of bladder cancer due to its β-naphthylamine content. Workers exposed to α-naphthylamine developed bladder tumors. The mean latent period was 22 years compared to 16 years for β-naphthylamine. One animal experiment demonstrated papillomata, but these results have never been confirmed.

Occupational exposure to commercial 1-naphthylamine containing 4% to 10% 2-naphthylamine is strongly associated with bladder cancer in man (1). However, it is not possible at present to determine unequivocally whether 1-naphthylamine free from the 2-isomer is carcinogenic to man (1).

The carcinogenicity of 1-naphthylamine in animals is equivocal. For example, no carcinogenic effect of 1-naphthylamine was found in the hamster following oral administration; inconclusive results were obtained in mice after oral and subcutaneous administration and in dogs. 1-Naphthylamine, if carcinogenic at all, was less so to the bladder than was the 2-isomer (1). The carcinogenicity of metabolites of 1-naphthylamine [e.g., N-(1-naphthyl)-hydroxylamine] in rodents has been reported (1)-(3). N-hydroxy-1-naphthylamine is a much more potent carcinogen than N-hydroxy-2-naphthylamine.

1-Naphthylamine is mutagenic in the *Salmonella*/microsome test (4).

Medical Surveillance: Placement and periodic examinations should include an evaluation of exposure to other carcinogens; use of alcohol, smoking, and medications; and family history. Special attention should be given on a regular basis to urine sediment and cytology. If red cells or positive smears are seen, cystoscopy should be done at once. The general health of exposed persons should also be evaluated in periodic examinations.

Special Tests: None commonly used. Some metabolites are known.

Personal Protective Methods: These are designed to supplement engineering controls and to prevent all skin or respiratory contact. Full body protective clothing and gloves should be used by those employed in handling operations. Fullface, supplied air respirators of continuous flow or pressure demand type should also be used. On exit from a regulated area, employees should shower and change into street clothes, leaving their protective clothing and equipment at the point of exit to be placed in impervious containers at the end of the work shift for decontamination or disposal. Effective methods should be used to clean and decontaminate gloves and clothing. Showers should be taken prior to dressing in street clothes.

References

(1) *IARC,* Monograph No. 4. International Agency, Lyon (1974) pp. 87-96.
(2) Boyland, E., Busby, E.R., Dukes, C.E., Grover, P.L., and Manson, D. Further experiments on implantation of materials into the urinary bladder of mice. *Brit. J. Cancer* 18 (1964) 575.
(3) Radomski, J.L., Brill, E., Deichmann, W.B., and Glass, E.M. Carcinogenicity testing of N-hydroxy and other oxidation and decomposition products of 1- and 2-naphthylamine. *Cancer Res.* 31 (1971) 1461.
(4) McCann, J., Choi, E., Yamasaki, E., and Ames, B.N. Detection of carcinogens as mutagens in the *Salmonella*/microsome test: Assay of 300 chemicals. *Proc. Nat. Acad. Sci.* 72 (1975) 5135.

BETA-NAPHTHYLAMINE

Description: $C_{10}H_7NH_2$, beta-naphthylamine, is a white to reddish crystal.

Synonyms: 2-Naphthylamine, 2-aminonaphthalene.

Potential Occupational Exposures: β-Naphthylamine is presently used only for research purposes. It is present as an impurity in α-naphthylamine. It was widely used in the manufacture of dyestuffs, as an antioxidant for rubber, and in rubber coated cables.

A partial list of occupations in which exposure may occur includes:

beta-Naphthylamine workers   Research workers

Permissible Exposure Limits: β-Naphthylamine is included in the Federal standard for carcinogens; all contact with it should be avoided.

Routes of Entry: Inhalation and percutaneous absorption.

Harmful Effects

Local—β-Naphthylamine is mildly irritating to the skin and has produced contact dermatitis.

Systemic—β-Naphthylamine is a known human bladder carcinogen with a latent period of about 16 years. The symptoms are frequent urination, dysuria, and hematuria. Acute poisoning leads to methemoglobinemia or acute hemorrhagic cystitis.

Epidemiological studies have shown that occupational exposure to 2-naphthylamine, either alone, or when present as an impurity in other compounds, is strongly associated with the occurrence of bladder cancer. (1)-(8).

2-Naphthylamine, administered orally, produced bladder carcinomas in the dog (9)-(11) and monkey (1)(12) and at high dosage levels, in the hamster (13) (14). Although oral administration of 2-naphthylamine increased the incidence of hepatomas in the mouse (15) it demonstrated little, if any, carcinogenic activity in the rat and rabbit (1)(15)(16).

Evidence to date suggests that several carcinogenic metabolites, rather than a single proximate carcinogen are responsible for the carcinogenic activity demonstrated by 2-naphthylamine (1)(17)(18). These include: 2-naphthylhydroxylamine and/or an ester thereof; bis(2-hydroxylamino-1-naphthyl)phosphate; bis(2-amino-1-naphthyl)phosphate; 2-amino-1-naphthol free and/or conjugated and 2-hydroxylamino-1-naphthol.

The demonstrated carcinogenicity spectrum of the metabolites is from highly active to weakly active in some species and testing system(s) (1)(17)-(22).

Twenty-four metabolites of 2-naphthylamine have been identified in the urine of rats, rabbits, dogs or monkeys by Boyland (19) and Boyland and Manson (20).

Recent reports have highlighted the potential problem of the metabolic conversion of industrial chemical precursors to 2-naphthylamine (23)(24). For example, phenyl-β-naphthylamine (PBNA) which is not currently regulated by OSHA, is widely used as an antioxidant in the rubber industry, as an antioxidant for grease and oils in the petroleum industry, as a stabilizer for the manufacture of synthetic rubber and as an intermediate in the synthesis of dyes as well as other antioxidants (23)(24).

In a recent study in the United States with volunteers, 3 to 4 micrograms of 2-naphthylamine were found in the urine of individuals who had 50 mg of PBNA (contaminated with 0.7 μg 2-naphthylamine) and from workers estimated to have inhaled 30 mg PBNA (24). These findings indicated that PBNA is at least partially metabolized by man to 2-naphthylamine and confirmed an earlier study in the Netherlands (23)(25) where volunteers who consumed 10 mg PBNA (containing 0.032 micrograms of 2-naphthylamine as an impurity) were found to have 3 to 8 micrograms of 2-naphthylamine in their urine samples.

It should be noted that 15,000 workers are at potential risk of exposure to phenyl-β-naphthylamine during its manufacture and use (24).

The carcinogenic potential of the metabolism of 2-nitronaphthalene (an unmarketed by-product produced during the commercial preparation of 1-naphthylamine) has also recently been stressed (23)(24). 2-Nitronaphthalene (analogous to PBNA) is metabolized by beagle dogs to 2-naphthylamine. In earlier studies, female dogs were fed 100 mg of 2-nitronaphthalene daily for 8 months; after 10.5 years, bladder papillomas were observed in various stages of malignancy in 3 of 4 dogs (23).

2-Naphthylamine as well as 2-naphthyl- and 1-naphthylhydroxylamines are mutagenic in the *Salmonella*/microsome test (29). However, N-hydroxy-1- and N-hydroxy-2-naphthylamines although toxic, were not mutagenic to intracellular T4 phage (26). Earlier studies indicated that N-hydroxy-1-naphthylamine (27) and N-hydroxy-2-naphthylamine caused mutations in bacteria (27)(28). However, the significance of the mutagenesis data reported in these studies (27)(28) was suggested (26) to be marginal (e.g., less than a 10 fold increase in the frequency of revertants in back-mutation experiments).

Medical Surveillance: Preplacement and periodic examinations should include an evaluation of exposure to other carcinogens; use of alcohol, smoking, and medications; and family history. Special attention should be given on a regular basis to urine sediment and cytology. If red cells or positive smears are seen, cystoscopy should be done at once. The general health of exposed persons should also be evaluated in periodic examinations.

Special Tests: None in common use; some metabolites are known.

Personal Protective Methods: These are designed to supplement engineering controls and to prevent all skin or respiratory contact. Full body protective clothing and gloves should be used by those employed in handling operations. Fullface, supplied air respirators of continuous flow or pressure demand type should also be used. On exit from a regulated area, employees should shower and change into street clothes, leaving their clothing and equipment at the point of exit to be placed in impervious containers at the end of the work shift for decontamination or disposal. Effective methods should be used to clean and decontaminate gloves and clothing. Showers should be taken prior to dressing in street clothes.

References

(1) *IARC,* Monograph No. 4. International Agency, Lyon (1974) pp. 87-96.
(2) Veys, C.A. Two epidemiological inquiries into the incidence of bladder tumors in industrial workers. *J. Nat. Cancer Inst.* 43 (1969) 219.
(3) Vigliani, E.D., and Barsotti, M. Environmental tumors of the bladder in some Italian dye-stuff factories. *Acta. Un. Int. Cancer* 18 (1961) 669.
(4) Temkin, I.S. *Industrial Bladder Carcinogenesis.* Pergamon Press, Oxford, London, New York, Paris (1963).
(5) Tsugi, I. Environmental and industrial cancer of the bladder in Japan. *Acta. Un. Int. Cancer* 18 (1963) 662.
(6) Gehrmann, G.H., Foulger, J.H., and Fleming, A.J. Occupational carcinoma of the bladder. In *Proceedings of the North International Congress of Industrial Medicine.* London (1948), Bristol, Wright, p. 427.
(7) Hueper, W.C. *Occupational Tumors and Allied Diseases.* Thomas Publ. Springfield, Ill.
(8) Billiard-Duchesne, J.L. Cas francais de tumeurs professionnelles de la vessie. *Acta. Un. Int. Cancer.* 16 (1960) 204.
(9) Boyland, E., Busby, E.R., Dukes, C.E., Grover, P.L., and Manson, D. Further experiments on implantation of materials into the urinary bladder of mice. *Brit. J. Cancer* 18 (1964) 575.
(10) Conzelman, G.M., Jr., and Moulton, J.E. Dose-response relationships of the bladder tumorigen 2-naphthylamine: A study in beagle dogs. *J. Nat. Cancer Inst.* 49 (1972) 193.
(11) Bonser, G.M., Clayson, D.B., Jull, J.W., and Pyrah, L.N. The carcinogenic activity of 2-naphthylamine. *Brit. J. Cancer* 10 (1956) 533.
(12) Conzelman, G.M. Jr., Moulton, J.E., Flanders, L.E. III, Springer, K., and Crout, D.W. Induction of transitional cell carcinomas of the urinary bladder in monkeys fed 2-naphthylamine. *J. Nat. Cancer Inst.* 42 (1969) 825.
(13) Saffiotti, U., Cefis, F., Montesano, R., and Sellakumar, A.R. Induction of bladder cancer in hamsters fed aromatic amines. In *Bladder Cancer. A Symposium.* eds. Deichmann, W., and Lampe, K.F., Aesculapius Press, Birmingham, Ala. (1967) p. 129.
(14) Sellakumar, A.R., Montesano, R., and Saffiotti, U. Aromatic amines carcinogenicity in hamsters. *Proc. Amer. Ass. Cancer Res.* 10 (1969) 78.
(15) Bonser, G.M., Clayson, D.B., Jull, J.W., and Pyrah, L.N. The carcinogenic properties of 2-amino-1-naphthol hydrochloride and its parent amine 2-naphthylamine. *Brit. J. Cancer* 6 (1952) 412.
(16) Hadidian, Z., Fredrickson, T.N., Weisburger, E.K., Weisburger, J.H., Glass, R.M., and Mantel, N. Tests for chemical carcinogens. Report on the activity of derivatives of aromatic amines, nitrosamines, quinolines, nitroalkanes, amides, epoxides, aziridines and purine anti-metabolites. *J. Nat. Can. In.* 41 (1968) 985.

(17) Arcos, J.C., and Argus, M.F. *Chemical Induction of Cancer.* Vol. IIB. Academic Press, NY (1974) 247-253.

(18) Arcos, J.C., and Argus, M.F. Molecular geometry and carcinogenic activity of aromatic compounds. New perspectives. *Adv. Cancer Res.* 11 (1968) 305.

(19) Boyland, E. The biochemistry of cancer of the bladder. *Brit. Med. Bull.* 14 (1958) 153.

(20) Boyland, E., and Manson, D. The metabolism of 2-naphthylamine and 2-naphthyl-hydroxylamine derivatives. *Biochem. J.* 101 (1966) 84.

(21) Radomski, J.L., and Brill, E. The role of N-oxidation products of aromatic amines in the induction of bladder cancer in the dog. *Arch. Toxicol.* 28 (1971) 159.

(22) Radomski, J.L., Brill, E., Deichmann, W.B., and Glass, E.M. Carcinogenicity testing of N-hydroxy and other oxidation and decomposition products of 1-, and 2-naphthylamine. *Cancer Res.* 31 (1971) 1461.

(23) Anon. NIOSH Issues alert on precursors of β-naphthylamine. *Occup. Hlth. Safety Letter* 6 (24) (1976) 4-5.

(24) Moore, R.M. Jr., Woolf, B.S., Stein, H.P., Thomas, A.W., and Finklea, J.F. Metabolic precursors of a known human carcinogen. *Science* 195 (1977) 344.

(25) Kummer, R., and Tordoir, W.F. *Tijdschr. Soc. Geneesk.* 53 (1975) 415.

(26) Corbett, T.H., Heidelberger, C., Dove, W.F. Determination of the mutagenic activity to bacteriophage T4 of carcinogenic and non-carcinogenic compounds. *Mol. Pharmacol.* 6 (1970) 667-679.

(27) Perez, G., and Radomski, J.L. The mutagenicity of the N-hydroxy naphthylamines in relation to their carcinogenicity. *Ind. Med. Surg.* 34 (1965) 714-716.

(28) Bellman, S. , Troll, W., Teebor, G., and Mukai, F. The carcinogenic and mutagenic properties of N-hydroxyaminonaphthalenes. *Cancer Res.* 28 (1968) 535-542.

(29) McCann, J., Choi, E., Yamasaki, E., and Ames, B.N. Detection of carcinogens as mutagens in the *Salmonella*/microsome test: Assay of 300 chemicals. *Proc. Nat. Acad. Sci.* 72 (1975) 5135.

NATURAL GAS

Description: Natural gas consists primarily of methane (85%) with lesser amounts of ethane (9%), propane (3%), nitrogen (2%), and butane (1%). Methane is a colorless, odorless, flammable gas.

Synonyms: Marsh gas.

Potential Occupational Exposures: Natural gas is used principally as a heating fuel. It is transported as a liquid under pressure. It is also used in the manufacture of various chemicals including acetaldehyde, acetylene, ammonia, carbon black, ethyl alcohol, formaldehyde, hydrocarbon fuels, hydrogenated oils, methyl alcohol, nitric acids, synthesis gas, and vinyl chloride. Helium can be extracted from certain types of natural gas.

A partial list of occupations in which exposure may occur includes:

Coal miners	Helium extractors
Electric power plant workers	Hydrogen makers
Gas fuel users	Nitric acid makers

> Organic chemical synthesizers   Synthetic gas makers
> Petroleum refinery workers   Vinyl chloride makers

Permissible Exposure Limits: There is no Federal standard for natural gas, methane, nitrogen, or butane. The Federal standard for propane is 1,000 ppm (1,800 mg/m^3). The ACGIH (1978) lists 600 ppm (1,430 mg/m^3) as a TWA for butane.

Route of Entry: Inhalation of gas.

Harmful Effects

Local—Upon escape from pressurized tanks, natural gas may cause frostbite.

Systemic—Natural gas is a simple asphyxiant. Displacement of air by the gas may lead to shortness of breath, unconsciousness, and death from hypoxemia. Incomplete combustion may produce carbon monoxide.

Medical Surveillance: No specific considerations are needed.

Special Tests: None are in use.

Personal Protective Method: Adequate ventilation should quite easily prevent any potential hazard.

NICKEL AND COMPOUNDS

Description: Ni, nickel, is a hard, ductile, magnetic metal with a silver-white color. It is insoluble in water and soluble in acids. It occurs free in meteorites and in ores combined with sulfur, antimony, or arsenic. Processing and refining of nickel is accomplished by either the Orford (sodium sulfide and electrolysis) or the Mond (nickel carbonyl) processes. In the latter, impure nickel powder is reacted with carbon monoxide to form gaseous nickel carbonyl which is then treated to deposit high purity metallic nickel.

Synonyms: None.

Potential Occupational Exposures: Nickel forms alloys with copper, manganese, zinc, chromium, iron, molybdenum, etc. Stainless steel is the most widely used nickel alloy. An important nickel-copper alloy is Monel metal, which contains 66% nickel and 32% copper and has excellent corrosion resistance properties. Permanent magnets are alloys chiefly of nickel, cobalt, aluminum, and iron.

Elemental nickel is used in electroplating, anodizing aluminum, casting operations for machine parts, and in coinage; in the manufacture of acid-resisting and magnetic alloys, magnetic tapes, surgical and dental instruments, nickel-cadmium batteries, nickel soaps in crankcase oils, and ground-coat enamels, colored ceramics, and glass. It is used as a catalyst in the hydrogenation of fats, oils, and other chemicals, in synthetic coal oil production, and as an intermediate in the synthesis of acrylic esters for plastics.

Exposure to nickel may also occur during mining, smelting, and refining operations.

A partial list of occupations in which exposure may occur includes:

Battery makers	Oil hydrogenators
Ceramic makers	Paint makers
Chemists	Pen point makers
Dyers	Spark plug makers
Enamelers	Textile dyers
Ink makers	Varnish makers

NIOSH estimates that 250,000 U.S. workers are potentially exposed to nickel.

Permissible Exposure Limits: The Federal standard for nickel metal and its soluble compounds is 1 mg/m^3 expressed as Ni. NIOSH recommends adherence to an exposure limit of 15 micrograms of nickel per cubic meter of air as a TWA for up to a 10-hour workday, 40-hour workweek. This limit differs greatly from the present Federal standard. ACGIH (1978) sets 1 mg/m^3 as a TWA for nickel metal and 0.1 mg/m^3 for soluble nickel compounds. In addition nickel sulfide roasting fume and dust are classified under "Human Carcinogens."

Route of Entry: Inhalation of dust or fume.

Harmful Effects

Local—Skin sensitization is the most commonly seen toxic reaction to nickel and nickel compounds and is seen frequently in the general population. This often results in chronic eczema "Nickel itch," with lichenification resembling atopic or neurodermatitis. Nickel and its compounds are also irritants to the conjunctiva of the eye and the mucous membrane of the upper respiratory tract.

Systemic—Elemental nickel (as deposited from inhalation of nickel carbonyl) and nickel salts are probably carcinogenic, producing an increased incidence of cancer of the lung and nasal passages. The average latency period for the induction of these cancers appears to be about 25 years (range 4 to 51 years). Effects on the heart muscle, brain, liver, and kidney have been seen in animal studies. Pulmonary eosinophilia (Loeffler's syndrome) has been reported in one study to be caused by the sensitizing property of nickel. Finely divided nickel has also shown some carcinogenic effects in rats by injection, and in guinea pigs by inhalation.

Medical Surveillance: Preemployment physical examinations should evaluate any history of skin allergies or asthma, other exposures to nickel or other carcinogens, smoking history, and the respiratory tract. Lung function should be studied and chest x-rays periodically evaluated. Special attention should be given to the nasal sinuses and skin.

Special Tests: Serum and urinary nickel can be determined, although opinions vary as to their value in monitoring exposures.

Personal Protective Methods: Full body protective clothing is advisable, as is the use of barrier creams to prevent skin sensitization and dermatitis. In areas of dust or fumes, masks or supplied air respirators are mandatory where concentrations exceed the standard limits. Clean work clothing should be provided daily; and showering should be required before changing to street clothes. No food should be eaten in work areas.

Bibliography

Kazantzis, G. 1976. Chromium and nickel. *Ann. Occup. Hyg.* 15:25.
Mastromatteo, E. 1967. Nickel: a review of its occupational health aspects. *J. Occup. Med.* 9:127.
McNeely, M.D., Nechay, M.W., and Sunderman, F.W. 1972. Measurement of nickel in serum and urine as indices of environmental exposure to nickel. *Clin. Chem.* 18:992.
National Institute for Occupational Safety and Health. *Criteria for a Recommended Standard: Occupational Exposure to Inorganic Nickel.* NIOSH Doc. No. 77-164 (1977).

NICKEL CARBONYL

Description: $Ni(CO)_4$, nickel carbonyl, is a colorless, highly volatile, flammable liquid with a musty odor. It decomposes above room temperature producing carbon monoxide and finely divided nickel. It is soluble in organic solvents.

Synonyms: Nickel tetracarbonyl.

Potential Occupational Exposures: The primary use for nickel carbonyl is in the production of nickel by the Mond process. Impure nickel powder is reacted with carbon monoxide to form gaseous nickel carbonyl which is then treated to deposit high purity metallic nickel and release carbon monoxide. Other uses include gas plating, the production of nickel products; in chemical synthesis as a catalyst, particularly for oxo reactions (addition reaction of hydrogen and carbon monoxide with unsaturated hydrocarbons to form oxygen-function compounds), e.g., synthesis of acrylic esters, and as a reactant.

A partial list of occupations in which exposure may occur includes:

Foundry workers   Gas platers

Organic chemical synthesizers   Petroleum refinery workers

Permissible Exposure Limits: The Federal standard for nickel carbonyl is 0.001 ppm (0.007 mg/m³), according to NIOSH. ACGIH (1978) quotes 0.05 ppm (0.35 mg/m³) as a TWA.

Routes of Entry: Inhalation of vapor. It may be possible for appreciable amounts of the liquid to be absorbed through the skin.

Harmful Effects

Local—Nickel dermatitis may develop. (See Nickel and Compounds.)

Systemic—Symptoms of exposure to the toxic vapors of nickel carbonyl are of two distinct types. Immediately after exposure, symptoms consist of frontal headache, giddiness, tightness of the chest, nausea, weakness of limbs, perspiring, cough, vomiting, cold and clammy skin, and shortness of breath. Even in exposures sufficiently severe to cause death, the initial symptoms disappear quickly upon removal of the subject to fresh air. Symptoms may be so mild during this initial phase that they go unrecognized.

Severe symptoms may then develop insidiously hours or even days after exposure. The delayed syndrome usually consists of retrosternal pain, tightness in the chest, dry cough, shortness of breath, rapid respiration, cyanosis, and extreme weakness. The weakness may be so great that respiration can be sustained only by oxygen support. Fatal cases are usually preceded by convulsion and mental confusion, with death occurring from 4 to 11 days following exposure. The syndrome represents a chemical pneumonitis with adrenal cortical suppression.

Nickel carbonyl is carcinogenic to the same degree as elemental nickel. (See Nickel and Compounds.)

Medical Surveillance: (See Nickel and Compounds.)

Special Tests: Urinary nickel levels for several days after acute exposures may be helpful.

Personal Protective Methods: (See Nickel and Compounds.)

Bibliography

Kazantzis, G. 1972. Chromium and nickel. *Ann. Occup. Hyg.* 15:25.
McDowell, R.S. 1971. Metal carbonyl vapors: rapid quantitative analysis by infrared spectrophotometry. *Am. Ind. Hyg. Assoc. J.* 32:621.
Sunderman, F.W., and Kincaid, J.F. 1954. Nickel poisoning II. Studies on patients suffering from acute exposure to vapors of nickel carbonyl. *J. Am. Med. Assoc.* 155:889.

Sunderman, F.W., and Sunderman, F.W. Jr. 1961. Loeffler's syndrome associated with nickel sensitivity. *Arch. Intern. Med.* 107:149.

NITRIC ACID

Description: Nitric acid, HNO_3, is a colorless liquid with a characteristic choking odor which fumes in moist air. It is a solution of nitrogen dioxide, NO_2, in water and so-called fuming nitric acid contains an excess of NO_2 and is yellow to brownish-red in color.

Synonyms: None.

Potential Occupational Exposures: NIOSH estimates that 27,000 workers are potentially exposed to nitric acid. Production of nitric acid is primarily by means of an ammonia oxidation process, and over 6 million tons of acid were produced in 1967. As of 1970, nitric acid was the second most important industrial acid and its production represented the sixth largest chemical industry in the United States. The largest use of nitric acid is in the production of fertilizers. Almost 15% of the production goes into the manufacture of explosives, with the remaining 10% distributed among a variety of uses such as etching, bright-dipping, electroplating, photoengraving, and production of rocket fuel.

Permissible Exposure Limits: NIOSH recommends adherence to the present Federal standard of 5 milligrams of nitric acid per cubic meter of air as a time-weighted average for up to a 10-hour workday, 40-hour workweek.

Routes of Entry: Skin contact or inhalation of vapors.

Harmful Effects

Exposure to nitric acid represents a dual health hazard; specifically, corrosion of the skin and other tissues from topical contact and acute pulmonary edema or chronic obstructive pulmonary disease from inhalation.

Medical Surveillance: NIOSH recommends that workers subject to nitric acid exposure have comprehensive preplacement and annual medical examinations including a 14" × 17" posterior-anterior chest x-ray, pulmonary function tests, and a visual examination of the teeth for evidence of dental erosion.

Special Tests: See Medical Surveillance above.

Personal Protective Methods: Engineering controls should be used wherever feasible to maintain airborne nitric acid concentrations below the recommended limit, and respirators should only be used in certain nonroutine or emergency situations.

Since many of the pulmonary disorders may result from the oxides of nitrogen, readily given off when nitric acid is dispersed in air, frequent reference is made in the criteria document to simultaneous environmental monitoring for airborne nitric acid and oxides of nitrogen.

Bibliography

National Institute for Occupational Safety and Health. *Criteria for a Recommended Standard: Occupational Exposure to Nitric Acid.* NIOSH Doc. No. 76-141 (1976).

NITROBENZENE

Description: $C_6H_5NO_2$, nitrobenzene, is a pale yellow liquid whose odor resembles bitter almonds.

Synonyms: Nitrobenzol, oil of mirbane, oil of bitter almonds.

Potential Occupational Exposures: Nitrobenzene is used in the manufacture of explosives and aniline dyes and as a solvent and intermediate. It is also used in shoe and floor polishes, leather dressings, and paint solvents, and to mask other unpleasant odors. Substitution reactions with nitrobenzene are used to form meta-derivatives.

A partial list of occupations in which exposure may occur includes:

Aniline dye makers	Paint makers
Explosive makers	Polish makers
Organic chemical synthesizers	

Permissible Exposure Limits: The Federal standard is 1 ppm (5 mg/m^3).

Routes of Entry: Inhalation and percutaneous absorption of liquid.

Harmful Effects

Local—Nitrobenzene may cause irritation of the eyes.

Systemic—There is a latent period of 1 to 4 hours before signs and symptoms appear. Nitrobenzene affects the central nervous system producing fatigue, headache, vertigo, vomiting, general weakness, and in some cases severe depression, unconsciousness, and coma. Nitrobenzene is a powerful methemoglobin former; cyanosis appears when methemoglobin reaches 15%. Sulfhemoglobin formation may also contribute to nitrobenzene toxicity. Chronic exposure may lead to spleen and liver damage, jaundice, liver impairments, and hemolytic icterus. Anemia and Heinz bodies in the red blood cells have been observed. Alcohol ingestion may increase the toxic effects.

Medical Surveillance: Preemployment and periodic examinations should be concerned particularly with a history of dyscrasias, reactions to medications, alcohol intake, eye disease, skin, and cardiovascular status. Liver and renal functions should be evaluated periodically, as well as blood and general health.

Special Tests: Follow methemoglobin levels until normal in all cases of suspected cyanosis. The metabolites in urine, p-nitro- and p-aminophenol, can be used as an evidence of exposure.

Personal Protective Methods: Impervious protective clothing should be worn in areas where risk of splash or spill exists. When splashed or spilled on ordinary work clothes, the clothes should be removed at once and the skin area washed thoroughly. In areas of vapor concentration fullface masks with organic vapor canisters or air supplied respirators should be used. Clean work clothing should be supplied daily, and showering made mandatory after each shift before workers change to street clothes.

Bibliography

Andreescheva, N.G. 1964. Substantiation of the maximum permissible concentration of nitrobenzene in atmospheric air. *Hyg. Sanit.* 29:4.

Myslak, A., Piotrowski, J.K., and Musialowicz, E. Acute nitrobenzene poisoning. A case report with data on urinary excretion of p-nitro-phenol and p-amino-phenol. *Arch. Toxikol.* 28:208.

Salmowa, J., Piotrowski, J. and Neuhorn, U. 1963. Evaluation of exposure to nitrobenzene. Absorption of nitrobenzene vapor through lungs and excretion of p-nitrophenol in urine. *Brit. J. Ind. Med.* 20:41.

4-NITROBIPHENYL

Description: $C_6H_5C_6H_4NO_2$, 4-nitrobiphenyl, exists as yellow plates or needles.

Synonyms: 4-Nitrodiphenyl, p-nitrobiphenyl, p-nitrodiphenyl, PNB.

Potential Occupational Exposures: 4-Nitrobiphenyl was formerly used in the synthesis of 4-aminodiphenyl. It is presently used only for research purposes; there are no commercial uses.

A partial list of occupations in which exposure may occur includes:

Research workers

Permissible Exposure Limits: 4-Nitrobiphenyl was included in the Federal standard for carcinogens; all contact with it should be avoided.

Routes of Entry: Inhalation and percutaneous absorption.

Harmful Effects

Local—None reported.

Systemic—4-Nitrobiphenyl is considered to be a human carcinogen. This is based on the evidence that it will induce bladder tumors in dogs and that human cases of bladder cancer were reported from a mixed exposure to 4-aminodiphenyl and 4-nitrobiphenyl. These human cases were attributed to 4-aminodiphenyl because the information available at the time showed that it produced bladder tumors in dogs. 4-Aminobiphenyl may be a metabolite.

Medical Surveillance: Placement and periodic examinations should include an evaluation of exposure to other carcinogens, as well as an evaluation of smoking, of use of alcohol and medications, and of family history. Special attention should be given on a regular basis to urine sediment and cytology. If red cells or positive smears are seen, cystoscopy should be done at once. The general health of exposed persons should also be evaluated in periodic examinations.

Special Tests: None commonly used. Can probably be determined in the urine as a metabolite.

Personal Protective Methods: These are designed to supplement engineering and to prevent all skin or respiratory contact. Full body protective clothing and gloves should be used by those employed in handling operations. Full-face, supplied air respirators of continuous flow or pressure demand type should also be used. On exit from a regulated area, employees should shower and change into street clothes, leaving their protective clothing and equipment at the point of exit to be placed in impervious containers at the end of the work-shift for decontamination or disposal. Effective methods should be used to clean and decontaminate gloves and clothing.

Bibliography

Deichmann, W.B. 1967. Introduction p. 3. In K.F. Lampe, ed. *Bladder Cancer, A Symposium.* Aesculapius Publishing Co., Birmingham, Alabama.

Melick, W.F., Escue, H.M., Naryka, J.J., Mezera, R.A., and Wheeler, E.P. 1955. The first reported cases of human bladder tumors due to a new carcinogen—xenylamine. *J. Urol.* 74:760.

NITROGEN OXIDES

Description: Nitrogen oxides include the following—

Nitrous oxide: N_2O

Nitric oxide: NO
Nitrogen dioxide: NO_2
Nitrogen trioxide: N_2O_3
Nitrogen tetroxide: N_2O_4
Nitrogen pentoxide: N_2O_5
Nitric acid: HNO_3
Nitrous acid: HNO_2

Nitrous oxide, N_2O, is a colorless, noncombustible gas, sweet-tasting, and slightly soluble in water. Nitric oxide is a colorless gas slightly soluble in water. Nitric oxide combines with oxygen to form nitrogen dioxide which is a reddish-brown gas with a characteristic odor. Nitrogen dioxide exists in equilibrium with nitrogen tetroxide, and these two compounds and oxygen are in equilibrium with the crystalline nitrogen pentoxide. However, nitrogen dioxide and nitric oxide are the dissociation products of nitrogen trioxide. When nitrogen dioxide comes in contact with water, nitrous acid and nitric acid are formed. Nitric acid is a colorless liquid when pure, but on exposure to light, the liquid may turn yellowish-brown as a result of nitrogen dioxide formation. Nitric acid mist almost always contains nitrogen oxide gases and is, therefore, included in this group. Nitrogen dioxide decomposes in water, nitrogen pentoxide is slightly soluble in water and nitric acid (70% aqueous solution) is soluble in water.

Synonyms

Nitrous oxide: Nitrogen monoxide
Nitric oxide: Mononitrogen monoxide
Nitrogen dioxide: None
Nitrogen trioxide: Dinitrogen trioxide, nitrous anhydride
Nitrogen tetroxide: Dinitrogen tetroxide
Nitrogen pentoxide: Nitric anhydride
Nitrous acid: None
Nitric acid: Aqua fortis, azotic acid, hydrogen nitrate

Potential Occupational Exposures: Exposure to nitrogen oxides is typically a mixed exposure to "nitrous fumes" which may evolve from various manufacturing processes and in many other industrial situations. Exposure to nitrogen oxides may occur during the manufacture of nitric and sulfuric acid, oxidized cellulose compounds, explosives, rocket propellants, fertilizers, dyes and dyestuffs, pharmaceuticals, and various other organic and inorganic chemicals such as nitrites, nitrates, and other nitro compounds, aqua regia, arsenic acid, oxalic acid, nitrous acid, phthalic acid, and phosphoric acid. Exposure may also occur during jewelry manufacturing, etching, brazing, lithographing, metal cleaning, textile (rayon) and food bleaching, glass blowing, electroplating, gas and electric arc welding, and during the nitration of chloroform. Nitrogen oxides also occur in garages from automobile exhaust, in silos from organic material decomposition, in tunnels following blasting, and when nitric acid comes in contact with organic materials.

A partial list of occupations in which exposure may occur includes:

Braziers	Medical technicians
Dentists	Metal cleaners
Dye makers	Nurses
Fertilizer makers	Organic chemical synthesizers
Food and textile bleachers	Photoengravers
Garage workers	Physicians
Gas and electric arc welders	Silo fillers
Jewelry makers	Sulfuric acid makers

Permissible Exposure Limits: The Federal standards are: nitric oxide, 25 ppm (30 mg/m^3); nitrogen dioxide, 5 ppm (9 mg/m^3); and nitric acid, 2 ppm (5 mg/m^3); determined as a TWA. Currently there are no standards for the other listed compounds. NIOSH has recommended a ceiling level of 1 ppm for nitrogen dioxide.

Routes of Entry: Inhalation of gas in the case of nitrogen oxide gases; inhalation of the mist or vapor, in the case of nitric and nitrous acids.

Harmful Effects

Local—Nitrogen oxide gases may produce irritation of the eyes and mucous membranes. Prolonged low level exposure may produce yellowish or brownish staining of the skin and teeth; however, this sign usually indicates nitric acid exposure. Nitric acid and nitrogen tetroxide are extremely corrosive liquids and may cause severe burns, ulcers, and necrosis of the skin, mucous membranes, and eye tissues.

Systemic—Exposure to high concentrations of nitrogen oxides may result in severe pulmonary irritation and methemoglobinemia. The former is believed to be caused by the nitrogen dioxide portion, while the latter is mainly caused by nitric oxide. It is postulated that nitric oxide is nonirritating but the distinction is of questionable importance since nitric oxide exposure generally includes other nitrogen oxides; moreover, nitric oxide at even moderate concentrations oxidizes rapidly and spontaneously in the presence of atmospheric oxygen.

Nitrogen dioxide at high concentrations has also been shown to cause methemoglobinemia in the dog. Typically, acute exposure may produce immediate malaise, cyanosis, cough, dyspnea, chills, fever, headache, nausea, and vomiting. Collapse and death may occur if exposure is sufficiently high. When lower concentrations are encountered, there may be only mild signs of bronchial irritation followed by a five to twelve-hour symptom-free period. Subsequently, the onset of signs and symptoms of acute pulmonary edema occur suddenly, which unfortunately may take place away from prompt medical aid.

Nitrogen oxides may be formed from green silage in amounts which, when restrained to the confines of a silo, may constitute a serious health hazard. "Silo-filler's disease" is the name used to designate the syndrome culminating in bronchiolitis fibrosa obliterans, caused by exposure to nitrogen oxides evolved in this way.

If the acute episode is survived, bronchiolitis fibrosa obliterans may develop usually within a few days but may be latent for as long as six weeks. Victims may develop severe and increasing dyspnea which is often accompanied by fever and cyanosis. Chest roentgenogram may reveal a diffuse, reticular, and fine nodular infiltration or numerous, uniform, scattered nodular densities ranging in size from 1 to 5 mm in diameter.

Chronic exposure may result in pulmonary dysfunction with decreased vital capacity, maximum breathing capacity and lung compliance, and increased residual volume. The most common complaint is of dyspnea upon exertion. Signs include moist rales and wheezes, sporadic cough with mucopurulent expectoration, a decrease in blood pH and serum proteins, and an increase in urinary hydroxyproline and acid mucopolysaccharides. These findings are suggestive of emphysema, although they are yet inconclusive.

The development of methemoglobinemia is typically mild and transient. In rare cases individuals may have a preexisting constantly high methemoglobin level due to a genetic defect. Such individuals are more susceptible to toxic methemoglobinemia.

Medical Surveillance: Preplacement and periodic examinations should be concerned particularly with the skin, eyes, and with significant pulmonary and heart diseases. Periodic chest x-rays and pulmonary function tests may be useful. Smoking history should be known. Methemoglobin studies may be of interest if exposure to nitric oxide is present. In the case of nitric acid vapor mist exposure, dental effects may be present.

Special Tests: None.

Personal Protective Methods: Workers should not enter confined areas where nitrogen oxides may accumulate (for example, silos) without appropriate eye and respiratory protection.

Individuals should be equipped with supplied air respirators with fullface piece or chemical goggles, and enclosed areas should be properly ventilated before entering. An observer equipped with appropriate respiratory protection should be outside the area and standing by to supply any aid needed.

Bibliography

Clutton-Brock, J. 1967. Two cases of poisoning by contamination of nitrous oxide with higher oxides of nitrogen during anesthesia. *Br. J. Anaesth.* 39:388.

Cooper, W.C., and Tabershaw, I.R. 1965. Biologic effects of nitrogen dioxide in relation to air quality standards. *Arch. Environ. Health* 10:455.

Kosmider, S., Ludyga, K., Musiewicz, A., Drozdz, M., and Sogan, J. 1972. Experimental and clinical investigations of the emphysematous effect of nitrogen oxides. *Zentralbl. Arbeitsmed.* 22:363.

Milne, J.E.H. 1969. Nitrogen dioxide inhalation and bronchiolitis obliterans. *J. Occup. Med.* 11:530.

Morley, R., and Silk, S.J. 1970. The industrial hazard from nitrous fumes. *Ann. Occup. Hyg.* 13:101.

National Institute for Occupational Safety and Health. *Criteria for a Recommended Standard: Occupational Exposure to Oxides of Nitrogen.* NIOSH Doc. No. 76-149 (1976).

Ramirez, R.J., and Dowell, A.R. 1971. Silo-filler's disease: nitrogen dioxide induced lung injury. *Ann. Intern. Med.* 74:569.

Scott, E.G., and Hunt, W.B. 1973. Silo-filler's disease. *Chest* 63:701.

NITROGLYCERIN AND ETHYLENE GLYCOL DINITRATE

Description: $C_3H_5(ONO_2)_3$, nitroglycerin. $O_2NOCH_2CH_2ONO_2$, ethylene glycol dinitrate. Both are oily, yellow liquids and are highly explosive. They may be detonated by mechanical shock, heat, or spontaneous chemical reaction.

Synonyms: Nitroglycerin, nitroglycerol; glyceryl trinitrate; trinitroglycerol; glonoin; trinitrin. Ethylene glycol dinitrate, nitroglycol; glycol dinitrate; ethylene dinitrate; EGDN.

Potential Occupational Exposures: Although ethylene glycol dinitrate is an explosive in itself, it is primarily used to lower the freezing point of nitroglycerin; together these compounds are the major constituents of commercial dynamite, cordite, and blasting gelatin. Occupational exposure generally involves a mixture of the two compounds. Ethylene glycol dinitrate is 160 times more volatile than nitroglycerin. Nitroglycerin is also used as a pharmaceutical.

A partial list of occupations in which exposure may occur includes:

Drug makers   Explosive makers

Permissible Exposure Limits: The Federal standard for nitroglycerin is 0.2 ppm (2 mg/m^3). The standard for ethylene glycol dinitrate and/or nitroglycerin is 0.2 ppm (2 mg/m^3) as a ceiling value, and, at concentrations greater than 0.02 ppm, personal protection may be necessary to avoid headache. These levels should be reduced when the substance is also absorbed percutaneously.

Routes of Entry: Inhalation of dust or vapor; ingestion of dust; percutaneous absorption.

Harmful Effects

Local—None reported.

Systemic—Exposure to small amounts of ethylene glycol dinitrate and/or nitroglycerin by skin exposure, inhalation, or swallowing may cause severe throbbing headaches. With larger exposure, nausea, vomiting, cyanosis, palpitations of the heart, coma, cessation of breathing, and death may occur. A temporary tolerance to the headache may develop, but this is lost after a few days without exposure. On some occasions a worker may have anginal pains a few days after discontinuing repeated daily exposure.

Medical Surveillance: Placement and periodic examinations should be concerned with central nervous system, blood, glaucoma, and especially history of alcoholism.

Special Tests: None commonly used, but urinary and blood ethylene glycol dinitrate may be determined by gas chromatography.

Personal Protective Methods: Both compounds are readily absorbed through the skin, lungs, and mucous membranes. It is, therefore, essential that adequate skin protection be provided for each worker: impervious clothing where liquids are likely to contaminate and full body clothing where dust creates the problem. All clothing should be discarded at the end of the shift and clean work clothing provided each day. Showers should be taken at the end of each shift and prior to changing to street clothing. In case of spill or splash that contaminates work clothing, the clothes should be changed at once and the skin area washed thoroughly. Masks of the dust type or organic vapor canister type may be necessary in areas of concentration of dust or vapors.

Bibliography

Bartalini, E., Cavagna, G., and Foa, V. 1967. Epidemiological and clinical features of occupational nitroglycol poisoning in Italy. *Med. Lavoro.* 58:618.

Carmichael, P., and Lieben, J. 1963. Sudden death in explosive workers. *Arch. Environ. Health* 7:424.

Lund, R.P., Haggendal, J. and Johnsson, G. 1968. Withdrawal symptoms in workers exposed to nitroglycerin. *Br. J. Ind. Med.* 25:136.

Munch. J.C., Friedland, B., and Shepard, M. 1965. Glyceryl trinitrate. II. Chronic toxicity. *Ind. Med. Surg.* 34:940.

NITROPARAFFINS

Description: Nitroparaffins are characterized by a $-C-NO_2$ group and may be either mono- or poly-substituted. Only certain mononitroparaffins are included in this section: nitromethane, nitroethane and 1- and 2-nitropropane. All of these are colorless liquids. Other mononitroparaffins are not commonly used, and use of the polynitroparaffins is limited almost entirely to fuels and fuel additives.

Synonyms: None.

Potential Occupational Exposures: Nitroparaffins are used as solvents for cellulose esters, vinyl copolymer, and other resins, oils, fats, waxes, and dyes. They are also used in various coating materials such as shellac, synthetic and processed rubber, paint and varnish removers, alkyl resins, and other high polymer coatings, and also in organic synthesis (1)-(5).

A partial list of occupations in which exposure may occur includes:

Cellulose workers	Resin makers
Dye makers	Rubber makers
Fat processors	Stainers
Organic chemical synthesizers	Wax makers
Plastic makers	

It is estimated that 100,000 workers in the United States are exposed to 2-nitropropane (2). 2-Nitropropane is not known to occur naturally but has been detected in tobacco smoke with other nitroalkanes, the levels were found to correlate with tobacco nitrate contents (5). The smoke content of a filterless 85 mm United States blend cigarette was found to contain (μg): 2-nitropropane, 1.1; 1-nitropropane, 0.13; 1-nitrobutane, 0.71; nitroethane, 1.1; and nitromethane, 0.53.

Permissible Exposure Limits: The Federal standards for these substances are: nitromethane 100 ppm (250 mg/m^3), nitroethane 100 ppm (310 mg/m^3), 1-nitropropane 25 ppm (90 mg/m^3), and 2-nitropropane 25 ppm (90 mg/m^3). ACGIH (1978) adds that 2-nitropropane is an "Industrial Substance Suspect of Carcinogenic Potential for Man."

Route of Entry: Inhalation of vapor.

Harmful Effects

Local—The nitroparaffins are irritants to the eyes and upper respiratory tract. There may be slight skin irritation due to solvent drying of skin.

Systemic—Only one report of occupational illness from nitroparaffins has been reported. The workers were exposed to 20 to 45 ppm of 2-nitropropane and complained of anorexia, nausea, vomiting, diarrhea, and occipital headache. Animal experiments indicate that high concentrations of nitroparaffins may produce light narcosis and central nervous system irritation. The lethal dose is generally lower than that producing significant narcosis. Liver and kidney damage have been observed in animals at lethal concentrations. Nitroparaffins release nitrate in vivo; however, methemoglobinemia and Heinz bodies have only been observed with 2-nitropropane. Experimental evidence indicates that nitroparaffin toxicity increases with molecule size.

NIOSH has recently reported that 2-nitropropane produced liver cancer in rats after 6 months exposure at about 200 ppm and suggested that "it would be prudent to handle 2-nitropropane as if it were a human carcinogen" (2).

Information as to the mutagenicity of 2-nitropropanes appears to be lacking.

Medical Surveillance: Based on animal data, preplacement and periodic examination should consider respiratory and central nervous system effects as well as liver and kidney function.

Special Tests: None commonly used. In the case of 2-nitropropane, Heinz bodies and methemoglobin levels would be of interest.

Personal Protective Methods: Barrier creams or gloves to protect exposed skin and, where vapor concentrations are excessive, fullface mask with organic vapor canister or air supplied respirators are advised.

References

(1) Martin, J.C., and Baker, P.J. Jr. Nitroparaffins. In *Kirk-Othmer Encyclopedia of Chemical Technology*, 2nd ed., Vol. 13. Interscience, New York, pp. 864-885.
(2) Anon. 2-Nitropropane causes cancer in rats. *Chem. Eng. News.* May 2 (1977) p. 10.
(3) Hara, K. Stabilized methyl chloroform. Japan Patent, 7455,606, 30 May (1974). *Chem. Abstr.* 82 (1975) 3756E.
(4) Sawabe, S., Genda, G., and Yamamoto, T. Stabilizing chlorohydrocarbons by addition of aliphatic alcohols, polyalkyl ethers and nitroalkanes. Japan Patent, 7403,963, 29 Jan. (1974). *Chem. Abstr.* 81 (1974) 151517X.
(5) Hoffman, D., and Rathkamp, G. Chemical studies on tobacco smoke. III. Primary and secondary nitroalkanes in cigaret smoke. *Beitr. Tabakforsch.* 4 (1968) 124-134.

Bibliography

Skinner, J.B. 1947. The toxicity of 2-nitropropane. *Ind. Med.* 16:441.

NITROPHENOL

Description: There are three isomers of nitrophenol $NO_2C_6H_4OH$. The meta-form is produced from m-nitroaniline, and the ortho and para isomers are produced by nitration of phenol. They are colorless to slightly yellowish crystals with an aromatic to sweetish odor.

Synonyms: None.

Potential Occupational Exposures: Nitrophenols are used in the synthesis of dyestuffs and other intermediates and as a chemical indicator.

A partial list of occupations in which exposure may occur includes:

Chemical indicator makers   Organic chemical synthesizers

Permissible Exposure Limits: There is no Federal standard for nitrophenol.

Routes of Entry: Inhalation and percutaneous absorption of liquid.

Harmful Effects

Local—Unknown.

Systemic—There is very little information available on the toxicity for humans of nitrophenols. Animal experiments have shown central and peripheral vagus stimulation, CNS depression, methemoglobinemia, and dyspnea. The p-isomer is the most toxic.

Medical Surveillance: Based on animal studies, individuals with cardiovascular, renal, or pulmonary disease and those with anemia are probably more subject to poisoning by nitrophenol. Liver and renal function and blood should be evaluated in placement or periodic examinations.

Special Tests: None commonly used. Nitrophenol is excreted rapidly in the urine as a conjugate. It may also be present as a metabolite of parathion.

Personal Protective Methods: Nitrophenols are readily absorbed through intact skin and by inhalation; full body protective clothing and appropriate type organic vapor canisters in areas of concentrations of dust or vapors should be provided. Spills on work clothing necessitate immediate clothing change and thorough washing of the skin area. Clean work clothes should be supplied daily; showers should be taken at the end of each shift prior to changing to street clothes.

N-NITROSODIMETHYLAMINE

Description: $(CH_3)_2NN=O$, N-nitrosodimethylamine, is a yellow liquid of low viscosity, soluble in water, alcohol, and ether.

Synonyms: Dimethylnitrosamine, DMN.

Potential Occupational Exposures: DMN is used in the manufacture of dimethylhydrazine. It has also been used as an industrial solvent and a nematocide.

A number of nitrosamines have been patented for use as gasoline and lubricant additives, antioxidants, and also pesticides. Dimethylnitrosamine (DMN)

[(CH$_3$)$_2$NN=O] is used primarily in the electrolytic production of the hypergolic rocket fuel 1,1-dimethylhydrazine (37)(38). Other areas of utility include the control of nematodes (39), the inhibition of nitrification in soil (40), use as plasticizer for acrylonitrile polymers (41), use in active metal anode-electrolyte systems (high-energy batteries) (42), in the preparation of thiocarbonyl fluoride polymers (43), in the plasticization of rubber (44), and in rocket fuels (45). Some N-nitroso compounds have been used as organic accelerators and antioxidants in the production of rubber, including N-nitrosodiphenylamine, N,N'-nitrosopentamethylenetetramine, polymerized N-nitroso-2,2,4-trimethyl-1,2-dihydroquinoline, and N-methyl-N-4-dinitrosoaniline (46).

A partial list of occupations in which exposure may occur includes:

Dimethylhydrazine makers   Solvent workers
Nematocide makers

Synthetic cutting fluids, semisynthetic cutting oils and soluble cutting oils may contain nitrosamines, either as contaminants in amines, or as products from reactions between amines and nitrite (45)-(50). Concentrations of nitrosamines have been found in certain synthetic cutting oils at levels ranging from 1 ppm to 1,000 ppm. It is believed that there are 8 to 12 additives that could be responsible for nitrosamine formation in cutting oils (49) and that approximately 750,000 to 780,000 persons employed by more than 1,000 cutting fluid manufacturing firms are endangered, in addition to an undetermined number of machine shop workers who use the fluids (49)(50).

N-nitroso compounds, primarily dimethylnitrosamine (DMN) have been found to be present as air pollutants in the ambient air of residential areas of Baltimore (51) with DMN levels varying from 16 to 760 ng/m^3 while on an industrial site in Baltimore, DMN levels reached 32,000 ng/m^3 (10.67 ppb) of ambient air in close proximity to a chemical factory which manufactured unsymmetrical dimethylhydrazine for which DMN was used as an intermediate (52). In Belle, West Virginia, DMN has been found near chemical factories which handle dimethylamine (53).

Another area of recent concern involves the finding of nitrosamines in a variety of herbicides (54)-(57). These ranged from less than 50 micrograms to 640,000 micrograms per liter of nitrosamines (e.g., N-nitrosodimethylamine and N-nitrosodipropylamine) (46). It has been estimated that 950 to 1,000 pesticide products may contain nitrosamines, and a sizeable number of these are available for use by homeowners (55). High levels of nitrosamines in soils (believed to arise from the use of triazine herbicide which can combine with nitrogen fertilizer) have been previously reported (58) as well as plant uptake and leaching of dimethylnitrosamine (59).

N-nitrosodiethanolamine has been found in amounts ranging from 1 ng/g (1 ppb) to a high of about 48,000 ppb in about 30 toiletry products (e.g., cosmetics, hand and body lotions and shampoos. The N-nitroso compound found probably results from nitrosation of di- and/or triethanolamine emulsifiers by a nitrite compound (59a).

Permissible Exposure Limits: DMN is included in the Federal standard for carcinogens; all contact with it should be avoided.

Routes of Entry: Inhalation of vapor and possibly percutaneous absorption.

Harmful Effects

Local—The liquid and vapor are not especially irritating to the skin or eyes, and warning properties are poor.

Systemic—DMN is a highly toxic substance in most species, including man. Systemic effects are characterized by onset in a few hours of nausea and vomiting, abdominal cramps, and diarrhea. Also headache, fever, weakness, enlargement of the liver, and jaundice may occur. Chronic exposures may lead to liver damage (central necrosis), with jaundice and ascites. There have been a number of reported cases, including severe liver injury in man and one death. Autopsy revealed an acute diffuse centrolobular necrosis. Recovery occurred in other cases.

In rats, guinea pigs, and other experimental animals, DMN is a highly potent carcinogen, producing malignant tumors, primarily of the liver and kidney, but also in the lung. Both ingestion and inhalation routes have produced tumors. These have not been reported in man, but in view of its potency in various other species, the material has been presumed to be carcinogenic in man also.

Nitrosamines possess considerable diversity of action, especially as carcinogens (1)-(5). Their occurrence, whether as direct emissions of N-nitroso compounds or via localized release of large amounts of precursor compounds (e.g., secondary amines, nitrogen oxides, nitrates, nitrites), effluent discharges from sewage treatment plants or runoff from feedlots or croplands treated with amine pesticides, ammonium fertilizers or nitrogenous organic materials (6)-(16) or accidental products in food processing and use, tobacco smoke (17)-(22), or via the body burden contributed by in vivo nitrosation (23)-(26) reactions, has sparked ever increasing intensive investigations as to the overall scope of the potential sources, mechanism of in vitro and in vivo formation, body burdens as well as to the need to develop a proper scientific foundation for a human health risk assessment (17)(22)(23)(27)-(36).

With regard to the mutagenicity of nitroso compounds, it is of note that the nitrosamines which are believed to require enzymatic decomposition before becoming active carcinogens (e.g., dimethyl- and diethylnitrosamines) are mutagenic in *Drosophila* (60)-(64) and *Arabidopsis thaliana* (65) and inactive in microorganisms such as *E. coli* (66)(67), *Serratia marcescens* (66), *Saccharomyces cerevisiae* (68)(69) and *Neurospora* (70) [but active in *Neurospora* (71) in the hydroxylating model system of Udenfriend (72), or in the presence of oxygen]. Eleven carcinogenic N-nitrosamines have been found mutagenic on *S. typhimurium* TA 100 (but not TA 98) when the bacteria, test substance were preincubated with rat liver S-9 mix, and then poured on a plate (72).

N-dealkylation of dimethylnitrosamine (DMN) and diethylnitrosamine (DEN) by tissue specific microsomal mixed-function oxidases is believed to generate alkylating intermediates that are responsible for the mutagenic (73)-(77), toxic and carcinogenic (3)(5)(78) effects of the parent compound in vivo and in vitro.

Medical Surveillance: Based on human experience and on animal studies, preplacement and periodic examinations should include a history of exposure to other carcinogens, alcohol and smoking habits, medications, and family history. Special attention should be given to liver size and function, and to any changes in lung symptoms or x-rays. Renal function should be followed. Sputum and urine cytology may be useful.

Special Tests: None commonly used.

Personal Protective Methods: These are designed to supplement engineering controls and to prevent all contact with the skin, eyes, or respiratory tract. Full body protective clothing and gloves should be provided and also appropriate type fullface supplied air respirators of continuous flow or pressure demand type.On exit from a regulated area, employees should be required to shower before changing into street clothes, leaving their protective clothing and equipment at the point of exit, to be placed in impervious containers at the end of the work shift for decontamination or disposal.

References

(1) Magee, P.N., and Barnes, J.M. The production of malignant primary hepatic tumors in the rat by feeding dimethylnitrosamine. *Br. J. Cancer* (1956) 10 114-122.
(2) Heath, D.F., and Magee, P.N. Toxic properties of dialkylnitrosamines and some related compounds. *Br. J. Ind. Med.* 19 (1962) 276-282.
(3) Magee, P.N., and Barnes, J.M. Carcinogenic nitroso compound. In Haddow, A., Weinhouse, S., (eds): *Advances In Cancer Research* 10 (1967) 163-246.
(4) Druckrey, H., et al. Erzeugung von krebs durch eine einmalige dosis von methylnitroso-harnstoff und verschiedenen dialkylnitrosaminen an ratten. *Z. Krebsforsch.* 66 (1964) 1-10.
(5) Druckrey, H., et al. Organotrope carcinogene wirkungen bei 65 verschiedenen N-nitroso-verbindungen an BD-ratten. *Z. Krebsforsch.* 69 (1967) 103-201.
(6) Magee, P.N. Possibilities of hazard from nitrosamines in industry. *Ann. Occup. Hyg.* 15 (1972) 19-22.
(7) Feuer, H. (ed). *The Chemistry of the Nitro and Nitroso Groups.* New York. Interscience Publishers Inc. (1969).
(8) Lijinsky, W., and Epstein, S.E. Nitrosamines as environmental carcinogens. *Nature* 225 (1970) 21-23.
(9) Shapley, D. Nitrosamines: Scientists on the trail of prime suspect in urban cancer. *Science* 191 (1976) 270.

(10) Wolfe, N.L., Zepp, R.G., Gordon, J.A., and Fincher, R.C. N-nitrosamine formation from atrazine. *Bull. Env. Contam. Toxicol.* 15 (1976) 342.

(11) Bogovski, P., and Walker, E.A. *N-Nitroso Compounds in the Environment.* IARC Scientific Publication No. 9. International Agency for Research on Cancer, Lyon (1974).

(12) Ayanaba, A., and Alexander, M. Transformation of methylamines and formation of a hazardous product, dimethylnitrosamine, in samples of treated sewage and lake water. *J. Environ. Qual.* 3 (1974) 83-89.

(13) Ayanaba, A., Verstraete, W., and Alexander, M. Formation of dimethylnitrosamine, a carcinogen and mutagen, in soils treated with nitrogen compounds. *Soil Sci. Soc. Am. Proc.* 37 (1973) 565-568.

(14) Tate, R.L., and Alexander, M. Formation of dimethylamine and diethylamine in soil treated with pesticides. *Soil Sci.* 118 (1974) 317-321.

(15) Tate, R.L., and Alexander, M. Stability of nitrosamines in samples of lake water, soil, and sewage. *J. Natl. Cancer Inst.* 54 (1975) 327-330.

(16) Elespuru, R. and Lijinsky, W. The formation of carcinogenic nitroso compounds from nitrite and some types of agricultural chemicals. *Food Cosmet. Toxicol.* 11 (1973) 807-817.

(17) Issenberg, P. Nitrite, nitrosamines, and cancer. *Fed. Proc.* 35 (1976) 1322.

(18) Fiddler, W., et al. Formation of N-nitrosodimethylamine from naturally occurring quaternary ammonium compounds and tertiary amines. *Nature* 236 (1972) 307.

(19) Scientific status summary, nitrites, nitrates, and nitrosamines in food: A dilemma. *J. Food Sci.* 37 (1972) 989-991.

(20) Serfontein, W.J., Smit, J.J. Evidence for the occurrence of N-nitrosamines in to-bacco. *Nature* 214 (1967) 169-170.

(21) Neurath, G. Zur frage des vorkommens von N-nitroso-vergindungen in tabakrauch. *Experientia* 23 (1967) 400-404.

(22) Wolff, I.A., and Wasserman, A.E. Nitrates, nitrites, and nitrosamines. *Science* 177 (1972) 15-19.

(23) Low, H. Nitroso compounds—safety and public health. *Arch. Env. Hlth.* 29 (1974) 256.

(24) Sander, J. Kann nitrit in der menschlichen nahrung ursache einer krebsentstehung durch nitrosaminebildung sein? *Arch. Hyg. Bakt.* 151 (1967) 22-24.

(25) Sander J., and Seif, F. Bakterielle reduktion von nitrat im magen des menschen als ursache einer nitrosamin-bildung. *Arzneim Forsch.* 19 (1969) 1091-1093.

(26) Hawksworth, G., and Hill, M.J. The formation of nitrosamines by human intestinal bacteria. *Biochem. J.* 122 (1971) 28P-29P.

(27) Magee, P.N. *Food Cosmet. Toxicol.* 9 (1971) 207.

(28) Wogan, G.N., and Tannenbaum, S.R. Environmental N-nitroso compounds: Implications for public health. *Toxicol. Appl. Pharmacol.* 31 (1975) 373.

(29) Scanlan, R.A. In *Critical Reviews in Food Technology.* Cleveland, CRC Press, (1975) Vol. 5, No. 4.

(30). Sen, N.P. In *Toxic Constituents of Animal Foodstuffs.* edited by I.E. Liener, New York, Academic (1974).

(31) Shank, R.C. Toxicology of N-nitroso compounds. *Toxicol. Appl. Pharmacol.* 31 (1975) 361.

(32) Mirvish, S.S. Formation of N-nitroso compounds: Chemistry, kinetics and in vitro occurrence. *Toxicol. Appl. Pharmacol.* 31 (1975) 325.

(33) Fiddler, W. The occurrence and determination of N-nitroso compounds. *Toxicol. Appl. Pharmacol.* 31 (1975) 352.

(34) Anon. Dupont to limit nitrosamine emissions: EPA studies nitrosamine problem. *Toxic Materials News* 3 (1976) 111.

(35) Anon. EPA advisory unit recommends no immediate regulatory action on nitro-samines. *Toxic Materials News* 3 (1975) 139.

(36) Anon. Science advisory board endorses closer look at nitrosamines. *Env. Hlth. Letter* 15 (1976) 1.

(37) Horvitz, D., and Cerwonka, E. U.S. Patent 2,916,426 (1959). *Chem. Abstr.* 54 (1960) 6370c.

(38) National Distillers and Chem. Corp. British Patent 817,523 (1959). *Chem. Abstr.* 54 (1960) 10601c.

(39) Maitlen, E.G. U.S. Patent 2,970,939 (1961). *Chem. Abstr.* 56 (1961) 11752f.

(40) Goring, C.A.I. U.S. Patent 3,256,083 (1966). *Chem. Abstr.* 65 (1966) 6253d.

(41) Lytton, M.R., Wielicki, E.A., and Lewis, E. U.S. Patent 2,776,946 (1957). *Chem. Abstr.* 51 (1957) 5466e.

(42) Elliot, W.E., Huff, J.R., Adler, R.W., and Towle, W.L. *Proc. Ann. Power Sources Conf.* 20 (1966) 67-70. *Chem. Abstr.* 66 (1967) 100955w.

(43) Middleton, W.J. U.S. Patent 3,240,765 (1966). *Chem. Abstr.* 64 (1966) 19826f.

(44) Lel'Chuck, Sh. L., and Sedlis, V.I. *Zh. Prikl. Khim.* 31 (1958) 128. *Chem. Abstr.* 52 (1958) 17787g.

(45) Klager, K. U.S. Patent 3,192,707 (1965).

(46) Boyland, E., Carter, R.L., Gorrod, J.W., and Roe, F.J.C. Carcinogenic properties of certain rubber additives. *Europ. J. Cancer* 4 (1968) 233.

(47) Anon. Nitrosamines warning. *Am. Ind. Hyg. Assoc. J.* 37 (1976) A-9.

(48) Anon. Nitrosamines reported in industrial cutting oils. *Occup. Hlth. Safety Letter.* 5 (1976) 3-4.

(49) Anon. Nitrosamines found in cutting oils. *Toxic Materials News* 3 (1976) 156-157.

(50) Anon. NIOSH issues nitrosamine alert. *Toxic Materials News* 3 (1967) 167.

(51) Fine, D.H., Roundbehler, D.P., Sawicki, E., Krost, K., and DeMarrais, G.A. N-nitroso compounds in the ambient community air of Baltimore, Maryland. *Anal. Letters* 9 (1976) 595-604.

(52) Pellizzari, E.D., Bunch, J.E., Bursey, J.T., Berkley, R.E., Sawicki, E., and Krost, K. Estimation of N-nitrosodimethylamine levels in ambient air by capillary gas-liquid chromatography/mass spectrometry. *Anal. Letters* 9 (1976) 579-594.

(53) Fine, D.H., Roundehler, D.P., Belcher, N.M., and Epstein, S.S. *Science* (1976) in press, cited in ref. 51.

(54) Anon. Nitrosamines found in herbicides: In vivo formation documented. *Toxic Materials News.* 3 (1976) 148.

(55) Anon. Nitrosamines: EPA refuses suspension request: NIOSH promises alert. *Toxic Materials News* 3 (1976) 156.

(56) Anon. Nitrosamines found in commercial pesticides. *Chem. Eng. News.* 54 (1976) 33.

(57) Anon. EPA investigating nitrosamine impurities in 24 herbicides. *Chem. Eng. News* 54 (1976) 12.

(58) Anon. High levels of nitrosamines in soils lead to government task force. *Pesticide Chem. News.* 4 (1976) 56.

(59) Dean-Raymond, D., and Alexander, M. Plant uptake and leaching of dimethylnitrosamine. *Nature* 262 (1976) 394-395.

(59a) Anon. N-nitrosamines found in toiletry products. *Chem. Eng. News* 55 (1977) 7, 8.

(60) Pasternak, L. Untersuchung uber die mutagene wirkung verschiedener nitrosamin und nitrosamid-verbindungen. *Arzneimittel-Forsch.* 14 (1964) 802.

(61) Pasternak, L. Mutagene wirkung von dimethyl nitrosamin bei *Drosophila melanogaster. Naturwissenschaften* 49 (1962) 381.

(62) Pasternak, L. The mutagenic effect of nitrosamines and nitroso methyl urea, untersuchungen uber die mutagene wirkung von nitrosaminen und nitrosomethyl harnstoff. *Acta. Biol. Med. Ger.* 10 (1963) 436-438.

(63) Fahmy, O.G., and Fahmy, M.J. Mutational mosaicism in relation to dose with the amine and amide derivatives of nitroso compounds in *Drosophila. Mutation Res.* 6 (1968) 139.

(64) Fahmy, O.G., Fahmy, M.J., Massasso, J., and Ondrej, M. Differential mutagenicity of the amine and amide derivatives of nitroso compounds in *Drosophila. Mutation Res.* 3 (1966) 201.

(65) Veleminsky, J., and Gichner, T. The mutagenic activity of nitrosamines in *Arabidopsis thaliana. Mutation Res.* 5 (1968) 429.

(66) Geisler, E. Uber die wirkung von nitrosaminen auf mikroorganismen. *Naturwissenschaften* 49 (1962) 380.

(67) Pogodina, O.N. O mutagennoy aktionosti kancerogenoviz gruppy nitrosaminov. *Cytologia* 8 (1966) 503.

(68) Marquardt, H., Zimmermann, F.K., and Schwaier, R. Die wirkung krebasuslösender nitrosamine und nitrosamide auf den adenin-6-45-ruck mutationssystem von *S. cerevisiae. Z. Verebungslehre* 95 (1964) 82.

(69) Marquardt, H., Zimmermann, F.K., and Schwaier, R. Nitrosamide als mutagene agentien. *Naturwiss.* 50 (1963) 625.

(70) Marquardt, H., Schwaier, R., and Zimmermann, F. Nicht-mutagenität von nitrosaminen bei *Neurospora crassa. Naturwiss.* 50 (1963) 135.

(71) Malling, H.V. Mutagenicity of two potent carcinogens, dimethylnitrosamine and diethylnitrosamine in *Neurospora crassa. Mutation Res.* 3 (1966) 537.

(72) Yahagi, I., Nagao, M., Seino, Y., Matsushima, T., Sugimura, T., and Okada, M. Mutagenicities of N-nitrosamines on *Salmonella. Mutation Res.* 48 (1977) 121-130.

(73) Malling, H.V. Dimethyl nitrosamine: Formation of mutagenic compounds by interaction with mouse liver microsomes. *Mutation Res.* 13 (1971) 425.

(74) Malling, H.V., and Frant, C.N. In vitro versus in vivo metabolic activation of mutagens. *Env. Hlth. Persp.* 6 (1973) 71.

(75) Czygan, P., Greim, H., Garro, A.J., Hutterer, F., Schaffner, F., Popper, H., Rosenthal, O., and Cooper, D.Y. Microsomal metabolism of dimethylnitrosamine and the cytochrom P-450 dependency of its activation to a mutagen. *Cancer Res.* 33 (1973) 2983.

(76) Bartsch, H., Macaveille, C., and Montesano, R. The predictive value of tissue-mediated mutagenicity assays to assess the carcinogenic risk of chemicals. In *Screening Tests in Chemical Carcinogenesis* (eds.) Montesano, R., Bartsch, H., and Tomatis, L., WHO/IARC Publ. No. 12, International Agency for Research on Cancer, Lyon (1976) pp. 467-491.

(77) Bartsch, H., Malaveille, C., and Montesano, R. In vitro metabolism and microsome-mediated mutagenicity and dialkylnitrosamines in rat, hamster, and mouse tissues. *Cancer Res.* 35 (1975) 644.

(78) Heath, D.F. The decomposition and toxicity of dialkylnitrosamines in rats. *Biochem. J.* 85 (1962) 72.

Bibliography

Jacobson, K.H., Wheelwright, H.G. Jr., Clem, J.H., and Shannon, R.N. 1955. Studies on the toxicology of N-nitrosodimethylamine vapor. *AMA Arch. Ind. Health* 2:617.

Le Page, R.N., and Christie, G.S. 1969. Induction of liver tumours in the rabbit by feeding dimethylnitrosoamine. *Br. J. Cancer* 23:125.

World Health Organization, International Agency for Research on Cancer. 1972. IARC Monographs on the Evaluation of Carcinogenic Risk of Chemicals to Man, Vol. I. International Agency for Research on Cancer, Lyon.

O

OSMIUM AND COMPOUNDS

Description: Os, osmium, is a blue-white metal. It is found in platinum ores and in the naturally occurring alloy osmiridium. Osmium when heated in air or when the finely divided form is exposed to air at room temperature oxidizes to form the tetroxide (OsO_4, osmic acid). It has a nauseating odor.

Synonyms: None.

Potential Occupational Exposures: Osmium may be alloyed with platinum metals, iron, cobalt, and nickel, and it forms compounds with tin and zinc. The alloy with iridium is used in the manufacture of fountain pen points, engraving tools, record player needles, electrical contacts, compass needles, fine machine bearings, and parts for watch and lock mechanisms. The metal is a catalyst in the synthesis of ammonia, and in the dehydrogenation of organic compounds. It is also used as a stain for histological examination of tissues. Osmium tetroxide is used as an oxidizing agent and as a fixative for tissues in electron microscopy. Other osmium compounds find use in photography. Osmium is no longer used in incandescent lights and in fingerprinting.

A partial list of occupations in which exposure may occur includes:

Alloy makers	Synthetic ammonia makers
Histology technicians	Platinum hardeners
Organic chemical synthesizers	

Permissible Exposure Limits: There is presently no Federal standard for osmium itself; the standard for osmium tetroxide is 0.002 mg/m³.

Route of Entry: Inhalation of vapor or fume.

Harmful Effects

Local—Osmium metal is innocuous, but persons engaged in the production of the metal may be exposed to acids and chlorine vapors. Osmium tetroxide vapors are poisonous and extremely irritating to the eyes; even in low concentrations, they may cause weeping and persistent conjunctivitis. Longer

340

exposure can result in damage to the cornea and blindness. Contact with skin may cause discoloration (green or black) dermatitis and ulceration.

Systemic—Inhalation of osmium tetroxide fumes is extremely irritating to the respiratory system, causing tracheitis, bronchitis, bronchial spasm, and difficulty in breathing which may last several hours. Longer exposures can cause serious inflammatory lesions of the lungs (bronchopneumonia with suppuration and gangrene). Slight kidney damage was seen in rabbits inhaling lethal concentrations of vapor for 30 minutes. Some fatty degeneration of renal tubules was seen in one fatal human case along with bronchopneumonia following an accidental overexposure.

Medical Surveillance: Consider the skin, eyes, respiratory tract, and renal function in preplacement or periodic examinations.

Special Tests: None in common use.

Personal Protective Methods: In areas where the concentration of osmium tetroxide fumes or vapors are excessive, fullface masks or supplied air respirators are necessary. Even low concentrations can cause severe irritation of the eyes. This can usually be prevented by proper ventilation (exhaust hoods, etc.) and the use of goggles. Gloves should be used to prevent burns of the skin and hands. Precautions should be taken to provide protection against acids and chlorine vapors in areas where the metal is produced. (See chlorine.)

Bibliography

McLoughlin, A.I.G., R. Milton, and K.M.A. Perry. 1946. Toxic manifestations of osmium tetroxide. *Br. J. Ind. Med.* 3:183.

OXALIC ACID

Description: HOOCCOOH·2H$_2$O, oxalic acid in solution, is a colorless liquid. Anhydrous oxalic acid is monoclinic in form and is produced by careful drying of the crystalline dihydrate.

Synonyms: Dicarboxylic acid, ethane-di-acid, ethanedioic acid.

Potential Occupational Exposures: Oxalic acid is used as an analytic reagent and in the manufacture of dyes, inks, bleaches, paint removers, varnishes, wood and metal cleansers, dextrin, cream of tartar, celluloid, oxalates, tartaric acid, purified methyl alcohol, glycerol, and stable hydrogen cyanide. It is also used in the photographic, ceramic, metallurgic, rubber, leather, engraving, pharmaceutical, paper, and lithographic industries.

A partial list of occupations in which exposure may occur includes:

Bleach makers	Laundry workers
Celluloid makers	Paper makers
Ceramic makers	Rubber makers
Cream of tartar makers	Tannery workers
Dye makers	Textile dyers
Glycerine makers	Printers
Ink makers	Wood bleachers

Permissible Exposure Limits: The Federal standard is 1 mg/m^3.

Route of Entry: Inhalation of mist and, occasionally, dust.

Harmful Effects

Local—Liquid has a corrosive action on the skin, eyes, and mucous membranes, which may result in ulceration. Local prolonged contact with extremities may result in localized pain, cyanosis, and even gangrenous changes probably resulting from localized vascular damage.

Systemic—Chronic exposure to mist or dust has been reported to cause chronic inflammation of the upper respiratory tract. Ingestion is of lesser importance occupationally. Symptoms appear rapidly and include shock, collapse, and convulsive seizures. Such cases may also have marked kidney damage with deposition of calcium oxalate in the lumen of the renal tubules.

Medical Surveillance: Evaluate skin, respiratory tract, and renal functions in placement or periodic examinations.

Special Tests: The presence of increased urinary oxalate crystals may be helpful in evaluating oral poisoning. Determination of blood calcium and oxalate levels may also be used for this purpose.

Personal Protective Methods: Protective clothing and goggles should be worn when working in areas where direct contact is possible. Respiratory protection from mist or dust may be needed.

Bibliography

Klauder, J.V., L. Shelanski, and K. Gabriel. 1955. Industrial uses of fluorine and oxalic acid. *AMA Arch. Ind. Health* 12:412.

OZONE

Description: O$_3$, ozone, is a bluish gas with a characteristic pungent odor, slightly soluble in water. Ozone is found naturally in the atmosphere as a result of the action of solar radiation and electrical storms. It is also formed around electrical sources such as x-ray or ultraviolet generators, electric arcs, mercury vapor lamps, linear accelerators, and electrical discharges.

Synonyms: None.

Potential Occupational Exposures: Ozone is used as an oxidizing agent in the organic chemical industry (e.g., production of azelaic acid); as a disinfectant for food in cold storage rooms and for water (e.g., public water supplies, swimming pools, sewage treatment); for bleaching textiles, waxes, flour, mineral oils and their derivatives, paper pulp, starch, and sugar; for aging liquor and wood; for processing certain perfumes, vanillin, and camphor; in treating industrial wastes; in the rapid drying of varnishes and printing inks; and in the deodorizing of feathers.

Industrial exposure often occurs around ozone generating sources, particularly during inert-gas shielded arc welding.

A partial list of occupations in which exposure may occur includes:

Air treaters	Organic chemical synthesizers
Arc welders	Sewage treaters
Cold storage food preservers	Textile bleachers
Industrial waste treaters	Water treaters
Liquor agers	Wax bleachers
Odor controllers	Wood agers
Oil bleachers	

Permissible Exposure Limits: The Federal standard is 0.1 ppm (0.2 mg/m^3).

Routes of Entry: Inhalation of the gas.

Harmful Effects

Local—Ozone is irritating to the eyes and all mucous membranes. In human exposures, the respiratory signs and symptoms in order of increasing ozone concentrations are: dryness of upper respiratory passages; irritation of mucous membranes of nose and throat; choking, coughing, and severe fatigue; bronchial irritation, substernal soreness, and cough. Pulmonary edema may occur, sometimes several hours after exposure has ceased. In severe cases, the pulmonary edema may be fatal.

Animal experiments demonstrate that ozone causes inflammation and congestion of respiratory tract and, in acute exposure, pulmonary edema, hemorrhage, and death.

Chronic exposure of laboratory animals resulted in chronic bronchitis, bronchiolitis, emphysematous and fibrotic changes in pulmonary parenchyma.

Systemic—Symptoms and signs of subacute exposure include headache, malaise, shortness of breath, drowsiness, reduced ability to concentrate, slowing of heart and respiration rate, visual changes, and decreased desaturation of oxyhemoglobin in capillaries. Animal experiments with chronic exposure showed aging effects and acceleration of lung tumorigenesis in lung-tumor susceptible mice.

Animal experiments further demonstrated that tolerance to acute pulmonary effects of ozone is developed and that this provided cross tolerance to other edemagenic agents. Antagonism and synergism with other chemicals also occur.

Ozone also has radiomimetic characteristics, probably related to its free-radical structure. Experimentally produced chromosomal aberrations have been observed.

Medical Surveillance: Preemployment and periodic physical examinations should be concerned especially with significant respiratory diseases. Eye irritation may also be important. Chest x-rays and periodic pulmonary function tests are advisable.

Special Tests: None.

Personal Protective Methods: In areas of excessive concentration, gas masks with proper canister and fullface-piece or goggles or the use of supplied air respirators is recommended.

Bibliography

Jaffe, L.S. 1967. The biological effects of ozone on man and animals. *Am. Ind. Hyg. Assoc. J.* 28:267.

Nasr, A.N.M. 1967. Biochemical aspects of ozone intoxication: A review. *J. Occup. Med.* 9:589.

Stokinger, H.E. 1954. Ozone toxicity—A review of the literature through 1963. *AMA Arch. Ind. Hyg. Occup. Med.* 9:366.

Stokinger, H.E. 1965. Ozone toxicology—A review of research and industrial experience: 1954-1964. *Arch. Environ. Health* 10:719.

P

PARAFFIN

Description: Paraffin is a white, somewhat translucent solid and consists of a mixture of solid aliphatic hydrocarbons. It may be obtained from petroleum.

Synonyms: Paraffin wax, hard paraffin.

Potential Occupational Exposures: Paraffin is used in the manufacture of paraffin paper, candles, food package materials, varnishes, floor polishes, and cosmetics. It is also used in waterproofing and extracting of essential oils from flowers for perfume.

A partial list of occupations in which exposure may occur includes:

Candle makers	Polish makers
Cosmetic makers	Varnish makers
Perfume makers	Wax paper makers

Permissible Exposure Limits: Paraffin wax fume has no established Federal standard; however, in 1978 the ACGIH recommended a TWA of 2.0 mg/m^3 for paraffin wax fume.

Route of Entry: Inhalation of fumes.

Harmful Effects

Local—Occasionally sensitivity reactions have been reported. Chronic exposure can produce chronic dermatitis, wax boils, folliculitis, comedones, melanoderma, papules, and hyperkeratoses.

Systemic—Carcinoma of the scrotum in wax pressmen exposed to crude petroleum wax has been documented. Other malignant lesions of an exposed area in employees working with finished paraffin are less well documented. Carcinoma of the scrotum, occurring in workmen exposed 10 years or more, began as a hyperkeratotic nevus-like lesion and developed into a squamous cell carcinoma. The lesions can metastasize to regional inguinal and pelvic lymph nodes. Paraffinoma has been reported from use of paraffin for cosmetic purposes.

Medical Surveillance: Medical examinations should be concerned especially with the skin. Surveillance should be continued indefinitely.

Special Tests: None appear useful.

Personal Protective Methods: Strict personal hygienic measures and protective clothing form the basis of a protective program.

Bibliography

Hendricks, N.V., C.M. Berry, J.G. Lione, and J.J. Thorpe. 1959. Cancer of the scrotum in wax pressmen. I. Epidemiology. *AMA Arch. Ind. Health* 19:524.

Hueper, W.C., and W.D. Conway. 1964. *Chemical Carcinogens and Cancers.* Charles C. Thomas, Springfield, Ill. p. 27.

Lione, J.G., and J.S. Denholm. 1959. Cancer of the scrotum in wax pressmen. II. Clinical observations. *AMA Arch. Ind. Health* 19:530.

Urbach, F., S.S. Wine, W.C. Johnson, and R.E. Davies. 1971. Generalized paraffinoma (sclerosing lipogranuloma). *Arch. Derm.* 103:277.

PARATHION

Description: Parathion,

$$(C_2H_5O)_2\overset{\overset{\textstyle S}{\|}}{P}OC_6H_4NO_2$$

is a pale yellow liquid boiling at 375°C, used as an insecticide.

Synonyms: O,O-diethyl-O-p-nitrophenyl phosphorothioate; diethyl-p-nitrophenyl monothiophosphate; DNTP.

Potential Occupational Exposures: NIOSH estimates that approximately 250,000 U.S. workers, including field workers, are potentially exposed to parathion in occupational settings.

Permissible Exposure Limits: NIOSH recommends the promulgation of a new Federal standard including, in part, an environmental limit of 0.05 mg of parathion per m^3 of air as a TWA for up to a 10-hour workday, 40-hour workweek. ACGIH (1978) sets 0.1 mg/m^3 as a TWA.

Routes of Entry: Skin contact, ingestion or inhalation.

Harmful Effects

Parathion, an organophosphorus insecticide, is converted in the environment and in the body to paraoxon, a potent inactivator of acetylcholinesterase, an enzyme responsible for terminating the transmitter action of acetylcholine at the junction of cholinergic nerve endings with their effector organs or post-synaptic sites.

The scientific basis for the recommended environmental limit is the prevention of medically significant inhibition of acetylcholinesterase. If functional acetylcholinesterase is greatly inhibited, a cholinergic crisis may ensue resulting in signs and symptoms of parathion poisoning including nausea, vomiting, abdominal cramps, diarrhea, involuntary defecation and urination, blurring of vision, muscular twitching, and difficulty in breathing.

Medical Surveillance: NIOSH recommends that medical surveillance, including preemployment and periodic examinations, shall be made available to workers who may be occupationally exposed to parathion. Biologic monitoring is also recommended as an additional safety measure.

Personal Protective Methods: Personal protective equipment and protective clothing are recommended for those workers occupationally exposed to parathion to further reduce exposure. In certain instances, such as emergency situations, NIOSH recommends that respirators be worn.

NIOSH also recommends that stringent work practices and engineering controls be adhered to in order to further reduce the likelihood of poisoning that may result from skin contact, ingestion, or from inhalation of parathion.

Bibliography

National Institute for Occupational Safety and Health. *Criteria for a Recommended Standard: Occupational Exposure to Parathion.* NIOSH Doc. No. 76-190 (1976).

PHENOL

Description: C_6H_5OH, phenol, is a white crystalline substance with a distinct aromatic, acrid odor.

Synonyms: Carbolic acid, phenic acid, phenylic acid, phenyl hydrate, hydroxybenzene, monohydroxybenzene.

Potential Occupational Exposures: Phenol is used in the production or manufacture of explosives, fertilizer, coke, illuminating gas, lampblack, paints, paint removers, rubber, asbestos goods, wood preservatives, synthetic resins, textiles, drugs, pharmaceutical preparations, perfumes, bakelite, and other plastics (phenol-formaldehyde resins). Phenol also finds wide use as a disinfectant in the petroleum, leather, paper, soap, toy, tanning, dye, and agricultural industries.

A partial list of occupations in which exposure may occur includes:

Coal tar workers	Paint and paint remover workers
Disinfectant makers	Paper makers
Dye workers	Rubber reclaimers
Explosive workers	Soap workers
Fertilizer makers	Tannery workers
Illuminating gas workers	Weed killer users
Lampblack makers	Wood preservers
Organic chemical synthesizers	

NIOSH estimates that 10,000 employees are potentially exposed to phenol in the United States.

Permissible Exposure Limits: The Federal standard is 5 ppm (19 mg/m^3). NIOSH recommends that exposure to phenol vapor, solid, or mists be limited to no more than 20 mg/m^3 expressed as a time-weighted average (TWA) concentration for up to a 10-hour workshift, 40-hour workweek. In addition, to protect employees from peak overexposures, NIOSH recommends that exposures be limited to no more than 60 mg/m^3 for any 15-minute period. There is thus no substantial change in the present Federal limit of 19 mg/m^3 (5 ppm), except for the inclusion of a ceiling value.

Routes of Entry: Inhalation of mist or vapor; percutaneous absorption of mist, vapor, or liquid.

Harmful Effects

Local—Phenol has a marked corrosive effect on any tissue. When it comes in contact with the eyes it may cause severe damage and blindness. On contact with the skin, it does not cause pain but causes a whitening of the exposed area. If the chemical is not removed promptly, it may cause a severe burn or systemic poisoning.

Systemic—Systemic effects may occur from any route of exposure. These include paleness, weakness, sweating, headache, ringing of the ears, shock, cyanosis, excitement, frothing of the nose and mouth, dark colored urine, and death. If death does not occur, kidney damage may appear.

Repeated or prolonged exposure to phenol may cause chronic phenol poisoning. This condition is very rarely reported. The symptoms of chronic poisoning include vomiting, difficulty in swallowing, diarrhea, lack of appetite, headache, fainting, dizziness, dark urine, mental disturbances, and possibly, skin rash. Liver and kidney damage and discoloration of the skin may occur.

Medical Suveillance: Consider the skin, eye, liver, and renal function as part of any preplacement or periodic examination.

Special Tests: Phenol can be determined in blood or urine.

Personal Protective Methods: In areas where there is likelihood of a liquid spill or splash, impervious protective clothing and goggles should be worn. In areas of heavy vapor concentrations, fullface mask with forced air supply should be used, as well as protective clothing, gloves, rubber boots, and apron.

Bibliography

American Industrial Hygiene Association. 1969. Community air quality guides. Phenol and cresol. *Am. Ind. Hyg. Assoc. J.* 30:425.

Evans, S.J. 1952. Acute phenol poisoning. *Br. J. Ind. Med.* 9:227.

National Institute for Occupational Safety and Health. *Criteria for a Recommended Standard: Occupational Exposure to Phenol.* NIOSH Doc. No. 76-196 (1976).

Piotrowski, J.K. 1971. Evaluation of exposure to phenol: absorption of phenol vapor in the lungs and through the skin and excretion of phenol in urine. *Br. J. Ind. Med.* 28:172.

PHENYLHYDRAZINE (See "Hydrazine")

PHOSGENE

Description: $COCl_2$, phosgene, is a colorless, noncombustible gas with a sweet, not pleasant odor in low concentrations. In higher concentrations, it is irritating and pungent. It decomposes in water but is soluble in organic solvents.

Synonyms: Carbonyl chloride, carbon oxychloride, carbonic acid dichloride, chloroformyl chloride, combat gas.

Potential Occupational Exposures: Phosgene is used in the manufacture of dyestuffs based on triphenylmethane, coal tar, and urea. It is also used in the organic synthesis of isocyanates and their derivatives, carbonic acid esters (polycarbonates), and acid chlorides. Occasional applications include its utilization in metallurgy, and in the manufacture of some insecticides and pharmaceuticals.

A partial list of occupations in which exposure may occur includes:

Chlorinated compound makers	Insecticide makers
Drug makers	Metallurgists
Dye makers	Organic chemical synthesizers
Firemen	

NIOSH estimates that 10,000 workers have potential occupational exposure to phosgene during its manufacture and use.

Permissible Exposure Limits: The Federal standard for concentrations of phosgene in air is 0.1 ppm (0.4 mg/m³). NIOSH recommends adherence to the present Federal standard of 0.1 ppm phosgene as a time-weighted average for up to a 10-hour workday, 40-hour workweek and proposes the addition of a 0.2 ppm ceiling value for any 15-minute period.

Route of Entry: Inhalation of gas.

Harmful Effects

Local—Conjunctivitis, lacrimation, and upper respiratory tract irritation may develop from gas. Liquid may cause severe burns.

Systemic—Acute exposure to phosgene may produce pulmonary edema frequently preceded by a latent period of 5-6 hours but seldom longer than 12 hours. The symptoms are dizziness, chills, discomfort, thirst, increasingly tormenting cough, and viscous sputum. Sputum may then become thin and foamy, and dyspnea, a feeling of suffocation, tracheal rhonci, and grey-blue cyanosis may follow. Death may result from respiratory or cardiac failure. The hazard of phosgene is increased because at low levels (205 mg/m³) it is lacking in warning symptoms.

Chronic exposure to phosgene may result in some tolerance to acute edemagenic doses, but may cause irreversible pulmonary changes of emphysema and fibrosis. Animal experimentation has shown an increased incidence of chronic pneumonitis and acute and fibrinous pneumonia from exposure to this agent.

Medical Surveillance: Preemployment medical examinations should include chest x-rays and baseline pulmonary function tests. The eyes and skin should be examined. Smoking history should be known. Periodic pulmonary function studies should be done. Workers who are known to have inhaled phosgene should remain under medical observation for at least 24 hours to insure that delayed symptoms do not occur.

Special Tests: None.

Personal Protective Methods: Where liquid phosgene is encountered, protective clothing should be supplied which is impervious to phosgene. Where gas is encountered above safe limits, fullface gas masks with phosgene canister or supplied air respirators should be used. Because of the potentially serious consequences of acute overexposure and the poor warning properties of the gas to the human senses, automatic continuous monitors with alarm systems are strongly recommended. Although such devices have only recently become commercially available and may present problems of optimum placement for appropriate warning of emergencies, it is felt that these devices will become increasingly useful tools in the effort to protect workers from loss of life or functional capacity on the job.

Bibliography

National Institute for Occupational Safety and Health. *Criteria for a Recommended Standard: Occupational Exposure to Phosgene.* NIOSH Doc. No. 76-137 (1976).
Potts, A.M., F.P. Simon, and R.W. Gerard. 1949. The mechanism of action of phosgene and diphosgene. *Arch. Biochem. Biophys.* 24:329.
Thiess, A.M., and P.J. Goldman. 1968. Ist die Phosgenvergiftung noch ein arbeitsmedizinisches problem. *Zentralbl. Arbeitsmed.* 18:132.

PHOSPHINE

Description: PH_3, phosphine, is a colorless gas with an odor of decaying fish. Phosphine presents an additional hazard in that it ignites at very low temperature. Phosphine is soluble in water, 26 ml/100 ml at 17°C, and in organic solvents.

Synonyms: Hydrogen phosphide, phosphoretted hydrogen, phosphorus trihydride.

Potential Occupational Exposures: Phosphine is only occasionally used in industry, and exposure usually results accidentally as a by-product of various processes. Exposures may occur when acid or water comes in contact with metallic phosphides (aluminum phosphide, calcium phosphide). These two phosphides are used as insecticides or rodenticides for grain, and phosphine is generated during grain fumigation. Phosphine may also evolve during the generation of acetylene from impure calcium carbide, as well as during metal shaving, sulfuric acid tank cleaning, rustproofing, and ferrosilicon, phosphoric acid and yellow phosphorus explosive handling.

A partial list of occupations in which exposure may occur includes:

Acetylene workers	Metal slag workers
Cement workers	Metallic phosphate workers
Firemen	Organic chemical synthesizers
Grain fumigators	Rustproofers
Metal refiners	Welders

Permissible Exposure Limits: 0.3 ppm (0.4 mg/m^3) is the Federal standard for occupational exposure to phosphine determined as TWA.

Route of Entry: Inhalation of vapor.

Harmful Effects

Local—Phosphine's strong odor may be nauseating. However, irritation to the eyes or skin is undocumented, and some authors indicate that lacrimation, if it occurs, results as a systemic effect rather than from local irritation.

Systemic—Acute effects are secondary to central nervous system depression, irritation of lungs, and damage to the liver and other organs. Most common effects include weakness, fatigue, headache, vertigo, anorexia, nausea, vomiting, abdominal pain, diarrhea, tenesmus, thirst, dryness of the throat, difficulty in swallowing, and sensation of chest pressure. In severe cases staggering gait, convulsions, and coma follow. Death may occur from cardiac arrest and, more typically, pulmonary edema, which may be latent in a manner similar to nitrogen oxide intoxication.

Chronic poisoning has been suggested by some authors and symptoms have been attributed to chronic phosphorus poisoning. However, there is evidence that phosphine may be metabolized to form nontoxic phosphates, and chronic exposure of animals has failed to produce toxic effects. Compounded with the lack of human experience and of extensive commercial usage, evidence indicates that chronic poisoning per se does not occur.

Medical Surveillance: No special considerations are necessary in placement or periodic examinations, other than evaluation of the respiratory system. If poisoning is suspected, workers should be observed for 48 hours due to the delayed onset of pulmonary edema.

Special Tests: None have been used.

Personal Protective Methods: In areas where vapors are excessive, workers should be supplied with fullface gas masks with proper canisters or supplied air respirators.

Bibliography

Courville, C.B. 1964. Confusion of presumed toxic gas poisoning for fatal granulomatous meningo-encephalitis resulting in a severe progressive arteritis and gross cerebral hemorrhages—report of fatal case assessed as hydrogen phosphide (phosphine) poisoning. *Bull. Los Angeles Neurol. Soc.* 29:76.

Hackenberg, V. 1972. Chronic ingestion by rats of standard diet treated with aluminum phosphide. *Toxicol. Appl. Pharmacol.* 23:147.

Harger, R.N., and L.W. Spolyar. 1958. Toxicity of phosphine, with a possible fatality from this poison. *AMA Arch. Ind. Health.* 18:497.

Jones, A.T., R.C. Jones, and E.O. Longley. 1964. Environmental and clinical aspects of bulk wheat fumigation with aluminum phosphide. *Am. Ind. Hyg. Assoc. J.* 25:376.

Mathew, G.G. 1961. The production of phosphine while machining spheroidal graphite iron. *Ann. Occup. Hyg.* 4:19.

PHOSPHORUS AND COMPOUNDS

Description: Phosphorus (white or yellow), P, is almost insoluble in water, but soluble in organic solvents. Phosphoric acid, H_3PO_4, is soluble in water and alcohol. Phosphorus trichloride, PCl_3, decomposes in cold water. Tetraphosphorus trisulfide, P_4S_3, is insoluble in water.

Red phosphorus is excluded in that it is a nontoxic allotrope, although it is frequently contaminated with a small amount of the yellow. White or yellow phosphorus is either a yellow or colorless, volatile, crystalline solid which darkens when exposed to light and ignites in air to form white fumes and greenish light. Phosphoric acid is also a crystal; however, it is typically encountered in a liquid form. Phosphorus pentachloride and phosphorus pentasulfide are white to pale yellow, fuming crystals, while tetraphosphorus trisulfide is a greenish-yellow crystal. Phosphorus trichloride and phosphorus oxychloride are also colorless, fuming liquids.

Elemental phosphorus does not occur free in nature, but is found in the form of phosphates. Phosphorus and phosphoric acid are prepared commercially from "phosphate rock" deposits of the Southern United States and, at one time, from bone in Europe. Phosphorus, once formed, is immediately converted to less toxic substances, such as phosphoric acid. The other compounds are prepared directly from red phosphorus and chlorine or sulfur respectively. Decomposition products of phosphorus compounds are also toxic and include hydrogen sulfide and phoshoric acid for sulfur-containing compounds.

Synonyms

Phosphorus: None.
Phosphoric acid: H_3PO_4, orthophosphoric acid.
Phosphorus trichloride: PCl_3, phosphorous chloride.
Phosphorus pentachloride: PCl_5, phosphoric chloride, phosphorus perchloride.
Tetraphosphorus trisulfide: P_4S_3, phosphorus sesquisulfide, trisulfurated phosphorus.

Phosphorus pentasulfide: P_2S_5, phosphoric sulfide, thio-
phosphoric anhydride, phosphorus persulfide.
Phosphorus oxychloride: $POCl_3$, phosphoryl chloride.

Potential Occupational Exposures: Yellow phosphorus is handled away from air so that exposure is usually limited. Phosphorus was at one time used for the production of matches or lucifers but has long since been replaced due to its chronic toxicity. Phosphorus is used in the manufacture of munitions, pyrotechnics, explosives, smoke bombs, and other incendiaries, in artificial fertilizers, rodenticides, phosphor bronze alloy, semiconductors, electroluminescent coating, and chemicals, such as, phosphoric acid and metallic phosphides.

Phosphoric acid is used in the manufacture of fertilizers, phosphate salts, polyphosphates, detergents, activated carbon, animal feed, ceramics, dental cement, pharmaceuticals, soft drinks, gelatin, rust inhibitors, wax, and rubber latex. Exposure may also occur during electropolishing, engraving, photoengraving, lithographing, metal cleaning, sugar refining, and water treating. Phosphorus trichloride and phosphorus pentachloride are used in the manufacture of agricultural chemicals, chlorinated compounds, dyes, gasoline additives, acetylcellulose, phosphorus oxychloride, plasticizers, saccharin, and surfactants.

Phosphorus pentasulfide and tetraphosphorus trisulfide are used in the manufacture of flotation agents, insecticides, lubricating oil, additives, ignition compounds, and matches. They are also used to introduce sulfur into agricultural, rubber, and organic chemicals.

A partial list of occupations in which exposure may occur includes:

Acetylcellulose makers	Metal refiners
Bronze alloy makers	Metallic phosphide makers
Chlorinated compound makers	Munitions workers
Electroluminescent coating makers	Pesticide workers
Fertilizer makers	Rat poison workers
Fireworks makers	Semiconductor makers
Hydraulic fluid makers	Smoke bomb makers
Incendiary makers	

Permissible Exposure Limits: Federal standards are: Phosphorus (yellow) 0.1 mg/m^3, phosphoric acid 1.0 mg/m^3, phosphorus trichloride 3.0 mg/m^3, phosphorus pentachloride 1.0 mg/m^3, phosphorus pentasulfide 1.0 mg/m^3.

Route of Entry: Inhalation of vapor or fumes or mist.

Harmful Effects

Local—Phosphorus, upon contact with skin, may result in severe burns, which are necrotic, yellowish, fluorescent under ultraviolet light, and have a garliclike odor. Other phosphorus compounds are potent irritants of the skin, eyes, and mucous membranes of nose, throat, and respiratory tract. At 1 ppm, the Federal standard, phosphoric acid mist is irritating to unacclimated workers but is easily tolerated by acclimated workers. Localized contact dermatitis, particularly of the thighs and eczema of the face and hands, have been observed in workers manufacturing the "strike anywhere" matches containing tetraphosphorus trisulfide.

Systemic—Acute phosphorus poisoning usually occurs as a result of accidental or suicidal ingestion. However, animal experiments indicate that acute systemic poisoning may follow skin burns. In acute cases, shock may ensue rapidly and the victim may succumb immediately. If acute attack is survived, an asymptomatic latency period of a few hours to a few days may follow. Death often occurs upon relapse from liver, kidney, cardiac, or vascular dysfunction or failure. Abnormal electrocardiograms, particularly of the QT, ST, or T wave phases, abnormal urinary and serum calcium and phosphate levels, proteinuria and aminoaciduria, and elevated serum SGPT are indicative signs. Vomitus, urine, and stools may be fluorescent in ultraviolet light, and a garlic odor of breath and eructations may be noted.

Inhalation of fumes produced by phosphorus compounds listed above may cause irritation of pulmonary tissues with resultant acute pulmonary edema. Chronic exposure may lead to cough, bronchitis, and pneumonia. The hazards of phosphorus pentasulfide are the same as for hydrogen sulfide to which it rapidly hydrolyzes in the presence of moisture.

Chronic phosphorus poisoning is a result of continued absorption of small amounts of yellow phosphorus for periods typically of 10 years; however, exposures of as short as 10 months may cause phosphorus necrosis of the jaw ("phossy jaw"). Chronic intoxication is characterized by periostitis with suppuration, ulceration, necrosis, and severe deformity of the mandible and, less often, maxilla. Sequestration of bone may occur. Polymorphic leukopenia, susceptibility to bone fracture, and failure of the alveolar bone to resorb following extractions are secondary clinical signs. Carious teeth and poor dental hygiene increase susceptibility.

Medical Surveillance: Special consideration should be given to the skin, eyes, jaws, teeth, respiratory tract, and liver. Preplacement medical and dental examination with x-ray of teeth is highly recommended in the case of yellow phosphorus exposure. Poor dental hygiene may increase the risk in yellow phosphorus exposures, and any required dental work should be completed before workers are assigned to areas of possible exposure.

Workers experiencing any jaw injury, tooth extraction, or any abnormal dental conditions should be removed from areas of exposure and observed. Roentgenographic examinations may show necrosis; however, in order to prevent full development of sequestra, the disease should be diagnosed in earlier stages. Liver function should be evaluated periodically. Pulmonary function tests may be useful when exposures are to the acid, chlorides, and sulfide compounds.

Special Tests: None commonly used.

Personal Protective Methods: Full body protective clothing including hat and face shield should be supplied to workers who may be exposed to spill, splashes, or spotters of phosphorus or phosphorus compounds. Inhalation of vapors or fumes can be prevented by proper ventilation in many cases, but in areas of higher concentration fullface mask respirators with proper canisters or supplied air respirators may be required. Continuing worker education of exposure risks for those in exposed areas is essential.

Bibliography

Burgess, J.F. 1951. Phosphorus sesquisulfide poisoning. *Can. Med. Assoc. J.* 65:567.
Fletcher, G.F., and J.T. Galambos. 1963. Phosphorus poisoning in humans. *Arch. Intern. Med.* 112:846.
Hughes, J.P.W., R. Baron, D.H. Buckland, M.A. Cooke, J.D. Craig, D.P. Duffield, A.W. Grosart, P.Q.J. Parkes, A. Porter, A.C. Frazer, J.W. Hallam, J.W.F. Snawdon, and R.W.H. Tavenner. 1962. Phosphorus necrosis of the jaw: A present-day study, with clinical and biochemical studies. *Br. J. Ind. Med.* 19:83.
Ive, F.A. 1967. Studies on contact dermatitis. XXI. *Trans. St. John's Hosp. Dermatol. Soc.* 53:135.
Matsumoto, S., Y. Kohri, K. Tanaka, and G. Tsuchiya. 1972. A case of acute phosphorus poisoning with various electrocardiographic changes. *Jap. Cir. J.* 36:963.
Salfelder, K., H.R. Doehnert, G. Doehnert, E. Sauerteig, T.R. De Liscano, and S.E. Fabrega. 1972. Fatal phosphorus poisoning: A study of forty-five autopsy cases. *Beitr. Pathol.* 147:321.

PHTHALIC ANHYDRIDE

Description: $C_8H_4O_3$, phthalic anhydride, is moderately flammable, white, lustrous, solid, with needle-like crystals.

Synonyms: Phthalic acid anhydride, benzene-o-dicarboxylic acid anhydride, phthalandione.

Potential Occupational Exposures: Phthalic anhydride is used in the manufacture of phthaleins, benzoic acid, alkyd and polyester resins, synthetic indigo, and phthalic acid, which is used as a plasticizer for vinyl resins. To a lesser extent, it is used in the production of alizarin dye, anthranilic acid, anthraquinone, diethyl phthalate, dimethyl phthalate, erythrosin, isophthalic acid, methyl aniline, phenolphthalein, phthalamide, sulfathalidine, and terephthalic acid. It has also found use in pesticides and herbicides, as well as perfumes.

A partial list of occupations in which exposure may occur includes:

Alizarin dye makers	Mylar plastic makers
Alkyd resin makers	Organic chemical synthesizers
Automobile finish makers	Phthalein makers
Cellulose acetate plasticizer makers	Resin makers
Dacron fiber makers	Vat dye makers
Erythrosin makers	Vinyl plasticizer makers
Insecticide makers	

Permissible Exposure Limits: The Federal standard is 2 ppm (12 mg/m^3). ACGIH (1978) sets 1 ppm (6 mg/m^3) as a TWA.

Route of Entry: Inhalation of dust, fume, or vapor.

Harmful Effects

Local—Phthalic anhydride, in the form of a dust, fume, or vapor, is a potent irritant of the eyes, skin, and respiratory tract. The irritant effects are worse on moist surfaces. Conjunctivitis and skin erythema, burning, and contact dermatitis may occur. If the chemical is held in contact with the skin, as under clothes or shoes, skin burns may develop. Hypersensitivity may develop in some individuals. Inhalation of the dust or vapors may cause coughing, sneezing, and a bloody nasal discharge. Impurities, naphthoquinone, as well as maleic anhydride, may also cause eye, skin, and pulmonary irritation.

Systemic—Repeated exposure may result in bronchitis, emphysema, allergic asthma, urticaria, and chronic eye irritation.

Medical Surveillance: Emphasis should be given to a history of skin or pulmonary allergy, and preplacement and periodic examinations should evaluate the skin, eye, and lungs, as well as liver and kidney functions. The hydrolysis product, phthalic acid, is rapidly excreted in the urine, although this has not been used in biological monitoring. Diagnostic patch testing may be helpful in evaluating skin allergy.

Special Tests: None in common use.

Personal Protective Methods: Proper ventilation, rubber gloves, protective clothing, head coverings, and goggles are recommended when repeated or prolonged contact is possible. Respiratory protection may be needed in dusty areas or where fumes or vapors are present.

Bibliography

Ghezzi, I., and P. Scott. 1965. Clinical contribution on the pathology induced by phthalic and maleic anhydride. *Med. Lav.* 56:746.

Merlevede, E., and J. Elskens. 1957. The toxicity of phthalic anhydride, maleic anhydride, and the phthalates. *Arch. Belg. Med. Soc.* 15:445.

PICRIC ACID

Description: $C_6H_2(NO_2)_3OH$, picric acid, is a pale yellow, odorless, intensely bitter crystal which is explosive upon rapid heating or mechanical shock.

Synonyms: Picronitric acid, trinitrophenol, nitroxanthic acid, carbazotic acid, phenol trinitrate.

Potential Occupational Exposures: Picric acid is used in the manufacture of explosives, rocket fuels, fireworks, colored glass, matches, electric batteries, and disinfectants. It is also used in the pharmaceutical and leather industries, and in dyes, copper and steel etching, forensic chemistry, histology, textile printing, and photographic emulsions.

A partial list of occupations in which exposure may ocur includes:

Battery makers	Explosive makers
Colored glass makers	Forsenic chemists
Copper etchers	Histology technicians
Disinfectant makers	Matchmakers
Drug makers	Photographic chemical workers
Dye makers	Tannery workers

Permissible Exposure Limits: The Federal standard for picric acid is 0.1 mg/m^3.

Routes of Entry: Inhalation and ingestion of dust; percutaneous absorption.

Harmful Effects

Local—Picric acid dust or solutions are potent skin sensitizers. In solid form, picric acid is a skin irritant, but in aqueous solution it irritates only hypersensitive skin. The cutaneous lesions which appear usually on exposed areas of the upper extremities consist of dermatitis with erythema, papular, and vesicular eruptions. Desquamation may occur following repeated or prolonged contact. Skin usually turns yellow upon contact, and areas around nose and mouth as well as the hair are most often affected. Dust or fume may cause eye irritation which may be aggravated by sensitization. Corneal injury may occur from exposure to picric acid dust and solutions.

Systemic—Inhalation of high concentrations of dust by one worker caused temporary coma followed by weakness, myalgia, anuria, and later polyuria. Following ingestion of picric acid, there may be headache, vertigo, nausea, vomiting, diarrhea, yellow coloration of the skin, hematuria, and albuminuria. High doses may cause destruction of erythrocytes, hemorrhagic nephritis, and hepatitis. High doses which cause systemic intoxication will color all tissues yellow, including the conjunctiva and aqueous humor, and cause yellow vision.

Medical Surveillance: Preplacement and periodic medical examinations should focus on skin disorders such as hypersensitivity atopic dermatitis, and liver and kidney function.

Special Tests: None commonly used. It is probably excreted as picric and picramic acid in the urine.

Personal Protective Methods: Skin protection by clothing and barrier creams can avoid the irritant and sensitizing action of picric acid. Masks of the dust type will prevent absorption by inhalation. Fullface masks are advisable or combination of chemical goggles with halfmask. Daily change of clean work clothes and showering after each shift before changing to street clothes are mandatory.

Bibliography

Chicago National Safety Council. 1969. *Picric Acid.* Data Sheet 351 (Revision A, Extensive). Chicago National Safety Council, Chicago, Ill.
Williams, R.T. 1959. *Detoxication Mechanism,* 2nd ed. J. Wiley and Sons, New York.

PLATINUM AND COMPOUNDS

Description: Pt, platinum, is a soft, ductile, malleable, silver-white metal, insoluble in water and organic solvents. It is found in the metallic form and as the arsenide, sperrylite. It forms complex soluble salts such as Na_2PtCl_6.

Synonyms: None.

Potential Occupational Exposures: Platinum and its alloys are utilized because of their resistance to corrosion and oxidation, particularly at high temperatures, their high electrical conductivity, and their excellent catalytic properties. They are used in relays, contacts and tubes in electronic equipment, in spark plug electrodes for aircraft, and windings in high temperature electrical furnaces. Platinum alloys are used for standards for weight, length, and temperature measurement. Platinum and platinum catalysts (e.g., hexachloroplatinic acid, H_2PtCl_6) are widely used in the chemical industry in persulfuric, nitric, and sulfuric acid production, in the synthesis of organic compounds and vitamins, and for producing higher octane gasoline. They are coming into use in catalyst systems for control of exhaust pollutants from automobiles. They are used in the equipment for handling molten glass and manufacturing fibrous glass; in laboratory, medical, and dental apparatus; in electroplating; in photography; in jewelry; and in x-ray fluorescent screens.

A partial list of occupations in which exposure may occur includes:

Alloy makers   Catalyst workers

Ceramic workers
Dental alloy makers
Drug makers
Electronic equipment makers
Electroplaters
Gasoline additive makers

Indelible ink makers
Jewelry makers
Laboratory ware makers
Mirror makers
Spark plug makers
Zinc etchers

Permissible Exposure Limits: The Federal standard for soluble salts of platinum is 0.002 mg/m^3 expressed as Pt.

Route of Entry: Inhalation of dust or mist.

Harmful Effects

Local—Hazards arise from the dust, droplets, spray, or mist of complex salts of platinum, but not from the metal itself. These salts are sensitizers of the skin, nasal mucosa, and bronchi, and cause allergic phenomena. One case of contact dermatitis from wearing a ring made of platinum alloy is recorded.

Systemic—Characteristic symptoms of poisoning occur after 2 to 6 months' exposure and include pronounced irritation of the throat and nasal passages, which results in violent sneezing and coughing; bronchial irritation, which causes respiratory distress; and irritation of the skin, which produces cracking, bleeding, and pain. Respiratory symptoms can be so severe that exposed individuals may develop status asthmaticus. After recovery, most individuals develop allergic symptoms and experience further asthma attacks when exposed to even minimal amounts of platinum dust or mists. Mild cases of dermatitis involve only erythema and urticaria of the hands and forearms. More severe cases affect the face and neck. All pathology is limited to allergic manifestations.

EPA research efforts indicate that platinum is more active biologically and toxicologically than previously believed. It methylates in aqueous media, establishing a previously unrecognized biotransformation and distribution mechanism. Because platinum complexes are used as antitumor agents, the potential for carcinogenic activity is present; tests to clarify this aspect should be completed within several months. While low levels of emissions of platinum particulate have been observed from some catalyst-equipped automobiles, the major potential source of Pt is from the disposal of spent catalysts.

Medical Surveillance: In preemployment and periodic physical examinations, the skin, eyes, and respiratory tract are most important. Any history of skin or pulmonary allergy should be noted, as well as exposure to other irritants or allergens, and smoking history. Periodic assessment of pulmonary function may be useful.

Special Tests: None commonly used.

Personal Protective Methods: In areas where dust or mist are excessive, masks or air-supplied respirators should be supplied. Where droplets, mist, or spray is encountered, impervious protective clothing, gloves, and goggles should be supplied.

Should the results of current EPA research efforts to document health hazards of Pt suggest a need to control these exposures, disposal control would be the most promising, since catalyst disposal is expected to be the largest contributor of Pt to the environment.

In addition, the value of the metal would help to offset the cost of reclaiming the Pt from discarded catalysts. If direct vehicular emissions of Pt are found to be significant, particulate traps, which are available at reasonable cost, may provide a technological solution. Other noble metals have been suggested as Pt substitutes; however, even less is known about their potential for hazard.

Bibliography

Annual Catalyst Research Program Report; EPA, Office of Research and Development (Publication No. EPA-650/3-75-010, Sept. 1975).

Brubaker, P.E., et al. Noble metals: A toxicological appraisal of potential new environmental contaminants. *Environmental Health Perspective.* Vol. 10 (1975).

Duffield, F.V.P., et al. Determination of human body burden baseline data of platinum through autopsy tissue analysis. EPA, Office of Research and Development (accepted for publication in *Environmental Health Perspectives,* March 1976).

Holbrook, D.J., Jr. *Assessment of Toxicity of Automotive Metallic Emissions, Vol. I, Assessment of Fuel Additives Emission Toxicity via Selected Assays of Nucleic Acid and Protein Synthesis.* Report EPA-600/1-76-0102, Research Triangle Park, N.C., U.S.E.P.A. Health Effects Research Laboratory, (Jan. 1976)

Mayer, R.A., Prehn, W.L., Jr. and Johnson, D.E. *A Literature Search and Analysis of Information Regarding Sources, Uses, Production, Consumption, Reported Medical Cases, and Toxicology of Platinum and Palladium.* Report EPA-650/1-74-008, Research Triangle Park, N.C., U.S.E.P.A. National Environmental Research Center (April 1974).

Occupational Safety and Health Standards: Subpart Z - Toxic and Hazardous Substances. 29 CFR 1910.1000, Table Z-1.

Parrot, J.L., R. Herbert, A. Saindelle, and F. Ruff. 1969. Platinum and platinosis—allergy and histamine release due to some platinum salts. *Arch. Environ. Health* 19:685.

Roberts, A.E. 1951. Platinosis—a five-year study of the effects of soluble platinum salts on employees in a platinum laboratory and refinery. *AMA Arch. Ind. Hyg. Occup. Med.* 4:549.

POLYCHLORINATED BIPHENYLS
(See "Chlorodiphenyls and Derivatives")

POLYNUCLEAR AROMATIC HYDROCARBONS

Description: The polynuclear aromatic hydrocarbons constitute a class of materials of which benzo[a]pyrene is one of the most common and also the most hazardous.

Benzo[a]pyrene, $C_{20}H_{12}$, is a yellowish crystalline solid, melting at 179°C. It consists of five benzene rings joined together.

Synonyms: PNAs, PAHs, PPAHs (Particulate polycyclic aromatic hydrocarbons) and POMs (Polynuclear organic materials). (Benzo[a]pyrene is also known as BaP.)

Potential Occupational Exposures: PNAs can be formed in any hydrocarbon combustion process and may be released from oil spills. The less efficient the combustion process, the higher the PNA emission factor is likely to be. The major sources are stationary sources, such as heat and power generation, refuse burning, industrial activity, such as coke ovens, and coal refuse heaps. While PNAs can be formed naturally (lightning-ignited forest fires), impact of these sources appears to be minimal. It should be noted, however, that while transportation sources account for only about 1% of emitted PNAs on a national inventory basis, transportation-generated PNAs may approach 50% of the urban resident exposures.

Because of the large number of sources, most people are exposed to very low levels of PNAs. BaP has been detected in a variety of foods throughout the world. A possible source is mineral oils and petroleum waxes used in food containers and as release agents for food containers. FDA studies have indicated no health hazard from these sources.

The air pollution aspects of the carcinogenic polynuclear aromatic hydrocarbons (PAH) and of benzo[a]pyrene (BaP) in particular have been reviewed in some detail by Olsen and Haynes (1). The total emissions of benzo[a]pyrene (BaP) and some emission factors for BaP are as presented by Goldberg (2).

Permissible Exposure Limits: A TLV of 0.2 mg/m^3 as benzene solubles has been assigned ty ACGIH. These materials are designated by ACGIH as human carcinogens.

Route of Entry: Inhalation of particulates, vapors.

Harmful Effects

Certain PNAs which have been demonstrated as carcinogenic in test animals at relatively high exposure levels are being found in urban air at very low levels. Various environmental fate tests suggest that PNAs are photo-oxidized, and react with oxidants and oxides of sulfur. Because PNAs are adsorbed on particulate matter, chemical half-lives may vary greatly, from a matter of a few hours to several days. One researcher reports that photo-oxidized PNA fractions of air extracts also appear to be carcinogenic. Environmental behavior/fate data have not been developed for the class as a whole.

It has been observed that PNAs are highly soluble in adipose tissue and lipids. Most of the PNAs taken in by mammals are oxidized and the metabolites excreted. Effects of that portion remaining in the body at low levels have not been documented.

Benzo[a]pyrene (BaP), one of the most commonly found and hazardous of the PNAs has been the subject of a variety of toxicological tests, which have been summarized by the International Agency for Research on Cancer. 50-100 ppm administered in the diet for 122-197 days produced stomach tumors in 70% of the mice studied. 250 ppm produced tumors in the forestomach of 100% of the mice after 30 days. A single oral administration of 100 mg to nine rats produced mammary tumors in eight of them. Skin cancers have been induced in a variety of animals at very low levels, and using a variety of solvents (length of application was not specified).

Lung cancer developed in 2 of 21 rats exposed to 10 mg/m³ BaP and 3.5 ppm SO_2 for 1 hour/day, five days a week, for more than one year. Five of 21 rats receiving 10 ppm SO_2 for 6 hours/day, in addition to the foregoing dosage, developed similar carcinomas. No carcinomas were noted in rats receiving only SO_2. No animals were exposed only to BaP. Transplacental migration of BaP has been demonstrated in mice. Most other PNAs have not been subjected to such testing.

Medical Surveillance: Preplacement and regular physical examinations are indicated for workers having contact with polynuclear aromatics in the workplace.

Special Tests: None in use.

Personal Protective Methods: Good particulate emission controls are the indicated engineering control scheme where polynuclear aromatics are encountered in the workplace.

References

(1) Olsen, D.A. and Haynes, J.L. *Air Pollution Aspects of Organic Carcinogens.* Report PB-188 090, Springfield, Virginia, Nat. Tech. Information Service (Sept. 1969).
(2) Goldberg, A.J. *A Suvey of Emissions and Controls for Hazardous and Other Pollutants.* Report PB-223 568, Springfield, Virginia, Nat. Tech. Information Service (Feb. 1973).

Bibliography

Begeman, C.R., and Colucci, J.M. *Polynuclear Aromatic Hydrocarbon Emissions from*

Automotive Engines. Warrendale, PA, Society of Automotive Engineers, Inc. (Publication No. 700469, 1971).

Gross, Herbert. *Hexane Extractables and PAH in the Black River (Ohio).* EPA, Office of Enforcement and General Counsel (Oct. 1974).

IARC Monographs on the Evaluation of Carcinogenic Risk of the Chemical to Man: Certain Polycyclic Aromatic Hydrocarbons and Heterocyclic Compounds. Volume 3, Lyon, International Agency for Research on Cancer (1973).

Moran, J.B. *Assuring Public Health Protection as a Result of the Mobile Source Emission Control Program.* Warrendale, PA, Society of Automotive Engineers, Inc. (Publication No. 740285, March 1974).

Moran, J.B. *Lead in Gasoline: Impact of Removal on Current and Future Automotive Emissions* for the Air Pollution Control Association (APCA) (June 1974).

Moran, J.B., Colucci, A., and Finklea, J.F. *Projected Changes in Polynuclear Aromatic Hydrocarbon Exposures from Exhaust and Tire Wear Debris of Light Duty Motor Vehicles.* Internal Report, ERC-RTP (May 1974).

Particulate Polycyclic Organic Matter. National Academy of Science. (1972).

Preferred Standards Path Report for Polycyclic Organic Matter. EPA, Office of Air Quality Planning and Standards (Oct. 1974).

Scientific and Technical Assessment Review of Particulate Polycyclic Organic Matter. EPA, Office of Research and Development, (1975).

PORTLAND CEMENT

Description: Portland cement is a class of hydraulic cements whose two essential constituents are tricalcium silicate and dicalcium silicate with varying amounts of alumina, tricalcium aluminate, and iron oxide. The quartz content of most is below 1%. The average composition of regular Portland cement is as follows:

	Percent
CaO	64.0
SiO_2	21.0
Al_2O_3	5.8
Fe_2O_3	2.9
MgO	2.5
Alkali Oxides	1.4
SO_3	1.7

Synonyms: None.

Potential Occupational Exposures: Cement is used as a binding agent in mortar and concrete (a mixture of cement, gravel, and sand). Potentially hazardous exposure may occur during both the manufacture and use of cement.

A partial list of occupations in which exposure may occur includes:

Asbestos cement workers   Brick masons

Bridge builders	Heat insulation makers
Building construction workers	Oil well builders
Burial vault builders	Silo builders
Cement workers	Storage tank builders
Concrete workers	Tunnel builders
Drain tile makers	Water pipe makers

Permissible Exposure Limits: The Federal standard for Portland cement is 50 mppcf.

Route of Entry: Inhalation of dust.

Harmful Effects

Local—Exposure may produce cement dermatitis which is usually due to primary irritation from the alkaline, hygroscopic, and abrasive properties of cement. Chronic irritation of the eyes and nose may occur. In some cases, cement workers have developd an allergic sensitivity to constituents of cement such as hexavalent chromate. It is not unusual for cement dermatitis to be prolonged and to involve covered areas of the body.

Systemic—No documented cases of pneumoconiosis or other systemic manifestations attributed to finished Portland cement exposure have been reported. Conflicting reports of pneumoconiosis from cement dust appear related to exposures that occurred in mining, quarrying, or crushing silica-containing raw materials.

Medical Surveillance: Preemployment and periodic medical examinations should stress significant respiratory problems, chest x-ray, pulmonary function tests, smoking history, and allergic skin sensitivities, especially to chromates. The eyes should be examined.

Special Tests: Patch test studies may be useful in dermatitis cases.

Personal Protective Methods: In areas exceeding safe dust levels, masks with proper cartridges should be provided. Gloves, barrier creams, and protective clothing (long-sleeved shirts, etc.) will help protect workers subject to dermatitis. Personal hygiene is very important, and all cement workers should be encouraged to shower following each shift before changing to street clothes. Freshly laundered work clothing should be supplied on a daily basis.

Bibliography

Calnan, C. 1960. Cement dermatitis. *J. Occup. Med.* 2:15.
Kalacic, I. 1973. Chronic nonspecific lung disease in cement workers. *Arch. Environ. Health* 26:78.
Kalacic, I. 1973. Ventilatory lung function in cement workers. *Arch. Environ. Health* 26:84.

Morris, G.E. 1960. The primary irritant nature of cement. *Arch. Environ. Health* 1:301.
Sander, O.A. 1968. Roentgen resurvey of cement workers. *AMA Arch. Ind. Health* 17:96.

POTASSIUM HYDROXIDE
[See "Sodium Hydroxide (Potassium Hydroxide)"]

BETA-PROPIOLACTONE

Description: $\overline{OCH_2CH_2CO}$, beta-propiolactone, is a colorless liquid which slowly hydrolyzes to hydracrylic acid and must be cooled to remain stable.

Synonyms: 2-Oxetanone, propiolactone, BPL, 3-hydroxy-beta-lactone-pro-panoic acid.

Potential Occupational Exposures: Beta-Propiolactone is used as a chemical intermediate in synthesis of acrylate plastics and as a vapor sterilizing agent, disinfectant, and a viricidal agent.

A partial list of occupations in which exposure may occur includes:

Acrylate plastic makers	Plastic makers
Chemists	Resin makers
Disinfectant workers	Viricidal agent makers

Permissible Exposure Limits: Beta-Propiolactone is included in the Federal standards for carcinogens; all contact with it should be avoided.

Routes of Entry: Inhalation of vapor and percutaneous absorption.

Harmful Effects

Local—Repeated or prolonged contact with liquid may cause erythema, vesication of the skin, and, as reported in animals, hair loss and scarring. In rodents, beta-propiolactone has also produced skin papilloma and sarcoma by skin painting, subcutaneous injection, and oral administration. Tumors of the connective tissue are also suspected. Direct eye contact with concentrated liquid may result in permanent corneal opacification. Skin cancer has not been reported in man.

Systemic—The systemic effect of beta-propiolactone in humans is unknown due to lack of reported cases. Acute exposure in animals has caused liver necrosis and renal tubular damage. Death has occurred following rapid development of spasms, dyspnea, convulsions, and collapse at relatively low levels (less than 5 ml/kg). Beta-propiolactone has been implicated as a carcinogen by a number of animal studies which produced a variety of skin tumors, stomach tumors, and hepatoma depending on the route of administration.

It should be noted that BPL is produced from formaldehyde and ketene which have been found to be mutagenic (1)-(4). Commercial grade BPL (97%) can contain trace quantities of the reactants.

Beta-propiolactone is a nucleophilic alkylating agent which reacts readily with acetate, halogen, thiocyanate, thiosulfate, hydroxyl and sulfhydryl ions (1)(5)-(7). 7-(2-Carboxyethyl)-guanine (in the enol form) has been suggested to be the major binding product of β-propiolactone with both DNA and RNA in vivo (6).

β-Propiolactone is carcinogenic in the mouse by skin application, subcutaneous or intraperitoneal application and in the rat by subcutaneous injection (5)(8)(9), while oral administration in the rat gave some indication of carcinogenicity (5)(10).

β-Propiolactone has been shown to be mutagenic in *Vicia faba* (11)(12), *Neurospora* (11), *E. coli* (13) and *Serratia marcescens* (13)(14), in the *Salmonella*/microsome test (15), causes chromosomal aberrations in *Vicia faba* (11), *Allium* (16) and *Neurospora* (11) and effects a decline in the transforming activity of DNA from *Bacillus subtilis* (17).

β-Propiolactone is mutagenic in bacteriophage T4 inducing primarily guanine//cytosine to adenine//thymine base pair transitions (18). This type of mispairing was suggested to be the most probable cause of BPL-induced mutagenesis (18).

Medical Surveillance: Based on its high toxicity and carcinogenic effects in animals, preplacement and periodic examinations should include a history of exposure to other carcinogens, alcohol and smoking habits, medication and family history. The skin, eye, lung, liver, and kidney should be evaluated. Sputum cytology may be helpful in evaluating the presence or absence of carcinogenic effects.

Special Tests: None in common use.

Personal Protective Methods: These are designed to supplement engineering controls and to prevent all contact with skin or respiratory tract. Full body protective clothing and gloves should be provided as well as fullface supplied air respirators of continuous flow or pressure demand type. Employees should remove and leave protective clothing and equipment at the point of exit, to be placed in impervious containers at the end of work shift for decontamination or disposal. Showers should be taken before dressing in street clothes.

References

(1) Fishbein, L., Flamm, W.G., and Falk, H.L. *Chemical Mutagens.* Academic Press, New York, (1970) pp. 215-216.

(2) Rapoport, I.A. Acetylation of gene proteins and mutations. *Dokl. Akad. Nauk. SSSR,* 58 (1947) 119.

(3) Auerbach, C. Drosophila tests in pharmacology. *Nature,* 210 (1966) 104.

(4) Khishin, A.F.E. The requirement of adenylic acid for formaldehyde mutagenesis. Mutation Res., 1 (1964) 202.

(5) IARC. β-propiolactone. In Vol. 4, *International Agency for Research on Cancer.* Lyon, France (1974) 259-269.

(6) Boutwell, R.L., Colburn, N.H., and Muckerman, C.C. In vivo reactions of β-propiolactone. *Ann. NY Acad. Sci.,* 163 (1969) 751.

(7) Fishbein, L. Degradation and residues of alkylating agents. *Ann. NY Acad. Sci.,* 163 (1969) 869.

(8) Van Duuren, B.L. Carcinogenic epoxides, lactones, and halo-ethers and their mode of action. *Ann. NY Acad. Sci.,* 163 (1969) 633.

(9) Dickens, F., and Jones, H.E.H. Further studies on the carcinogenic action of certain lactones and related substances in the rat and mouse. *Brit. J. Cancer* 19 (1965) 392.

(10) Van Duuren, B.L., Langseth, L., Orris, L., Teebor, G., Nelson, N., and Kuschner, M. Carcinogenicity of epoxides, lactones and peroxy compounds. IV. Tumor response in epithelial and connective tissue in mice and rats. *J. Natl. Cancer Inst.,* 37 (1966) 825.

(11) Smith, H.H., and Srb, A.M. Induction of mutations with β-propiolactone. *Science,* 114 (1951) 490.

(12) Swanson, C.P., and Merz, T. Factors influencing the effect of β-propiolactone on chromosomes of *Vicia faba. Science,* 129 (1959) 1364.

(13) Mukai, F., Belman, S., Troll, W., and Hawryluk, I. The mutagenicity of aryl-hydroxylamines. *Proc. Am. Assoc. Cancer Res.,* 8 (1967) 49.

(14) Kaplan, R.W. Probleme der prüfüng von pharmaka, zusatzstoffen und chemikaller auf ihre mutationsauslösende wirkung. *Naturwiss,* 49 (1962) 457.

(15) McCann, J., Choi, E., Yamasaki, E., and Ames, B.N. Detection of carcinogens as mutagens in *Salmonella*/microsome test: Assay of 300 chemicals. *Proc. Natl. Acad. Sci.,* 72 (1975) 5135.

(16) Smith, H.H., and Lofty, T.A. Effects of β-propiolactone and creepryn on chromosomes of *Vicia* and *Allum. Am. J. Botany,* 42 (1955) 750.

(17) Kubinski, H., and Kubinski, Z.O. Effects of monoalkylating carcinogens and mutagens on transforming DNA, *Fed. Proc.,* 35 (1976) 1551.

(18) Corbett, T.H., Heidelberger, C., and Dove, W.F. Determination of the mutagenic activity to bacteriophage T4 of carcinogenic and noncarcinogenic compounds. *Mol. Pharmacol.,* 6 (1970) 667-679.

Bibliography

Palmes, E.O., L. Orris, and N. Nelson. 1962. Skin irritation and skin tumor production by beta-propiolactone (BPL). *Am. Ind. Hyg. Assoc. J.* 23:257.

Van Duuren, B.L., L. Langseth, B.M. Goldschmidt, and L. Orris. 1967. Carcinogenicity of epoxides, lactones, and peroxy compounds. VI. Structure and carcinogenic activity. *J. Natl. Cancer Inst.* 39:1217.

PROPYL ALCOHOL

Description: There are two isomers of propyl alcohol—n-propyl alcohol and isopropyl alcohol. Both are colorless, volatile liquids.

Synonyms: n-Propyl alcohol: 1-Propanol, propylic alcohol. Isopropyl alcohol: Isopropanol, 2-propanol, secondary propyl alcohol, dimethylcarbinol.

Potential Occupational Exposures: Isopropyl alcohol is the more widely used of the two isomers. In the pharmaceutical industry, it has replaced ethyl alcohol in liniments, skin lotions, cosmetics, permanent wave preparations, hair tonics, mouth washes, and skin disinfectants and is widely used as a rubbing alcohol. Isopropyl alcohol is used in the manufacture of acetone, isopropyl derivatives, and safety glass, as a solvent in perfumes, resins and plastics, dye solutions, nitrocellulose lacquers, and in many extraction processes and is an ingredient of antifreezes, deicing agents, liquid soaps, and window cleaners.

Further applications include use as a preservative and dehydrating agent, a coupling agent in oil emulsions, an extracting agent for sulfonic acids from petroleum oils, and for coatings in textiles. n-Propyl alcohol is used in lacquers, dopes, cosmetics, dental lotions, cleaners, polishes, and pharmaceuticals and as a surgical antiseptic. It is a solvent for vegetable oils, natural gums and resins, rosin, shellac, certain synthetic resins, ethylcellulose, and butyral.

A partial list of occupations in which exposure may occur includes:

Disinfectant makers	Resin makers
Drug makers	Soap makers
Gum processors	Stainers
Metal degreasers	Vegetable oil processors
Nurses	Wax makers
Perfume makers	Window cleaning fluid makers
Polish makers	

Permissible Exposure Limits: The Federal standard for n-propyl alcohol is 200 ppm (500 mg/m^3) and for isopropyl alcohol, 400 ppm (980 mg/m^3).

Routes of Entry: Isopropyl alcohol: Inhalation of vapor. n-Propyl alcohol: Inhalation of vapor, percutaneous absorption.

Harmful Effects

Local—The two isomers are similar in physical and in most physiological properties. The vapors are mildly irritating to the conjunctiva and mucous membranes of the upper respiratory tract.

Systemic—No cases of poisoning from industrial exposure have been recorded for either isomer. n-Propyl alcohol can produce mild central nervous system depression; isopropyl alcohol is potentially narcotic in high concentrations.

Medical Surveillance: No specific considerations are needed.

Special Tests: Isopropyl alcohol and its metabolite, acetone, may be detected in blood, urine, and body tissues.

Personal Protective Methods: Clothing and barrier creams are recommended.

Bibliography

Henson, E.V. 1960. The toxicology of some aliphatic alcohols, parts I and II. *J. Occup. Med.* 2:442 and 497.
Von Oettingen, W.F. 1943. The aliphatic alcohols: their toxicity and potential dangers in relation to their chemical constitution and their fate in metabolism. *Public Health Bulletin* No. 281. U.S. Public Health Service, p. 112.

PROPYLENE DICHLORIDE

Description: $CH_3CHClCH_2Cl$, propylene dichloride, is a colorless liquid with a characteristic unpleasant odor.

Synonyms: 1,2-Dichloropropane, propylene chloride.

Potential Occupational Exposures: Propylene dichloride is widely used for degreasing and dry cleaning; it is also used as a soil fumigant and in the manufacture of cellulose plastics, rubber, waxes, scouring compounds and other manufacture of organic synthetics.

A partial list of occupations in which exposure may occur includes:

Cellulose plastic makers	Organic chemical synthesizers
Dry cleaners	Rubber makers
Fat processors	Scouring compound makers
Fumigant workers	Solvent workers
Gum processors	Stain removers
Metal degreasers	Wax makers
Oil processors	

Permissible Exposure Limits: The Federal standard is 75 ppm ($350 \, mg/m^3$).

Route of Entry: Inhalation of vapor.

Harmful Effects

Local—Propylene dichloride may cause dermatitis by defatting the skin. More severe irritation may occur if it is confined against the skin by clothing. Undiluted, it is moderately irritating to the eyes, but does not cause permanent injury.

Systemic—In animal experiments, acute exposure to propylene dichloride produced central nervous system narcosis, fatty degeneration of the liver and kidneys.

Medical Surveillance: Evaluate the skin and liver and renal function on a periodic basis, as well as cardiac and respiratory status and general health.

Special Tests: None in common use. Propylene dichloride can be determined in expired air.

Personal Protective Methods: Barrier creams or gloves and protective clothing. Masks in areas of vapor concentration.

Bibliography

Heppel. L.A., P.A. Neal, B. Highman, and V.T. Porterfield. 1946. Toxicology of 1,2-dichloropropane (propylene dichloride). I. Studies on effects of daily inhalation. *J. Ind. Hyg. Toxicol.* 28:1.

PYRIDINE

Description: C_5H_5N, pyridine, is a colorless liquid with an unpleasant odor. It is both flammable and explosive when exposed to a flame and decomposes on heating to release cyanide fumes. Pyridine is soluble in water, alcohol, and ether. The odor can be detected well below 1 ppm.

Synonyms: Azine.

Potential Occupational Exposures: Pyridine is used as a solvent in the chemical industry and as a denaturant for ethyl alcohol. It is used in the manufacture of paints, explosives, dyestuffs, rubber, vitamins, sulfa drugs, and disinfectants.

A partial list of occupations in which exposure may occur includes:

Alcohol denaturant makers	Dye makers
Alcohol denaturers	Explosive workers
Drug makers	Organic chemical synthesizers

Paint makers
Rubber workers
Resin workers

Solvent workers
Vitamin makers

Permissible Exposure Limits: The Federal standard is 5 ppm (15 mg/m^3).

Routes of Entry: Inhalation of vapor and percutaneous absorption of liquids.

Harmful Effects

Local—Irritation of the conjunctiva of the eye and cornea and mucous membranes of the upper respiratory tract and skin may occur. It occasionally causes skin sensitization, and photosensitizatiion has been reported.

Systemic—Very high concentrations may cause narcosis. Repeated, intermittent, or continuous low level exposure may lead to transient effects on the central nervous system and gastrointestinal tract. The symptoms include headache, dizziness, insomnia, nervousness, anorexia, nausea, vomiting, and diarrhea. Low back pain and urinary frequency with no changes in urine sediment or liver or renal function and complete recovery have been reported to follow exposures to about 100 ppm. Liver and kidney injury have been reported from its use as an oral medication.

Medical Surveillance: Placement and periodic examinations should consider possible effects on skin, central nervous system, and liver and kidney function.

Special Tests: None in common use. Metabolites are known and can be determined in blood and urine.

Personal Protective Methods: Rubber and plastic gloves should not be relied upon to prevent contact with pyridine as its salts penetrate the material. The odor is detectable at less than 1 ppm but cannot be relied upon as a preventative. In areas of elevated vapor concentration, workers should be supplied with fullface supplied air masks and protective clothing. Clothing that is contaminated by spills or splashes should be immediately changed and discarded and the area of involved skin thoroughly washed. Clean work clothes should be supplied daily with the worker showering after his shift before changing to street clothes.

Bibliography

Baldi, D. 1953. Patologia professional da piridina. *Med. Lav.* 44:244.

Q

QUINONE

Description: $C_6H_4O_2$, quinone, exists as large yellow, monoclinic prisms; the vapors have a pungent, irritating odor.

Synonyms: Benzoquinone, chinone, p-benzoquinone, 1,4-benzoquinone.

Potential Occupational Exposures: Because of its ability to react with certain nitrogen compounds to form colored substances, quinone is widely used in the dye, textile, chemical, tanning, and cosmetic industries. It is used as an intermediate in chemical synthesis for hydroquinone and other chemicals.

A partial list of occupations in which exposure may occur includes:

Chemical laboratory workers	Organic chemical synthesizers
Cosmetic makers	Photographic film developers
Dye makers	Protein fiber makers
Gelatin makers	Tannery workers
Hydrogen peroxide makers	Textile workers

Permissible Exposure Limits: The Federal standard is 0.1 ppm (0.4 mg/m³).

Route of Entry: Inhalation of vapor.

Harmful Effects

Local—Solid quinone in contact with skin or the lining of the nose and throat may produce discoloration, severe irritation, swelling, and the formation of papules and vesicles. Prolonged contact with the skin may cause ulceration. Quinone vapor is highly irritating to the eyes. Following prolonged exposure to vapor, brownish conjunctival stains may appear. These may be followed by corneal opacities and structural changes in the cornea and loss of visual acuity. The early pigmentary stains are reversible, while the corneal dystrophy tends to be progressive.

Systemic—No systemic effects have been found in workers exposed to quinone vapor over many years.

Medical Surveillance: Careful examination of the eyes, including visual acuity and slit lamp examinations, should be done during placement and periodic examinations. Also evaluate skin.

Special Tests: No useful laboratory tests for monitoring exposure have been developed.

Personal Protective Methods: In areas of high vapor concentrations, protection must be aimed at the eyes and respiratory tract. Fullface mask with organic vapor canisters or respirators with forced air afford protection. The skin can be damaged by contact with solid quinone, solutions, or vapor condensing on the skin, so protective clothing, gloves and boots are indicated. Personal hygiene is encouraged, with clothes being changed after each shift or after becoming damp from contact with the liquid. Workers should shower before changing to street clothes.

Bibliography

Anderson, B., and F. Oglesby. 1959. Corneal changes from quinone-hydroquinone exposure. *Arch. Ophthalmol.* 59:495.

S

SELENIUM AND COMPOUNDS

Description: Se, selenium, exists in three forms: a red amorphous powder, a grey form, and red crystals. Selenium, along with tellurium, is found in the sludges and sediments from electrolytic copper refining. It may also be recovered in flue dust from burning pyrites in sulfuric acid manufacture.

Synonyms: None.

Potential Occupational Exposures: Most of the selenium produced is used in the manufacture of selenium rectifiers. It is utilized as a pigment for ruby glass, paints, and dyes, as a vulcanizing agent for rubber, a decolorizing agent for green glass, a chemical catalyst in the Kjeldahl test, and an insecticide; in the manufacture of electrodes, selenium photocells, selenium cells, and semiconductor fusion mixtures; in photographic toning baths; and for dehydrogenation of organic compounds. Se is used in radioactive scanning of the pancreas and for photostatic and x-ray xerography. It may be alloyed with stainless steel, copper, and cast steel.

Hydrogen selenide (selenium hydride, H_2Se) is a colorless gas with a very disagreeable odor which is soluble in water. It is not used commercially. However, it may be produced by the reaction of acids or water and metal selenides or hydrogen and soluble selenium compounds. Selenium hexafluoride (SeF_6) is a gas and is utilized as a gaseous electric insulator. Other selenium compounds are used as solvents, plasticizers, reagents for alkaloids, and flameproofing agents for textiles and wire-cable coverings.

Selenium is a contaminant in most sulfide ores of copper, gold, nickel, and silver, and exposure may occur while removing selenium from these ores.

A partial list of occupations in which exposure may occur includes:

Arc light electrode makers	Pesticide makers
Copper smelters	Photographic chemical makers
Electric rectifier makers	Pigment makers
Glass makers	Plastic workers
Organic chemical synthesizers	Pyrite roasters

375

Rubber makers   Sulfuric acid makers
Semiconductor makers   Textile workers

Permissible Exposure Limits: The Federal standards are: Selenium compounds (as Se): 0.2 mg/m^3. Selenium hexafluoride: 0.05 ppm, 0.4 mg/m^3. Hydrogen selenide: 0.05 ppm, 0.2 mg/m^3.

Routes of Entry: Inhalation of dust or vapor; percutaneous absorption of liquid; ingestion.

Harmful Effects

Local—Elemental selenium is considered to be relatively nonirritating and is poorly absorbed. Some selenium compounds (particularly selenium dioxide and selenium oxychloride) are strong vesicants and can cause destruction of the skin. They are strong irritants to the upper respiratory tract and eyes, and may cause irritation of the mucous membrane of the stomach. Selenium compounds also may cause dermatitis of exposed areas. Allergy to selenium dioxide has been reported in the form of an urticarial generalized rash, and may cause a pink discoloration of the eyelids and palpebral conjunctivitis ("rose-eye"). Selenium oxide also may penetrate under the free edge of the nail, causing excruciatingly painful nail beds and painful paronychia. Selenium compounds may be absorbed through intact skin to produce systemic effects (Se sulfide in shampoo).

Selenium is considered to be an essential trace element for rats and chickens, and there is strong evidence of its essentiality in man. It is capable of antagonizing the toxic effects of certain other metals, e.g., As and Cd.

Systemic—The effects of hydrogen selenide intoxication are similar to those caused by other irritating gases in industry: irritation of the mucous membranes of the nose, eyes, and upper respiratory tract, followed by slight tightness in the chest. These symptoms clear when the worker is removed from the exposed area. In some cases, however, pulmonary edema may develop suddenly after a latent period of six to eight hours following exposure. Selenium dioxide inhaled in large quantities may also produce pulmonary edema.

The first and most characteristic sign of selenium absorption is a garlic odor of the breath. This may be related to the excretion in the breath of small amounts of dimethyl selenide. This odor dissipates completely in seven to ten days after the worker is removed from the exposure. It cannot be relied upon as a certain guide to selenium absorption. A more subtle and earlier sign is a metallic taste in the mouth, but many workers accept this without complaint.

Other systemic effects are less specific: pallor, lassitude, irritability, vague gastrointestinal symptoms (indigestion), and giddiness. Vital organs appear to escape harm from selenium absorption, but, based on the results of animal experimentation, liver and kidney damage should be regarded as possible.

Liver damage and other effects have been long recognized in livestock grazing on high selenium soils. Selenium has been mentioned for its carcinogenic, anticarcinogenic, and teratogenic effects, but, to date, these effects have not been seen in man.

Medical Surveillance: Preemployment and periodic examinations should consider especially the skin and eyes as well as liver, respiratory and kidney disease and function. The fingernails should be examined.

Special Tests: Urinary selenium excretion has been used to indicate exposure in the environment and also occupational exposure. It varies with the Se content of the diet and geographic location. Dimethyl selenide can be determined in breath.

Personal Protective Methods: Protective clothing with special emphasis on personal hygiene (showering and care of fingernails) should help prevent skin exposure and sensitization. Masks and supplied air respirators are needed in areas where concentrations of dust and vapors exceed the allowable standards. These should be equipped with fullface plates. Work clothing should be changed daily and showering encouraged prior to change to street clothing.

Bibliography

Environmental Protection Agency. *Selenium.* Report EPA-600/1-76-014. Research Triangle Park, N.C., Health Effects Research Laboratory (Jan. 1976).

Glover, J.R. 1970. Selenium and its industrial toxicology. *Ind. Med. Surg.* 30:50.

Harr, J.R., and O.H. Muth. 1972. Selenium poisoning in domestic animals and its relationship to man. *Clin. Toxicol.* 5:175.

Hausknecht, D., and Ziskind, R. (Science Applications, Inc.) *Health Effects of Selenium.* Report EPRI 571-1, Palo Alto, Calif., Electric Power Research Institute (Jan. 1976).

Nelson, A.A., Fitzhugh, O.G., and Calvery, H.O. 1943. Liver tumors following cirrhosis caused by selenium in rats. *Cancer Res.* 3:230.

Ransone, J.W., Scott, N.M. Jr., and Knoblock, E.C. 1961. Selenium sulfide intoxication. *N. Engl. J. Med.* 264:384.

Robertson, D.S.F. 1970. Selenium—a possible teratogen? *Lancet* 1:518.

Shapiro J.R. 1972. Selenium and carcinogenesis: A review. *Ann. N.Y. Acad. Sci.* 192:215.

SILICA, CRYSTALLINE

Description: Crystalline silica, SiO_2, is a crystalline material which melts to a glass at very high temperatures.

Synonyms: Silicon dioxide, silicic anhydride. It occurs in nature as agate, amethyst, chalcedony, cristobalite, flint, quartz, sand, tridymite.

Potential Occupational Exposure: NIOSH estimates that 1,200,000 workers are potentially exposed to crystalline silica in such industries as granite quarrying and cutting, foundry operations, metal, coal, and nonmetallic mining, and manufacture of clay and glass products.

Permissible Exposure Limits: NIOSH has concluded that adherence to an environmental limit of 50 micrograms (μg) crystalline silica per cubic meter of air as a time-weighted average, together with a program of work practices and medical examinations, will protect the worker from adverse effects of silica exposure for up to a 10-hour workday, 40-hour workweek, over a working lifetime.

The current Federal standard is based on a formula which is 10 milligrams total respirable dust divided by the sum of the percent of silica in the sample plus 2. The effect of the recommended change would be to reduce the environmental limit to about half of the existing limit for dusts of high quartz content; however, the reduction would be slight for dusts of very low quartz content. The recommended limit is based on sampling for respirable dust and analysis for free silica content.

The coal mine standard for silica promulgated under the Federal Mine Health and Safety Act of 1969 is approximately half as stringent as the proposed silica standard. While some of this difference may be explained by the specialized worker exposures found in the coal mining industry, the coal mine standard will be reviewed by NIOSH in the light of the conclusions in the criteria document.

Route of Entry: Inhalation of dust.

Harmful Effects

Crystalline silica can cause silicosis, a progressive and frequently incapacitating pneumoconiosis evident on x-ray and in pulmonary function testing, as well as in subjective respiratory complaints.

Medical Surveillance

(a) Medical examinations shall be made available to all workers subject to "exposure to free silica" prior to employee placement and at least once each 3 years thereafter. Examinations shall include as a minimum:

   1) A medical and occupational history to elicit data on worker exposure to free silica and signs and symptoms of respiratory disease.
   2) A chest roentgenogram (posteroanterior 14" × 17" or 14" × 14") classified according to the 1971 ILO International Classification of Radiographs of Pneumoconioses. [ILO U/C International Classification of

Radiographs of Pneumoconioses 1971, Occupational Safety and Health Series 22 (rev). Geneva, International Labor Office, 1972.]

3) Pulmonary function tests including forced vital capacity (FVC) and forced expiratory volume at one second (FEV_1) to provide a baseline for evaluation of pulmonary function and to help determine the advisability of the workers using negative- or positive-pressure respirators. It should be noted that pulmonary function tests may vary significantly in various ethnic groups. For example, in black persons, the test values for the FVC should be divided by 0.85 before the percentage value is compared with normal figures.

4) Body weight.

5) Height.

6) Age.

7) Initial medical examinations for presently employed workers shall be offered within 6 months of the promulgation of a standard incorporating these recommendations.

(b) Medical Management. An employee with or without roentgenographic evidence of silicosis who has respiratory distress and/or pulmonary functional impairment should be fully evaluated by a physician qualified to advise the employee whether he should continue working in a dusty trade.

(c) These records shall be available to the medical representatives of the Secretary of Health, Education and Welfare, of the Secretary of Labor, of the employee or former employee, and of the employer.

(d) Medical records shall be maintained for at least 30 years following the employee's termination of employment.

Special Tests: See "Medical Surveillance" above.

Personal Protective Methods: Engineering controls shall be used to maintain free silica dust exposures below the prescribed limit. Subsection (a) below shall apply whenever a variance from the standard recommended is granted under provisions of the Occupational Safety and Health Act, or in the interim period during the application for a variance. When the limits of exposure to free silica prescribed cannot be met by limiting the concentration of free silica in the work environment, an employer must utilize, as provided in subsection (a) of this section, a program of respiratory protection to effect the required protection of every worker exposed.

(a) Respiratory Protection—Appropriate respirators, as prescribed by NIOSH (see table below) shall be provided and used when a variance has been granted to allow respirators as a means of control of exposure to routine operations and while the application is pending. Administrative controls may also be used to reduce exposure. Respirators shall also be provided and used for non-routine operations (occasional brief exposures above the environmental standard and for emergencies); however, for these instances a variance is not required but the requirements set forth below continue to apply. Appropriate respirators as described in the table which follows shall only be used pursuant to the following requirements:

1) For the purpose of determining the type of respirator to be used, the employer shall measure the atmospheric concentration of free silica in the workplace when the initial application for variance is made and thereafter whenever process, worksite, climate, or control changes occur which are likely to affect the free silica concentration. This requirement shall not apply when only atmosphere-supplying positive-pressure respirators are used. The employer shall ensure that no worker is exposed to free silica in excess of the standard because of improper respirator selection, fit, use, or maintenance.

2) Employees experiencing breathing difficulty while using respirators shall be evaluated by a physician to determine the ability of the worker to wear a respirator.

3) A respiratory protective program meeting the requirements of Section 1910.134 of the Occupational Safety and Health Standards shall be established and enforced by the employer. (29 CFR 1910.134 published in the *Federal Register,* vol. 39, page 23671, dated June 27, 1974, as amended.)

4) The employer shall provide respirators in accordance with the table and shall ensure that the employee uses the appropriate respirator.

5) Respiratory protective devices in the table below shall be those approved either under 30 CFR 11, published March 25, 1972, or under the following regulations:

 (A) Filter-type dust, fume, and mist respirators—30 CFR 14 (Bureau of Mines Schedule 21B).

 (B) Supplied air respirator—30 CFR 12 (Bureau of Mines Schedule 19B).

6) A respirator specified for use in higher concentrations of free silica may be used in atmospheres of lower concentrations.

7) Employees shall be given instruction on the use of respirators assigned to them, on cleaning respirators, and on testing for leakage.

(b) Work Clothing—Where exposure to free silica is above the recommended environmental limit, work clothing shall be vacuumed before removal. Clothes shall not be cleaned by blowing or shaking.

Requirements for Respirator Usage at Concentrations Above the Standard

Concentrations of Free Silica in Multiples of the Standard	Respirator Type*
Less than or equal to 5X	Single use (valveless type) dust respirator.
Less than or equal to 10X	Quarter or half mask respirator with replaceable dust filter or single use (with valve) dust respirator.
	Type C, demand type (negative pressure), with quarter or half mask facepiece.

(continued)

Concentrations of Free Silica in Multiples of the Standard	Respirator Type*
Less than or equal to 100X	Full facepiece respirator with replaceable dust filter.
	Type C, supplied air respirator, demand type (negative pressure), with full facepiece.
Less than or equal to 200X	Powered air-purifying (positive-pressure) respirator, with replaceable applicable filter.**
Greater than 200X	Type C, supplied air respirator, continuous flow type (positive pressure), with full facepiece, hood, or helmet.

*Where a variance has been obtained for abrasive blasting with silica sand, use only Type C continuous flow, supplied air respirator with hood or helmet.

**An alternative is to select the standard high efficiency filter which must be at least 99.97% efficient against 0.3-micrometer dioctyl phthalate (DOP).

Bibliography

National Institute for Occupational Safety and Health. *Criteria for a Recommended Standard: Occupational Exposure to Crystalline Silica.* NIOSH Doc. No. 75-120, Wash., D.C. (1974).

SILVER AND COMPOUNDS

Description: Ag, silver, is a white metal and is extremely ductile and malleable, insoluble in water but soluble in hot sulfuric and nitric acids.

Synonyms: None.

Potential Occupational Exposures: Silver may be alloyed with copper, aluminum, cadmium, lead, or antimony; the alloys are used in the manufacture of silverware, jewelry, coins, ornaments, plates, commutators, scientific instruments, automobile bearings, and grids in storage batteries. Silver is used in chrome-nickel steels, in solders and brazing alloys, in the application of metallic films on glass and ceramics, to increase corrosion resistance to sulfuric acid, in photographic films, plates and paper, as an electroplated undercoating for nickel and chrome, as a bactericide for sterilizing water, fruit juices, vinegar, etc., in busbars and windings in electrical plants, in dental amalgams, and as a chemical catalyst in the synthesis of aldehydes. Because of its resistance to acetic and other food acids, it is utilized in the manufacture of pipes, valves, vats, pasteurizing coils and nozzles for the milk, vinegar, cider, brewing, and acetate rayon silk industries.

Silver compounds are used in photography, silver plating, inks, dyes, coloring glass and porcelain, etching ivory, in the manufacture of mirrors, and as analytical chemical reagents and catalysts. Some of the compounds are also of medical importance as antiseptics or astringents, and in the treatment of certain diseases, particularly in veterinary medicine.

A partial list of occupations in which exposure may occur includes:

Alloy makers	Glass makers
Bactericide makers	Hair dye makers
Ceramic makers	Hard solder workers
Coin makers	Ivory etchers
Chemical laboratory workers	Mirror makers
Dental alloy makers	Organic chemical makers
Drug makers	Photographic workers
Electric equipment makers	Water treaters

Permissible Exposure Limits: The Federal standard for silver metal and soluble compounds is 0.01 mg/m^3. The ACGIH (1978) proposes a TWA of 0.1 mg/m^3 for silver metal.

Routes of Entry: Inhalation of fumes or dust; ingestion of solutions or dust.

Harmful Effects

Local—The only local effect from metallic silver derives from the implant of small particles in the skin of the workmen (usually hands and fingers) which causes a permanent discoloration equivalent to the process of tattooing (local argyria). Silver nitrate dust and solutions are highly corrosive to the skin, eyes, and intestinal tract. The dust of silver nitrate may cause local irritation of the skin, burns of the conjunctiva, and blindness. Localized pigmentation of the skin and eyes may occur. The eye lesions are seen first in the caruncle, and then in the conjunctiva and cornea. The nasal septum and tonsillar pillars also are pigmented.

Systemic—All forms of silver are extremely cumulative once they enter body tissues, and very little is excreted. Studies on the occurrence of argyria following injection of silver arsphenamine indicate that the onset of visible argyria begins at a total dose of about 0.9 grams of silver. Generalized argyria develops when silver oxide or salts are inhaled or possibly ingested by workmen who handle compounds of silver (nitrate, fulminate, or cyanide). The condition produces no constitutional symptoms, but it may lead to permanent pigmentation of the skin and eyes. The workman's face, forehead, neck, hands, and forearms develop a dark, slate-grey color, uniform in distribution and varying in depth depending on the degree of exposure. Fingernails, buccal mucosa, toe nails, and covered parts of the body to a lesser degree, can also be affected by this discoloration process. The dust is also deposited in the lungs and may be regarded as a form of pneumoconiosis, although it carries no hazard of fibrosis. The existence of kidney lesions of consequence to renal function is improbable from occupational exposure.

Medical Surveillance: Special attention should be given to other sources of silver exposure, e.g., medications or previous occupational exposure. Inspection of the nasal septum, eyes, and throat will generally give incidence of pigmentation before generalized argyria occurs. This will usually be seen first in the ear lobes, face and hands.

Special Tests: Silver is excreted principally in the feces. Urine and blood levels have not been found useful in monitoring.

Personal Protective Methods: Workers involved with silver nitrate solution should be protected from spills and splashes by impervious protective clothing and chemical goggles. In areas of excessive dust levels, masks with full-face plates should be worn. Clean clothing should be provided daily and meals eaten in noncontaminated areas. Showers should be taken after each shift before change to street clothes.

Bibliography

Heimann, H. 1943. Toxicity of metallic silver. *N.Y. State Dept. Lab. Ind. Bull.* 22:81.
Holden, R.F. 1950. Observations in argyria. *J. Lab. Clin. Med.* 36:837.
Montandon, M.A. 1959. Argyrose des voies respiratoires. *Arch. Mal. Prof.* 20:419.

SODIUM HYDROXIDE/POTASSIUM HYDROXIDE

Description: NaOH, sodium hydroxide, is a white, deliquescent material sold as pellets, flakes, lumps, or sticks. It is soluble in water, alcohol, and glycerine. Aqueous solutions are known as soda lye.

KOH, potassium hydroxide, exists as white or slightly yellow deliquescent lumps, rods, or pellets. It is soluble in water. Aqueous solutions are known as lye.

Synonyms: Sodium hydroxide: caustic soda, caustic alkali, caustic flake, sodium hydrate, soda lye, white caustic. Potassium hydroxide: potassium hydrate, caustic potash, potassa, caustic alkali.

Potential Occupational Exposures: Sodium hydroxide is utilized to neutralize acids and make sodium salts in petroleum refining, viscose rayon, cellophane, and plastic production, and in the reclamation of rubber. It hydrolyzes fats to form soaps, and it precipitates alkaloids and most metals from aqueous solutions of their salts. It is used in the manufacture of mercerized cotton, paper, explosives, and dyestuffs, in metal cleaning, electrolytic extraction of zinc, tin plating, oxide coating, laundering, bleaching, and dishwashing, and it is used in the chemical industries. Total United States production of sodium hydroxide in 1974 was 10.9 million tons and NIOSH estimates that 150,000 workers are potentially exposed to the alkali.

Potassium hydroxide is used in the manufacture of liquid soap, as a mordant for wood, as a carbon dioxide absorber, in mercerizing cotton, in electroplating, photoengraving, and lithography, in printing inks, in paint and varnish removers, and in analytical chemistry, organic synthesis, and the production of other potassium compounds.

A partial list of occupations in which exposure may occur includes:

Bleachers	Paint removers
Bleach makers	Paper makers
Cellophane makers	Photoengravers
Chemical laboratory workers	Printers
Dye makers	Printing ink makers
Electroplaters	Rayon makers
Etchers	Rubber reclaimers
Explosive makers	Soap makers
Laundry workers	Textile bleachers
Lithographers	Tin platers
Mercerizers	Varnish removers
Organic chemical synthesizers	Zinc extractors

Permissible Exposure Limits: The Federal standard for sodium hydroxide is 2 mg/m^3. There is no standard for potassium hydroxide. NIOSH recommends a ceiling concentration of 2.0 milligrams sodium hydroxide/m^3 as determined by a sampling period of 15 minutes. The present Federal standard is 2.0 milligrams sodium hydroxide/m^3 as an 8-hour time-weighted average concentration. ACGIH (1978) gives 2 mg/m^3 as a TWA for both NaOH and KOH.

The recommendation for a limit on airborne workplace sodium hydroxide concentrations serves to protect against the irritation of the respiratory tract from sodium hydroxide aerosols.

Route of Entry: Inhalation of dust or mist.

Harmful Effects

Local—Both compounds are extremely alkaline in nature and are very corrosive to body tissues. Dermatitis may result from repeated exposure to dilute solutions in the form of liquids, dusts, or mists.

Extensive work practices are recommended to protect workers from local contact with sodium hydroxide. Local contact of sodium hydroxide with eyes, skin, and the alimentary tract has resulted in extensive damage to tissues, with resultant blindness, cutaneous burns, and perforations of the alimentary tract.

Systemic—Systemic effects are due entirely to local tissue injury. Extreme pulmonary irritation may result from inhalation of dust or mist. During the tissue regeneration process in the alimentary tract, some squamous cell carcinomas have developed.

Medical Surveillance: The skin, eyes, and respiratory tract should receive special attention in any placement or periodic examination. NIOSH recommends that workers subject to sodium hydroxide exposure have comprehensive preplacement medical examinations. Medical examinations shall be made available promptly to all workers with signs or symptoms of skin, eye, or upper respiratory tract irritation resulting from exposure to sodium hydroxide.

Special Tests: None.

Personal Protective Methods: Protection should be provided by impervious protective clothing, rubber boots, face and eye shields, and dust respirators. All skin area burns, especially of the eyes, demand immediate care by flooding with large quantities of water for 15 minutes or longer and specialized medical care.

Bibliography

National Institute for Occupational Safety and Health. *Criteria for a Recommended Standard: Occupational Exposure to Sodium Hydroxide.* NIOSH Doc. No. 76-105, Wash., D.C. (1976).

STIBINE

Description: SbH_3, stibine, is a colorless gas with a characteristic disagreeable odor. It is produced by dissolving zinc-antimony or magnesium-antimony in hydrochloric acid.

Synonyms: Antimony hydride.

Potential Occupational Exposures: Stibine is used as a fumigating agent. Exposure to stibine usually occurs when stibine is released from antimony-containing alloys during the charging of storage batteries, when certain antimonial drosses are treated with water or acid, or when antimony-containing metals come in contact with acid. Operations generally involved are metallurgy, welding or cutting with blow torches, soldering, filling of hydrogen balloons, etching of zinc, and chemical processes.

A partial list of occupations in which exposure may occur includes:

Etchers	Storage battery workers
Solderers	Welders

Permissible Exposure Limits: The Federal standard is 0.1 ppm (0.5 mg/m^3).

Route of Entry: Inhalation of gas.

Harmful Effects

Local—No local effects have been noted.

Systemic—Stibine is a powerful hemolytic and central nervous system poison. In acute poisoning, the symptoms are severe headache, nausea, weakness, abdominal and lumbar pain, slow breathing, and weak, irregular pulse. One of the earliest signs of overexposure may be hemoglobinuria. Laboratory studies may show a profound hemolytic anemia. Death is preceded by jaundice and anuria. Chronic stibine poisoning in man has not been reported.

Medical Surveillance: In preemployment and periodic examinations special attention should be given to significant blood, kidney, and liver diseases. The general health of exposed workmen should be evaluated periodically. Blood hemoglobin and urine tests for hemoglobin on persons suspected of stibine overexposure are indicated. Workers should also be advised to immediately report any red or dark urinary discoloration to the medical department. This frequently is the initial sign of stibine poisoning. (See Arsine.)

Special Tests: None in common use.

Personal Protective Methods: In areas where stibine gas is suspected, all persons entering or working in the area should be provided with fullface gas masks or supplied air respirators.

Bibliography

Dernehl, C.V., Stead, F.M., and Nau, C.A. 1944. Arsine, stibine, and H$_2$S. Accidental generation in a metal refinery. *Ind. Med. Surg.* 13:361.

Nau, C.A., Anderson, W., and Cone, R.E. 1944. Arsine, stibine, and H$_2$S—accidental industrial poisoning by a mixture. *Ind. Med. Surg.* 13:308.

Webster, S.H. 1946. Volatile hydrides of toxicological importance. *Ind. Hyg. and Toxicol.* 28:167.

STYRENE/ETHYL BENZENE

Description: $C_6H_5CH=CH_2$, styrene, is a colorless to yellowish, very refractive, oily liquid with a penetrating odor. $C_6H_5C_2H_5$, ethyl benzene, is a colorless flammable liquid with a pungent odor.

Synonyms: Styrene: Cinnamene, cinnemenol, cinnamol, phenethylene, phenylethylene, styrene monomer, styrol, styrolene, vinyl benzene. Ethyl benzene: Ethylbenzol, phenylethane, EB.

Potential Occupational Exposures: Upon heating to 200°C, styrene polymerizes to form polystyrene, a plastic. It is also used in combination with 1,3-butadiene or acrylonitrile to form copolymer elastomers, butadiene-styrene rubber, and acrylonitrile-butadiene-styrene (ABS). It is also used in the manufacture of resins, polyesters, and insulators.

Ethyl benzene is used in the manufacture of cellulose acetate, styrene, and synthetic rubber. It is also used as a solvent or diluent and as a component of automotive and aviation gasoline.

A partial list of occupations in which exposure may occur includes:

Adhesive makers
Aviation fuel blenders
Emulsifier agent makers
Fibrous glass molders
Insulator makers
Lacquer workers
Organic chemical synthesizers
Petroleum refinery workers

Polyester resin laminators
Polystyrene makers
Potting compound workers
Protective coating workers
Resin makers
Rubber makers
Solvent workers
Varnish makers

Permissible Exposure Limits: The Federal standard for styrene for an 8-hour TWA is 100 ppm (420 mg/m^3). The acceptable ceiling concentration is 200 ppm with an acceptable maximum peak of 600 ppm for a maximum duration of 5 minutes in any 3 hours. The Federal standard for ethyl benzene is 100 ppm (435 mg/m^3).

Routes of Entry: Inhalation of vapor; percutaneous absorption.

Harmful Effects

Local—Liquid and vapor are irritating to the eyes, nose, throat, and skin. The liquids are low-grade cutaneous irritants, and repeated contact may produce a dry, scaly, and fissured dermatitis.

Systemic—Acute exposure to high concentrations may produce irritation of the mucous membranes of the upper respiratory tract, nose, and mouth, followed by symptoms of narcosis, cramps, and death due to respiratory center paralysis. Effects of short-term exposure to styrene under laboratory conditions include prolonged reaction time and decreased manual dexterity.

Medical Surveillance: Consider possible irritant effects on the skin, eyes, and respiratory tract in any preplacement or periodic examinations, as well as blood, liver, and kidney function.

Special Tests: None in common use. Mandelic acid in urine has been used as a measure of the intensity of styrene exposure.

Personal Protective Methods: Barrier creams or gloves and protective clothing may be all that are needed where the vapor concentrations do not exceed existing standards. Where vapor concentration exists above allowable standards, masks with organic vapor canisters and face plates or respirators with air supply are recommended. Clothing saturated with styrene or ethyl benzene should be changed at once. Personal hygiene is encouraged with frequent changes of work clothes.

Bibliography

Stewart, R.D., Dodd, H.C., Baretta, E.D., and Schaffer, A.W. 1968. Human exposure to styrene vapor. *Arch. Environ. Health* 16:656.
Wilson, R.H. 1944. Health hazards encountered in the manufacture of synthetic rubber. *J. Am. Med. Assoc.* 124:701.

SULFUR CHLORIDE

Description: S_2Cl_2, sulfur chloride, is a fuming, oily liquid with a yellowish-red to amber color and a suffocating odor. It has an added hazard since it oxidizes and hydrolyzes to sulfur dioxide and hydrogen chloride.

Synonyms: Sulfur monochloride, sulfur subchloride, disulfur dichloride.

Potential Occupational Exposures: Sulfur chloride finds use as a chlorinating agent and an intermediate in the manufacture of organic chemicals, e.g., carbon tetrachloride, and sulfur dyes, insecticides, synthetic rubber, and pharmaceuticals. Exposure may also occur during the extraction of gold, purification of sugar juice, finishing and dyeing textiles, processing vegetable oils, hardening wood, and vulcanization of rubber.

A partial list of occupations in which exposure may occur includes:

Carbon tetrachloride makers	Insecticide makers
Drug makers	Rubber workers
Gold extractors	Sugar juice purifiers

Sulfur dye makers
Synthetic rubber makers
Textile dye makers and finishers

Vegetable oil processors
Wood hardeners

Permissible Exposure Limits: The Federal standard for sulfur chloride (sulfur monochloride) is 1 ppm (6 mg/m^3).

Route of Entry: Inhalation of vapor.

Harmful Effects

Local—Fumes, in sufficient quantity, may cause severe irritation to eyes, skin, and mucous membranes of the upper respiratory tract.

Systemic—Although this compound is capable of producing severe pulmonary irritation, very few serious cases of industrial exposure have been reported. This is probably because the pronounced irritant effects of sulfur chloride serve as an immediate warning signal when concentration of the gas approaches a hazardous level.

Medical Surveillance: Preemployment and periodic examinations should give special emphasis to the skin, eyes, and respiratory system. Pulmonary function tests may be useful. Exposures may also include sulfur dioxide and hydrochloric acid. (See these compounds.)

Special Tests: None are known to be useful.

Personal Protective Methods: In areas where vapor levels are excessive, workers should be supplied with fullface gas masks with proper canister or supplied air respirators with fullface-piece. Skin protection can usually be afforded by work clothes and barrier creams, but under certain instances (spills, etc.), full impervious protective suits may be necessary.

SULFUR DIOXIDE

Description: SO_2, sulfur dioxide, is a colorless gas at ambient temperatures with a characteristic strong suffocating odor. It is soluble in water and organic solvents.

Synonyms: Sulfurous anhydride, sulfurous oxide.

Potential Occupational Exposures: NIOSH estimates that 500,000 workers are potentially exposed to sulfur dioxide, which is encountered in many industrial operations. Sulfur dioxide is used in the manufacture of sodium sulfite, sulfuric acid, sulfuryl chloride, thionyl chloride, organic sulfonates, disinfectants, fumigants, glass, wine, ice, industrial and edible protein, and vapor pressure thermometers. It is also used in the bleaching of beet sugar, flour, fruit, gelatin, glue, grain, oil, straw, textiles, wicker ware, wood pulp, and wool; in the tanning of leather; in brewing and preserving; and in the refrigeration industry. Exposure may also occur in various other industrial processes as it is a by-product of ore smelting, coal and fuel oil combustion, paper manufacturing, and petroleum refining.

A partial list of occupations in which exposure may occur includes:

Beet sugar bleachers	Ore smelter workers
Boiler water treaters	Paper makers
Brewery workers	Petroleum refinery workers
Disinfectant makers	Protein makers
Diesel engine operators and repairmen	Refrigeration workers
Firemen	Sodium sulfite makers
Fumigant makers	Sulfuric acid makers
Furnace operators	Tannery workers
Gelatin bleachers	Thermometer makers (vapor)
Glass makers	Wine makers
Ice makers	Wood bleachers

Permissible Exposure Limits: The Federal standard is 5 ppm (13 mg/m^3). NIOSH has recommended lowering this standard to 2 ppm as a TWA. NIOSH has concluded from experimental and epidemiologic studies that exposure to sulfur dioxide at the existing Federal standard of 5 ppm can cause adverse respiratory effects by increasing airway resistance in a significant number of workers. Some workers are especially sensitive to these effects. In addition, there is evidence that the effects produced by sulfur dioxide are enhanced by airborne particulate matter and that sulfur dioxide may promote the carcinogenic action of other airborne substances. Compliance with the NIOSH recommended standard of 2 ppm, however, should prevent adverse effects of sulfur dioxide on the health and safety of workers. ACGIH (1978) also has set a TWA of 2 ppm (5 mg/m^3).

Routes of Entry: Inhalation of gas. Direct contact of gas or liquid phase on skin and mucous membranes.

Harmful Effects

Local—Gaseous sulfur dioxide is particularly irritating to mucous membranes of the upper respiratory tract. Chronic effects include rhinitis, dryness of the throat, and cough. Conjunctivitis, corneal burns, and corneal opacity may occur following direct contact with liquid.

Systemic—Acute over-exposure may result in death from asphyxia. Survivors may later develop chemical bronchopneumonia with bronchiolitis obliterans. Bronchoconstriction with increased pulmonary resistance, high-pitched rales, and a tendency to prolongation of the expiratory phase may result from moderate exposure, though bronchoconstriction may be asymptomatic. The effects on pulmonary function are increased in the presence of respirable particles.

Chronic exposure may result in nasopharyngitis, fatigue, altered sense of smell, and chronic bronchitis symptoms such as dyspnea on exertion, cough, and increased mucous excretion. Transient stimulation of erythropoietic activity of the bone marrow has been reported. Slight tolerance, at least to the odor threshold, and general acclimatization are common. Sensitization in a few individuals, particularly young adults, may also develop following repeated exposures. There is some evidence that some individuals may be innately hypersusceptible to SO_2. Animal experimentation has also indicated that sulfur dioxide may be a possible co-carcinogenic agent.

Medical Surveillance: Preplacement and periodic medical examinations should be concerned especially with the skin, eye, and respiratory tract. Pulmonary function should be evaluated, as well as smoking habits, and exposure to other pulmonary irritants.

Special Tests: None commonly used.

Personal Protective Methods: In areas where levels of sulfur dioxide gas are excessive, the worker should be supplied with fullface-piece cartridge or canister respirator or with supplied air respirators. Goggles, protective clothing, and gloves should be worn if splashes with liquid are likely. Work clothing should be changed at least twice a week to freshly laundered work clothes. Showering following each work shift should be encouraged. In areas of splash or spill, impervious clothing should be supplied, but if work clothes are wetted by sulfur dioxide, they should be promptly removed and the skin area thoroughly washed.

Bibliography

Ferris, B.G. Jr., Burgess, W.A., and Worcester, J. 1967. Prevalence of chronic respiratory disease in a pulp mill and a paper mill in the United States. *Br. J. Ind. Med.* 24:26.
National Institute for Occupational Safety and Health. *Criteria for a Recommended Standard: Occupational Exposure to Sulfur Dioxide.* NIOSH Doc. No. 74-111 (1974).
Skalpe, I.O. 1964. Long-term effects of sulphur dioxide exposure in pulp mill. *Br. J. Ind. Med.* 21:69.
Snell, R.E., and Luchsinger, P.C. 1969. Effects of sulfur dioxide on expiratory flow rates and total respiratory resistance in normal human subjects. *Arch. Environ. Health* 18:693.

SULFURIC ACID

Description: H_2SO_4, concentrated sulfuric acid, is a colorless, odorless, oily liquid which is commercially sold at 93% to 98% H_2SO_4, the remainder being water. Fuming sulfuric acid (oleum) gives off free sulfur trioxide and is a colorless or slightly colored, viscous liquid. Sulfuric acid is soluble in water and alcohol.

Synonyms: Oil of vitriol, spirit of vitriol, spirit of sulfur, hydrogen sulfate.

Potential Occupational Exposures: Sulfuric acid is used as a chemical feedstock in the manufacture of acetic acid, hydrochloric acid, citric acid, phosphoric acid, aluminum sulfate, ammonium sulfate, barium sulfate, copper sulfate, phenol, superphosphates, titanium dioxide, as well as synthetic fertilizers, nitrate explosives, artificial fibers, dyes, pharmaceuticals, detergents, glue, paint, and paper. It finds use as a dehydrating agent for esters and ethers due to its high affinity for water, as an electrolyte in storage batteries, for the hydrolysis of cellulose to obtain glucose, in the refining of mineral and vegetable oil, and in the leather industry. Other uses include fur and food processing, carbonization of wool fabrics, gas drying, uranium extraction from pitchblende, and laboratory analysis. NIOSH estimates that 200,000 workers are potentially exposed to sulfuric acid.

A partial list of occupations in which exposure may occur includes:

Aluminum sulfate makers	Food processors
Battery makers	Glue makers
Cellulose workers	Jewelers
Chemical synthesizers	Leather workers
Copper sulfate makers	Metal cleaners
Detergent makers	Paint makers
Dye makers	Paper makers
Explosive makers	Phenol makers

Permissible Exposure Limits: The Federal standard for sulfuric acid is 1 mg/m^3. NIOSH has concluded that adherence to the present Federal standard of one milligram per cubic meter of air ($1\ mg/m^3$), in conjunction with a strong program of work practices, will protect the worker from sulfuric acid exposure for up to a 10-hour workday, 40-hour workweek over a working lifetime.

Harmful Effects

Local—Burning and charring of the skin are a result of the great affinity for, and strong exothermic reaction with, water. Concentrated sulfuric acid will effectively remove the elements of water from many organic materials with which it comes in contact. It is even more rapidly injurious to mucous membranes and exceedingly dangerous to the eyes. Ingestion causes serious burns of the mouth or perforation of the esophagus or stomach. Dilute sulfuric acid does not possess this property, but is an irritant to skin and mucous membranes due to its acidity and may cause irreparable corneal damage and blindness as well as scarring of the eyelids and face.

Systemic—Sulfuric acid mist exposure causes irritation of the mucous membranes, including the eye, but principally the respiratory tract epithelium. The mist also causes etching of the dental enamel followed by erosion of the enamel and dentine with loss of tooth substance. Central and lateral incisors are mainly affected. Breathing high concentrations of sulfuric acid causes tickling in the nose and throat, sneezing, and coughing. At lower levels sulfuric acid causes a reflex increase in respiratory rate and diminution of depth, with reflex bronchoconstriction resulting in increased pulmonary air flow resistance. A single overexposure may lead to laryngeal, tracheobronchial, and pulmonary edema. Repeated excessive exposures over long periods have resulted in bronchitic symptoms, and rhinorrhea, lacrimation, and epistaxis. Long exposures are claimed to result in conjunctivitis, frequent respiratory infections, emphysema, and digestive disturbances.

Medical Surveillance: Preplacement and periodic medical examinations should give special consideration to possible effects on the skin, eyes, teeth, and respiratory tract. Pulmonary function tests should be performed.

Special Tests: None commonly used.

Personal Protective Methods: In all areas where liquid sulfuric acid is handled, impervious clothing should be provided, including gloves, goggles or face mask, rubber suits, and rubber shoes. Any work clothing wetted by sulfuric acid should be immediately changed and the skin area thoroughly washed and flooded with water. In areas where mist or gas is excessive, gas masks with appropriate canister or supplied air respirators should be provided. In either instance the worker should be supplied with fullface protection.

Bibliography

Malcolm, D., and Paul, E. 1961. Erosion of the teeth due to sulphuric acid in the battery industry. *Br. J. Ind. Med.* 18:63.
National Institute for Occupational Safety and Health. *Criteria for a Recommended Standard: Occupational Exposure to Sulfuric Acid.* NIOSH Doc. No. 74-128 (1974).
Williams, M.K. 1970. Sickness absence and ventilatory capacity of workers exposed to sulfuric acid mist. *Br. J. Ind. Med.* 27:61.

T

TELLURIUM AND COMPOUNDS

Description: Te, tellurium, is a semimetallic element with a bright luster which is insoluble in water and organic solvents. It may exist in a hexagonal crystalline form or an amorphous powder. It is found in sulfide ores and is produced as a by-product of copper or bismuth refining.

Synonyms: Aurum paradoxum, metallum problematum.

Potential Occupational Exposures: The primary use of tellurium is in the vulcanization of rubber. It is also used as a carbide stabilizer in cast iron, a chemical catalyst, a coloring agent in glazes and glass, a thermocoupling material in refrigerating equipment, and as an additive to selenium rectifiers; in alloys of lead, copper, steel, and tin for increased resistance to corrosion and stress, workability, machinability, and creep strength, and in certain culture media in bacteriology. Since tellurium is present in silver, copper, lead, and bismuth ores, exposure may occur during purification of these ores.

A partial list of occupations in which exposure may occur includes:

Alloy makers	Lead refinery workers
Ceramic makers	Porcelain makers
Copper refinery workers	Rubber workers
Electronic workers	Semiconductor makers
Enamel makers	Silverware makers
Foundry workers	Stainless steel makers

Permissible Exposure Limits: The applicable Federal standards are: Tellurium: 0.1 mg/m^3. Tellurium hexafluoride: 0.02 ppm (0.2 mg/m^3).

Routes of Entry: Inhalation of dust or fume; percutaneous absorption from dust.

Harmful Effects

Local—The literature contains no indication of any local effect from tellurium.

Systemic—The toxicity of tellurium and its compounds is of a low order.

There is no indication that either tellurium dust or fume is damaging to the skin or lungs. Inhalation of fumes may cause symptoms, however, some of which are particularly annoying socially to the worker. The most common signs of exposure are foul (garliclike) breath and perspiration, metallic taste in the mouth, and dryness. This is probably due to the presence of dimethyl telluride. These symptoms may appear after relatively short exposures at high concentrations, or longer exposures at lower concentrations, and may persist for long periods of time after the exposure has ended. Workers also complain of afternoon somnolence and loss of appetite.

Exposure to hydrogen telluride produces symptoms of headache, malaise, weakness, dizziness, and respiratory and cardiac symptoms similar to those caused by hydrogen selenide. Pulmonary irritation and the destruction of red blood cells have been reported in studies of laboratory animals exposed to hydrogen telluride.

In other animal studies, tellurium hexafluoride was found to be a respiratory irritant which caused pulmonary edema, and metallic tellurium was shown to have a teratogenic effect on the fetus of rats.

Medical Surveillance: Oral hygiene and the respiratory tract should receive special attention in preplacement or periodic examinations.

Special Tests: Urinary tellurium excretion has been studied in relation to exposure, but is of uncertain value.

Personal Protective Methods: Clean change of work clothes is necessary for hygienic purposes, and showering after each shift before change to street clothes should be encouraged. Respiratory protection is indicated in areas where exposure to hydrogen telluride and tellurium hexafluoride fumes and dust are above the allowable limits.

Bibliography

Agnew,W.F., and Curry, E. 1972. Period of teratogenic vulnerability of rat embryo to induction of hydrocephalus by tellurium. *Experientia.* 28:1444.

Blackadder, E.S., and Manderson, W.G. 1975. Occupational absorption of tellurium: a report of two cases. *Brit. J. Ind. Med.* 32:59.

Cerwenka, E.A., Jr., and Cooper. W.C. 1961. Toxicology of selenium and tellurium and their compounds. *Arch. Environ. Health* 3:189.

Duckett, S. 1972. Teratogenesis caused by tellurium. *Ann. N.Y. Acad. Sci.* 192:220.

Steinberg, H.H., Massari, S.C., Miner, A.G., and Rink, R. 1942. Industrial exposure to tellurium—atmospheric studies and clinical evaluation. *J. Ind. Hyg. Toxicol.* 24:183.

TETRACHLORODIBENZO-p-DIOXIN

Description: Polychlorinated dibenzo-p-dioxins are formed in the manufacturing process of all chlorophenols. However, the amount formed is dependent on the degree to which the temperature and pressure are controlled during production (1)-(5).

An especially toxic dioxin, 2,3,7,8-tetrachlorodibenzo-p-dioxin (TCDD), is formed during the production of 2,4,5-TCP (trichlorophenol) by the alkaline hydrolysis of 1,2,4,5-tetrachlorobenzene. Tetrachlorodibenzo-p-dioxin has the formula $C_{12}H_4Cl_4O_2$.

As can be anticipated, TCDD has been associated with all synthetic compounds derived from 2,4,5-TCP (6). This includes the widely used herbicide and defoliant 2,4,5-T (2,4,5-trichlorophenoxyacetic acid) (7).

Synonyms: TCDD.

Potential Occupational Exposures: In the cases of human exposure to 2,4,5-TCP, the only adverse effects reported were caused by occupational exposure or accidents that occurred during the manufacture of chlorinated phenols or products derived from them.

In 1949, intermediary chemicals of the manufacturing process were released in a U.S. 2,4,5-T plant. This accident led to 117 cases of chloracne among exposed workers (8).

In 1953 there was an accident in a Middle Rhine factory manufacturing 2,4,5-TCP from 1,2,4,5-tetrachlorobenzene. In addition to contracting chloracne (9), many workers had liver cirrhosis, heart complaints, and nervous system disorders, and were depressed (10).

In 1958, 31 employees of a Hamburg, Germany, plant in which 2,4,5-T was made from technical 2,4,5-TCP contracted chloracne and suffered the physical and psychological symptoms associated with it (11). In 1961 Bauer et al (10), conclusively identified TCDD as the cause of the chloracne.

An explosion occurred in a 2,4,5-T plant in Amsterdam in 1963. Six months later, 9 of 18 men, who were attempting to decontaminate the plant, developed chloracne. All of the men had worn deep sea diving suits, and all but one wore face masks with goggles while working in the plant. Of these men, three died within 2 years. The man without the face mask or goggles was severely affected. He was unable to walk and is still undergoing treatment (8).

In 1964, workers in a 2,4,5-T plant in the United States developed chloracne from exposure to TCDD (11).

There was an explosion at the Coalite Co.'s 2,4,5-TCP plant in Great Britain in 1968. TCDD had accidentally been produced as the result of an exothermic reaction (2). Seventy-nine cases of chloracne were reported; many of them were severe.

In 1971 there was an accidental poisoning episode in the United States that affected humans, horses, and other animals. Waste oil contaminated with TCDD had been sprayed on a riding arena to control dust. Later analyses showed that the arena contained TCDD in concentrations of 31.8 to 33.0 $\mu g/g$ (12). Commoner and Scott (13) found that the most important route of entry of dioxin into the body was the skin. (This does not preclude the effects of ingesting food contaminated with dioxin from handling.)

A 6-year-old girl was the most severely affected. She had an inflammatory reaction of the kidneys and bladder bleeding that was diagnosed as acute hemorrhagic cystitis with signs of focal pyelonephritis. Nine less severely affected persons developed diarrhea, headaches, nausea, polyarthralgias, and persistent skin lesions (14). The girl most affected was thoroughly reexamined in 1976. Results indicated that all of her original symptoms had completely disappeared. She had grown normally and all tests, including a detailed neurological examination, were normal (15).

In July 1976, 2 to 10 pounds of TCDD were accidentally released in the Seveso region of Italy (16). Most of the inhabitants were adversely affected. Reports of immediate symptoms and indications of many long-term effects are just becoming available. The first overt reaction was the appearance of numerous burn-like lesions on many of the inhabitants. These lesions generally receded. Whiteside (8) believes that they were probably caused by direct contact with the sodium hydroxide and phenolic components in the fallout. However, $2\frac{1}{2}$ months after the explosion, an increasing number of children and young people in the zone most affected began to develop symptoms of chloracne on their faces and bodies, a definite mark of dioxin poisoning. By November 28 people had confirmed cases of chloracne. This number rose to 38 by December and to 130 a year after the explosion. A number of the victims exposed underwent a "complete change of character": they became extremely nervous, tired, moody, and irritable, and had a marked loss of appetite.

There were a number of Seveso women who were pregnant at the time of the accident. Whiteside (8) reported that the total number of legal and illegal abortions performed as a result of the explosion probably totaled 90. There were 51 spontaneous (as distinct from induced) abortions. A survey conducted by an epidemiological commission has shown that 183 babies were delivered in the 2 months following the accident. Eight cases of birth abnormalities have been noted among babies born to women in the Seveso area who were pregnant at the time of the explosion. However, local physicians have had difficulty relating these abnormalities directly to the explosion because the incidence of birth abnormalities was not significantly higher than the normal incidence of abnormal births (8).

Permissible Exposure Limits: There are no numerical limits; in view of its effects, all contact should be avoided.

Routes of Entry: Skin absorption, inhalation of vapors.

Harmful Effects

See "Potential Occupational Exposures" above. The toxicity of a dioxin varies with the position and number of chlorines attached to the aromatic rings. Generally, the toxicity increases with increased chlorine substitution. Those dioxins that have halogens at the 2, 3, and 7 positions are particularly toxic. TCDD, which has chlorine atoms at the 2, 3, 7, and 8 positions, is considered the most toxic of the dioxins.

Medical Surveillance: In short, contact with TCDD should be avoided but obviously careful preplacement and regular physical exams should be carried out in those cases where worker exposure cannot be avoided.

Special Tests: To obtain a meaningful assessment of the levels of TCDD present in the environment and to determine the amount that could be accumulated in the food chain, a sensitive analytical method had to be developed that could accurately identify TCDD in parts per trillion. During the past 10 years considerable advances have been made in this regard. The analytical procedure that is currently considered the most sensitive is gas-liquid chromatography coupled with high resolution mass spectrometry.

Personal Protective Methods: As stated above, contact with TCDD should be avoided. When it cannot be avoided, extreme worker protection should be provided.

References

(1) Fishbein, L. 1973. Mutagens and potential mutagens in the biosphere I. DDT and its metabolites, polychlorinated biphenyls, chlorodioxins, polycyclic aromatic hydrocarbons, haloethers. *Sci. Total Environ.* 4:305-340.
(2) Milnes, M.H. 1971. Formation of 2,3,7,8-tetrachlorodibenzodioxin by thermal decomposition of sodium 2,4,5-trichlorophenate. *Nature* 232:395-396.
(3) Schulz, K.H. 1968. On the clinical aspects and etiology of chloracne. *Arbeitsmed. Socialmed. Arbeitshyg.* 3(2):25-29.
(4) Higginbotham, G.R., Huang, A., Firestone, D., Verrett, J., Ress, J., and Campbell, A.D. 1968. Chemical and toxicological evaluations of isolated and synthetic chloro derivatives of dibenzo-p-dioxin. *Nature* 220:702-703.
(5) Muelder, W.W., and Shadoff, L.A. 1973. The preparation of uniformly labeled ^{14}C-2,7-dichlorodibenzo-p-dioxin and ^{14}C-2,3,7,8-tetrachlorodibenzo-p-dioxin. In *Chlorodioxins—Origin and Fate.* Advances in Chemistry Series No. 120. American Chemical Society, pp. 1-6.
(6) Kearney, P.C., Woolson, E.A., Isensee, A.R. and Helling. C.S. 1973. Tetrachlorodibenzodioxin in the environment: sources, fate, and decontamination. *Environ. Health Perspect.* 5:273-277.
(7) *Federal Register* 43, No. 149, 34026-34054 (Aug. 2, 1978).
(8) Whiteside, T. July 25, 1977. The pendulum and the toxic cloud. *New Yorker.* 30-39, 40-41, 44-46, 48-55.

(9) Goldmann, P.J. 1972. Extremely severe acute chloracne due to trichlorophenol decomposition products. *Ind. Med. Soc. Med. Ind. Hyg.* 7(1):12-18.

(10) Bauer, H., Schulz, K.H., and Spiegelberg, U. 1961. Occupational intoxication in the production of chlorinated phenol compounds. *Arch. fur Gewerbepathol. Gewerbehyg.* 18:538-555.

(11) Poland, A.P., Smith, D., Metter, G., and Possick, P. 1971. A health survey of workers in a 2,4-D and 2,4,5-T plant: with special attention to chloracne, porphyria cutanea tarda, and psychologic parameters. *Arch. Environ. Health.* 22:316-327.

(12) Carter, C.D., Kimbrough, R.D., Liddle, J.A. Cline, R.E., Zack, M.M., Jr., Barthel, W.F., Koehler, R.E., and Phillips, P.E. 1975. Tetrachlorodibenzodioxin: an accidental poisoning episode in horse arenas. *Science* 188(4189):738-740.

(13) Commoner, B., and Scott, R.E. 1976. Accidental contamination of soil with dioxin in Missouri: Effects and countermeasures.

(14) U.S. Environmental Protection Agency. 1975. *Hazardous Waste Disposal Damage Reports.* Document No. 2 EPA/530/SW-151.2 pp. iii-iv, 1-9, references.

(15) Beale, M.G., Shearer, W.T., Karl, M.M., and Robson, A.M. 1977. Long-term effects of dioxin exposure. *Lancet* 1(8014):748.

(16) Dewse, C.D. 1976. Dangers of TCDD. *Lancet* 11(7981):363.

TETRACHLOROETHANE

Description: $CHCl_2CHCl_2$, tetrachloroethane, is a heavy, volatile liquid which is nonflammable and has a sweetish, chloroform-like odor. Oxidative decomposition of tetrachloroethane by ultraviolet radiation or by contact with hot metal results in the formation of small quantities of phosgene, hydrochloric acid, carbon monoxide, carbon dioxide, or dichloroacetyl chloride.

Synonyms: 1,1,2,2-Tetrachloroethane, sym-tetrachloroethane, acetylene tetrachloride, ethanetetrachloride.

Potential Occupational Exposures: Tetrachloroethane is used as a drycleaning agent, as a fumigant, in cement, and in lacquers. It is used in the manufacture of tetrachloroethylene, artificial silk, artificial leather, and artificial pearls. Recently, its use as a solvent has declined due to replacement by less toxic compounds. It is also used in the estimation of water content in tobacco and many drugs, and as a solvent for chromium chloride impregnation of furs.

A partial list of occupations in which exposure may occur includes:

Biologists	Mineralogists
Drycleaners	Oil processors
Fat processors	Paint makers
Fumigators	Phosphorus processors
Gasket makers	Resin makers
Herbicide workers	Soil treaters
Insecticide workers	Solvent workers
Lacquer workers	Varnish workers
Metal cleaners	Waxers

Permissible Exposure Limits: The Federal standard is 5 ppm (35 mg/m^3). NIOSH has recommended a time-weighted average limit of 1 ppm.

Routes of Entry: Inhalation of vapor and absorption of liquid through the skin. There is some evidence that tetrachloroethane absorbed through the skin affects the central nervous system only.

Harmful Effects

Local—Repeated or prolonged contact with this chemical can produce a scaly and fissured dermatitis.

Systemic—Early effects brought on by tetrachloroethane narcotic action include tremors, headache, a prickling sensation and numbness of limbs, loss of kneejerk, and excessive sweating. Paralysis of the interossei muscles of the hands and feet and disappearance of ocular and pharyngeal reflexes have also occurred due to peripheral neuritis which may develop later. Blood changes include increases in mononuclear leukocytes, progressive anemia, and a slight thrombocytosis.

Clinical symptoms following these changes are fatigue, headache, constipation, insomnia, irritability, anorexia, and nausea. Later on, liver dysfunction may result in complaints of general malaise, drowsiness, loss of appetite, nausea, and unpleasant taste in the mouth, and abdominal discomfort. This may be followed by jaundice, mental confusion, stupor or delirium, hematemesis, convulsions, and purpuric rashes.

Pulmonary edema ascribed to capillary injury has been noted in severe cases, along with renal damage, though it is not known to what extent this contributes to the total toxic picture. Nephritis may develop and the urine may contain albumin and casts. Fatty degeneration of the myocardium has been reported only in animal experiments.

Medical Surveillance: Preplacement and periodic examination should be comprehensive because of the possible involvement of many systems. Special attention should be given to liver, kidney, and bone marrow function, as well as to the central and peripheral nervous systems. Alcoholism may be a predisposing factor.

Special Tests: None commonly used. Blood or breath analyses may be useful.

Personal Protective Methods: Gloves and protective clothing should be worn, and appropriate respirators or masks should be used in areas of elevated vapor concentration.

Bibliography

Lobo-Medonca, R. 1963. Tetrachloroethane—a survey. *Br. J. Ind. Med.* 20:50.

National Institute for Occupational Safety and Health. *Criteria for a Recommended Standard: Occupational Exposure to 1,1,2,2-Tetrachloroethane.* NIOSH Doc. No. 77-121, Wash., D.C. (1977).

Towe, V.L., Wukjkowski, T., Wolf, M.A., Sadek, S.E., and Steward, R.E. 1963. Toxicity of a solvent of 1,1,1-trichloroethane and tetrachloroethylene as determined by experiments on laboratory animals and human subjects. *Am. Ind. Hyg. Assoc. J.* 24:541.

TETRACHLOROETHYLENE

Description: $Cl_2C=CCl_2$, tetrachloroethylene, is a clear, colorless, nonflammable liquid with a characteristic odor. The odor is noticeable at 50 ppm, though after a short period it may become inconspicuous, thereby becoming an unreliable warning signal.

Synonyms: Perchloroethylene, carbon dichloride, ethylene tetrachloride, perclene.

Potential Occupational Exposures: Tetrachloroethylene is a widely used solvent with particular use as a drycleaning agent, a degreaser, a chemical intermediate, a fumigant, and medically as an anthelmintic.

A partial list of occupations in which exposure may occur includes:

Cellulose ester processors	Metal degreasers
Degreasers	Printers
Dope processors	Rubber workers
Drug makers (anthelmintics)	Soap workers
Drycleaners	Solvent workers
Electroplaters	Tar processors
Ether processors	Vacuum tube makers
Fumigant workers	Wax makers
Gum processors	Wool scourers

Permissible Exposure Limits: The Federal standard is 100 ppm (670 mg/m³), as an 8-hour TWA with an acceptable ceiling concentration of 200 ppm; acceptable maximum peaks above the ceiling of 300 ppm are allowed for 5 minutes duration in a 3-hour period. NIOSH has recommended a time-weighted average limit of 50 ppm and a ceiling limit of 100 ppm determined by 15-minute samples, twice daily. Neither of these levels may provide adequate protection from potential carcinogenic effects because they were selected to prevent toxic effects other than cancer.

Routes of Entry: Inhalation of vapor and percutaneous absorption of liquid.

Harmful Effects

Local—Repeated contact may cause a dry, scaly, and fissured dermatitis. High concentrations may produce eye and nose irritation.

Systemic—Acute exposure to tetrachloroethylene may cause central nervous system depression, hepatic injury, and anesthetic death. Cardiac arrhythmias and renal injury have been produced in animal experiments. Signs and symptoms of overexposure include malaise, dizziness, headache, increased perspiration, fatigue, staggering gait, and slowing of mental ability. These usually subside quickly upon removal into the open air.

Perchloroethylene was very recently reported to be carcinogenic in NCI studies (1), producing liver hepatocellular tumors in $B_6C_3F_1$ hybrid male and female mice when tested at MTD and ½ MTD dose levels in corn oil solution by gavage (2)(3). No carcinogenic activity was observed in analogously treated Osborne-Mendel rats of both sexes (2).

Perchloroethylene has not been found to be carcinogenic in inhalation studies with rabbits, mice (4), rats, guinea pigs and monkeys.

Perchloroethylene, as well as the *cis-* and *trans-*isomers of 1,2-dichloroethylene were found to be nonmutagenic when tested in the metabolizing in vitro system with *E. coli* K12 (5). The mutagenicity of vinyl chloride, vinylidene chloride, and trichloroethylene, in the above test system was attributed to their initially forming unstable oxiranes, whereas halocarbons such as perchloroethylene and *cis-* and *trans-*1,2-dichloroethylene which form much more stable oxiranes were nonmutagenic (5)(6).

Metabolic pathways of tetrachloroethylene have been proposed by Bonse and Henschler (6). The metabolic formation of trichloroacetic acid can be explained by the primary formation of the oxirane and subsequent rearrangement to trichloroacetyl chloride and its subsequent hydrolysis.

Medical Surveillance: Evaluate skin, and liver and kidney function, as well as central nervous system. Alcoholism may be a predisposing factor.

Special Tests: Breath analyses may be helpful in evaluating exposures. Workers with preemployment histories of liver, kidney, or nervous disorders should be advised as to possible increased risk.

The *NIOSH Occupational Exposure Sampling Strategy Manual,* NIOSH publication #77-173, may be helpful in developing efficient programs to monitor employee exposures to tetrachloroethylene. The manual discusses determination of the need for exposure measurements, selection of appropriate employees for sampling, and selection of sampling times.

Personal Protective Methods: Skin protection in the form of barrier creams, gloves, and protective clothing should be used. In areas of vapor concentration, full face masks should be worn.

Employers should provide impervious gloves, face shields (8-inch minimum) and other appropriate clothing necessary to prevent repeated or prolonged skin contact with liquid tetrachloroethylene.

Employers should see that employee clothing wet with liquid tetrachloroethylene is placed in closed containers for storage until it can be discarded or until the employer provides for the removal of tetrachloroethylene from the clothing. If the clothing is to be laundered or otherwise cleaned to remove the tetrachloroethylene, the employer should inform the person performing the operation of the hazardous properties of tetrachloroethylene including the fact that it is a possible human carcinogen.

Employers should see that permeable clothing which becomes contaminated with liquid tetrachloroethylene be removed promptly and not reworn until the tetrachloroethylene is removed from the clothing.

Employers should see that employees who handle liquid tetrachloroethylene wash their hands thoroughly with soap or mild detergent before eating, smoking, or using toilet facilities.

Employers should see that employees whose skin becomes contaminated with liquid tetrachloroethylene promptly wash or shower with soap and mild detergent and water to remove any tetrachloroethylene from the skin.

Engineering and work practice controls should be used to minimize employee exposure to tetrachloroethylene.

To ensure that ventilation equipment is working properly, it is advised that effectiveness be checked at least every three months (e.g., air velocity, static pressure or air volume). System effectiveness should also be checked within five days of any change in production, process, or control which might result in significant increases in airborne exposures to tetrachloroethylene.

Exposure to tetrachloroethylene should not be controlled with the use of respirators except:

> During the time period necessary to install or implement engineering or work practice controls; or

> In work situations in which engineering and work practice controls are technically not feasible; or

> To supplement engineering and work practice controls when such controls fail to adequately control exposure to tetrachloroethylene; or

For operations which require entry into tanks or closed vessels; or

In emergencies.

References

(1) *Bioassay of Tetrachloroethylene for Possible Carcinogenicity.* DHEW Publication No. (NH) 77-813. U.S. Department of Health, Education, and Welfare, Public Health Service, National Institutes of Health, National Cancer Institute, October 1977.

(2) Weisburger, E.K. Carcinogenicity studies of halogenated hydrocarbons. NIEHS Conference on Comparative Metabolism and Toxicity of Vinyl Chloride and Related Compounds, Bethesda, Md. May 2-4 (1977).

(3) Anon. NCI clearinghouse subgroup finds tris tetrachloroethylene carcinogenic. *Toxic Materials News* 4 (9) (1977) p. 60.

(4) Kylin, B., Sumegi, I., and Yllner, S. Hepatotoxicity of inhaled trichloroethylene, and tetrachloroethylene, long-term exposure. *Acta. Pharmacol. Toxicol.* 22 (1965) 379-385.

(5) Greim, H., Bonse, G., Radwan, Z., Reichert, D., and Henschler, D. Mutagenicity in vitro and potential carcinogenicity of chlorinated ethylenes as a function of metabolic oxirane formation. *Biochem. Pharmacol.* 24 (1975) 2013.

(6) Bonse, G., Henschler, D. Chemical reactivity, biotransformation, and toxicity of polychlorinated aliphatic compounds. *CRC Crit. Revs. Toxicology* 4 (1976) 395-409.

Bibliography

Rampy, L.W., Quast, J.F., Leong, B.K.J., and Gehring, P.J. Results of longterm inhalation toxicity studies on rats of 1,1,1-trichloroethane and perchloroethylene formulations. Toxicology Research Laboratory, Dow Chemical, U.S.A., Poster presentation, International Congress of Toxicology, Toronto, Canada, April, 1977.

National Institute for Occupational Safety and Health. *Criteria for a Recommended Standard: Occupational Exposure to Tetrachloroethylene.* NIOSH Doc. No. 76-185 (1976).

Fishbein, L. Industrial mutagens and potential mutagens I. Halogenated aliphatic derivatives. *Mutat. Res.* 32:267-308, 1976.

Fujii, T. Variation in the liver function of rabbits after administration of chlorinated hydrocarbons. *Jap. J. Ind. Hlth.* 17:81-88, 1975.

Duprat, P., Delsaut, L., and Gradiski, D. Irritant potency of the principal aliphatic chloride solvents on the skin and ocular mucous membranes of rabbits. *Europ. J. Toxicol.* 3:171-177, 1976.

Brancaccio, A., Mazza, V., and Di Paolo, R. Renal function in experimental tetrachloroethylene poisoning. *Folia Med.* 54:233-237, 1971.

Mazza, V. Enzymatic changes in experimental tetrachloroethylene poisoning. *Folia Med.* 55:373-381, 1972.

Korn, J. How many more? Perchloroethylene intoxication in coin drycleaning establishments. *Ugeskr. Laeg.* 139:303-304, 1977.

Weichardt, H., and Lindner, J. Health hazards due to perchloroethylene in chemical drycleaning enterprises, from the viewpoint of occupational medicine and toxicology. *Staub Reinhalt Luft.* 35:416-420, 1975.

Medek, V. and Kovarik, J. The effect of perchloroethylene on the health of workers. *Pracovni lekarstvi* 25:339-341, 1973.

Larsen, N., Nielsen, B., and Rayn Nielsen, A. Perchloroethylene intoxication. A hazard in the use of coin laundries. *Ugeskr. Laeg.,* 139:270-275, 1977.

Schwetz, B.A., Leong, K.J., and Gehring, P.J. The effect of maternally inhaled trichloroethylene, perchloroethylene, methyl chloroform and methylene chloride on embryonal and fetal development in mice and rats. *Toxicol. Appl. Pharmacol.* 32:84-96, 1975.

Ikeda, M., and Imamura, T. Biological halflife of trichloroethylene and tetrachloroethylene in human subjects. *Int. Arch. Arbeitsmed.* 31:209-224, 1973.

Gobbato, F., and Slavich, G. 1968. Intossicazione acuta da tetrachloroethano. *Med. Lavoro.* 59:667.

Gold, J.H. 1969. Chronic perchloroethylene poisoning. *Can. Psychiatr. Assoc. J.* 14:627.

Mecker, L.C., and Phelps, D.K. 1966. Liver disease secondary to tetrachloroethylene exposure. *J. Am. Med. Assoc.* 1971:662.

Meunzer, M., and Heder, K. 1972. Results of an industrial hygiene survey and medical examinations in drycleaning firms. *Zentralbl. Arbeitsmed.* 22:133.

Stewart, R.D., Baretta, E.D., Dodd, H.C., and Torkelson, T.R. 1970. Experimental human exposure to tetrachloroethylene. *Arch. Environ. Health* 20:224.

National Institute for Occupational Safety and Health. Current Intelligence Bulletin No. 20: *Tetrachloroethylene,* Wash., D.C. (Jan. 20, 1978).

TETRAMETHYLTHIURAM DISULFIDE

Description: $C_6H_{12}N_2S_4$, tetramethylthiuram disulfide, is a white or yellow crystal insoluble in water, but soluble in organic solvents.

Synonyms: Thiram, bis-(dimethylthiocarbamyl) disulfide, TMTD, thirad, thiuram.

Potential Occupational Exposures: Tetramethylthiuram disulfide is used as a rubber accelerator and vulcanizer; a seed, nut, fruit, and mushroom disinfectant; a bacteriostat for edible oils and fats; and as an ingredient in sun-tan and antiseptic sprays and soaps. It is also used as a fungicide, rodent repellent, wood preservative, and may be used in the blending of lubricant oils.

A partial list of occupations in which exposure may occur includes:

Food disinfectant makers	Rubber makers
Fungicide workers	Soap makers
Lubricating oil blenders	Wood preservative makers
Rat repellent makers	

Permissible Exposure Limits: The Federal standard for thiram (tetramethylthiuram disulfide) is 5 mg/m^3.

Route of Entry: Inhalation of dust, spray, or mist.

Harmful Effects

Local—Irritation of mucous membranes, conjunctivitis, rhinitis, sneezing, and cough may result from excessive exposures. Skin irritation with erythema and urticaria may also occur. Allergic contact dermatitis has been reported in workers who wore rubber gloves containing tetramethylthiuram disulfide.

Systemic—Systemic effects have not been reported in the U.S. literature. Bronchitis was mentioned in one European report in workers exposed to thiram or other products during synthesis. Intolerance to alcohol has been observed in workers exposed to thiram, manifested by flushing of face, palpitation, rapid pulse, dizziness, and hypotension. These effects are thought to be due to the blocking of the oxidation of acetaldehyde. It should be noted in this connection that the diethyl homologue of this compound, tetraethylthiuram disulfide, is marketed as the drug "Antabuse" and that severe and disagreeable symptoms ensue immediately in subjects who ingest the smallest amount of ethyl alcohol after they have been "premedicated" with the drug.

Medical Surveillance: Preplacement and periodic medical examinations should give special attention to history of skin allergy, eye irritation, and significant respiratory, liver, or kidney disease. Workers should be aware of the potentiating action of alcoholic beverages when working with tetramethylthiuram disulfide.

Special Tests: None in common use.

Personal Protective Methods: Skin and eye protection should be provided by protective clothing, gloves, and goggles. Employees should be encouraged to shower following each shift and to change to clean work clothes at the start of each shift. In areas where dust, spray, or mist are excessive, respiratory protection by dust masks or gas mask respirators with proper canister or supplied air respirators should be provided.

Bibliography

Finulli, M., and Magistretti, M. 1961. Antabuse-like toxic manifestations in workmen employed in the manufacture of a synthetic anticryptogamic: TMTD (tetramethylthiuram disulfide). *Med. Lav.* 52:132.
Gleason, M.N., Gosselin, R.E., and Hodge, H.C. 1963. *Clinical Toxicology of Commercial Products.* William and Wilkins, Baltimore.

TETRYL

Description: Tetryl, $(NO_2)_3C_6H_2N(NO_2)CH_3$, is a yellow solid.

Synonyms: Trinitrophenylmethylnitramine, nitramine, tetranitromethylaniline, pyrenite, picrylmethylnitramine, picrylnitromethylamine, N-methyl-N,2,4,6-tetranitroaniline, tetralite.

Potential Occupational Exposures: Tetryl is used in explosives as an intermediary detonating agent and as a booster charge; it is also used as a chemical indicator.

A partial list of occupations in which exposure may occur includes:

Chemical indicator makers   Explosive makers

Permissible Exposure Limits: The Federal standard is 1.5 mg/m^3.

Routes of Entry: Inhalation and skin absorption.

Harmful Effects

Local—Tetryl is a potent sensitizer, and allergic dermatitis is common. Dermatitis first appears on exposed skin areas, but can spread to other parts of the body in fair skinned individuals or those with poor personal hygiene. The severest forms show massive generalized edema with partial obstruction of the trachea due to swelling of the tongue, and these cases require hospitalization. Contact may stain skin and hair yellow or orange. Tetryl is acutely irritating to the mucous membranes of the respiratory tract and the eyes, causing coughing, sneezing, epistaxis, conjunctivitis, and palpebral and periorbital edema.

Systemic—Tetryl exposure may cuase irritability, easy fatigability, malaise, headaches, lassitude, insomnia, nausea, and vomiting. Anemia either of the marrow depression or deficiency type has been observed among tetryl workers. Tetryl exposure has produced liver and kidney damage in animals.

Medical Surveillance: Preplacement physical examination should give special attention to those individuals with a history of allergy, blood dyscrasias, or skin, liver, or kidney disease. Periodic examinations should be directed primarily to the control of dermatitis and allergic reactions, plus any effects on the respiratory tract, eyes, central nervous system, blood, liver, or kidneys.

Special Tests: None in common use.

Personal Protective Methods: Skin protection is necessary by means of protective clothing and gloves. Where significant air concentration of dusts or vapors exist, masks to prevent inhalation are necessary. Daily change to clean work clothes is strongly advised, with showers after each shift mandatory, before dressing in street clothes.

Bibliography

Bergman, B.B. 1952. Tetryl toxicity: a summary of ten years' experience. *AMA Arch. Ind. Hyg. Occup. Med.* 5:10.

Hardy, H.L., and Maloof, C.C. 1950. Evidence of systemic effect of tetryl with summary of available literature. *AMA Arch. Ind. Med. Occup. Med.* 1:545.

Norwood, W.D. 1943. Trinitrotoluene (TNT); its effective removal from the skin by a special liquid soap. *Ind. Med.* 12:206.

THALLIUM AND COMPOUNDS

Description: Tl, thallium, is a soft, heavy metal insoluble in water and organic solvents. It is usually obtained as a by-product from the flue dust generated during the roasting of pyrite ores in the smelting and refining of lead and zinc.

Synonyms: None.

Potential Occupational Exposures: Thallium and its compounds are used as rodenticides, fungicides, insecticides, catalysts in certain organic reactions, in phosphor activators, in bromoiodide crystals for lenses, plates, and prisms in infrared optical instruments, in photoelectric cells, in mineralogical analysis, alloyed with mercury in low temperature thermometers, switches and closures, in high-density liquids, in dyes and pigments, and in the manufacture of optical lenses, fireworks, and imitation precious jewelry. It forms a stainless alloy with silver and a corrosion-resistant alloy with lead. Its medicinal use for epilation has been almost discontinued.

A partial list of occupations in which exposure may occur includes:

Alloy makers	Glass makers
Artificial diamond makers	High refractive index makers
Chlorinated compound makers	Infrared instrument makers
Dye makers	Optical glass makers
Fireworks makers	Photoelectric cell makers
Gem makers	Rodenticide workers

Permissible Exposure Limits: The Federal standard for thallium (soluble compounds) is 0.1 mg Tl/m^3.

Routes of Entry: Inhalation of dust and fume. Ingestion and percutaneous absorption of dust.

Harmful Effects

Local—Thallium salts may be skin irritants and sensitizers, but these effects occur rarely in industry.

Systemic—Thallium is an extremely toxic and cumulative poison. In nonfatal occupational cases of moderate or long term exposure, early symptoms usually include fatigue, limb pain, metallic taste in the mouth and loss of hair, although loss of hair is not always present as an early symptom. Later, peripheral neuritis, proteinuria, and joint pains occur.

Occasionally, neurological signs are the presenting factor, especially in more severe poisonings. Long term exposure may produce optic atrophy, paresthesia, and changes in pupillary and superficial tendon reflexes (slowed responses). Acute poisoning rarely occurs in industry, and is usually due to ingestion of thallium. When it occurs, gastrointestinal symptoms, abdominal colic, loss of kidney function, peripheral neuritis, strabismus, disorientation, convulsions, joint pain, and alopecia develop rapidly (within 3 days). Death is due to damage to the central nervous system.

Medical Surveillance: Preplacement and periodic examinations should give special consideration to the central nervous system, gastrointestinal symptoms, and liver and kidney function. Hair loss may be a significant sign. Urine examinations may be helpful.

Special Tests: Thallium has been determined in the urine, but the levels do not relate to degree, exposure, or to symptoms.

Personal Protective Methods: Eating, gum chewing, and smoking should not be allowed in production areas. Strict enforcement of high standards of personal hygiene is recommended. Appropriate respiratory protection should be used. Protective clothing, hats, goggles, and gloves may be needed to prevent dust absorption through the skin. Daily change of work clothes and showers at the end of the shift will reduce the chances of significant absorption.

Bibliography

Bank, W.J., Pleasure, D.E., Suzuki, K., et al. 1972. Thallium poisoning. *Arch. Neurol.* 26:456.
Jacobs, M.B. 1962. The determination of thallium in urine. *Am. Ind. Hyg. Assoc. J.* 23:411.
Richeson, E.M. 1958. Industrial thallium intoxication. *Ind. Med. Surg.* 27:607.

THORIUM AND COMPOUNDS

Description: Th, thorium, is a natural radioactive element insoluble in water and organic solvents. It occurs in the minerals monazite, thorite, and thorinite, usually mixed with its distintegration products.

Synonyms: None.

Potential Occupational Exposures: Metallic thorium is used in nuclear reactors to produce nuclear fuel, in the manufacture of incandescent mantles, as an alloying material, especially with some of the lighter metals, e.g., magnesium, as a reducing agent in metallurgy, for filament coatings in incandescent lamps and vacuum tubes, as a catalyst in organic synthesis, in ceramics, and in welding electrodes.

Exposures may occur during production and use of thorium-containing materials, in the casting and machining of alloy parts, and from the fume produced during welding with thorium electrodes.

A partial list of occupations in which exposure may occur includes:

Ceramic makers	Metal refiners
Gas mantle makers	Nuclear reactor workers
Incandescent lamp makers	Organic chemical synthesizers
Magnesium alloy makers	Vacuum tube makers

Permissible Exposure Limits: Maximum permissible concentration for thorium under the Federal standard (see 20 CFR Part 20-Table 1) is $1 \times 10^{-6} \mu Ci/ml$ (air).

Routes of Entry: Ingestion of liquid, inhalation of dust or gas, and percutaneous absorption.

Harmful Effects

Local—Thorium and thorium compounds are relatively inert, but some irritant effect may occur depending on the anion present. Gas and aerosols can penetrate the body by way of the respiratory system, the digestive system, and the skin.

Systemic—Thorium and its compounds are toxicologically inert on the basis of its chemical toxicity. Only 0.001% of an ingested dose is retained in the body. Thorium, once deposited in the body, remains for long periods of time. It has a predilection for bones, lungs, lymphatic glands, and parenchymatous tissues. Characteristic effects of the activity of thorium and its disintegration products are changes in blood forming, nervous and reticuloendothelial systems, and functional and morphological damage to lung and bone tissue. Only much later do illness and symptoms characteristic of chronic radiation disease appear. After a considerable time, neoplasms may occur and the immunological activity of the body may be reduced. External radiation with gamma rays can occur from contact with material containing mesothorium, with thorium in large quantities, and with by-products that contain disintegration products of thorium. Thorium dioxide (thorotrast) is known to cause severe radiation damage and cancer of bone, blood vessels, liver, and other organs when administered to patients for diagnostic purposes. Its use is now forbidden for introduction into body tissues. Workers in plants where thorium dioxide is produced have not experienced either chemical or radiation injury.

Medical Surveillance: Monitoring of personnel for early symptoms and changes such as abnormal leukocytes in the blood smear may be of value.

Special Tests: In cases of chronic or acute exposure, the determination of thorium in the urine or the use of whole body radiation counts and breath radon are useful methods of monitoring the exposure dose and excretion rates.

Personal Protective Methods: Protection of the worker is afforded by respiratory protection with either dust masks, special canister gas masks, or supplied air respirators. Protective clothing and gloves to prevent dust settling on the skin, with daily change of work clothes, and showering after each shift before change to street clothes should be routine.

Bibliography

Albert, R., Klevin, P., Fresco, J., Harley, J., Harris, W., and Eisenbud, M. 1955. Industrial hygiene and medical survey of a thorium refinery. *Arch. Ind. Health* 11:234.
Baker, W.J., Bulkley, J.B., Dudley, R.A., Evans, R.D., McCluskey, H.B., Reeves, J.D., Jr., Ryder, R.H., Salter, L.P., and Shanaham, M.M. 1961. Observations on the late effects of internally deposited mixtures of mesothorium and radium in twelve dial painters. *N. Engl. J. Med.* 265:1023.
Saragoca, A., Tabares, M.H., Barros, R.B., and Horta, J.D. 1972. Some clinical and laboratory findings in patients injected with thorium dioxide—study of 155 cases. *Am. J. Gastroenterol.* 57:301.

TIN AND COMPOUNDS

Description: Sn, tin, is a soft, silvery-white metal insoluble in water. The primary commercial source of tin is cassiterite (SnO_2, tinstone).

Synonyms: Stannum.

Potential Occupational Exposures: The most important use of tin is as a protective coating for other metals such as in the food and beverage canning industry, in roofing tiles, silverware, coated wire, household utensils, electronic components, and pistons. Common tin alloys are phosphor bronze, light brass, gun metal, high tensile brass, manganese bronze, die-casting alloys, bearing metals, type metal, and pewter. These are used as soft solders, fillers in automobile bodies, and as coatings for hydraulic brake parts, aircraft landing gear and engine parts. Metallic tin is used in the manufacture of collapsible tubes and foil for packaging.

Organic and inorganic tin compounds are important industrially in the production of drill-glass, ceramics, porcelain, enamel, glass, and inks; as a mordant it is important in the production of fungicides, anthelmintics, insecticides; as a stabilizer it is used in polyvinyl plastics and chlorinated rubber paints; and it is used in plating baths.

Organotin compounds are used as additives in a variety of products and processes. Diorganotins find application as heat stabilizers in plastics, as catalysts in the production of urethane foams, in the cold curing of rubber, and as scavengers for halogen acids. Tri- and tetraorganotins are used as preservatives for wood, leather, paper, paints, and textiles and as biocides. Production of organotins has increased from a few tons in the 1940s to 25,000 tons in 1975. NIOSH estimates that 30,000 employees in the United States may be exposed to organotin compounds.

Exposures to tin may occur in mining, smelting, and refining, and in the production and use of tin alloys and solders.

A partial list of occupations in which exposure may occur includes:

> Babbitt metal (tin, copper, antimony) makers
> Brass (essentially copper and zinc) founders
> Britannia metal (tin, copper, antimony) makers
> Bronze (tin, copper) founders
> Dye workers
> Fungicide workers
> Pewter makers
> Pigment workers
> Plastic makers
> Solder makers
> Textile workers
> Type metal (lead, antimony, tin) makers

Permissible Exposure Limits: The Federal standard for organic tin compounds is 0.1 mg/m^3 and for inorganic compounds excluding the oxides it is 2.0 mg/m^3.

Routes of Entry: Inhalation of dust. Ingestion, inhalation, or percutaneous absorption of organotins.

Harmful Effects

Local—Certain inorganic tin salts are mild irritants to the skin and mucous membranes. They may be strongly acid or basic depending on the cation or anion present. Organic tin compounds, especially tributyl and dibutyl compounds, may cause acute burns to the skin. The burns produce little pain but may itch. They heal without scarring. Clothing contaminated by vapors or liquids may cause subacute lesions and diffuse erythematoid dermatitis on the lower abdomen, thighs, and groin of workmen who handle these compounds. The lesions heal rapidly on removal from contact. The eyes are rarely involved, but accidental splashing with tributyl tin has caused lacrimation and conjunctival edema which lasted several days; there was no permanent injury.

Systemic—Exposure to dust or fumes of inorganic tin is known to cause a benign pneumoconiosis (stannosis). This form of pneumoconiosis produces distinctive progressive x-ray changes of the lungs as long as exposure persists, but there is no distinctive fibrosis, no evidence of disability, and no special complicating factors. Because tin is so radio-opaque, early diagnosis is possible.

Certain organic tin compounds, especially alkyltin compounds, are highly toxic when ingested. The trialkyl and tetraalkyl compounds cause damage to the central nervous system with symptoms of headaches, dizziness, photophobia, vomiting, and urinary retention, some weakness and flaccid paralysis of the limbs in the most severe cases. Percutaneous absorption of these compounds has been postulated, but to date, deaths and serious injury have resulted only from ill-advised attempts at therapeutic use by mouth. The mechanism of action of the organotins is not clearly understood, although triethyltin is an extremely potent inhibitor of oxidative phosphorylation. Occasionally, mild organotin intoxication is seen in chemical laboratories with headache, nausea, and EEG changes.

Medical Surveillance: In the case of inorganic tin compounds, the skin and eyes are of particular interest. Chest x-rays may reveal that exposures have occurred. For organotins, preplacement and periodic examinations should include the skin, eyes, blood, central nervous system, and liver and kidney function.

Special Tests: None in use.

Personal Protective Methods: It is important that employees be trained in the correct use of personal protective equipment. Skin contact should be prevented by protective clothing, and, especially in the case of organic tin compounds, clean work clothes should be supplied daily and the worker required to shower following the shift and prior to change to street clothes. In all areas of dust concentration, dust masks should be provided, and in the case of fumes, masks with proper canisters or supplied air respirators should be used.

Bibliography

Barnes, J.M., and Stoner, H.B. 1958. Toxic properties of some dialkyl and trialkyl tin salts. *Br. J. Ind. Med.* 15:15.
Lyle, W.H. 1958. Lesion of the skin in process workers caused by contact with butyltin compounds. *Br. J. Ind. Med.* 15:193.
National Institute for Occupational Safety and Health. *Criteria for a Recommended Standard: Occupational Exposure to Organotin Compounds.* NIOSH Doc. No. 77-115 (1977).
Pendergrass, E.P., and Pryde, A.W. 1948. Benign pneumoconiosis due to tin oxide—a case report with experimental investigation of the radiographic density of the tin oxide dust. *J. Ind. Hyg. Toxicol.* 30:119.

Prull, G., and Rompel, K. 1970. EEG changes in acute poisoning with organic tin compounds. *Electroencephalogr. Clin. Neurophysiol.* 29:215.

TITANIUM AND COMPOUNDS

Description: Ti, titanium, is a dark-grey, lustrous metal insoluble in water. It is brittle when cold and malleable when hot. The most important minerals containing titanium are ilmenite, rutile, perovskite, and titanite or sphene.

Synonyms: None.

Potential Occupational Exposures: Titanium metal, because of its low weight, high strength, and heat resistance, is used in the aerospace and aircraft industry as tubing, fittings, fire walls, cowlings, skin sections, and jet compressors, and it is also used in surgical appliances. It is used, too, as control-wire casings in nuclear reactors, as a protective coating for mixers in the pulp-paper industry and in other situations in which protection against chlorides or acids is required, in vacuum lamp bulbs and x-ray tubes, as an addition to carbon and tungsten in electrodes and lamp filaments, and to the powder in the pyrotechnics industry. It forms alloys with iron, aluminum, tin, and vanadium of which ferrotitanium is especially important in the steel industry.

Titanium dioxide (TiO_2, rutile, anatase, titania) is a white pigment in the rubber, plastics, ceramics, paint, and varnish industries, in dermatological preparations, and is used as a starting material for other titanium compounds, as a gem, in curing concrete, and in coatings for welding rods.

Other titanium compounds are utilized in smoke screens, as mordants in dyeing, in the manufacture of cemented metal carbides, as thermal insulators, and in heat resistant surface coatings in paints and plastics.

A partial list of occupations in which exposure may occur includes:

Ceramic makers	Paper makers
Glass makers	Plastic makers
Incandescent lamp makers	Rayon makers
Ink makers	Smoke screen makers
Lacquer makers	Steel workers
Nuclear steel makers	Vacuum tube makers
Paint makers	Welding rod makers

Permissible Exposure Limits: The Federal standard for titanium dioxide is 15 mg/m³. There is no standard for titanium itself or other titanium compounds.

Route of Entry: Inhalation of dust or fume.

Harmful Effects

Local—Titanium and titanium compounds are, for the most part, virtually inert and not highly toxic to man. Titanium tetrachloride, which is released into the air during maintenance of chlorinating and rectifying operations, is an exception. Titanium tetrachloride and its hydrolysis products are highly toxic and irritating. Skin exposure may cause irritation and burns, and even brief contact with the eyes may cause suppurating conjunctivitis and keratitis, followed by clouding of the cornea.

Systemic—During the production of titanium metal, it is possible that the air may be contaminated with chlorine, hydrogen chloride, titanium tetrachloride, and similar harmful constituents. Reports of severe lung injury caused by such exposures have been recorded; in some cases the condition resembles silicotic lungs. Reports of pulmonary fibrosis due to titanium carbide are now mostly discounted, but precautions are still recommended. Titanium tetrachloride may cause injury to the upper respiratory tract and acute bronchitis.

Medical Surveillance: Preemployment and periodic physical examinations should give special attention to lung disease, especially if irritant compounds are involved. Chest x-rays should be included in both examinations and pulmonary function evaluated periodically. Smoking history should be taken. Careful attention should be given to the eyes and the skin.

Personal Protective Methods: Employees exposed to titanium tetrachloride should wear protective clothing and respirators. In areas of dust or fumes of titanium tetrachloride, all workers should be provided with goggles and dust masks, fullface gas masks, or supplied air respirators. Clothing should be changed daily to avoid dust inhalation from clothing, and employees should be encouraged to shower before changing to street clothes.

Bibliography

Elo, R., Maatta, K., Uksila, E., and Arstila, A.U. 1972. Pulmonary deposits of titanium dioxide in man. *Arch. Pathol.* 94:417.
Joseph, M. 1969. Hard metal pneumoconiosis. *Australas. Radiol.* 12:92.
Lawson, J.J. 1961. The toxicity of titanium tetrachloride. *J. Occup. Med.* 3:7.

TOLUENE

Description: $C_6H_5CH_3$, toluene, is a clear, colorless, noncorrosive liquid with a sweet, pungent, benzene-like odor.

Synonyms: Toluol, methylbenzene, phenylmethane, methylbenzol.

Potential Occupational Exposures: Toluene may be encountered in the manufacture of benzene. It is also used as a chemical feed for toluene diisocyanate, phenol, benzyl and benzoyl derivatives, benzoic acid, toluene sulfonates, nitrotoluenes, vinyl toluene, and saccharin; as a solvent for paints and coatings; or as a component of automobile and aviation fuels.

A partial list of occupations in which exposure may occur includes:

Aviation fuel blenders	Perfume makers
Benzene makers	Petrochemical workers
Chemical laboratory workers	Rubber cement makers
Coke oven workers	Saccharin makers
Gasoline blenders	Solvent workers
Lacquer workers	Toluene diisocyanate makers
Paint thinner makers	Vinyl toluene makers

It is estimated that 100,000 workers are potentially exposed to toluene.

Permissible Exposure Limits: The Federal standard is 200 ppm as an 8-hour TWA with an acceptable ceiling concentration of 300 ppm; acceptable maximum peaks above the ceiling of 500 ppm are allowed for 10 minutes duration. NIOSH has recommended a limit of 100 ppm (TWA) with a ceiling of 200 ppm for a ten minute sampling period. ACGIH (1978) cites a TWA of 100 ppm (375 mg/m^3).

Routes of Entry: Inhalation of vapor and percutaneous absorption of liquid.

Harmful Effects

Local—Toluene may cause irritation of the eyes, respiratory tract, and skin. Repeated or prolonged contact with liquid may cause removal of natural lipids from the skin, resulting in dry, fissured dermatitis. The liquid splashed in the eyes may cause irritation and reversible damage.

Systemic—Acute exposure to toluene predominantly results in central nervous system depression. Symptoms and signs include headache, dizziness, fatigue, muscular weakness, drowsiness, incoordination with staggering gait, skin paresthesia, collapse, and coma.

Medical Surveillance: Preplacement and periodic examinations should evaluate possible effect on skin, central nervous system, as well as liver and kidney function. Hematologic studies should also be done if there is significant contamination of the solvent with benzene.

Special Tests: Hippuric acid levels above 5 g/l of urine may result from exposure greater than 200 ppm determined as a TWA. Blood levels can also be determined for toluene.

Personal Protective Methods: Where vapor concentration exists above allowable standards, employees should be provided with respirators (air supplied) or gas masks with organic vapor canister and fullface plate. Impervious clothing, gloves, or other coverings to protect potentially exposed areas of the body should be supplied to employees in operations requiring continued exposure to liquid toluene. Toluene-wet clothing should be immediately removed unless impervious, and work clothing changed at least twice a week. Safety glasses or goggles should be worn in areas where splash or spill is likely.

Bibliography

Jenkins, L.J., Jones, R.A., and Siegel, J. 1970. Long-term inhalation screening studies of benzene, toluene, o-xylene, and cumene on experimental animals. *Toxicol. Appl. Pharmacol.* 16:818.

National Institute for Occupational Safety and Health. *Criteria for a Recommended Standard: Occupational Exposure to Toluene.* NIOSH Doc. No. 73-11023 (1973).

TOLUENE DIISOCYANATE (See "Isocyanates")

1,1,1-TRICHLOROETHANE

Description: CH_3CCl_3, 1,1,1-trichloroethane, is a colorless, nonflammable liquid with an odor similar to chloroform. Upon contact with hot metal or exposure to ultraviolet radiation, it will decompose to form the irritant gases hydrochloric acid, phosgene, and dichloroacetylene.

Symptoms: Methyl chloroform.

Potential Occupational Exposures: In recent years, 1,1,1-trichloroethane has found wide use as a substitute for carbon tetrachloride. In liquid form it is used as a degreaser and for cold cleaning, dip-cleaning, and bucket cleaning of metals. Other industrial applications of 1,1,1-trichloroethane's solvent properties include its use as a drycleaning agent, a vapor degreasing agent, and a propellant.

A partial list of occupations in which exposure may occur includes:

Degreasers	Metal degreasers
Drycleaners	Propellant makers
Machinery cleaners	Stain removers

Permissible Exposure Limits: The Federal standard is 350 ppm (1,900 mg/m³). NIOSH has recommended a 350 ppm ceiling as determined by a 15-minute sampling period.

Routes of Entry: Inhalation of vapor and moderate skin absorption.

Harmful Effects

Local—Liquid and vapor are irritating to eyes on contact. This effect is usually noted first in acute exposure cases. Mild conjunctivitis may develop but recovery is usually rapid. Repeated skin contact may produce a dry, scaly, and fissured dermatitis, due to the solvent's defatting properties.

Systemic—1,1,1-Trichloroethane acts as a narcotic and depresses the central nervous system. Acute exposure symptoms include dizziness, incoordination, drowsiness, increased reaction time, unconsciousness, and death.

Medical Surveillance: Consider the skin, liver function, cardiac status, especially arrythmias, in preplacement or periodic examinations.

Special Tests: Expired air analyses may be useful in monitoring exposure.

Personal Protective Methods: 1,1,1-Trichloroethane attacks natural rubber; therefore, protective clothing of leather, polyvinyl alcohol, or neoprene is recommended. In areas of high concentrations, fullface masks should be worn.

Bibliography

Hatfield, T.R., and Maykoski, R.T. 1970. A fatal methyl chloroform (trichloroethane) poisoning. *Arch. Environ. Health* 20:279.
Manufacturing Chemists Association, Inc. 1965. *Properties and Essential Information for Safe Handling and Use of 1,1,1-Trichloroethane.* Chemical Safety Data Sheet SD-90. MCA, Washington, D.C.
National Institute for Occupational Safety and Health. *Criteria for a Recommended Standard: Occupational Exposure to 1,1,1-Trichloroethane (Methyl Chloroform).* NIOSH Doc. No. 76-184, Wash., D.C. (1976).
Stahl, C.J., Fatteh, A.V., and Dominquez, A.M. 1969. Trichloroethane poisoning: observations on the pathology and toxicology in six fatal cases. *J. Forensic Sci.* 14:393.
Stewart, R.D. 1968. The toxicity of 1,1,1-trichloroethane. *Ann. Occup. Hyg.* 11:71.
Torkelson, T.R., Oyen, F., McCollister, D.D., and Rowe, V.R. 1958. Toxicity of 1,1,1-trichloroethane as determined on laboratory animals and human subjects. *Am. Ind. Hyg. Assoc. J.* 19:353.

1,1,2-TRICHLOROETHANE

Description: $CH_2ClCHCl_2$, 1,1,2-trichloroethane, is a colorless, nonflammable liquid. It is an isomer of 1,1,1-trichloroethane but should not be confused with it toxicologically. 1,1,2-Trichloroethane is comparable to carbon tetrachloride and tetrachloroethane in toxicity.

Synonyms: Vinyl trichloride.

Potential Occupational Exposures: 1,1,2-Trichloroethane is used as a chemical intermediate and as a solvent, but is not as widely used as its isomer 1,1,1-trichloroethane.

A partial list of occupations in which exposure may occur includes:

        Organic chemical synthesizers   Solvent makers

Permissible Exposure Limits: The Federal standard is 10 ppm (45 mg/m^3) as a TWA.

Routes of Entry: Inhalation of vapor and absorption through the skin.

Harmful Effects

Local—Irritation to eyes and nose, and infection of the conjunctiva have been shown in animals.

Systemic—Little is known of the toxicity of 1,1,2-trichloroethane since no human toxic effects have been reported. Animal experiments show 1,1,2-trichloroethane to be a potent central nervous system depressant. The injection of anesthetic doses in animals was associated with both liver and renal neurosis.

Medical Surveillance: Consider the skin, central nervous system, and liver and kidney function. Alcoholism may be a synergistic factor.

Special Tests: None commonly used, but expired air analyses may be useful in monitoring exposure.

Personal Protective Methods: Protective clothing and gloves should be worn. Respirators should be used in areas of high vapor concentration.

Bibliography

Carpenter, C.P., Smith, H.F., and Pozzani, V.C. 1949. Assay of acute vapor toxicity and grading and interpretation of results in 96 chemical compounds. *J. Ind. Hyg.* 31:343.
Lazarew, N.W. 1929. Uber die Narkotische Wirkungstraft der dampfe der chlorderivaten des methans, des athans und des athylens. *Archiv. Exp. Pathol. Pharmakol.* 141:19-24.

TRICHLOROETHYLENE

Description: ClCH=CCl$_2$, trichloroethylene, a colorless, nonflammable, noncorrosive liquid has the "sweet" odor characteristic of some chlorinated hydrocarbons. Decomposition of trichloroethylene, due to contact with hot metal or ultraviolet radiation, forms products including chlorine gas, hydrogen chloride, and phosgene. Dichloroacetylene may be formed from the reaction of alkali with trichloroethylene.

Synonyms: Ethylene trichloride, ethinyl trichloride, trichloroethene, tri.

Potential Occupational Exposures: Trichloroethylene is primarily used as a solvent in vapor degreasing. It is also used for extracting caffeine from coffee, as a drycleaning agent, and as a chemical intermediate in the production of pesticides, waxes, gums, resins, tars, paints, varnishes, and specific chemicals such as chloroacetic acid.

A partial list of occupations in which exposure may occur includes:

Anesthetic makers	Metal cleaners
Caffeine processors	Oil processors
Cleaners	Perfume makers
Disinfectant makers	Printers
Degreasers	Resin workers
Drug makers	Rubber cementers
Drycleaners	Shoe makers
Dye makers	Soap makers
Electronic equipment cleaners	Solvent workers
Fat processors	Textile cleaners
Glass cleaners	Tobacco denicotinizers
Mechanics	Varnish workers

It was estimated by NIOSH in 1973 that 200,000 workers are potentially exposed to trichloroethylene and that many of these exposures are in small workplaces. In 1978, this estimate was revised to 100,000 full-time exposures to TCE with up to 3.5 million more workers subjected to continuous low levels or to brief exposures of various levels.

Permissible Exposure Limits: The Federal standard is 100 ppm (535 mg/m^3) as an 8-hour TWA with an acceptable ceiling concentration of 200 ppm; acceptable maximum peaks above the ceiling of 300 ppm are allowed for 5 minutes duration in a 2-hour period. The NIOSH Criteria for a Recommended Standard recommends limits of 100 ppm as a TWA and a peak of 150 ppm determined by a sampling time of 10 minutes.

It was recommended by NIOSH in 1978 that the permissible limit for occupational exposure to trichloroethylene be reduced and that TCE be controlled as an occupational carcinogen. Current information regarding engineering feasibility indicates that personnel exposures of 25 ppm, on a time-weighted-average, can be readily attained using existing engineering control technology. However, NIOSH does not feel that this should serve as a final goal. Rather, industry should pursue further reductions in worker exposure as advancements in technology research allow.

Routes of Entry: Inhalation and percutaneous absorption.

Harmful Effects

Local—Exposure to trichloroethylene vapor may cause irritation of the eyes, nose, and throat. The liquid, if splashed in the eyes, may cause burning irritation and damage. Repeated or prolonged skin contact with the liquid may cause dermatitis.

Systemic—Acute exposure to trichloroethylene depresses the central nervous system exhibiting such symptoms as headache, dizziness, vertigo, tremors, nausea and vomiting, irregular heart beat, sleepiness, fatigue, blurred vision, and intoxication similar to that of alcohol. Unconsciousness and death have been reported. Alcohol may make the symptoms of trichloroethylene overexposure worse. If alcohol has been consumed, the overexposed worker may become flushed. Trichloroethylene addiction and peripheral neuropathy have been reported.

The National Cancer Institute (NCI) in the United States has recently issued a "state of concern" alert, warning producers, users, and regulatory agencies that trichloroethylene administered by gastric intubation to B6C3F mice induced predominantly hepatocellular carcinomas with some metastases to the lungs, e.g., 30 of 98 (30.6%) of the mice given the low dose (1,200 mg/kg and 900 mg/kg for male and female respectively) and 41 of 95 (43.2%) of the mice given the higher dose (2,400 mg/kg and 1,800 mg/kg for male and female respectively). Only one of 40 (2.5%) control mice developed these carcinomas (1)(2).

No hepatocellular carcinomas were observed in both sexes of Osborne-Mendel rats administered trichloroethylene at levels of 1.0 or 0.5 g/kg by gastric intubation 5 times weekly for an unspecified period (1).

No liver lesions or hepatomas were found in NLC mice given oral doses by gavage of 0.1 ml of a 40% solution of trichloroethylene in oil twice weekly for an unspecified period (3).

Trichloroethylene (3.3 mM) in the presence of a metabolic activating microsomal system induced reverse mutations in *E. coli* strain K12 (4). It has also been shown to induce frameshift as well as base substitution mutation in *S. cerevisiae* strain XV185-14C in the presence of mice liver homogenate (5).

Trichloroethylene is metabolized in rats to trichloroacetic acid and trichloroethanol which are proposed to have been derived from a primary metabolite, trichloroethylene oxide (6)-(8). Trichloroethylene is also metabolized to trichloroethanol and trichloroacetic acid in dogs (9).

Epoxides are now recognized as obligatory intermediates in the metabolism of olefins by hepatic microsomal mixed-function oxidases (10)(11). The formation of metabolites such as trichloroethanol and trichloroacetic acid implies rearrangement of the transient trichloroethylene oxide intermediate into chloral. This has been confirmed in studies involving: (a) the rearrangement of the oxides belonging to a series of chlorinated ethylenes (12), and (b) the identification of chloral as a trichloroethylene metabolite in vitro (13) and in vivo (14). Chloral hydrate has been shown to be mutagenic in *Antirrhinum* (15).

Medical Surveillance: Preplacement and periodic examinations should include the skin, respiratory, cardiac, central, and peripheral nervous systems, as well as liver and kidney function. Alcohol intake should be evaluated.

Special Tests: Expired air analysis and urinary metabolites have been used to monitor exposure.

Personal Protective Methods: Gloves and protective clothing should be worn, and fullface mask should be used in areas of excessive vapor concentrations.

References

(1) Lloyd, J.W., Moore, R.M., and Breslin, P. Background information on trichloroethylene. *J. Occup. Med.* 17 (1975) 603-605.
(2) National Institutes of Health. *HEW News.* June 14 (1976).
(3) Rudali, G. A propos de l'activite oncogene de quelques hydrocarbures halogenes utilises en therapeutique. *UICC Monograph.* 7 (1967) 138-143.
(4) Greim, H., Bonse, G., Radwan, Z., Reichert, D., and Henschler, D. Mutagenicity in vitro and potential carcinogenicity of chlorinated ethylenes as a function of metabolic oxirane formation. *Biochem. Pharmacol.* 24 (1975) 2013.
(5) Shahin, M.M., and Von Borstel, R.C. Mutagenic and lethal effects of α-benzene hexachloride, dibutyl phthalate and trichloroethylene in *Saccharomyces cerevisiae. Mutation Res.* 48 (1977) 173-180.
(6) Daniel, J.W. Metabolism of [36]Cl-labelled trichloroethylene and tetrachloroethylene in the rat. *Biochem. Pharmacol.* 12 (1963) 795.
(7) Parke, D.V. In *The Biochemistry of Foreign Compounds.* Vol. 5 (eds) Campbell, P.N., Datta, S.D., and Engel, L.L. Pergamon Press, Oxford (1968).
(8) Powell, J.F. Trichloroethylene: Absorption, elimination and metabolism. *Brit. J. Ind. Med.* 2 (1945) 142-145.
(9) Butler, T.C. Metabolic transformation of trichloroethylene. *J. Pharmacol. Exp. Ther.* 97 (1949) 84-92.
(10) Anon. Vinylidene chloride linked to cancer. *Chem. Eng. News.* Feb. 28 (1977) pp. 6-7.
(11) Maynert, E.W., Foreman, R.L., and Watanabe, T. Epoxides as obligatory intermediates in the metabolism of olefins to glycols. *J. Biol. Chem.* 245 (1970) 5234.
(12) Bonse, G., Urban, T., Reichert, D., and Henschler, D. Chemical reactivity, metabolic oxirane formation and biological reactivity of chlorinated ethylenes in the isolated perfused rat liver preparation. *Biochem. Pharmacol.* 24 (1975) 1829-1834.
(13) Byington, K.H., and Leibman, K.C. Metabolism of trichloroethylene by liver microsomes. II. Identification of the reaction product as chloral hydrate. *Molec. Pharmacol.* 1 (1965) 247-254.

(14) Scansetti, G., Rubino, G.F., and Trompeo, G. Studio sully intossicazione cronica da trielina III. Metabolismo del trichloroetilene. *Med. Lavoro.* 50 (1959) 743-754.

(15) Barthelmess, A. Mutagene arzneimittel. *Arzneimittelforsch.* 6 (1956) 157-168.

Bibliography

Bauer, M., and S.F. Rabene. 1974. Cutaneous manifestations of trichloroethylene toxicity. *Arch. Derm.* 110:886.

Feldman, R.G., Mayer, R.M., and Traub, A. 1970. Evidence for peripheral neurotoxic effect of trichloroethylene. *Neurology* 20:599.

Lloyd, J.W., Moore, R.M. Jr., and Breslin, P. 1975. Background information on trichloroethylene. *J. Occup. Med.* 17:603.

Lowry, L.K., Vandervort, R., and Polakoff, P.L. 1974. Biological indicators of occupational exposure to trichloroethylene. *J. Occup. Med.* 16:98.

Pardys, S., and Brotman, M. 1974. Trichloroethylene and alcohol: a straight flush. *J. Am. Med. Assoc.* 229:521.

Air Pollution Assessment of Trichloroethylene. EPA, Office of Air Quality Planning and Standards (September, 1975, Draft).

Criteria Document: Occupational Exposure to Trichloroethylene. HEW, National Institute for Occupational Safety and Health (June 1973).

Memorandum of Alert: Trichloroethylene. HEW, National Cancer Institute (March 21, 1975).

Mitchell, A.B.S., and Parsons-Smith, B.G. Trichloroethylene neuropath. *Br. Med. J.* 1:422-23 (1969).

Preliminary Study of Selected Potential Environmental Contaminants—Optical Brighteners, Methyl Chloroform, Trichloroethylene, Tetrachloroethylene, Ion-Exchange Resins. EPA, Office of Toxic Substances (Publication No. EPA-560/2-75-002, July 1975).

Stuber, K. Injuries to health in the industrial use of trichloroethylene and the possibility of their prevention. *Arch Gewerbepathol Gewerbehyg* (Ger.) 2:398-456 (1932).

Trichloroethylene (Background Report). HEW, National Institute for Occupational Safety and Health (June 6, 1975).

Trichloroethylene (data sheet 389) (revised). Chicago, National Safety Council (1964).

National Institute for Occupational Safety and Health. *Criteria for a Recommended Standard: Occupational Exposure to Trichloroethylene.* NIOSH Doc. No. 73-11025 (1973).

National Institute for Occupational Safety and Health. *Special Occupational Hazard Review with Control Recommendations: Trichloroethylene.* NIOSH Doc. No. 78-130, Wash., D.C. (Jan. 1978).

TRICRESYL PHOSPHATES

Description: Tricresyl phosphates, $(CH_3C_6H_4O)_3PO$, are available as the ortho-isomer (TOCP), the meta-isomer (TMCP), and the para-isomer (TPCP). The ortho-isomer is the most toxic of the three; the meta- and para-isomers are relatively inactive. The commercial product may contain the ortho-isomer as a contaminant unless special precautions are taken during manufacture. Pure tri-para-cresyl phosphate is a solid, and ortho- and meta- are colorless, oily, odorless liquids.

Synonyms: Tritolyl phosphate, TCP.

Potential Occupational Exposures: Tricresyl phosphate is used as a plasticizer for chlorinated rubber, vinyl plastics, polystyrene, polyacrylic, and polymethacrylic esters, as an adjuvant in milling of pigment pastes, as a solvent and as a binder in nitrocellulose and various natural resins, and as an additive to synthetic lubricants and gasoline. It is also used as hydraulic fluid, fire retardant and in the recovery of phenol in coke-oven waste waters.

A partial list of occupations in which exposure may occur includes:

Gasoline additive makers	Polystyrene makers
Gasoline blenders	Polyvinyl chloride makers
Hydraulic fluid workers	Solvent workers
Lead scavenger makers	Surgical instrument sterilizers
Nitrocellulose workers	Waterproofing makers
Plasticizer workers	

Permissible Exposure Limits: The Federal standard for tri-ortho-cresyl phosphate is 0.1 mg/m^3; there is no standard for the meta- and para-isomers.

Routes of Entry: Inhalation of ortho-isomer vapor or mist, especially when heated; ingestion and percutaneous absorption of liquids. The widespread epidemics of poisoning that have occurred have been due to ingested ortho-isomer as a contaminant of foodstuff. There have been relatively few reports of neurological symptoms in workers handling these substances. Experimental human studies with labeled phosphorus derivatives show only 0.4% of the applied dose was absorbed.

Harmful Effects

Local—None reported.

Systemic—The major effects from inhaling, swallowing, or absorbing tricresyl phosphate through the skin are on the spinal cord and peripheral nervous system, the poison attacking the anterior horn cells and pyramidal tract as well as the peripheral nerves. Gastrointestinal symptoms on acute exposure (nausea, vomiting, diarrhea, and abdominal pain) are followed by a latent period of 3 to 30 days with the progressive development of muscle soreness and numbness of fingers, calf muscles, and toes, with foot and wrist drop. In chronic intoxication, the g.i. symptoms pass unnoticed, and after a long latent period, flaccid paralysis of limb and leg muscles appear. There are minor sensory changes and no loss of sphincter control.

Medical Surveillance: Preplacement and periodic examinations should include evaluation of spinal cord and neuromuscular function, especially in the extremities, and a history of exposure to other organo-phosphate esters, pesticides, or neurotoxic agents. Periodic cholinesterase determination may relate to exposure, but not necessarily to neuromuscular effect.

Special Tests: None used except for determination of serum of red cell choline or acetylcholine esterases.

Personal Protective Methods: Protective clothing should be worn to prevent skin absorption and, where dust or vapor concentrates, masks should be supplied to employees.

Bibliography

Hodge, H.C., and Sterner, J.H. 1943. Skin absorption of triorthocresyl phosphate as shown by radioactive phosphorus. *J. Pharmacol. Exp. Ther.* 79:225.
Hunter, D., Perry, K.M.A., and Evans. R.B. 1944. Toxic polyneuritis arising during the manufacture of tricresyl phosphate. *Br. J. Ind. Med.* 1:227.
Prineas, J. 1969. Triorthocresyl phosphate myopathy. *Arch. Neurol.* 21:150.
Tabershaw, I.R., and Kleinfield, M. 1957. Manufacture of tricresyl phosphate and other alkyl phenyl phospates: an industrial hygiene study. II. Clinical effects of tricresyl phosphate. *AMA Arch. Ind. Health* 15:541.

TRIMELLITIC ANHYDRIDE

Description: Trimellitic anhydride, $C_9H_4O_5$, is a crystalline solid melting at 161° to 163.5°C. It is the anhydride of trimellitic acid (1,2,4-benzenetricarboxylic acid).

Synonyms: 1,3-Dihydro-1,3-dioxo-5-isobenzofurancarboxylic acid; anhydrotrimellitic acid; 4-carboxyphthalic anhydride, TMA.

Potential Occupational Exposure: NIOSH estimates approximately 20,000 American workers are currently at risk of exposure to trimellitic anhydride in its various applications. TMA is used as a curing agent for epoxy and other resins, in vinyl plasticizers, paints and coatings, polymers, polyesters, agricultural chemicals, dyes and pigments, pharmaceuticals, surface active agents, modifiers, intermediates, and specialty chemicals. The sole domestic producer of trimellitic anhydride is Amoco Chemicals Corporation which has a 50 million pound-per-year plant at Joliet, Illinois.

Permissible Exposure Limits: There is no current Occupational Safety and Health Administration (OSHA) exposure standard for trimellitic anhydride. The Amoco Chemicals Corporation, the sole domestic producer, suggests a limit of "0.05 mg/m^3 or less for susceptible individuals" (1).

Routes of Entry: Skin contact or inhalation of dusts or vapors.

Harmful Effects

The National Institute for Occupational Safety and Health (NIOSH) recommends that trimellitic anhydride (TMA) be handled as an extremely toxic agent in the workplace. Exposure to this compound may result in noncardiac pulmonary edema (apparently without benefit of a pulmonary irritation warning), immunological sensitization, and irritation of the pulmonary tract, eyes, nose, and skin.

The ability for trimellitic anhydride to cause pulmonary edema (excessive fluid in the lungs) has been demonstrated by Rice et al (2). Two workers had been employed by the same company for only a short period of time (3 and 6 weeks). They received multiple inhalation exposures to an epoxy resin containing trimellitic anhydride when it was sprayed on heated pipes. The levels of trimellitic anhydride were not available to the authors. No mention was made of severe irritation of the upper respiratory tract while they were receiving their exposures, suggesting little or no warning of subsequent damage to the lungs. The possibility that the pulmonary edema was the result of a hypersensitivity reaction must therefore be considered. Resins can be sensitizers (e.g., toluene diisocyanate or TDI), though most of the reported effects have been those of direct irritation (3).

Sensitization to trimellitic anhydride was reported by Zeiss et al (4). Respiratory symptoms were observed in fourteen workers employed in the synthesis of trimellitic anhydride. The authors suggest three distinct syndromes induced by inhalation of TMA. The first, rhinitis and/or asthma, developed over an industrial exposure period of weeks to years. After this period, the sensitized worker exhibited symptoms immediately following exposure to trimellitic anhydride dust or fume, which abated after the work exposure had stopped. The second syndrome, termed "TMA-flu" by the workers, also required a sensitization period of exposure and was characterized by delayed onset cough, wheezing, and labored breathing starting 4 to 8 hours after a work shift and peaking at night. These respiratory symptoms were usually accompanied by malaise, chills, fever, muscle and joint aches, and appeared to be associated with relatively high exposures to trimellitic anhydride during particular work shifts. The third syndrome, which followed initial high exposure to TMA, was primarily an irritant effect. It was characterized by a "running" nose without itching or sneezing, occasional nosebleed, cough, labored breathing, and occasional wheezing. Symptoms usually abated after 8 hours and rarely lasted into the night.

The above studies suggest harmful respiratory effects of trimellitic anhydride at relatively high concentrations, but even at lower concentrations some workers may develop an immunological sensitization over a period of time.

Fawcett et al (5) also observed sensitization in a worker exposed to trimellitic anhydride in the production of tubular steel shop fittings coated with an epoxy resin. The chemical agent responsible for asthma symptoms of six workers was identified by careful inhalation challenge testing, simulating work exposure. Typical attacks which began one year after onset of TMA exposure consisted of cough and breathlessness lasting for 30 minutes, which subsided, only to be followed the same evening by sneezing which persisted for about 24 hours. Subsequent attacks were prevented by avoiding exposure.

Data on occupational exposures to trimellitic anhydride were also obtained during a NIOSH Health Hazard Evaluation of a paint and varnish company during the manufacture of an epoxy paint (6). The Health Hazard Evaluation was conducted at the request of employees who were concerned about possible harmful effects of trimellitic anhydride exposure during processing and decontamination operations. The occupational airborne exposure levels averaged 1.5 mg/m^3 TMA (with a range from "none detected" to 4.0 mg/m^3) during processing operations and 2.8 mg/m^3 TMA (ranging from "none detected" to 7.5 mg/m^3) during decontamination operations. A total of 13 employees (5 present and 8 former employees) were interviewed and briefly examined. Employees' symptoms and complaints were: eye irritation, nasal irritation, shortness of breath, wheezing, cough, heartburn, nausea, headache, skin irritation, and throat irritation. Three of the former workers stated that they had left that department for health reasons. Complaints subsided when non-TMA-containing products were being formulated.

The Occupational Health and Safety Division, Department of Labour, Alberta, Canada, has reported to NIOSH that they are aware of employee reactions in two plants using TMA-epoxy powder pipe coatings. The one plant, started in 1971, had a number of employees with an immediate reaction. After instituting engineering and administrative controls, there has been no further incidence. In the second plant, begun in 1974, the first adverse reaction occurred in 1975. There have been 9 cases of adverse reactions reported to date. Most of these employees were kept in intensive care while they recuperated and were advised by their physicians to seek new jobs. However, some returned to their previous jobs and became ill again. Due to the unavailability of a good analytical method for trimellitic anhydride, occupational levels could not be documented until November of 1977. The TMA concentrations found ranged from 0.11 mg/m^3 to 0.27 mg/m^3.

Medical Surveillance: The *NIOSH Occupational Exposure Sampling Strategy Manual,* NIOSH Publication #77-173, may be helpful in developing efficient programs to monitor employee exposures to trimellitic anhydride. The manual discusses determination of the need for exposure measurements, selection of appropriate employees for sampling, and selection of sampling times.

Special Tests: None in common use.

Personal Protective Methods: Engineering and work practice controls should be used to minimize employee exposure to trimellitic anhydride.

To ensure that ventilation equipment is working properly, effectiveness (e.g., air velocity, static pressure or air volume) should be checked at least every three months. System effectiveness should also be checked within five days of any change in production, process, or control which might result in significant increases in airborne exposures to trimellitic anhydride.

Employers should provide appropriate protective clothing and equipment necessary to prevent repeated or prolonged skin contact with trimellitic anhydride (7).

Employers should see that employees whose clothing may have become contaminated with trimellitic anhydride change into uncontaminated clothing before leaving the work premises.

Employers should see that nonimpervious clothing which becomes contaminated with trimellitic anhydride be promptly removed and not reworn until the trimellitic anhydride is removed from the clothing.

Employers should see that clothing contaminated with trimellitic anhydride is placed in closed containers for storage until it can be discarded or cleaned. If the clothing is to be laundered or otherwise cleaned to remove the trimellitic anhydride, the employer should tell the person performing the cleaning operation of the hazardous properties of trimellitic anhydride.

Employers should provide dust-resistant safety goggles where there is any possibility of trimellitic anhydride dust contacting the eyes.

Exposure to trimellitic anhydride should not be controlled with the use of respirators (7) except:

>During the time period necessary to install or implement engineering or work practice controls; or

>In work situations in which engineering and work practice controls are technically not feasible; or

>To supplement engineering and work practice controls when such controls fail to adequately control exposure to trimellitic anhydride; or

>For operations which require entry into tanks or closed vessels; or

>In emergencies.

Respirators should be approved by the National Institute for Occupational Safety and Health (NIOSH) or by the Mining Enforcement and Safety Administration (MESA). Refer to *NIOSH Certified Equipment, December 15, 1975,* NIOSH Publication #76-145 and *Cumulative Supplement June 1977, NIOSH Certified Equipment,* NIOSH Publication #77-195. The use of faceseal coverlets or socks with any respirator voids NIOSH/MESA approvals.

Quantitative faceseal fit test equipment (such as sodium chloride, dioctyl phthalate, or equivalent) should be used. Refer to *A Guide to Industrial Respiratory Protection,* NIOSH Publication #76-189 for guidelines on appropriate respiratory protection programs.

Where respirators are needed and NIOSH recommendations allow their use to reduce employee exposure, the following types of respirators may be used. They are listed in order of increasing protection factors.

Protection factor of 50: any chemical cartridge respirator with a full facepiece, organic vapor cartridge(s), and high efficiency filter(s) (30 CFR 11.150 and 11.130); any gas mask with a chin-style organic vapor canister and high efficiency filter [30 CFR 11.90(a) and 11.130]; any supplied-air respirator with a full facepiece, helmet, or hood [30 CFR 11.110(a)]; any self-contained breathing apparatus with a full facepiece [30 CFR 11.70(a)].

Protection factor of 1,000: any powered air purifying chemical cartridge respirator with full facepiece, organic vapor cartridge(s), and high efficiency particulate filter(s).

Protection factor of 2,000: any type C supplied-air respirator with a full facepiece operated in positive pressure-demand or other positive pressure mode or with a full facepiece, hood, or helmet operated in continuous flow mode [30 CFR 11.110(a)].

Protection factor of 10,000+: self-contained breathing apparatus with a full facepiece operated in positive pressure-demand or other positive pressure mode [30 CFR 11.70(a)]; any combination respirator which includes a type C supplied-air respirator with a full facepiece operated in positive pressure-demand or other positive pressure or continuous flow mode and an auxilary self-contained breathing apparatus operated in positive pressure-demand or positive pressure mode [30 CFR 11.70(b)].

References

(1) Amoco-Industrial Hygiene Toxicology and Safety Data Sheet. Environmental Health Services. Medical and Health Services Department. July 8, 1976.

(2) Rice, D.L., Jenkins, D.E., Gray, J.M., and Greenberg, S.D. Chemical pneumonitis secondary to inhalation of epoxy pipe coating. *Archives of Environmental Health* 32(4):173-8, July-August 1977.

(3) Patty, F.A., ed. *Industrial Hygiene and Toxicology.* Second Revised Edition. Volume II. Fawcett, D.W., Irish, D.D., eds. John Wiley & Sons, Inc., New York.

(4) Zeiss, C.R., Patterson, R., Pruzansky, J.J., Miller, M.M., Rosenberg, M., Levitz, D., Trimellitic anhydride-induced airway syndromes: chemical and immunologic studies. *J. Allergy and Clinical Immunology* 60(2):96-103, August 1977.

(5) Fawcett, D.W., Taylor, A.J., Pepys, J. Asthma due to inhaled chemical agents-epoxy

resin systems containing phthalic acid anhydride, trimellitic acid anhydride and tri-ethylene tetramine. *Clinical Allergy* 7(1):14, January 1977.

(6) National Institute for Occupational Safety and Health. Health Hazard Evaluation Determination Report No. 74-111-283.

(7) National Institute for Occupational Safety and Health. *Trimellitic Anhydride (TMA).* Current Intelligence Bulletin No. 21, Wash., D.C. (Feb. 3, 1978).

TRINITROTOLUENE

Description: TNT exists in 5 isomers; 2,4,6-trinitrotoluene is the most commonly used. All are crystalline solids in pure form. TNT is a relatively stable high explosive.

Synonyms: TNT, sym-trinitrotoluol, methyltrinitrobenzene.

Potential Occupational Exposures: TNT is used as an explosive, i.e, as a bursting charge in shells, bombs, and mines.

A partial list of occupations in which exposures may occur includes:

Explosives workers

Permissible Exposure Limits: The Federal standard is 1.5 mg/m^3, according to NIOSH. ACGIH (1978) quotes 0.5 mg/m^3 as a TWA.

Routes of Entry: Inhalation of dust, fume, or vapor; ingestion of dust; percutaneous absorption from dust.

Harmful Effects

Local—Exposure to trinitrotoluene may cause irritation of the eyes, nose, and throat with sneezing, cough, and sore throat. It may cause dermatitis and may stain the skin, hair, and nails a yellowish color.

Systemic—Numerous fatalities have occurred in workers exposed to TNT from toxic hepatitis or aplastic anemia. TNT exposure may also cause methemoglobinemia with cyanosis, weakness, drowsiness, dyspnea, and unconsciousness. In addition it may cause muscular pains, heart irregularities, renal irritation, cataracts, menstrual irregularities, and peripheral neuritis.

Medical Surveillance: Placement or periodic examinations should give special considerations to history of allergic reactions, blood dyscrasias, reactions to medications, and alcohol intake. The skin, eye, blood, and liver and kidney function should be followed.

Special Tests: Urine may be examined for TNT by the Webster test or for the urinary metabolite 2,6-dinitro-4-aminotoluene; however, both may be negative if there is a liver injury.

Personal Protective Methods: Protective clothing should be worn. The Webster skin test (colorimetric test with alcoholic sodium hydroxide) or indicator soap should be used to make sure workers have washed all TNT off their skins. Daily change of clean work clothes should be provided, and showers made compulsory at the end of each shift prior to changing to street clothes.

Bibliography

Goodwin, J.W., 1972. Twenty years of handling TNT in a shell loading plant. *Am. Ind. Hyg. Assoc. J.* 33:41.
McConnell, W.J., and Flinn, R.H. 1946. Summary of twenty-two trinitrotoluene fatalities in World War II. *J. Ind. Hyg. and Toxicol.* 20:76.
Morton, A.R., Ranadive, M.V., and Hathaway, J.A. 1976. Biological effects of trinitrotoluene from exposure below the threshold limit value. *Am. Ind. Hyg. Assoc. J.* 37:56.
Norwood, W.D. 1943. Trinitrotoluene (TNT), its effective removal from the skin by a special liquid soap. *Ind. Med.* 12:206.

TUNGSTEN AND CEMENTED TUNGSTEN CARBIDE

Description: Tungsten, symbol W, is element atomic number 74. It is a steel-gray to tin-white metal with a melting point of 3410°C. Tungsten carbide, WC, is a gray powder melting at 2780°C. Cemented tungsten carbide is a mixture consisting of 85% to 95% WC and 5% to 15% cobalt.

Synonyms: Cemented tungsten carbide is also known as "hard metal."

Potential Occupation Exposure: NIOSH estimates that at least 30,000 employees in the United States are potentially exposed to tungsten and its compounds, based on actual observations in the National Occupational Hazards Survey.

It has been stated that the principal health hazards from tungsten and its compounds arise from inhalation of aerosols during mining and milling operations. The principal compounds of tungsten to which workers are exposed are ammonium-p-tungstate, oxides of tungsten (WO_3, W_2O_5, WO_2), metallic tungsten, and tungsten carbide. In the production and use of tungsten carbide tools for machining, exposure to the cobalt used as a binder or cementing substance may be the most important hazard to the health of the employees. Since the cemented tungsten carbide industry uses such other metals as tantalum, titanium, niobium, nickel, chromium, and vanadium in the manufacturing process, the occupational exposures are generally to mixed dusts.

Potential occupational exposures to sodium tungstate are found in the textile industry, where the compound is used as a mordant and fireproofing agent, and in the production of tungsten from some of its ores, where sodium tungstate is an intermediate product. Potential exposures to tungsten and its compounds are also found in the ceramics, lubricants, plastics, printing inks, paint, and photographic industries.

A partial list of occupations in which exposure may occur includes:

Alloy makers	Melting, pouring, casting
Carbonyl workers	workers
Ceramic workers	Metal sprayers
Cemented tungsten carbide	Ore-refining and foundry
workers	workers
Cement makers	Paint and pigment makers
Dyemakers	Papermakers
Dyers	Penpoint makers
Flameproofers	Petroleum refinery workers
High-speed tool steel-	Photographic developers
workers	Spark-plug makers
Incandescent-lamp makers	Textile dyers
Industrial chemical synthesizers	Tool grinders
Ink makers	Tungsten and molybdenum
Lamp-filament makers	miners
Lubricant makers	Waterproofing makers
	Welders

Permissible Exposure Limits:

(a) Occupational exposure to insoluble tungsten shall be controlled so that employees are not exposed to insoluble tungsten at a concentration greater than 5 mg/m^3 of air, measured as tungsten, determined as a TWA concentration for up to a 10-hour work shift in a 40-hour workweek.

(b) Occupational exposure to soluble tungsten shall be controlled so that employees are not exposed to soluble tungsten at a concentration greater than 1 mg/m^3 of air, measured as tungsten, determined as a TWA concentration for up to a 10-hour work shift in a 40-hour workweek.

(c) Occupational exposure to dust of cemented tungsten carbide which contains more than 2% cobalt shall be controlled so that employees are not exposed at a concentration greater than 0.1 mg/m^3 of air, measured as cobalt, determined as a TWA concentration for up to a 10-hour work shift in a 40-hour workweek.

(d) Occupational exposure to dust of cemented tungsten carbide which contains more than 0.3% nickel shall be controlled so that employees are not exposed at a concentration greater than 15 μg nickel/m^3 air determined as a TWA concentration, for up to a 10-hour workshift in a 40-hour workweek as specified in NIOSH's *Criteria for a Recommended Standard for Occupational Exposure to Inorganic Nickel.*

Routes of Entry: Skin contact and inhalation of dusts.

Harmful Effects

Insoluble tungsten compounds and cemented tungsten carbide may cause transient or permanent lung damage and skin irritation, while soluble tungsten compounds have the potential to cause systemic effects involving the gastrointestinal tract and CNS. No carcinogenic, mutagenic, teratogenic, or reproductive effects in humans have been reported. Compliance with the appropriate recommended environmental limits should eliminate the hazards associated with tungsten compounds and cemented tungsten carbide, except for a few individuals who may become sensitized to cobalt or nickel and have adverse reactions upon exposure to extremely small amounts of cemented tungsten carbide.

Medical Surveillance: Preplacement medical screening is recommended to identify any preexisting pulmonary conditions that might make a worker more susceptible to exposures in the work environment. Periodic medical examinations will aid in early detection of any occupationally related illnesses which might otherwise go undetected because of either delayed toxic effects or subtle changes. Maintenance of medical records for a period of 30 years is recommended.

Special Tests: None in common use.

Personal Protective Methods: Exposures to tungsten and its compounds in occupational environments can best be prevented by engineering controls and good work practices. Since tungsten compounds and dusts from cemented tungsten carbide affect chiefly the respiratory system, measures are recommended that will reduce the atmospheric concentrations of tungsten in the work atmosphere. Adoption of these measures during normal operations will also minimize the possibility of skin contact or accidental ingestion.

In addition to using sound engineering controls, employers should institute a program of work practices which emphasizes good sanitation and personal hygiene. These practices are important in preventing skin and respiratory irritation caused by tungsten compounds or cemented tungsten carbide.

Respirators should not be used as a substitute for proper engineering controls in normal operations. However, during emergencies and during non-routine repair and maintenance activities, exposures to airborne dusts or mists of tungsten compounds or cemented tungsten carbide might not be reduced either by engineering controls or by administrative measures to the levels specified. If this occurs, then respiratory protection may be used only: during the time necessary to install or test the required engineering controls; for operations such as maintenance and repair activities causing brief exposure at concentrations above the TWA concentration limits; and during emergencies when airborne concentrations may exceed the TWA concentration limits.

Eye protection shall be provided in accordance with 29 CFR 1910.133 for operations, such as grinding, which produce and scatter particulates into the air.

While most workers do not experience skin irritation as a result of exposure to tungsten compounds or cemented tungsten carbide, there are some who develop sensitivity. Fingerless gloves may be used during grinding of hard metal to protect the hands from abrasion. Protective sleeves of dustproof material may be worn to prevent impact of hard-metal dust on the skin of the arms. In the absence of such gloves and sleeves, creams protective against abrasion may be applied liberally to the hands and arms to minimize contact of the skin with hard-metal dust. When skin irritation occurs, these workers should be referred to a physician for appropriate protective and therapeutic measures. When abrasive dust of tungsten carbide is likely to contact major parts of an employee's body, the employee should wear closely woven coveralls provided by the employer. The coveralls should be laundered frequently to minimize mechanical irritation from dust in the cloth.

Bibliography

National Institute for Occupational Safety and Health. *Criteria for a Recommended Standard: Occupational Exposure to Tungsten and Cemented Tungsten Carbide.* NIOSH Doc. No. 77-127 (Sept. 1977).

TURPENTINE

Description: Turpentine is the oleoresin from species of *Pinus pinacea* trees. The crude oleoresin (gum turpentine) is a yellowish, sticky, opaque mass and the distillate (oil of turpentine) is a colorless, volatile liquid. Chemically, it contains: alpha pinene, beta pinene, camphene, monocyclic terpenes, and terpene alcohols.

Synonyms: Gum turpentine, oil of turpentine, spirit of turpentine, gum spirit, gum (derived from pine resin), wood turpentine (derived from pine stumps), sulfate wood pulp waste.

Potential Occupational Exposures: Turpentines have found wide use as chemical feedstock for the manufacture of floor, furniture, shoe, and automobile polishes, camphor, cleaning materials, inks, putty, mastics, cutting and grinding fluids, paint thinners, resins, and degreasing solutions. Recently, alpha and beta pinenes, which can be extracted, have found use as volatile bases for various compounds.

A partial list of occupations in which exposure may occur includes:

Art glass workers	Oil additive makers
Belt dressing makers	Paint workers
Camphor makers	Pine oil makers
Drug makers	Resin makers
Furniture polish makers	Rubber workers
Ink makers	Solvent workers
Insecticide makers	Stain makers
Lacquer makers	Varnish workers
Lithographers	Wax makers

Permissible Exposure Limits: The Federal standard for turpentine is 100 ppm (560 mg/m^3).

Routes of Entry: Inhalation of vapor and percutaneous absorption of liquid are the usual paths of occupational exposure. However, symptoms have been reported to develop from percutaneous absorption alone.

Harmful Effects

Local—High vapor concentrations are irritating to the eyes, nose, and bronchi. Aspiration of liquid may cause direct lung irritation resulting in pulmonary edema and hemorrhage. Turpentine liquid may produce contact dermatitis. Eczema from turpentine is quite common and has been attributed to the auto-oxidation products of the terpenes (formic acid, formaldehyde, and phenols). This hypersensitivity usually develops in a small portion of the working population. Liquid turpentine splashed in the eyes may cause corneal burns and demands emergency treatment.

Systemic—Turpentine vapor in acute concentrations may cause central nervous system depression. Symptoms include headache, anorexia, anxiety, excitement, mental confusion, and tinnitus. Convulsions, coma, and death have been reported in animal experiments.

Turpentine vapor also produces kidney and bladder damage. Chronic nephritis with albuminuria and hematuria has been reported as a result of repeated exposures to high concentrations. Predisposition to pneumonia may also occur from such exposures. Recovery usually takes from a few days to a few weeks. Several animal experiments of chronic low level exposure have produced no ill effects to the central nervous system, kidneys, bladder, or blood.

Medical Surveillance: Consideration should be given to skin disease or skin allergies in any preplacement or periodic examinations. Liver, renal, and respiratory disease should also be considered.

Special Tests: None in common use.

Personal Protective Methods: Rubber gloves, protective clothing, masks for high concentrations.

U

URANIUM AND COMPOUNDS

Description: U, uranium, is a hard, silvery-white amphoteric metal and is a radioactive element. In the natural state, it consists of three isotopes: ^{238}U (99.28%), ^{234}U (0.006%), and ^{235}U (0.714%). There are over one hundred uranium minerals; those of commercial importance are the oxides and oxygenous salts. The processing of uranium ore generally involves extraction then leaching either by an acid or a carbonate method. The metal may be obtained from its halides by fused salt electrolysis.

Synonyms: None.

Potential Occupational Exposures: The primary use of natural uranium is in nuclear energy as a fuel for nuclear reactors, in plutonium production, and as feeds for gaseous diffusion plants. It is also a source of radium salts. Uranium compounds are used in staining glass, glazing ceramics, and enamelling, in photographic processes, for alloying steels, and as a catalyst for chemical reactions, radiation shielding, and aircraft counterweights.

Uranium presents both chemical and radiation hazards, and exposures may occur during mining, processing of the ore, and production of uranium metal.

A partial list of occupations in which exposure may occur includes:

Atomic bomb workers	Hydrogen bomb workers
Ceramic makers	Nuclear reactor workers
Glass makers	Photographic chemical makers

Permissible Exposure Limits: The Federal standards are: uranium, soluble compounds, 0.05 mg/m^3; and uranium, insoluble compounds, 0.25 mg/m^3 according to NIOSH. ACGIH (1978) cites 0.2 mg/m^3 for both soluble and insoluble natural U compounds.

Routes of Entry: Inhalation of fume, dust, or gas. The following uranium salts are reported to be capable of penetrating intact skin:

Uranyl nitrate, $UO_2(NO_3)_2 \cdot 6H_2O$.
Uranyl fluoride, UO_2F_2.

Uranium pentachloride, UCl_5.
Uranium trioxide (uranyl oxide), UO_3.
Sodium diuranate [sodium uranate(VI), $Na_2U_2O_7 \cdot H_2O$].
Ammonium diuranate [ammonium uranate(VI), $(NH_4)_2U_2O_7$].
Uranium hexafluoride, UF_6.

Harmful Effects

Local—No toxic effects have been reported, but prolonged contact with skin should be avoided to prevent radiation injury.

Systemic—Uranium and its compounds are highly toxic substances. The compounds which are soluble in body fluids possess the highest toxicity. Poisoning has generally occurred as a result of accidents. Acute chemical toxicity produces damage primarily to the kidneys. Kidney changes precede in time and degree the effects on the liver. Chronic poisoning with prolonged exposure gives chest findings of pneumoconiosis, pronounced blood changes, and generalized injury.

It is difficult to separate the toxic chemical effects of uranium and its compounds from their radiation effects. The chronic radiation effects are similar to those produced by ionizing radiation. Reports now confirm that carcinogenicity is related to dose and exposure time. Cancer of the lung, osteosarcoma, and lymphoma have all been reported.

Medical Surveillance: Special attention should be given to the blood, lung, kidney, and liver in preemployment physical examinations. In periodic examinations, tests for blood changes, changes in chest x-rays, or for renal injury and liver damage are advisable.

Special Tests: Uranium excretion in the urine has been used as an index of exposure. Whole body counting may also be useful.

Personal Protective Methods: It is important that a formal monitoring system be established to measure each employee's exposure to uranium. This industry has an excellent record of safety to this hazardous material because of good industrial hygiene practices and monitoring of work practices. Protective clothing, gloves, and respirators are necessary in cases of spills and accidents, and must be worn when dealing with soluble compounds in open systems. Closed systems are essential because of the carcinogenic effects.

Bibliography

Archer, V.E., J.K. Wagoner, and F.E. Lundin, Jr. 1973. Cancer mortality among uranium mill workers. *J. Occup. Med.* 15:11.
Voegtlin, C., and H.C. Hodge. 1949. *Pharmacology and Toxicology of Uranium Compounds.* McGraw-Hill Book Company, New York.

V

VANADIUM AND COMPOUNDS

Description: V, vanadium, is a light grey or white, lustrous powder or fused hard lump insoluble in water. It is produced by roasting the ores, thermal decomposition of the iodide, or from petroleum residues, slags from ferrovanadium production, or soot from oil burning.

Synonyms: None.

Potential Occupational Exposures: Most of the vanadium produced is used in ferrovanadium and of this, the majority is used in high speed and other alloy steels with only small amounts in tool or structural steels. It is usually combined with chromium, nickel, manganese, boron, and tungsten in steel alloys.

Vanadium pentoxide (V_2O_5) is an industrial catalyst in oxidation reactions, is used in glass and ceramic glazes, is a steel additive, and is used in welding electrode coatings. Ammonium metavanadate (NH_4VO_3) is used as an industrial catalyst, a chemical reagent, a photographic developer, and in dyeing and printing. Other vanadium compounds are utilized as mordants in dyeing, in insecticides, as catalysts, and in metallurgy.

Since vanadium itself is considered nontoxic, there is little hazard associated with mining; however, exposure to the more toxic compounds, especially the oxides, can occur during smelting and refining. Exposure may also occur in conjunction with oil-fired furnace flues.

A partial list of occupations in which exposure may occur includes:

Alloy makers	Glass makers
Ceramic makers	Organic chemical synthesizers
Dye makers	Photographic chemical makers
Ferrovanadium workers	Textile dye workers

Permissible Exposure Limits: The Federal standards are: V_2O_5 dust, 0.5 mg/m^3, and V_2O_5 fume, 0.1 mg/m^3, according to NIOSH. The ACGIH (1978) cites a lower limit of 0.05 mg/m^3 for vanadium fume.

439

Route of Entry: Inhalation of dust or fume.

Harmful Effects

Local—Vanadium compounds, especially vanadium pentoxide, are irritants to the eyes and skin. The initial eye symptoms are profuse lacrimation and a burning sensation of the conjunctiva. Skin lesions are of the eczematous type which itch intensely. In some cases there may be generalized urticaria. Workers may also exhibit greenish discoloration of the tongue. This same discoloration may be detectable on the butts of cigarettes smoked by vanadium workers.

Systemic—Vanadium compounds are irritants to the respiratory tract. Entrance to the body is through inhalation of dusts or fumes. Serous or hemorrhagic rhinitis, sore throat, cough, tracheitis, bronchitis, expectoration, and chest pain, may result after even a brief exposure. More serious exposure may result in pulmonary edema and pneumonia which may be fatal. Individuals who recover may experience persistent bronchitis resembling asthma, and bouts of dyspnea; however, no chronic lung lesions have been described.

The results of experimental biochemical studies show that vanadium compounds inhibit cholesterol synthesis and the activity of the enzyme cholinesterase. A variety of other biochemical effects have been noted experimentally, but these have not been reported in relation to occupational exposures. Slightly lower cholesterol levels in blood were noted in one report, but this seems of doubtful significance.

Medical Surveillance: Preemployment and periodic physical examinations should emphasize effects on the eyes, skin, and lungs.

Special Tests: Urinary vanadium excretion may be useful as an index of exposure.

Personal Protective Methods: Employees should receive training in personal hygiene and in the use of personal protective equipment. In certain areas, masks or respirators may be necessary to prevent inhalation of dust and fumes. Protective clothing and gloves will be helpful in preventing dermatitis. Showering after each shift before changing to street clothes is very important. Clean work clothes should be supplied daily.

Bibliography

Lewis, C.E. 1959. The biological actions of vanadium—I. Effects upon serum cholesterol levels in man. *AMA Arch. Ind. Health* 19:419.
Mountain, J.T., Stockell, F.R., and Stockinger, H.E. 1955. Studies in vanadium toxicology. *AMA Arch. Ind. Health* 12:494.
National Institute for Occupational Safety and Health. *Criteria for a Recommended Standard: Occupational Exposure to Vanadium.* NIOSH Doc. No. 77-222 (1977).

Smith, R.G. Vanadium. 1972. In: D.K. Lee, *Metallic Contaminants and Human Health*. Academic Press, New York.

Zenz, C., and Berg, B.A. 1967. Human responses to controlled vanadium pentoxide exposure. *Arch. Environ. Health* 14:709.

VINYL CHLORIDE

Description: $CH_2=CHCl$, vinyl chloride, is a flammable gas at room temperature and is usually encountered as a cooled liquid. The colorless liquid forms a vapor which has a pleasant ethereal odor.

Synonyms: Chloroethylene, chloroethene, monochloroethylene.

Potential Occupational Exposures: Vinyl chloride is used as a vinyl monomer in the manufacture of polyvinyl chloride and other resins. It is also used as a chemical intermediate and as a solvent (1).

A partial list of occupations in which exposure may occur includes:

Polyvinyl resin makers   Organic chemical synthesizers
Rubber makers

The hazard of vinyl chloride was originally believed to primarily concern workers employed in the conversion of VCM to PVC who may receive a particularly high exposure of VCM in certain operations (e.g., cleaning of polymerization kettles) or a long-term exposure to relatively low concentrations in air of VCM at different factory sites. Much larger populations are now believed to be potentially at risk including: (1) producers of VCM, (2) people living in close proximity to VCM or PVC producing industries, (3) users of VCM as propellant in aerosol sprays, (4) persons in contact with resins made from VCM, (5) consumers of food and beverage products containing leachable amounts of unreacted VCM from PVC packaged materials and (6) ingestion of water containing unreacted VCM leached from PVC pipes.

NIOSH (38) estimates definite worker exposure to vinyl chloride at 27,000 and probable worker exposure at 2,200,000.

Permissible Exposure Limits: The Federal standard for exposure to vinyl chloride sets a limit of 1 ppm over an 8-hour period, and a ceiling of 5 ppm averaged over any period not exceeding 15 minutes. ACGIH lists vinyl chloride as a human carcinogen and NIOSH (38) has documented the designation of vinyl chloride as a demonstrated carcinogen in humans.

Routes of Entry: Vinyl chloride gas is absorbed by inhalation. Skin absorption has been suggested but experimental evidence is presently lacking.

Harmful Effects

Local—Vinyl chloride is a skin irritant, and contact with the liquid may cause frostbite upon evaporation. The eyes may be immediately and severely irritated.

Systemic—Vinyl chloride depresses the central nervous system causing symptoms which resemble mild alcohol intoxication. Lightheadedness, some nausea, and dulling of visual and auditory responses may develop in acute exposures. Death from severe vinyl chloride exposure has been reported.

Chronic exposure of workers involved in reactor vessel entry and hand cleaning may result in the triad of acro-osteolysis, Raynaud's phenomenon, and sclerodermatous skin changes. Chronic exposure may also cause hepatic damage.

Vinyl chloride is regarded as a human carcinogen (38) and a cause of angiosarcoma of the liver. Excess cancer of the lung and lymphatic and nervous systems has also been reported. Experimental evidence of tumor induction in a variety of organs, including liver, lung, brain, and kidney, as well as non-malignant alterations, such as fibrosis and connective tissue deterioration, indicate the multisystem oncogenic and toxicologic effects of vinyl chloride.

Vinyl chloride monomer contains some ethylene dichloride in many cases (2). As has been previously noted in this volume, ethylene dichloride is mutagenic in *Drosophila* (3)(4), in *S. typhimurium* TA 1530, TA 1535 and TA 100 tester strains (without metabolic activation) (5)(6) and in *E. coli* (DNA polymerase deficient pol A⁻ strain) (7).

Vinyl chloride has been shown to produce tumors of different types, (especially angiosarcomas of the liver) as well as lung adenomas, brain neuroblastoma lymphomas in mice, rats and hamsters (8)-(13). Extensive worldwide epidemiological studies to date have indicated about 50 cases of angiosarcomas of the liver associated with VCM exposure among workers employed in the manufacture of PVC resins (14)-(18). Infante et al (18) cited a significant excess of mortality from cancer of the lung and brain in addition to cancer of the liver, among workers occupationally exposed to VCM. The risk of dying from cancer of the lymphatic and hematopoietic system also appears to increase with an increase in latency. A study of cancer mortality among populations residing proximate to VCM polymerization facilities also demonstrated an increased risk of dying from CNS and lymphatic cancer (18). However, it was noted by Infante et al (18) that although these findings raise cause for concern about out-plant emissions of VCM, without further study these cancers cannot be unequivocally interpreted as being related to out-plant exposure to VCM (18).

Vinyl chloride has been shown to produce chromosome breaks in exposed workers (19)-(22). The mode of mutagenic action of VCM in special tester strains of *Salmonella typhimurium* (17)(23)-(26) appears to be multifaceted. For example, it could be mutagenic per se (25), or could be active via its microsomal metabolites such as chloroethylene oxide and 2-chloroacetaldehyde (17)(23)(26) produced in the presence of liver microsomes from mice (17)(23), rats (17)(23)(24) and humans (17)(23). It was also reported that the stimulatory effect of hepatic extracts is not due to microsomal activation but rather it is due to a nonenzymatic reaction and that mutagenic activity of VCM in Salmonella might involve a free radical mechanism (27). VCM is mutagenic in *Drosophila melanogaster* (28), and produced gene mutations in *S. pombe* (forward mutations) and gene conversions in two loci of a diploid *S. cerevisiae* in the presence of liver microsomes (29) and in the host-mediated assay when mice were treated with an oral dose of 700 mg/kg (29). In the absence of metabolic activation, vinyl chloride was not mutagenic or recombinogenetic in *S. cerevisiae* strains D5 and XV185-14C (30). VCM (in gaseous form or in ethanol solution) was also nonmutagenic in two strains of *Neurospora crassa* (31) and not mutagenic in male CD-1 mice at inhalation levels of 3,000, 10,000, and 30,000 ppm for 6 hours/day for 5 days as measured by the dominant lethal test (32).

Elmore et al (33) screened without exogenous activation seven potential metabolites of vinyl chloride in their pure forms as well as the related epichlorohydrin in tester strains of *Bacillus* and *Salmonella*. Chlorooxirane (chloroethylene oxide), chloroacetaldehyde, chloroacetaldehyde monomer hydrate, chloroacetaldehyde dimer hydrate, chloroacetaldehyde trimer and epichlorohydrin produced significant mutagenic activity in *S. typhimurium* strains sensitive to base-pair mutation. A recombination repair deficient strain of *B. subtilis* was inhibited in growth by these compounds, whereas excision repair deficient and wild type strains of *B. subtilis* were relatively unaffected.

Vinyl chloride at a concentration of 0.0106 M (723 ppm) in nutrient broth was negative in both the *Salmonella* and *Bacillus* cultures. [High concentrations of VCM (20% v/v in air-200,000 ppm) produced mutagenic action in previous assays with *Salmonella* tester strains] (23)(26).

Chloroethanol and chloroacetic acid which probably are metabolic intermediates were nonmutagenic at 1 mM concentrations in the above mutagenicity assays of Elmore et al (33). The metabolic pathways of vinyl chloride have been postulated by Hefner et al (34). Among the compounds tested by Elmore et al (33) in this scheme, chloroacetaldehyde and chlorooxirane were the most mutagenic with the lowest toxic side effects. Hence, it was suggested that they may be the active carcinogenic derivatives of vinyl chloride (33). However, chloroacetaldehyde monomer hydrate was considered a more realistic choice as the ultimate carcinogen than the monomer compound which reacts immediately with water. The lower mutagenic activity of chlorooxirane compared to chloroacetaldehyde monomer hydrate may reflect the unstable nature of chlorooxirane as an α-chloroether (24). One mode of action of chlorooxirane is a rearrangement to chloroacetaldehyde (35) via the NIH shift (36); another is a homolytic ring cleavage to yield a stabilized diradical intermediate ClCH−CH₂O, with both being capable of reacting with DNA to account for the mutagenicity of chlorooxirane (24).

Bartsch and Montesano (37) proposed a possible biotransformation of VCM in rats and an alternative biotransformation of VCM involving mixed function oxidase. Products obtained by reaction of 2-chloroacetaldehyde with adenosine or cytidine are also described (37). Base alterations of this type in the bacterial DNA may explain the activity of 2-chloroacetaldehyde as a bacterial mutagen since this compound induced base-pair substitutions in *S. typhimurium* TA 1530 (37).

Medical Surveillance: Preplacement and periodic examinations should emphasize liver function and palpation. Liver scans and grey-scale ultrasonography have been useful in detecting liver tumors. Medical histories should include alcoholic intake; past hepatitis; exposure to hepatotoxic agents, drugs and chemicals; past blood transfusions; past hospitalizations. Radiographic examinations of the hands may be helpful if acroosteolysis is suspected. Long term followup of exposed persons is essential as in the case of other carcinogens.

Special Tests: None in common use. Metabolism is being studied.

Personal Protective Methods: Where vinyl chloride levels cannot meet the standard, workers should be required to wear respiratory protection, either air supplied respirator or, if the level does not exceed 25 ppm, a chemical cartridge or cannister type gas mask. In hazard areas, proper protective clothing to prevent skin contact with the vinyl chloride or polyvinyl chloride residue should be worn.

References

(1) EPA. Scientific and technical assessment report on vinyl chloride and polyvinyl chloride. Office of Research and Development. Washington, D.C. June, 1975.
(2) IARC. Some anti-thyroid and related substances, nitrofurans and industrial chemicals. Monograph No. 7. International Agency for Research on Cancer. Lyon, 1974.
(3) Sakarnis, V.F. 1,2-Dichloroethane-induced chromosome non-divergence of the X-chromosome and recessive sex-linked lethal mutation in *Drosophila melanogaster. Genetika* 5 (1969) 89-95.
(4) Rapoport, I.A. The reaction of genic proteins with 1,2-dichloroethane. *Akad. Nauk. SSR Dokl. Biol. Sci.* 134 (1960) 745-747.
(5) McCann, J., Spingarn, N.E., Kobori, J., and Ames, B.N. Carcinogens as mutagens: Bacterial tester strains with R factor plasmids. *Proc. Natl. Acad. Sci.* 72 (1975) 979-983.
(6) McCann, J., Simon, V., Streitweiser, D., and Ames, B.N. Mutagenicity of chloroacetaldehyde, a possible metabolic product of 1,2-dichloroethane, chloroethanol, and cyclophosphamide. *Proc. Natl. Acad. Sci.* 72 (1975) 3190-3193.

(7) Brem, H., Stein, A.B., and Rosenkranz, H.S. The mutagenicity and DNA-modifying effect of haloalkanes. *Cancer Res.* 34 (1974) 2576-2579.

(8) Maltoni, C. The value of predictive experimental bioassays in occupational and environmental carcinogenesis. *Ambio* 4 (1975) 18.

(9) Maltoni, C., and Lefemine, G. Carcinogenicity bioassays of vinyl chloride, I. Research plans and early results. *Environ. Res.* 7 (1974) 387-405.

(10) Viola, P.L., Bigotti, A., and Caputo, A. Oncogenic response of rat skin, lungs, and bones to vinyl chloride. *Cancer Res.* 31 (1971) 516-519.

(11) Maltoni, C. La ricerca sperimentale ei suoa risultati, *Sapere* 776 (1974) 42-45.

(12) Maltoni, C., and Lefemine, G. La potenzialita dei saggi sperimentali nella predizione dei rischi oncogeni ambientali. Un esempio: Il cloruro di vinile. *Rend. Sci. Fis. Mat. Nat.* 66 (1974) 1-11.

(13) Maltoni, C., and Lefemine, G. Carcinogenicity bioassays of vinyl chloride, current results. *Ann. NY Acad. Sci.* 246 (1975) 195-218.

(14) IARC. Report of a working group on vinyl chloride. Lyon. IARC Internal Technical Report No. 74/005, 1974.

(15) Creech, J.L., and Johnson, M.N. Angiosarcoma of liver in the manufacture of polyvinyl chloride. *J. Occup. Med.* 16 (1974) 150.

(16) Holmberg, B., and Molina, M.J. The industrial toxicology of vinyl chloride: A review. *Work Environ. Health.* 11 (1974) 138-144.

(17) Bartsch, H., and Montesano, R. Mutagenic and carcinogenic effects of vinyl chloride. *Mutation Res.* 32 (1975) 93-114.

(18) Infante, P.F., Wagoner, J.K., and Waxweiler, R.J. Carcinogenic, mutagenic and teratogenic risks associated with vinyl chloride. *Mutation Res.* 41 (1976) 131-142.

(19) Purchase, I.F.H., Richardson, C.R., and Anderson, D. Chromosomal and dominant lethal effects of vinyl chloride. *Lancet* 11 (1976) 410-411.

(20) Hansteen, I.L., Hillestad, L., and Thiis-Evensen, E. Chromosome studies on workers exposed to vinyl chloride. *Mutation Res.* 38 (1976) 112.

(21) Funes-Cravioto, F., Lambert, B., et al. Chromosome aberrations in workers exposed to vinyl chloride. *Lancet* 1 (1975) 459.

(22) Ducatman, V., Hirschhorn, K., and Selikoff, I.J. Vinyl chloride exposure and human chromosome aberrations. *Mutation Res.* 31 (1975) 163-169.

(23) Bartsch, H., Malaveille, C., and Montesano, R. Human, rat and mouse liver-mediated mutagenicity of vinyl chloride in *S. typhimurium* strains. *Int. J. Cancer* 15 (1975) 429-437.

(24) Rannug, U., Johansson, A., Ramel, C., and Wachtmeister, C.A. The mutagenicity of vinyl chloride after metabolic activation. *Ambio* 3 (1974) 194-197.

(25) McCann, J., Simon, V., Streitsweiser, D., and Ames, B.N. Mutagenicity of chloroacetaldehyde, a possible metabolic product of 1,2-dichloroethane, chloroethanol, and cyclophosphamide. *Proc. Natl. Acad. Sci.* 72 (1975) 3190-3193.

(26) Malaveille, C., Bartsch, H., Barbin, A., Camus, A.M., and Montesano, R. Mutagenicity of vinyl chloride, chloroethylene oxide, chloroacetaldehyde and chloroethanol. *Biochem. Biophys. Res. Comm.* 63 (1975) 363-370.

(27) Garro, A.J., Guttenplan, J.B., and Milvy, P. Vinyl chloride dependent mutagenesis: Effects of liver extracts and free radicals. *Mutation Res.* 38 (1976) 81-88.

(28) Magnusson, J., and Ramel, C. Mutagenic effects of vinyl chloride in *Drosophila melanogaster.* *Mutation Res.* 38 (1976) 115.

(29) Loprieno, N., Barale, R., Baroncelli, S., Bauer, C., Bronzetti, G., Cammellini, A., et al. Evaluation of the genetic effects induced by vinyl chloride monomer (VCM) under mammalian metabolic activation: studies in vitro and vivo. *Mutation Res.* 40 (1976) 85-96.

(30) Shahin, M.M. The non-mutagenicity and recombinogenicity of vinyl chloride in the absence of metabolic activation. *Mutation Res.* 40 (1976) 269-272.

(31) Drozdowicz, B.Z., and Huang, P.C. Lack of mutagenicity of vinyl chloride in two strains of *Neurospora crassa. Mutation Res.* 48 (1977) 43-50.

(32) Anderson, D., Hodge, M.F.E., and Purchase, I.F.H. Vinyl chloride: Dominant lethal studies in male CD-1 mice. *Mutation Res.* 40 (1976) 359-370.

(33) Elmore, J.D., Wong, J.L., Laumbach, A.D., and Streips, U.N. Vinyl chloride mutagenicity via the metabolites chlorooxirane and chloroacetaldehyde monomer hydrate. *Biochim. Biophys. Acta.* 442 (1976) 405-419.

(34) Hefner, J.R.E., Watanabe, P.G., and Gehring, P.J. Preliminary studies on the fate of inhaled vinyl chloride monomer in rats. *Ann. NY Acad. Sci.* 246 (1975) 135-148.

(35) Walling, C., and Fredericks, P.S. Positive halogen compounds, IV. Radical reactions of chlorine and t-butyl hypochlorite with some small ring compounds. *J. Am. Chem. Soc.* 84 (1962) 3326-3331.

(36) Daly, J.W., Jerina, D.M., and Witkop, B. Arene oxides and the NIH shift. The metabolism, toxicity and carcinogenicity of aromatic compounds. *Experientia* 28 (1972) 1129.

(37) Bartsch, H., and Montesano, R. Mutagenic and carcinogenic effects of vinyl chloride. *Mutation Res.* 32 (1975) 93-114.

(38) National Institute for Occupational Safety and Health. *Vinyl Halides—Carcinogenicity.* Current Intelligence Bulletin No. 28, Wash., D.C. (Sept. 21, 1978).

Bibliography

Berk, P.D., Martin, J.F., Young, R.S., Creech, J., Selikoff, I.J., Falk, H., Watanabe, P., Popper, H., and Thomas, L. 1976. Vinyl chloride-associated liver diseases. *Ann. Intern. Med.* 84:717.

Dodson, V.N., Dinman, B.D., Whitehouse, W.M., Nasr, A.H.M., and Magnuson, H.J. Occupational acro-osteolysis: III. A clinical study. *Arch. Environ. Health.* 22:83.

Duck, B.W. 1976. Medical surveillance of vinyl chloride workers. *Proc. R. Soc. Med.* 69:307.

Editorial: Vinyl chloride: the carcinogenic risk. 1976. *Brit. Med. J.* 2:134.

Falk, H., Creech, J.L. Jr., Heath, C.W. Jr., Johnson, M.N., and Key, M.M. 1974. Hepatic disease among workers at a chloride polymerization plant. *J. Am. Med. Assoc.* 230:59.

Fox, A.J., and Collier, P.F. 1977. Mortality experience of workers exposed to vinyl chloride monomer in the manufacture of polyvinyl chloride in Great Britain. *Brit. J. Ind. Med.* 34:1.

Haley, T.J. 1975. Vinyl chloride: How many unknown problems? *J. Toxicol. and Environ. Health.* 1:47.

Makk, L., Creech, J.L., Whelan, J.G., and Johnson, M.N. 1974. Liver damage and angiosarcoma in vinyl chloride workers: a systematic detection program. *J. Am. Med. Assoc.* 230:64.

Preston, B.J., Jones, K. Lloyd, and Grainer, R.G. 1976. Clinical aspects of vinyl chloride disease. *Proc. R. Soc. Med.* 69:284.

Selikoff, I.J., and Hammond, E.C., eds. 1975. Toxicity of vinyl chloride-polyvinyl chloride. *Ann. NY Acad. Sci.* 246:1.

Taylor, K.J.W., Williams, D.M.J., Smith, P.M., and Duck, B.W. 1975. Grey-scale ultrasonography for monitoring industrial exposure to hepatotoxic agents. *Lancet* 1:1222.

Viola, P.L. 1970. Pathology of vinyl chloride. *Med. Lav.* 61:174.

Waxweiller, R.J., Stringer, W., Wagoner, J.K., and Jones, J. 1976. Neoplastic risk among workers exposed to vinyl chloride. *Ann. NY Acad. Sci.* 271:40.

VINYLIDENE CHLORIDE (VDC)

Description: Vinylidene chloride is a volatile liquid, boiling at 31.7°C at 760 mm, with a mild, sweet odor resembling that of chloroform. It has a density D_4^{20} of 1.2129 and a flash point of -15°C.

Synonyms: 1,1-dichloroethylene, DCE, VDC.

Potential Occupational Exposures: Dow Chemical and PPG Industries annually produce 270 million pounds of VDC monomer in three Gulf Coast plants. About 50% is used in the production of methyl chloroform by PPG. The remainder is polymerized to plastic resins at 12 facilities owned by a number of companies throughout the country. The resin is then fabricated into plastics at 60 to 75 plants.

These copolymers (in latex, fiber, film and resin forms) are referred to as "Saran" and have wide utility mainly for film wraps for food. Saran production is estimated at about 150 million pounds per year (1).

It has been estimated that about four million pounds of VDC were lost to the air in 1974. One EPA-funded report estimates that as much as 25% of the VDC used in any given Saran production run is disposed of in landfill, primarily in polymerized form, but there are no estimates of the levels of unreacted monomer.

In the past, worker exposure has generally not been monitored. Tests demonstrate that 20,000 ppm can easily be attained in the immediate vicinity of a spill. In some cases, past worker exposures to VDC may have exceeded those to vinyl chloride (which were measured at 300 to 1,000 ppm before OSHA limits were imposed). The odor threshold of VDC is 500 ppm.

The number of workers engaged in the production of vinylidene chloride monomer per se in the United States (compared to vinyl chloride monomer) appears to be small, e.g., 75 and 12 to 15 at two major vinylidene chloride production facilities (2). NIOSH (19) estimates definite worker exposure to vinylidene chloride at 6,500 and probable worker exposure to vinylidene chloride at 58,000.

Vinylidene chloride has also been reported to be a trace impurity in vinyl chloride monomer (3)-(5). Workers involved in manufacturing facilities using VCM monomer in polymerization processes (e.g., PVC) can be exposed to vinylidene chloride concentrations in amounts of less than 5 ppm (3)(4) and more frequently to trace amounts (4).

Although the widespread use of vinylidene polymers as food wraps could result in the release of unreacted monomer into the food chain (1)(6), and vinylidene chloride copolymers containing a minimum of 85% vinylidene chloride have been approved for use with irradiated foods (7), information is scant as to the migration of unreacted monomers from these sources either into food or via disposal of the polymeric material per se. One report states that no more than 10 ppm of unreacted vinylidene chloride is contained in Dow's product Saran Wrap and that within detectable limits, no more than 10 ppb could get into food, even under severe conditions of use (8).

Permissible Exposure Limits: The American Conference of Governmental Industrial Hygienists has published a threshold limit value of 10 ppm (40 mg/m^3) as of 1978. NIOSH has gone further and recommended (19) an exposure standard of 1 ppm, the same as for vinyl chloride.

EPA is preparing an assessment of the air pollution problems associated with VDC production and use. Fetotoxicity and embryotoxicity have been demonstrated under EPA-funded contracts. Data on environmental effects of VDC are also being obtained.

Route of Entry: Inhalation of vapor.

Harmful Effects

Aspects of the reported carcinogenicity of vinylidene chloride appear conflicting and indicate sex, species, and strain specificity. Maltoni (9) reported that Swiss male mice exposed to 25 ppm of vinylidene chloride in air for 4 hours daily, 4 to 5 weeks, for 52 weeks, developed adenocarcinoma of the kidney. No effects were found in Balb/C, C56B1 or C$_3$H mice or Sprague-Dawley rats and hamsters similarly exposed to vinylidene chloride (9). However, Viola (10) reported that male and female Wistar rats exposed to 100 ppm of vinylidene chloride by inhalation developed abdominal lymphomas and subcutaneous fibromas.

Two year studies at Dow Chemical Co., involving both vinylidene chloride administered in the drinking water (60,100 and 120 ppm) and repeated inhalation (10 or 40 ppm 6 hours/day; 5 days/week; after 5 weeks, 75 ppm for up to 18 months) to male and female Sprague-Dawley rats have been carried out (11)(12) and indicated no dose-related clinical differences or cumulative mortality differences or findings of neoplasia. Reproduction studies with vinylidene chloride administered to Sprague Dawley rats by inhalation or ingestion in the drinking water showed the compound to be neither a teratogen nor mutagen nor one adversely affecting reproductivity (12). The vinylidene chloride (99.5%) tested in the Dow studies contained trace amounts (ppm) of the following impurities: vinyl bromide, 4; vinyl chloride, 3 to 50; trans-1,2-dichloroethylene, 138 to 1,300; cis-1,2-dichloroethylene, 0.013 to 0.16%; 1,1,1-trichloroethane, 0.03; and 1,1,2-trichloroethane (11).

Winston et al (13) and Lee (14) reported the only tumor in rats exposed to 55 ppm vinylidene chloride for 9 months was a subcutaneous hemangiosarcoma of the skin in one of the 9 rats. Exposure of mice to 55 ppm of vinylidene chloride for up to 9 months resulted in the development of one hepatic hemangiosarcoma. Bronchiolar adenoma and/or acinar proliferation in the lung also occurred in 5 to 42 mice.

Vinylidene chloride induces point mutation in the histidine-auxotroph strains of *Salmonella typhimurium* TA 1530 and TA 100 when tested in the presence of rat or mouse liver in vitro (15) (its mutagenic activity was higher than that of VCM). The mutagenic response, which was greater in TA 100 strain increased in both strains after exposure to 2% vinylidene chloride in air. The lower mutagenic response observed with a concentration of 20% vinylidene chloride may have resulted from an inhibitory action of vinylidene chloride and or its metabolite(s) on the microsomal enzymes responsible for its metabolic activation. It was postulated by Bartsch et al (15) that 1,1-dichloroethylene oxide (in analogy with chloroethylene oxide, the suggested primary metabolite of vinyl chloride) (16) may be a primary reactive metabolite of vinylidene chloride. It is also considered possible that partial dechlorination of vinylidene chloride by microsomal enzymes results in vinyl chloride and its metabolic products (15).

Vinylidene chloride has also been found to be mutagenic when tested in a metabolizing in vitro system with *E. coli* K12 (17). In contrast to the results in *S. typhimurium* (15), the mutagenicity of VCM was several times higher than that of vinylidene chloride when tested in *E. coli* (back-mutation system, arg$^+$) (17).

No chromosomal aberrations have been found in Sprague-Dawley rats exposed to 75 ppm vinylidene chloride 6 hours/day, 5 days/week for 26 weeks (12).

Hathaway (18) recently proposed a scheme for the in vivo metabolism of vinylidene chloride in rats. Of the compounds included, chloroacetic acid, thiodiglycollic acid, thioglycollic acid, dithioglycollic acid and lactam compounds have been isolated.

Medical Surveillance: See Vinyl Chloride.

Special Tests: None.

Personal Protective Methods: The primary requirement for reduction of exposure to VDC would be to limit emissions through improved housekeeping procedures in the industry. Beyond that, the use of gloves and protective clothing is indicated and a fullface mask should be used in areas of excessive vapor concentrations.

References

(1) Anon. Vinylidene chloride linked to cancer. *Chem. Eng. News,* Feb. 28 (1977) p. 6-7.
(2) Anon. Rap for film wrap. *Chem. Week,* Oct. 16 (1974) p. 20.
(3) Ott, M.G., Langner, R.R., and Holder, B.B. Vinyl chloride exposure in a controlled industrial environment. *Arch. Env. Hlth.* 30 (1975) 333-339.
(4) Jaeger, R.J., Conolly, R.B., Reynolds, E.S., and Murphy, S.D. Biochemical toxicology of unsaturated halogenated monomers. *Env. Hlth. Persp.* 11 (1975) 121-123.

(5) Kramer, C.D., and Mutchler, J.E. The correlation of clinical and environmental measurement for workers exposed to vinyl chloride. *Am. Ind. Hyg. Assoc. J.* 33 (1975) 19-30.

(6) Lehman, A.J. Chemicals in foods: a report to the Association of Food and Drug Officials on Current Developments. *Assoc. Food and Drug Officials, U.S. Quart. Bull.* 15 (1951) 82-89.

(7) Anon. Food additives, packaging material for use during irradiation of pre-packaged foods. *Fed. Register* 33 (1968) 4659.

(8) Anon. Vinylidene chloride: No trace of cancer at Dow. *Chem. Eng. News* March 14 (1977) pp. 21-22.

(9) Maltoni, C. Recent findings on the carcinogenicity of chlorinated olefins. NIEHS Conference on Comparative Metabolism and Toxicity of Vinyl Chloride and Related Compounds, Bethesda, MD May 2-4 (1977).

(10) Viola, P.L. Carcinogenicity studies on vinylidene chloride (VDC). NIEHS Conference on Comparative Metabolism and Toxicity of Vinyl Chloride and Related Compounds, Bethesda, MD May 2-4 (1977).

(11) Rampy, L.W. Toxicity and carcinogenicity studies on vinylidene chloride. NIEHS Conference on Comparative Metabolism and Toxicity of Vinyl Chloride and Related Compounds, Bethesda, MD May 2-4 (1977).

(12) Norris, J.M. The MCA-Toxicology Program for Vinylidene Chloride. Presented at 1977 European Tappi Meeting, Hamburg, Germany, Jan. 26 (1977).

(13) Winston, J.M., Lee, C.C., Bhandari, J.C., Dixon, R.C., and Woods, J.S. A study of the carcinogenicity of inhaled vinyl chloride and vinylidene chloride (VDC) in rats and mice. *Abstracts Int. Congress of Toxicology.* Toronto, March 3-April 2 (1977) p. 32.

(14) Lee, C.C. Toxicity and carcinogenicity of vinyl chloride compared to vinylidene chloride. NIEHS Conference on Comparative Metabolism and Toxicity of Vinyl Chloride and Related Compounds. Bethesda, MD May 2-4 (1977).

(15) Bartsch, H., Malaveille, C., Montesano, R., and Tomatis, L. Tissue-mediated mutagenicity of vinylidene chloride and 2-chlorobutadiene in *Salmonella typhimurium. Nature* 255 (1975) 641-643.

(16) Bartsch, H., and Montesano, R. Mutagenic and carcinogenic effects of vinyl chloride. *Mutation Res.* 32 (1975) 93-114.

(17) Greim, H., Bonse, G., Radwan, Z., Reichert, D., and Henschler, D. Mutagenicity in vitro and potential carcinogenicity of chlorinated ethylenes as a function of metabolic oxirane formation. *Biochem. Pharmacol.* 24 (1975) 2013.

(18) Hathaway, D.E. Comparative mammalian metabolism of vinylidene and vinyl chloride in relation to outstanding oncogenic potential. NIEHS Conference on Comparative Metabolism and Toxicity of Vinyl Chloride Related Compounds, Bethesda, MD May 2-4 (1977).

(19) National Institute for Occupational Safety and Health. *Vinyl Halides—Carcinogenicity.* Current Intelligence Bulletin No. 28, Wash., D.C. (Sept. 21, 1978).

Bibliography

Air Pollution Assessment Report: Vinylidene Chloride. EPA, Office of Air Quality Planning and Standards. (Prepared under contract.)

Jaeger, R.J., Trabulus, M.J., and Murphy, S.D. Biochemical effects of 1,1-dichloroethylene in rats: dissociation of its hepatotoxicity from a liperoxidative mechanism. *Toxicology and Applied Pharmacology.* 24:457-567 (1973).

The Merck Index, 8th edition. Rahway, NJ, Merck and Co., Inc. (1968).

Prendergast, J.A., Jones, R.A., Jenkins, L.J. Jr., and Siegel, J. Effects on experimental animals of long-term inhalation of trichloroethylene, carbon tetrachloride, 1,1,1-trichloroethane, dichlorofluoromethane and 1,1-dichloroethylene. *Toxicology and Applied Pharmacology.* 10:270-289 (1967).

Vinylidene Chloride Monomer Emissions from the Monomer, Polymer, and Polymer Processing Industries. EPA, Office of Air Quality Planning and Standards (Prepared under contract 68-02-1332, Task 13).

Wessling, R., and Edwards, F.G. Poly(vinylidene chloride). *Kirk-Othmer Encyclopedia of Chemical Technology.* 2nd edition, (21:275-303), NY Interscience Publishers (1967).

X

XYLENE

Description: $C_6H_4(CH_3)_2$, xylene, exists in three isomeric forms, ortho-, meta- and para-xylene. Commercial xylene is a mixture of these three isomers and may also contain ethylbenzene as well as small amounts of toluene, trimethylbenzene, phenol, thiophene, pyridine, and other nonaromatic hydrocarbons. meta-Xylene is predominant in commercial xylene and shares physical properties with ortho-xylene in that both are mobile, colorless, flammable liquids. para-Xylene, at a low temperature (13°-14°C), forms colorless plates or prisms.

Synonyms: Xylol, dimethylbenzene.

Potential Occupational Exposures: Xylene is used as a solvent; as a constituent of paint, lacquers, varnishes, inks, dyes, adhesives, cements, cleaning fluids and aviation fuels; and as a chemical feedstock for xylidines, benzoic acid, phthalic anhydride, isophthalic, and terephthalic acids, as well as their esters (which are specifically used in the manufacture of plastic materials and synthetic textile fabrics). Xylene is also used in the manufacture of quartz crystal oscillators, hydrogen peroxide, perfumes, insect repellants, epoxy resins, pharmaceuticals, and in the leather industry.

A partial list of occupations in which exposure may occur includes:

Adhesive workers	Phthalic anhydride makers
Aviation gasoline workers	Polyethylene terephthalate film makers
Benzoic acid makers	Quartz crystal oscillator makers
Cleaning fluid makers	Solvent workers
Histology technicians	Synthetic textile makers
Lacquer workers	Terephthalic acid makers
Leather workers	Varnish makers
Paint workers	

NIOSH estimates that 140,000 workers are potentially exposed to xylene.

Permissible Exposure Limits: The Federal standard is 100 ppm (435 mg/m³). NIOSH recommends adherence to the present Federal Standard of 100 ppm as a time-weighted average for up to a 10-hour workday, 40-hour workweek. NIOSH also recommends a ceiling concentration of 200 ppm as determined by a sampling period of 10 minutes.

Routes of Entry: Inhalation of vapor and, to a small extent, percutaneous absorption of liquid.

Harmful Effects

Local—Xylene vapor may cause irritation of the eyes, nose, and throat. Repeated or prolonged skin contact with xylene may cause drying and defatting of the skin which may lead to dermatitis. Liquid xylene is irritating to the eyes and mucous membranes, and aspiration of a few milliliters may cause chemical pneumonitis, pulmonary edema, and hemorrhage. Repeated exposure of the eyes to high concentrations of xylene vapor may cause reversible eye damage.

Systemic—Acute exposure to xylene vapor may cause central nervous system depression and minor reversible effects upon liver and kidneys. At high concentrations xylene vapor may cause dizziness, staggering, drowsiness, and unconsciousness. Also at very high concentrations, breathing xylene vapors may cause pulmonary edema, anorexia, nausea, vomiting, and abdominal pain.

Medical Surveillance: Preplacement and periodic examinations should evaluate possible effects on the skin and central nervous system, as well as liver and kidney functions. Hematologic studies should be done if there is any significant contamination of the solvent with benzene.

Special Tests: Although metabolites are known, biologic monitoring has not been widely used. Hippuric acid or the ether glucuronide of ortho-toluic acid may be useful in diagnosis of meta-, and para- and ortho-xylene exposure, respectively.

Personal Protective Methods: When vapor concentrations exceed allowable standards, fullface masks with organic vapor canisters or air-supplied respirators should be furnished. Impervious protective clothing and gloves should be worn to cover exposed portions of the body of employees exposed to liquid xylene. Xylene-wet clothing should be changed quickly. Personal hygiene, as well as appropriate changes of work clothes, is necessary. Goggles or safety glasses in areas of spill or splash, or in areas where vapors concentrate, are advised. Barrier creams may be useful.

Bibliography

Matthaus, W. 1964. Beitrag zur hornhauterkrankung von oberflachenbearbeiten in der mobelindustrie. *Klin. Monatsbl. Augenheilkd.* 144:713.

Morley, R., D.W. Eccleston, C.P. Douglas, W.E.J. Greville, D.J. Scott, and J. Anderson. 1970. Xylene poisoning: a report of one fatal case and two cases of recovery after prolonged unconsciousness. *Br. Med. J.* 3:442.

National Institute for Occupational Safety and Health. *Criteria for a Recommended Standard: Occupational Exposure to Xylene,* NIOSH Doc. No. 75-168 (1975).

Z

ZINC CHLORIDE

Description: $ZnCl_2$, zinc chloride, consists of white hexagonal, deliquescent crystals, soluble in water (1 g/0.5 ml) and in organic solvents. It may be produced from zinc sulfide ore, zinc oxide, or zinc metal.

Synonyms: Butter of zinc.

Potential Occupational Exposures: Zinc chloride is used as a wood preservative, for dry battery cells, as a soldering flux, and in textile finishing, in vulcanized fiber, reclaiming rubber, oil and gas well operations, oil refining, manufacture of parchment paper, dyes, activated carbon, chemical synthesis, dentists' cement, deodorants, disinfecting and embalming solutions, and taxidermy. It is also produced by military screening-smoke devices.

A partial list of occupations in which exposure may occur includes:

Activated carbon makers	Military personnel
Dental cement makers	Paper makers
Deodorant makers	Petroleum refinery workers
Disinfectant makers	Rubber workers
Dry cell battery makers	Solderers
Dye makers	Taxidermists
Embalmers	Textile finishers

Permissible Exposure Limits: The Federal standard for zinc chloride fume is 1 mg/m³.

Routes of Entry: Inhalation of dust and fumes; ingestion.

Harmful Effects

Local—Solid zinc chloride is corrosive to the skin and mucous membranes. Aqueous solutions of 10% or more are also corrosive and cause primary dermatitis and chemical burns, especially at sites of minor trauma. Aqueous solutions are also extremely dangerous to the eyes, causing extreme pain, inflammation, and swelling, which may be followed by corneal ulceration.

Zinc chloride may produce true sensitization of the skin in the form of eczematoid dermatitis. Ingestion of zinc chloride may cause serious corrosive effects in the esophagus and stomach, often complicated by pyloric stenosis.

Systemic—There are no reports of inhalation of zinc chloride from industrial exposure. All reported experience with inhaled zinc chloride is based on exposures caused by military accidents. In all of those cases, there was severe irritation of the respiratory tract. In the more severe cases, acute pulmonary edema developed within two to four days following exposure. The fatalities reported were due to severe lung injury with hemorrhagic alveolitis and bronchopneumonia. In human experimentation with concentrations of 120 mg/m^3, there were complaints of irritation of the nose, throat, and chest after 2 minutes. With exposure to 80 mg/m^3 for 2 minutes, the majority of subjects experienced slight nausea, all noticed the smell, and one or two coughed.

Medical Surveillance: In preemployment and periodic physical examinations, special attention should be given to the skin and to the history of allergic dermatitis, as well as to exposed mucous membranes, the eyes, and the respiratory system. Chest x-rays and periodic pulmonary function studies may be helpful. Smoking history should be known.

Special Tests: Urinary zinc excretion may be useful.

Personal Protective Methods: Employees exposed to zinc chloride should be given instruction in personal hygiene, and in the use of personal protective equipment. Goggles should be provided in areas where splash or spill of liquid is possible. In areas with excessive dust or fume levels, respiratory protection by use of filter type dust masks or air supplied respirators with fullface pieces should be required. In areas where danger of spills or splashes exists, skin protection should be provided with rubber gloves, face shields, rubber aprons, gantlets, suits, and rubber shoes.

Bibliography

Johnson, F.A., and R.B. Stonehill. 1961. Chemical pneumonitis from inhalation of zinc chloride. *Dis. Chest.* 40:619.

ZINC OXIDE

Description: ZnO, zinc oxide, is an amorphous, odorless, white or yellowish-white powder, practically insoluble in water. It is produced by oxidation of zinc or by roasting of zinc oxide ore.

Synonyms: Zinc white, flowers of zinc.

Potential Occupational Exposures: Zinc oxide is primarily used as a white pigment in rubber formulations and as a vulcanizing aid. It is also used in photocopying, paints, chemicals, ceramics, lacquers, and varnishes, as a filler for plastics, in cosmetics, pharmaceuticals, and calamine lotion. Exposure may occur in the manufacture and use of zinc oxide and products, or through its formation as a fume when zinc or its alloys are heated.

A partial list of occupations in which exposure may occur includes:

Alloy makers	Lacquer makers
Brass foundry workers	Paint makers
Ceramic makers	Pigment makers
Chemical synthesizers	Plastic makers
Cosmetic makers	Rubber workers
Electroplaters	Welders
Galvanizers	

NIOSH estimates that 50,000 workers have potential exposure to zinc oxide.

Permissible Exposure Limits: The Federal standard for zinc oxide fume is 5 mg/m^3. NIOSH recommends adherence to the present Federal standard of 5 mg/m^3 as a time-weighted average for up to a 10-hour workday, 40-hour workweek. NIOSH also recommends a ceiling concentration of 15 mg/m^3 as determined by a sampling time of 15 minutes.

Route of Entry: Inhalation of dust or fumes.

Harmful Effects

Local—When handled under poor hygienic conditions, zinc oxide powder may produce a dermatitis called "oxide pox." This condition is due primarily to clogging of the sebaceous glands with zinc oxide and produces a red papule with a central plug. The area rapidly becomes inflamed and the central plug develops into a pustule which itches intensely. Lesions occur in areas of the skin that are exposed or subject to heavy perspiration. These usually clear, however, in a week to ten days with good hygiene and proper care of secondary infections.

Systemic—The syndrome of metal fume fever is the only important effect of exposure to freshly formed zinc oxide fumes and zinc oxide dusts of respirable particle size. The fumes are formed by subjecting either zinc or alloys containing zinc to high temperatures. Typically, the syndrome begins four to twelve hours after sufficient exposure to freshly formed fumes of zinc oxide. The worker first notices the presence of a sweet or metallic taste in the mouth, accompanied by dryness and irritation of the throat. Cough and shortness of breath may occur, along with general malaise, a feeling of weakness, fatigue, and pains in the muscles and joints. Fever and shaking chills then develop. Fever can range from 102° to 104°F. Profuse sweats develop and the fever subsides. The entire episode runs its course in 24 to 48 hours.

During the acute period, there is an elevation of the leukocyte count (rarely above 20,000/mm^3), and the serum LDH may be elevated. Chest x-rays are not diagnostic.

Metal fume fever produces rapid development of tolerance or short-lived relative immunity. This may be lost, however, over a weekend or holiday, and the worker may again develop the complete syndrome when he returns to work if fume levels are sufficiently high. There are no sequelae to the attacks.

Other possible systemic effects of zinc oxide are in doubt. Cases of gastro-intestinal disturbance have been reported, but most authorities agree there is no evidence of chronic industrial zinc poisoning.

Medical Surveillance: Preemployment and periodic physical examinations should be made to assess the status of the general health of the worker. Examinations are also recommended following episodes of metal fume fever or intercurrent illnesses.

Special Tests: Zinc excretion in urine can be used as an index of exposure.

Personal Protective Methods: Employees should receive instruction in personal hygiene and in the causes and effects of metal fume fever. Workers exposed to zinc oxide powder should be supplied wth daily clean work clothes and should be required to shower before changing to street clothes. In cases of accident or where excessive fume concentrations are present, gas masks with proper canister or supplied air respirators should be provided.

Bibliography

National Institute for Occupational Safety and Health, *Criteria for a Recommended Standard: Occupational Exposure to Zinc Oxide.* NIOSH Doc. No. 76-104 (1976).

ZIRCONIUM AND COMPOUNDS

Description: Zr, zirconium, is a greyish-white, lustrous metal in the form of platelets, flakes, or a bluish-black, amorphous powder. It is never found in the free state; the most common sources are the ores zircon and baddeleyite. It is generally produced by reduction of the chloride or iodide. The metal is very reactive, and the process is carried out under an atmosphere of inert gas. The powdered metal is a fire and explosive hazard.

Synonyms: None.

Potential Occupational Exposures: Zirconium metal is used as a "getter" in vacuum tubes, a deoxidizer in metallurgy, and a substitute for platinum; it is used in priming of explosive mixtures, flashlight powders, lamp filaments, flash bulbs, and construction of rayon spinnerets. Zirconium or its alloys (with nickel, cobalt, niobium, tantalum) are used as lining materials for pumps and pipes, for chemical processes, and for reaction vessels. Pure zirconium is a structural material for atomic reactors, and alloyed, particularly with aluminum, it is a cladding material for fuel rods in water-moderated nuclear reactors. A zirconium-columbium alloy is an excellent super-conductor.

Zircon ($ZrSiO_4$) is utilized as a foundry sand, an abrasive, a refractory in combination with zirconia, a coating for casting molds, a catalyst in alkyl and alkenyl hydrocarbon manufacture, a stabilizer in silicone rubbers, and as a gem stone; in ceramics it is used as an opacifier for glazes and enamels and in fritted glass filters. Both zircon and zirconia (zirconium oxide, ZrO_2) bricks are used as linings for glass furnaces. Zirconia itself is used in die extrusion of metals and in spout linings for pouring metals as a substitute for lime in oxyhydrogen light, as a pigment, and an abrasive; it is used, too, in incandescent lights, as well as in the manufacture of enamels, white glass, and refractory crucibles.

Other zirconium compounds are used in metal cutting tools, thermocouple jackets, waterproofing textiles, ceramics, and in treating dermatitis and poison ivy.

A partial list of occupations in which exposure may occur includes:

Abrasive makers	Incandescent lamp makers
Ceramic workers	Metallurgists
Crucible makers	Pigment makers
Deodorant makers	Rayon spinneret makers
Enamel makers	Refractory material makers
Explosives workers	Textile waterproofers
Foundry workers	Vacuum tube makers
Glass makers	

Permissible Exposure Limits: The Federal standard for zicronium compounds is 5 mg/m^3 as Zr.

Route of Entry: Inhalation of dust or fume.

Harmful Effects

Local—No ill effects from industrial exposure to zirconium have been proven. A recent study from the USSR, however, reports that some workers exposed to plumbous titanate zirconate developed a mild occupational dermatitis associated with hyperhydrosis of the hands. This condition was accompanied by subjective complaints of vertigo, sweet taste in the mouth, and general indisposition. These workers were also said to have elevated thermal and pain sensitivity, and electric permeability of the horny layer, along with increased sweating, and reduced capillary resistance.

Zircon granulomas were reported in the U.S. as early as 1956. This condition arose from the use of deodorant sticks in the axillae, but it was resolved when use was stopped. Zircon is no longer used as a deodorant. Because of a possible allergic sensitivity reaction, individuals who have experienced granulomas from zirconium should avoid dust and mist.

Systemic—Inhalation of zirconium dust and fumes has caused no respiratory or other pathological problems. Animal experiments, however, have produced interstitial pneumonitis, peribronchial abscesses, peribronchiolar granuloma, and lobular pneumonia.

Medical Surveillance: No special considerations are needed.

Special Tests: None in common use.

Personal Protective Methods: Employees should be trained in the correct use of personal protective equipment. In areas of dust accumulation or high fume concentrations, respiratory protection is advised either by dust mask or supplied air respirators. Skin protection is not generally necessary, but where there is a history of zircon granuloma from deodorants, it is probably advisable.

Bibliography

Bukharovich, M.N., N.N. Speransky, I. Zakharov, and A.F. Malitsky. 1972. Skin diseases in workers of a department engaged in the production of titanate zirconate. *Gig. Tr. Prof. Zabol.* 16:35.

Sheard, C., Jr., F.E. Cormia, S.C. Atkinson, and E.L. Worthington. 1957. Granulomatous reactions to deodorant sticks. *J. Am. Med. Assoc.* 164:1085.

Shelley, W.B., and H.J. Hurley, Jr. 1971. The immune granuloma: late delayed hypersensitivity to zirconium and beryllium. In: M. Samter, ed. *Immunological Diseases.* Vol. II. Little, Brown and Company, Boston.

Shelly, W.B. 1973. Chondral dysplasia induced by zirconium and hafnium. *Cancer Res.* 33:287.

DRINKING WATER DETOXIFICATION 1978

Edited by M.T. Gillies

Pollution Technology Review No. 49

This book reviews the methods used to remove toxic chemicals from drinking water including toxic substances formed by chlorination. It also concerns itself with new and proposed amendments to the *Safe Drinking Water Act* of 1974 and therewith presents a comprehensive exposition of the whole drinking water topic. While the primary aim is to remove potential carcinogens, a serendipitous note is introduced by the fact that the proposed filter systems will also greatly improve the taste and odor of any potable water.

At issue is EPA's controversial plan to require waterworks serving more than 75,000 persons to filter their water through beds of granular activated carbon in order to remove traces of chemicals suspected of causing cancer. Most suspect are the trihalomethanes such as chloroform, which are formed from natural organic matter whenever water is chlorinated. Also implicated are many other organic chemicals that enter water supplies as a result of industrial activity, and from municipal and agricultural runoffs.

It should be noted that adsorption on activated carbon is a nonspecific process which will remove not only selected key contaminants, but many other poisons, some presently undetected by current analytical capability. This may result in greater protection of human health than the arbitrary setting of maximum contaminant levels for selected substances. This book, based mostly on federally funded studies and official publications, consists of 10 chapters.

A summarized table of contents follows here:

ISBN 0-8155-0723-2 $48   348 pages

TRACE CONTAMINANTS FROM COAL
1978

Edited by S. Torrey

Pollution Technology Review No. 50

During the past ten years man has become increasingly aware of the long range adverse health effects of most air and water pollutants. It has become apparent that interactions between air pollutants and lung tissue can undermine pulmonary functions long before there is any pathological evidence.

Airborne particulate matter is the most conspicuous, because of its ubiquity in urban atmospheres, its abundance in the emissions of vital industrial plants including coal-fired electric power generating stations, its effect on climate, visibility, materials and vegetation, and its apparent capacity to enhance the damaging effects of other air pollutants.

There is a paucity of information about the far-reaching impact of trace elements which become ecological contaminants upon their release by the burning of coal. These trace elements not only contaminate the air as particulates and gases but are also in effluents from ash and sludge disposal systems.

This book assesses their potential impact by assembling data on (1) the kinds and amounts of trace elements emitted during the coal combustion process; (2) their transport and deposition in the environment; and (3) their availability, accumulation, and toxicity in ecosystems. As is the case with other energy-producing methods, coal combustion is not a panacea, but requires critical evaluation and comparison with other means of obtaining energy.

The book is based on federally-funded studies and the bibliography at the end of the volume lists the important sources. A partial and condensed table of contents follows here:

ISBN 0-8155-0724-0

$39

294 pages

INORGANIC CHEMICAL INDUSTRY PROCESSES, TOXIC EFFLUENTS AND POLLUTION CONTROL 1978

by Marshall Sittig

Chemical Technology Review No. 118
Pollution Technology Review No. 52

While industrial inorganic chemistry may seem less glamorous than the complex chemistry of the life sciences, the progress in wonder drugs, miracle fibers and recombinant DNA is basically dependent on the uninterrupted flow of "heavy" chemicals, such as sulfuric acid, chlorine and sodium hydroxide.

The present book provides helpful directions for making these inorganic chemicals. The arrangement is encyclopedic starting with alumina and ending with various zinc compounds. The process details provide a wide choice of raw materials due to a changing economy and partial cutoff of the customary supplies, e.g., natural gas, which in inorganic processing is a most convenient source of hydrogen, carbon, carbon monoxide and carbon dioxide. Careful perusal of this book will reveal many alternate sources and energy-sparing processes.

Contaminated wastewater and air pollution problems are considerable, but new commercial processes are shown, as well as detailed technology from the U.S. patent literature. Because of the stable nature of this industry, established processes and obsolete equipment units remain sufficiently profitable to continue in use. Still, considerable progress in pollution control has been achieved during the past 5 or 6 years and is reflected in this book. The following chemicals have received detailed treatment:

Alumina	Cuprous Oxide	Potassium Nitrate
Aluminum Chloride	Ferric Chloride	Potassium Perchlorate
Aluminum Fluoride	Ferrous Sulfate	Potassium Permanganate
Aluminum Sulfate	Fluorine	Potassium Sulfate
Ammonia	Helium	Silver Nitrate
Ammonium Chloride	Hydrochloric Acid	Sodium Bicarbonate
Ammonium Diuranate	Hydrogen	Sodium Bisulfite
Ammonium Hydroxide	Hydrogen Fluoride	Sodium Borohydride
Ammonium Nitrate	Hydrogen Peroxide	Sodium Bromide
Ammonium Phosphate	Iodine	Sodium Carbonate
Ammonium Sulfate	Iron Blue Pigments	Sodium Chlorate
Antimony Oxide	Iron Oxide Pigments	Sodium Chloride
Arsenic Acid	Lead Carbonate	Sodium Chromate
Arsenic Oxides	Lead Nitrate	Sodium Cyanide
Barium Carbonate	Lead Oxides	Sodium Dichromate
Barium Sulfate	Lead Sulfate	Sodium Fluoride
Beryllium Hydroxide	Lithium Carbonate	Sodium Sulfide
Beryllium Oxide	Lithium Hydroxide	Sodium Hydrosulfite
Borax	Magnesium Chloride	Sodium Hydroxide
Boric Acid	Magnesium Sulfate	Sodium Metal
Boron Trichloride	Manganese Sulfate	Sodium Perchlorate
Bromine	Mercuric Chloride	Sodium Phosphates
Cadmium Pigments	Mercuric Oxide	Sodium Silicate
Cadmium Sulfide	Mercuric Sulfide	Sodium Silicofluoride
Calcium Arsenate	Mercurous Chloride	Sodium Sulfite
Calcium Carbide	Molybdate Chrome Pigments	Sodium Thiosulfate
Calcium Carbonate	Nickel Sulfate	Stannic Oxide
Calcium Chloride	Nitric Acid	Stannous Chloride
Calcium Hydroxide	Nitrogen & Oxygen	Strontium Carbonate
Calcium Oxide	Nitrous Oxide	Sulfur Chlorides
Calcium Phosphate	Phosphate Rock	Sulfuric Acid
Carbon Dioxide	Phosphoric Acid	Sulfuryl Chloride
Carbon Monoxide	Phosphorus	Superphosphoric Acid
Chlorine & Caustic	Phosphorus Oxychloride	Supported Catalysts
Chlorosulfonic Acid	Phosphorus Pentasulfide	Thallium Carbonate
Chrome Green Pigment	Phosphorus Pentoxide	Thionyl Chloride
Chrome Yellow Pigment	Phosphorus Trichloride	Titanium Dioxide
Chromic Acid	Potassium Chlorate	Ultramarine Pigments
Chromic Oxide Pigment	Potassium Chloride	Zinc Chloride
Cobalt Compounds	Potassium Dichromate	Zinc Oxide
Cobalt Oxide	Potassium Iodide	Zinc Sulfate
Copper Sulfate	Potassium Metal	Zinc Yellow Pigment

ISBN 0-8155-0726-7

351 pages

TOXIC ORGANIC CHEMICALS DESTRUCTION AND WASTE TREATMENT 1978

by E. Ellsworth Hackman III

Pollution Technology Review No. 40
Chemical Technology Review No. 107

This book has been written with several types of readers in mind. Manufacturing personnel are given a basic understanding of a wide variety of potential methods, followed by an exposition of current methods considered best at present.

Pollution control engineers, consultants, R&D investigators, and facilities planners or designers are given approaches for waste treatment and separation of difficult-to-destroy toxic organic chemicals.

Local, regional and national environmental control personnel in government are furnished insights into the scope and magnitude of toxic and refractory organic chemicals wastes problems and their refractoriness.

At the outset it will become obvious that there is no way in which modern life can function without the vast majority of these compounds which play an important part in materials, food, and energy technology. Bans on use and precipitous searches for substitutes were found to be poor alternatives in most cases. This author maintains that effective pollution control and waste treatment is much to be preferred.

It is necessary to walk a very fine line to achieve success in protecting the general public, the workers, the environment and future generations from toxic organics while at the same time trying to expand the job market in chemical products.

This book concentrates on those toxic organic compounds which are considered the most toxic and refractory substances, but having many useful properties.

Part I (chapters 1 to 5) identifies and describes the problems of toxic organic chemical pollution, and then shows all the technical avenues of approach to solving these problems, whether they are already in practice or forecast for use in the future.

Part II (chapters 6 to 12) categorizes the industrial sources of toxic organic chemical pollution, and presents the waste treatment systems found most practical and effective. Part II also places emphasis on cost factors.

Each compound that has been examined in this book has very powerful reasons for its existence and use. When substitutes for certain chemicals are sought, the virtues of the original compound come into clear focus. At times the substitutes are found to be hazardous. Some very effective fire retardants and noncombustible fluids are being removed from use because of their toxic hazards. Benefits and costs of substitution are weighed carefully.

Therefore emissions control should be the keystone of all future industrial efforts. In general, the air, water and solid waste emissions from the plant processes we design will have to be such that none of us would mind having the plant in our own neighborhood.

Part I: Toxic Organic Pollutant Problems and Potential Methods for Waste Treatment

1. RISK ASSESSMENT, HAZARDOUS INCIDENTS & LEGISLATIVE RESPONSES

2. TOXIC ORGANIC CHEMICALS DEFINED

3. METHODS OF DESTRUCTION AND DETOXIFICATION

4. SEPARATIONS TECHNOLOGY APPLICABLE TO WASTE TREATMENT

5. HOW WASTES ARE GENERATED & DISSEMINATED

Part II: Industrial Sources of Toxic Organics. Typical Waste Treatment Methods & Associated Capital & Operating Costs

6. PESTICIDES PRODUCTION

7. ORGANIC CHEMICALS PRODUCTION

8. PETROLEUM REFINING

9. POLYMERS AND PLASTICS PRODUCTION

10. TEXTILES MANUFACTURING

11. COAL CONVERSION

12. DETAILED WASTE TREATMENT DESIGNS AND COSTS: THE PCB EXAMPLE

ISBN 0-8155-0700-3

317 pages

UNIT OPERATIONS FOR TREATMENT OF HAZARDOUS INDUSTRIAL WASTES 1978

Edited by D. J. De Renzo

Pollution Technology Review No. 47

This book describes ways of combating pollution from hazardous waste materials through the use of basic unit operations common to chemical engineering processes. These unit operations are either physical, such as neutralization and precipitation, or biological, e.g. treatment with suitable enzymes.

There exists an urgent and immediate need for treatment processes which can detoxify, destroy or apply recovery processes to industrial wastes. This study examines over 40 unit engineering processes for their applicability to treating hazardous wastes.

For each potentially applicable method there are given, in depth, the:

Technical characteristics
Underlying physical, chemical, or biological principles
Operating characteristics
 • Physical and chemical properties of suitable feed streams
 • Mode of operation
 • Physical and chemical properties of the output streams
Operating experience
 • Principal current applications
 • Potential applications to hazardous wastes
Capital and operating costs
Environmental impacts
Energy requirements
Outlook & development needs

A condensed table of contents follows here:

In order to make all of the information readily available and useful, the section on TREATMENT TECHNIQUES is preceded by a series of reference tables to be used when screening processes for application to an individual waste stream. These tables are so organized as to provide a method for screening out processes that do not have any potential for the waste in question.

ISBN 0-8155-0717-8

920 pages